T0388664

Euphorbia in Southern Africa

Peter V. Bruyns

Euphorbia in Southern Africa

Volume 2

Peter V. Bruyns
University of Cape Town
Rondebosch, South Africa

ISBN 978-3-030-49398-1 ISBN 978-3-030-49399-8 (eBook)
https://doi.org/10.1007/978-3-030-49399-8

© Springer Nature Switzerland AG 2022
This work is subject to copyright. The Author has marked certain illustrations in the manuscript in the figure legends as follows: '© PVB'. whether the whole or part of the material is concerned, specifically the rights of translation, reprinting, reuse of illustrations, recitation, broadcasting, reproduction on microfilms or in any other physical way, and transmission or information storage and retrieval, electronic adaptation, computer software, or by similar or dissimilar methodology now known or hereafter developed.
The use of general descriptive names, registered names, trademarks, service marks, etc. in this publication does not imply, even in the absence of a specific statement, that such names are exempt from the relevant protective laws and regulations and therefore free for general use.
The publisher, the authors, and the editors are safe to assume that the advice and information in this book are believed to be true and accurate at the date of publication. Neither the publisher nor the author give a warranty, expressed or implied, with respect to the material contained herein or for any errors or omissions that may have been made. The publisher remains neutral with regard to jurisdictional claims in published maps and institutional affiliations.

This Springer imprint is published by the registered company Springer Nature Switzerland AG
The registered company address is: Gewerbestrasse 11, 6330 Cham, Switzerland

Contents

Volume 2

3 *Euphorbia* **subg.** *Chamaesyce* 475
    3.1   Sect. Anisophyllum 477
            3.1.1   Subsect. Hypericifoliae 477
    3.2   Sect. Articulofruticosae 482
    3.3   Sect. Espinosae 559
    3.4   Sect. Frondosae 569
    3.5   Sect. Gueinziae 577
    3.6   Sect. Tenellae 581

4 *Euphorbia* **subg.** *Esula* 589
    4.1   Sect. Aphyllis 591
            4.1.1   Subsect. Africanae 591
    4.2   Sect. Esula 607

5 **Euphorbia subg. Euphorbia** 639
    5.1   Sect. Euphorbia 641
    5.2   Sect. Monadenium 863
    5.3   Sect. Tirucalli 868

6 **Addenda** 893
    6.1   Names of Uncertain Application or Excluded from Euphorbia
           and Naturally Occurring Hybrids 893
            6.1.1   Names of Uncertain Application or Excluded from Euphorbia 893
            6.1.2   Naturally Occurring Hybrids 894
    6.2   The Species of Sect. *Euphorbia* and Sect. *Monadenium*
           in Moçambique 897
            6.2.1   Sect. Euphorbia 899
            6.2.2   Sect. Monadenium 957

References 965

Name Index 973

Subject Index 977

New synonyms published in this work: *Euphorbia grandialata* R.A. Dyer is reduced to synonymy under *E. grandicornis* A. Blanc, *E. halipedicola* L.C. Leach under *E. bougheyi* L.C. Leach, *E. decliviticola* L.C. Leach under *E. graniticola* L.C. Leach, *E. stenocaulis* Bruyns under *E. plenispina* S. Carter.

Lectotypes are designated here for *Euphorbia benguelensis* Pax, *E. caerulescens* Haw., *E. cucumerina* Willd., *E. enopla* Boiss., *E. fleckii* Pax, *E. genistoides* var. *leiocarpa* Boiss., *E. genistoides* var. *major* Boiss., *E. grandicornis* K.I. Goebel, *E. grandicornis* J.E. Weiss, *E. involucrata* var. *megastegia* Boiss., *E. latimammillaris* Croizat, *E. melanosticta* E. Mey. ex Boiss., *E. nodosa* N.E. Br., *E. platymammillaris* Croizat, *E. polygonata* G. Lodd., *E. proteifolia* Boiss., *E. trichadenia* var. *gibbsiae* N.E. Br. and *Tithymalus zeyheri* Klotzsch & Garcke.

A type is also designated for the genus *Tirucalia* Raf.

Euphorbia subg. *Chamaesyce*

Euphorbia subg. **Chamaesyce** Raf., *Amer. Monthly Mag. & Crit. Rev.* 2: 119 (1817). Type (Wheeler 1939): *Euphorbia supina* Raf. (= *Euphorbia maculata* L.)
Chamaesyce Gray (see under sect. *Anisophyllum*).

Bisexual (sometimes unisexual), annual herbs, succulent shrubs with slender cylindrical stem and branches (rarely with spines developing from tips of branches), rarely geophytes or small trees. *Leaves* alternate or opposite, ovate to lanceolate and sometimes reduced to minute caducous rudiments, rarely arising on raised tubercles (these irregularly arranged along branches); small stipules often present and glandular or filiform. *Synflorescences* in axils of leaves towards apices of stem and branches or terminal on peduncle, peduncles usually simple, often with alternating or opposite bracts often differently shaped from leaves and sometimes larger just beneath cyathium, further peduncles with terminal cyathia rarely developing from axils of uppermost bracts. *Cyathia* bisexual or unisexual (then males larger than females); glands 4–5, usually elliptic and flat with entire, crenulate or few-toothed outer margins, sometimes with petaloid appendages; male florets with glabrous or pubescent pedicels. *Capsule* obtusely 3-lobed, smooth, glabrous to pubescent, 2–10 mm diam., sessile to exserted, dehiscing explosively. *Seeds* 3 per capsule, 4-angled to ellipsoidal or oblong, smooth to tuberculate, with or without caruncle.

Of the 16 sections (15 recognized by Yang et al. (2012), with one added by Tian et al. (2018)), only six are represented naturally in southern Africa and most of them have only a few species in the region (see Fig. 3.1). These range from small, prostrate annuals to shrubby succulents and small deciduous trees (Table 3.1). *Euphorbia heterophylla* L. of sect. *Poinsettia* (with a single, cupped gland on each cyathium and native to the Americas) is an occasional weed in the northern parts of Botswana and South Africa.

Key to the sections of subg. *Chamaesyce* in southern Africa:

1. Annuals or non-succulent (non-geophytic) ephemerals usually with aborting main axis, cyathia 1–2.2 mm diam., glands often 4 and frequently with entire white or pink petaloid appendages on outer margins.................................2.
1. Plant not ephemeral (geophytes or succulent to woody shrubs or small trees), cyathia > 2.5 mm diam, glands (4) 5 and without petaloid appendages.................................3.
2. Seeds with caruncle... sect. **Tenellae**
2. Seeds without caruncle.. sect. **Anisophyllum**
3. Plant a unisexual geophyte, above-ground branches and leaves deciduous.................................sect. **Gueinziae**
3. Plant not geophytic, with perennial above-ground branches, woody with deciduous leaves or succulent and sometimes with very reduced and ephemeral leaves, plant unisexual or cyathia bisexual.................................4.
4. Woody shrub to small tree with shiny peeling bark on trunk or towards bases of branches (lacking green branches), cyathia bisexual, seeds with caruncle,.................................sect. **Espinosae**
4. Shrubby to dwarf succulent with fleshy green stem and branches without peeling bark, plant unisexual or cyathia bisexual, seeds without caruncle.................................5.
5. Plant mostly unisexual, branches not tuberculate, frequently slender and cylindrical, leaves usually opposite and sometimes with small globular stipules, cyathial glands usually 5, with entire margins.................................sect. **Articulofruticosae**
5. Plant and most cyathia bisexual, branches with scattered to densely distributed tubercles, stipules minute and ± linear, cyathial glands often 4, bilobed or with wavy outer margins.................................sect. **Frondosae**

3 Euphorbia subg. Chamaesyce

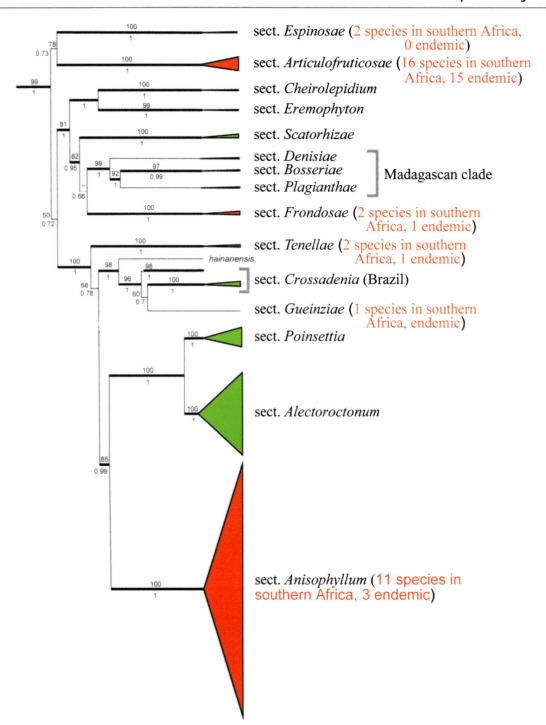

Fig. 3.1. Relationships between southern African members of subg. *Chamaesyce*. The sections represented in southern Africa are red (those not represented are green). The length of the vertical side of each of the triangles is proportional to the number of species in that section (adapted from Yang et al. 2012).

3.1 Sect. Anisophyllum

Table 3.1. The sections and subsections of *Euphorbia* subg. *Chamaesyce* occurring naturally in southern Africa.

Section	Sub-section	Namibia (endemic)	South Africa (endemic)	Southern Africa (endemic)	Elsewhere (total)	Annuals	Perennial, non-succulent herbs	Woody shrub to tree	Succulents	Geophytes
Articulofruticosae		10 (2)	13 (5)	16 (15)	1 (16)	0	0	0	16	0
Espinosae		2 (0)	2 (0)	2 (0)	2 (2)	0	0	2	0	0
Frondosae		2 (1)	1 (0)	2 (1)	6 (7)	0	0	0	2	0
Tenellae		2 (0)	2 (0)	2 (1)	3 (4)	2	0	0	0	0
Gueinziae		0 (0)	1 (0)	1 (1)	0 (1)	0	0	0	0	1
Anisophyllum		4 (2)	3 (1)	11 (3)	362 (365)	8	3	0	0	0
	Hypericifoliae	4 (2)	3 (1)	11 (3)	362 (365)	8	3	0	0	0

3.1 Sect. Anisophyllum

Euphorbia sect. **Anisophyllum** Roep. in Duby, *Bot. Gall.*, ed. 2, 1: 412 (1828). *Anisophyllum* Haw., *Syn. Pl. Succ.*: 159 (1812), *nom. illegit.*, *non* Jacq. (1763). *Chamaesyce* Gray, *Nat. Arr. Brit. Pl.* 2: 260 (1821). *Euphorbia* subg. *Chamaesyce* (Gray) Caesalp. ex. Rchb., *Deut. Bot. Herb.-Buch.*: 193 (1841), *nom. illegit.*, *non* Raf. (1817). Type (Wheeler 1941): *Euphorbia peplis* L.

The 368 species of this section were split into two subsections, one with three species found only in America and the other, subsect. *Hypericifoliae*, containing the remaining 365 species (Yang et al. 2012), of which 11 occur naturally in southern Africa.

3.1.1 Subsect. Hypericifoliae

Euphorbia subsect. **Hypericifoliae** Boiss. in DC., *Prodr.*15 (2): 20 (1862). Type: *Euphorbia hypericifolia* L.

Bisexual non-succulent annual or perennial herbs, with short and soon aborting stem bearing fibrous roots and growth continuing from dichotomously forking branches arising in axils of few leaves on stem. *Branches* prostrate to ascending or erect. *Leaves* to 30 mm long, opposite, orbicular-ovate and sometimes asymmetrically rounded at base, usually sessile, stipules filamentous or absent. *Synflorescences* many, each a solitary bisexual cyathium terminating branchlets, sometimes branchlet then forking around it, with minute sessile bracts, glands 4 (5), entire, green, with pale petaloid appendages along outer margins. *Capsule* 2–3 mm diam., exserted on decurved and later erect pedicel. *Seeds* minutely tuberculate or ridged to wrinkled, ± 4-angled, without caruncle.

This subsection has 365 species. Most species are found in the Americas and it is poorly represented in Africa as a whole and in southern Africa too, where 11 have been recorded. These are *E. chamaesycoides* B.Nord., *E. eylesii* Rendle, *E. inaequilatera* Sond., *E. livida* E.Mey. ex Boiss., *E. mossambicensis* (Klotzsch & Garcke) Boiss., *E. neopolycnemoides* Pax & K.Hoffm., *E. pergracilis* P.G.Mey., *E. rubriflora* N.E.Br., *E. schlechteri* Pax, *E. tettensis* Klotzsch and *E. zambesiana* Benth. Several cosmopolitan weeds such as *E. prostrata* and *E. hirta* are also common in southern Africa. Apart from *E. chamaesycoides* and *E. pergracilis* from Namibia, most of the southern African species occur on the moister eastern side of the subcontinent. A key to the species is presented below. To give an idea of how the species look, photographs of the very widespread *E. inaequilatera* and an unnamed species (perhaps *E. lupatensis*) from northern Botswana are shown. The distributions of *E. livida* (endemic and restricted to the Eastern Cape and Kwazulu-Natal of South Africa) and *E. neopolycnemoides* (widespread, also known in Moçambique and Zimbabwe) are also shown (in Figs. 3.2 and 3.3).

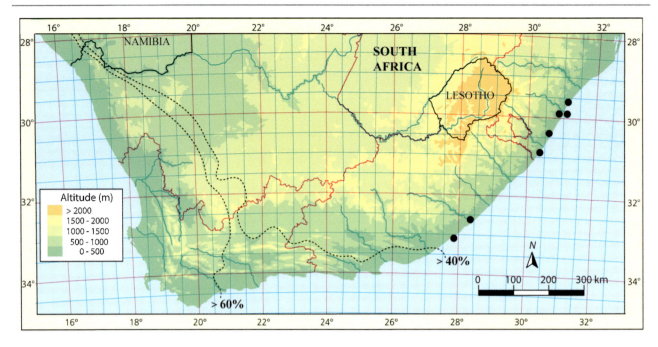

Fig. 3.2. Distribution of *Euphorbia livida* (© PVB).

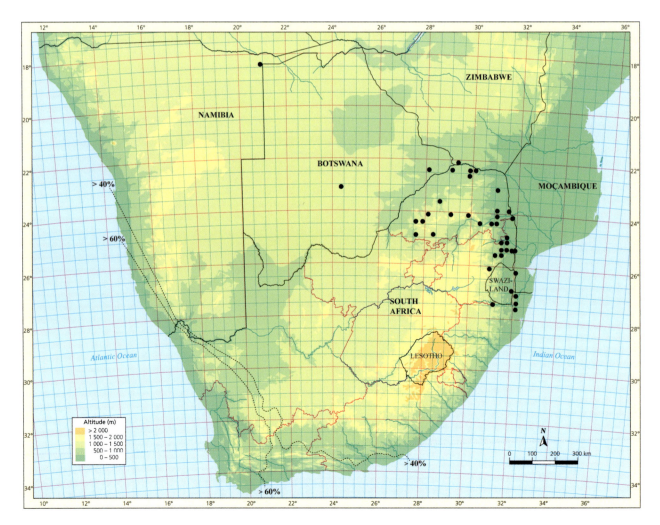

Fig. 3.3. Distribution of *Euphorbia neopolycnemoides* (© PVB).

3.1 Sect. Anisophyllum

Note: *Euphorbia austro-occidentalis* Thell. was included in Bruyns (2012) as if it occurred naturally in Namibia. There are two problems with this. Firstly, Meyer (1967) placed it in synonymy under *E. forskalii* Gay along with *E. aegyptiaca* Boiss. There are no grounds for questioning Meyer's treatment of it. This is especially since Brown (1911: 508) included several of the collections listed by Thellung (1916) for *E. austro-occidentalis* under *E. aegyptiaca*. Secondly, Thellung said that it agreed in most features with *E. forskalii,* except for some minute details of the stipules. Thirdly, there is no evidence that it occurs naturally in Namibia, since *E. forskalii* is a very widespread weed and Dinter mentioned on some of his specimens that it was a weed of cultivated land. This species is therefore not included in the key below.

Several species of this section are common weeds in disturbed areas in southern Africa. These include *E. maculata* (which occurs naturally in the Americas), *E. prostrata* (native to tropical America) and *E. hirta* (Africa to India) and they are not included in the key that follows.

Fig. 3.4. *Euphorbia inaequilatera*, prostrate herb ± 15 cm diam., southern outskirts of Zeerust, South Africa, Feb. 2019 (© PVB).

Fig. 3.5. *Euphorbia inaequilatera*, ± 7 cm diam., hills south of Kuruman, South Africa, Feb. 2019 (© PVB).

Fig. 3.6. *Euphorbia inaequilatera*, flowering, hills south of Kuruman, South Africa, Feb. 2019 (© PVB).

Fig. 3.7. *Euphorbia inaequilatera*, flowering, hills south of Kuruman, South Africa, Feb. 2019 (© PVB).

Fig. 3.8. *Euphorbia inaequilatera*, fruiting, Matsap Pan, Postmasburg, South Africa, Mar. 2016 (© PVB).

Fig. 3.9. *Euphorbia maculata*, leaves ± 6 mm long, in crevices in a driveway in Cape Town, South Africa, Dec. 2019 (© PVB).

3.1 Sect. Anisophyllum

Key to the species of subsect. *Hypericifoliae* that occur naturally in southern Africa

1. Annuals to short-lived perennials with slender fibrous roots, occasionally with main root to ± 4 mm thic..........................2.
1. Perennials with a woody rootstock 5–10 mm thick from which fibrous roots arise..9.
2. Cyathia grouped closely together in densely leafy cymes...**E. tettensis**
2. Cyathia laxly grouped or solitary...3.
3. Branches pubescent on their upper surface ....**E. mossambicensis**
3. Branches glabrous except possibly near base...4.
4. Margins of leaves finely to boldly toothed...**E. inaequilatera**
4. Margins of leaves entire...5.
5. Plant prostrate, leaves mostly at least 5 mm broad, leathery to slightly fleshy..**E. livida**
5. Plant to 0.4 m tall with ascending to erect branches, leaves mostly 3–4 (7) mm broad, mostly herbaceous or only slightly leathery..6.
6. Petaloid appendages on two cyathial glands longer than other two...7.
6. Petaloid appendages on all 4 (5) cyathial glands ± equal in length...8.
7. Stipules linear and entire, 0.5–1.5 mm long..**E. eylesii**
7. Stipules ± 0.5 mm long, divided above broad base into 3–5 linear teeth..**E. neopolycnemoides**
8. Stipules short and shortly ciliate, petaloid appendages on cyathial glands ± equalling glands......................**E. pergracilis**
8. Stipules divided into 2–4 lanceolate-subulate lobules, petaloid appendages on cyathial glands narrow and rim-like on glands...**E. chamaesycoides**
9. Petaloid appendages on cyathial glands forming inconspicuous narrow white margin on glands...............**E. schlechteri**
9. Petaloid appendages on cyathial glands conspicuous, often pink or red...10.
10. Leaves with minute point at apex, cyathia borne on peduncle 1–25 mm long, petaloid appendages on cyathial glands all ± equal in size...**E. zambesiana**
10. Leaves rounded at apex, cyathia ± sessile (on peduncle ± 0.5 mm long), petaloid appendages on two cyathial glands next to pedicel of exserted capsule longer than others...**E. rubriflora**

Fig. 3.10. *Euphorbia lupatensis?*, ± 15 cm across, north of Pandamatenga, Botswana, Feb. 2019 (© PVB).

Fig. 3.11. *Euphorbia lupatensis?*, north of Pandamatenga, Botswana, Feb. 2019 (© PVB).

3.2 Sect. Articulofruticosae

Euphorbia sect. **Articulofruticosae** Bruyns, *Taxon* 55: 416 (2006). Type: *Euphorbia aequoris* N.E.Br. (= *Euphorbia juttae* Dinter).

Mostly unisexual, succulent shrubs or dwarf succulents often dichotomously branched. *Branches* cylindrical or slightly longitudinally ridged, grey-green to dark green (to brown towards bases), apices drying into spines in *E. spinea*. *Leaves* rudimentary and caducous, rarely more than 5 mm long, opposite, without petiole, occasionally with small globular stipules. *Synflorescences* often many, each a unisexual (occasionally bisexual) cyathium terminating shoot with cymes sometimes branching repeatedly, surrounded by bracts slightly shorter than leaves, glands 5, entire, often yellow-green, without appendages. *Capsule* < 5 mm diam., ± sessile to exserted on erect to decurved and later erect pedicel. *Seeds* roughened, ± 4-angled, without caruncle.

This section has 16 species, which are found only in Angola, Botswana and South Africa. In southern Africa it is especially associated with the western side of the subcontinent and is most diverse in the parts of the Namib Desert that receive rainfall in winter. It reaches its greatest diversity on the Knersvlakte of Namaqualand (Fig. 3.12).

Key to the species of sect. *Articulofruticosae*

1. Two small to quite prominent globular stipules present alongside bases of leaf-rudiments (and remaining on branches after leaf-rudiments fall off)..2.
1. Stipules absent alongside bases of leaf-rudiments..4.
2. Ultimate branches 1–2 mm thick, stipules ± 0.3 mm diam., capsules glabrous and shiny green with bright red in grooves..**E. exilis**
2. Ultimate branches 3–5 mm thick, stipules 1–1.5 mm diam., capsules pubescent and dull grey-green to dull reddish green. ....3.
3. Free-standing shrub (usually), branches smooth, stipules smooth and shiny, capsule exserted from cyathium on maturing ....**E. burmanni**
3. Scandent shrublet inside other bushes, branches papillate, stipules tuberculate and dull, capsule remaining included in cyathium on maturing...**E. suffulta**
4. Leaf-rudiments with small to prominent, horn-like outgrowths at base of 'lamina' (usually still visible when leaf-rudiment dried out) giving them ± triangular outline...5.
4. Leaf-rudiments without small, horn-like outgrowths at base of 'lamina' and elliptical in outline....................................10.
5. Plant with underground tuber and branches spreading underground as slender cylindrical rhizomes (often rebranching beneath ground and developing subsidiary tubers), becoming abruptly thicker above surface where irregularly cylindrical and erect but frequently only 5–50 mm long..**E. stapelioides**

3.2 Sect. Articulofruticosae

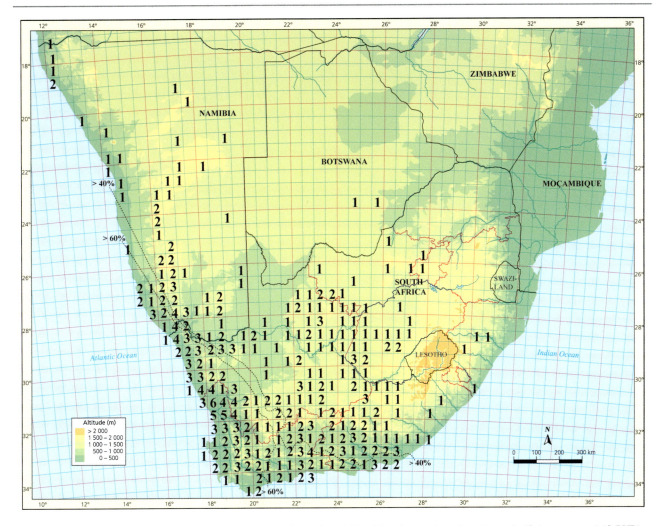

Fig. 3.12. Distribution of *Euphorbia* sect. *Articulofruticosae* in southern Africa (showing number of species per half-degree square) (© PVB).

5. Plant shrub-like (though sometimes very small and with branches spreading close to ground) from fibrous to sometimes swollen base, without rhizomes, branches spreading or ascending to erect ....6.
6. Surface of branches coarsely asperulous or convoluted with raised irregular warts and papillae, branches mostly spreading close to (but above) surface of ground and forming small, often very diffuse shrublet..7.
6. Surface of branches finely asperulous to smooth, branches generally ascending to erect (if spreading close to ground then with smooth epidermis)..8.
7. Rootstock a slender and wiry taproot, branches not knobbly-tuberculate (when turgid), leaf-rudiments raised on small petiole above tubercle, branches (when turgid) not distinctly articulated and not easily breaking at joints, capsule glabrous to sparsely pubescent..**E. muricata**
7. Rootstock somewhat swollen, branches knobbly-tuberculate (when turgid), leaf-rudiments not raised on small petiole above tubercle and remaining ± embedded in apex of tubercle, branches (when turgid) distinctly articulated and easily breaking at joints, capsule densely white-pubescent..**E. verruculosa**
8. Ultimate branchlets usually somewhat downwardly curved and developing into sharp spines at their tips, low but dense hemispherical shrublet, capsule on pedicel 3–5 mm long..**E. spinea**
8. Tips of branches ascending to erect (occasionally horizontally spreading close to ground) and not (or very rarely) developing into spines, variable in shape from small horizontally spreading shrublet to larger robust hemispherical or V-shaped shrub, capsule sessile or on pedicel to 1.5 mm long..9.
9. Slender shrub (often narrowly V-shaped in outline) with ultimate branches often only 2 mm thick (in *fynbos* or in *renosterveld* in SW Cape)..**E. tenax**

9. Slender to stout shrub (often hemispherical in outline) with ultimate branches usually at least 3 mm thick (outside SW Cape and particularly in Succulent Karoo)...................**E. rhombifolia**
10. Cyathium exceeded by bracts (usually about half their length), capsules ± spherical and sessile (or exserted to 1 mm) when mature, nodes thickened towards apex but remaining circular in cross-section there (so that leaf-rudiments do not arise on slight tubercles), stem (generally clearly visible at base of plant) and lowest parts of branches (at base of plant) usually becoming blackish...................**E. ephedroides**
10. Cyathium much longer than bracts (rarely equalling them), capsules distinctly 3-lobed and usually distinctly to far-exserted when mature, nodes often thickened towards apex but then distinctly flattened there (so that leaf-rudiments arise on slight tubercles), stem (if visible) and lowest parts of branches (at base of plant) remaining green or becoming covered with grey bark...................11.
11. Dwarf plants rarely exceeding 0.1 m (to 0.22 m) tall, branches swollen and usually highly succulent, capsule exserted on pedicel 3–6 mm long when mature, leaf-rudiments mostly borne on distinctly swollen tubercles with branch often strongly articulated at nodes, leaf-rudiments abruptly narrowing towards base into short petiole...................12.
11. Shrubs (often large) 0.3–1 m tall, branches slender and usually only slightly succulent, capsule sessile to exserted to 1.5 mm when mature, leaf-rudiments borne on low and indistinct tubercles or tubercles absent with branch not articulated at nodes, leaf-rudiments gradually narrowing towards base and without petiole...................14.
12. Leaf-rudiments with toothed margins, internodes with 1–2 distinct ridges running downwards from leaf-rudiments, rootstock slender and rapidly tapering off into fibrous roots...................**E. herrei**
12. Leaf-rudiments with entire margins, internodes without distinct ridges running downwards from leaf-rudiments, rootstock usually swollen before tapering off into fibrous roots...................13.
13. Plant densely branched, branches bearing many small cymes each with 1–3 cyathia around their tips...................**E. gentilis**
13. Plant sparingly branched (often with only 1–3 branches), branches usually bearing solitary cyme of 1–3 cyathia at their tips...................**E. juttae**
14. Plant with erect branches, capsule glabrous...................**E. spartaria**
14. Plant with some lower branches spreading or wholly prostrate, capsule pubescent...................15.
15. Plant erect with lower branches spreading, male cyathia almost cylindrical...................**E. giessii**
15. Plant ± prostrate, male cyathia swollen around middle, urn-shaped...................**E. negromontana**

Euphorbia burmanni E.Mey. ex Boiss. in DC., *Prodr.* 15 (2): 75 (1862). Lectotype (Bruyns 2012): South Africa, Cape, towards Blauwberg, *Drège 2920* (P).

Euphorbia biglandulosa Willd., *Enum. Pl., Suppl.*: 27 (1814), *nom. illegit. non* Desf. (1808).

Arthrothamnus burmanni E.Mey. ex Klotzsch & Garcke, *Abh. Königl. Akad. Wiss. Berlin* 1859: 62 (1860). *Tirucalia burmanni* (E.Mey. ex Klotzsch & Garcke) P.V.Heath, *Calyx* 5: 88 (1996). Type: South Africa, Cape, *Drège* (missing). Neotype (Bruyns 2012): South Africa, Cape, *Drège 2920* (P).

Arthrothamnus bergii Klotzsch & Garcke, *Abh. Königl. Akad. Wiss. Berlin* 1859: 63 (1860). Type: South Africa, Cape, *Bergius* (missing).

Euphorbia corymbosa N.E.Br., *Fl. Cap.* 5 (2): 279 (1915). *Tirucalia corymbosa* (N.E.Br.) P.V.Heath, *Calyx* 5: 89 (1996). Type: South Africa, Cape, near Albertinia, 16 Nov. 1910, *Muir* (K, holo.; PRE, iso.).

Euphorbia karroensis (Boiss.) N.E.Br., *Fl. Cap.* 5 (2): 290 (1915). *Euphorbia burmanni* var. *karroensis* Boiss. in DC., *Prodr.* 15 (2): 75 (1862). *Tirucalia karroensis* (Boiss.) P.V.Heath, *Calyx* 5: 91 (1996). Type: South Africa, Cape, Karoo between Hol River and Mierenkasteel, 500–1000', 5 Aug. 1830, *Drège 2947* (P, holo.; K, iso.).

Euphorbia macella N.E.Br., *Fl. Cap.* 5 (2): 288 (1915). *Tirucalia macella* (N.E.Br.) P.V.Heath, *Calyx* 5: 91 (1996). Type: South Africa, Cape, near Little Brak River, 10 Oct. 1814, *Burchell 6197/2* (K, holo.).

Unisexual spineless and glabrous free-standing succulent shrub (rarely growing inside other shrubs) 0.2–0.7 m tall (more rarely to 1–2 m tall when sheltered), branching and rebranching extensively from similar stem to 300 mm tall with slender woody rootstock bearing fibrous roots. *Branches* ascending to erect, repeatedly rebranching, cylindrical, 50–500 × 3–5 mm, without distinct tubercles, smooth, soft and succulent when young and becoming woody towards base, dark green becoming grey with age towards base; *leaf-rudiments* towards apices of branches, opposite, 2–5 × 1–3 mm, ascending and spreading towards tip, fleeting, spathulate, glabrous, slightly channelled above by upwardly folded somewhat toothed margins, obtuse, tapering below into very short petiole 1 mm long, subtended on either side at base by conspicuous ellipsoidal brown smooth and shiny stipule 1–1.5 mm diam. *Synflorescences* terminal on branches (sometimes racemose), finely pubescent, each of 1–3 (–many) unisexual crowded cyathia on short peduncles 1–3 mm long each subtended by 2 obovate to spathulate bracts very like leaf-rudiments 1–3 × 1 mm; *cyathia* conical, minutely pubescent, 2.5–3.5 mm broad (± 2 mm long below insertion of glands), with 5 obovate lobes with deeply dissected margins, green; *glands* 5, transversely elliptic, 1 mm broad, spreading, yellow-green, outer margins flat and entire; *stamens* with glabrous pedicels, bracteoles filiform and pubescent; *ovary* ellipsoidal, pubescent, raised on pedicel 0.5 mm long; *styles* ± 0.5 mm long, divided nearly to base, often horizontally spreading. *Capsule* 3–3.5 mm diam., obtusely 3-lobed, pubescent, dull green to red, slightly exserted from cyathium on short erect pedicel.

Distribution & Habitat

Euphorbia burmanni is a widespread species. Most records have been made within the winter-rainfall area of the southwestern side of southern Africa. Here it is recorded from just north of Oranjemund in Namibia southwards along the coastal plain to around Cape Town and Paarl. A few collections have been made in Namaqualand on the slopes of some of the higher mountains west of the escarpment (as, for example, around Komaggas) and only on the Knersvlakte, from Nuwerus southwards, has it been recorded a little further away from the sea. Further south, the distribution begins to broaden out, with a few collections from the mountainous western margin of the Tanqua Karoo. It is also common on the Worcester-Robertson

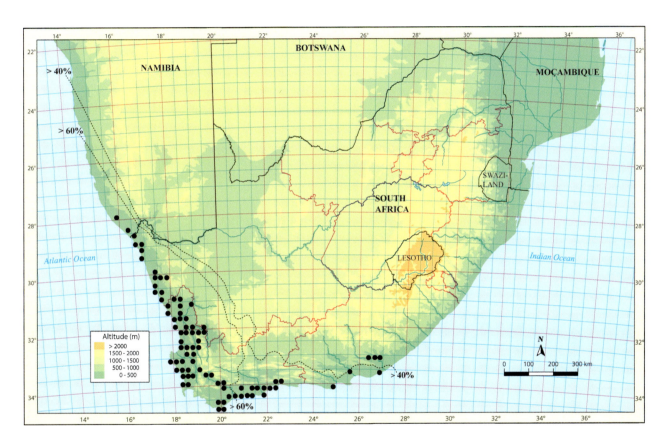

Fig. 3.13. Distribution of *Euphorbia burmanni* (© PVB).

Karoo to Kogman's Kloof near Montagu. Although it is found around Montagu, it is unknown further east on the Klein Karoo (material figured as *E. burmanni* in Vlok and Schutte-Vlok (2010) is a slender form of *E. mauritanica*). *Euphorbia burmanni* also occurs along the southern coastal plain from Bredasdorp to Mossel Bay and George. East of this it is sporadic around Port Elizabeth and Addo and is once again more common in the Fish River Valley, north and east of Grahamstown.

Euphorbia burmanni generally grows in flat areas or on the lower rocky slopes of hills, very often among other shrubs of a similar size. Consequently, although it is often common and is an important element of the winter-rainfall Succulent Karoo, it is not a dominant feature of the landscape, like *E. mauritanica*. It is mostly a denizen of karroid scrub and is especially common in the *strandveld* of the West Coast on grey to red sands with a wide variety of other succulents. From Addo eastwards it grows in dense, predominantly non-succulent bush. It also occurs in low *fynbos* on the coastal limestones from Bredasdorp to Mossel Bay, though here too it will be encountered occasionally in the denser bush that develops on steep slopes.

Diagnostic Features & Relationships
In the western part of its distribution, specimens of *E. burmanni* are usually not more than 0.3 m tall, branching from the base and repeatedly rebranching above to form a neat, conical to almost spherical shrub. Specimens of *E. burmanni* growing near the coast in Namaqualand and southern Namibia tend to have noticeably stouter and somewhat shorter branches than those, for example, from the Clanwilliam district or from near Cape Town and have sometimes been thought to represent a distinct species (and sometimes referred to as '*E. karooensis*'). From Addo to around Grahamstown, *E. burmanni* occurs in dense bush and often clambers in other shrubs, reaching heights of 1–2 m on occasion and rarely forming discrete shrubs.

In *E. burmanni* the branches are smooth and cylindrical, with a dark green colour. On young growth, the branches bear small, opposite pairs of spathulate leaf-rudiments.

Fig. 3.14. *Euphorbia burmanni*, shrub ± 30 cm tall, south of Bitterfontein, Namaqualand, South Africa, 10 Jul. 2005 in dry conditions (© PVB).

Alongside the base of each of these leaf-rudiments, two quite conspicuous, flattened, globular, brown stipules will be found. These stipules persist alongside the scars that are left once the leaf-rudiments fall off and are usually clearly visible in dried material, making them an important feature for identification.

In years of good rain, *Euphorbia burmanni* is generously floriferous between July and early September, when the whole bush becomes covered with cyathia. Although the cyathia are small and inconspicuous, they are paler than the branches so that the plants can readily be seen to be flowering. Afterwards, the female plants become covered with the small, pubescent capsules which are initially dull green and then change to reddish green before ripening.

There are three species in this section in which the bases of the leaves are subtended by rounded stipules, namely *E. burmanni*, *E. exilis* and *E. suffulta*. *Euphorbia burmanni* never occurs together with *E. suffulta*, which has similarly shaped and similarly sized stipules. However, it will be found sometimes in the company of *E. exilis*, where similar stipular structures are also present. Although *E. exilis* is also a shrub-forming plant like *E. burmanni*, the stems are considerably more slender and are noticeably brittle, while the stipules are far smaller than those of *E. burmanni*. The capsules of *E. burmanni* are always finely pubescent and are consequently somewhat dully coloured, while those of *E. exilis* are a brightly shiny green with red markings.

Euphorbia burmanni is remarkable for being edible, both to stock and to man and, as a consequence, it is known as *soetmelkbos*. Although there is no burning sensation caused by eating them, the young branches have a rather astringent taste and so this epithet is not especially well-earned.

Fig. 3.15. *Euphorbia burmanni*, male shrub ± 40 cm tall and broad, Sandberg, SE of Worcester, South Africa, 13 Sep. 2014 in a year of good rains (© PVB).

Fig. 3.16. *Euphorbia burmanni*, male cyathia, Sandberg, SE of Worcester, South Africa, 13 Sep. 2014 (© PVB).

Fig. 3.17. *Euphorbia burmanni*, female cyathia, 20 km east of Redelinghuys, South Africa, 30 Sep. 2007 (© PVB).

3 Euphorbia subg. Chamaesyce

Fig. 3.18. *Euphorbia burmanni*, capsules, east of Katdoringvlei, Namaqualand, South Africa, 8 Sep. 1994 (© PVB).

History

The earliest collection of *E. burmanni* is that of Thunberg, made around 1774, but he called this *Euphorbia tirucalli*. White et al. (1941) believed that '*Euphorbia viminalis*' of N.L. Burman (1768) referred to *E. burmanni*. However, here Burman copied his references out of Linnaeus (1753) and so he was not describing a new species but was listing Linnaeus' *E. viminalis*. It is unlikely that Linnaeus' *E. viminalis* refers to *E. burmanni*. In addition, Linnaeus had made some errors here since he referred to t. 190 of Alpinus (1735) and t. 368 of Dillenius (1732). Both of these are actually pages of text, not plates: the first one is accompanied by figure 53, while there is no t. 368 in Dillenius' *Hortus Ethamensis*. The figure printed opposite page 34 in Veslingius (1638) and as figure 53 next to page 190 in the yet later edition, Alpinus (1735), was present in a much simpler form in Alpinus'

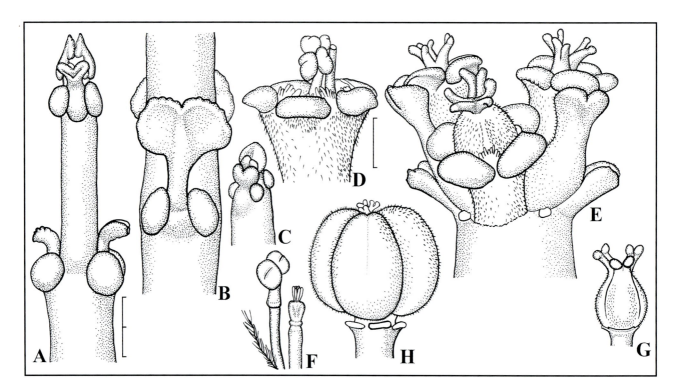

Fig. 3.19. *Euphorbia burmanni*. **A–C**, young branches with leaf-rudiments and stipules (scale 2 mm, as for **B, C, H**). **D**, male cyathium from side (scale 1 mm, as for **E–G**). **E**, female cyathia from side. **F**, anther, bracteole and rudimentary female from male cyathium. **G**, female floret. **H**, capsule. Drawn from: **A–C**, De Wet, near Worcester, South Africa. **D–H**, Koekenaap, near Vredendal, South Africa (© PVB).

original book of 1592 (about plants that Alpinus saw in Egypt between 1580 and 1583). Alpinus called this plant '*Felfel tavil, piper longum Aegyptium*' and the modifications in the illustration of 1638 from the original of 1592 by Alpinus may have originated from Veslingius' observations during his time in Egypt around 1620. Figure 53 in Alpinus (1735) has been taken as the lectotype of *Cynanchum viminale* (Liede and Meve 1993). However, its height of up to 1.5 m, its alternate branching, the fine striations on the branches, the slender alternating leaves lasting only a month and the white sap produced when it is cut that 'stings like pepper or burns the tongue like flames if tasted', suggests much more that it represents a poor drawing of the widely cultivated *Euphorbia tirucalli*.

The synonyms *E. karroensis* and *E. macella* refer to somewhat stouter-branched plants, the first from southern Namaqualand and the latter from near the present-day Mossel Bay. *Euphorbia karroensis* was first described as a variety of *E. burmanni* and it is not clear why N.E. Brown considered it to represent a separate species.

Euphorbia ephedroides E.Mey. ex Boiss. in DC., *Prodr.* 15 (2): 75 (1862). *Tirucalia ephedroides* (E.Mey. ex Boiss.) P.V.Heath, *Calyx* 5: 89 (1996). Lectotype (Bruyns 2012): South Africa, Cape, Karoo at Goedemanskraal, 2500', 8 Sept. 1830, *Drège 2949* (P; MO, K, S, iso.).

Unisexual spineless and glabrous free-standing succulent shrub 0.05–0.6 (1) × 0.03–0.4 m, branching and rebranching from base and above from short, slightly woody, grey-green later becoming grey to black stem 30–300 × 10–30 (50) mm bearing fibrous roots from base. *Branches* erect, repeatedly rebranching, cylindrical, 50–500 × 1–5 mm, without distinct tubercles, smooth, soft and succulent and yellow- to grey-green when young, becoming smooth and shiny grey and slightly woody with age and often more coarse and blackish near base; *leaf-rudiments* on new growth towards apices of branches, opposite, 1.5–20 × 1–5 mm, initially ascending and adpressed to branch then soon spreading, fleeting, spathulate to linear-spathulate, glabrous, flat above, margins entire, obtuse to mucronate, tapering below into petiole 0.5–2 mm long or with slightly swollen base, estipulate. *Synflorescences* terminal on branches, glabrous except towards bases of bracts and bases of cyathia, initially with 1 unisexual sessile cyathium subtended by 2 obovate to spathulate pale green bracts very like leaf-rudiments 2–4 × 1 mm, further solitary cyathia developing in axils of bracts on peduncles 5–12 mm long eventually forming irregularly branched structure to 80 mm long or more; *cyathia* conical-cupular, sparsely pubescent towards base otherwise glabrous, 2–3 mm broad (± 1 mm long below insertion of glands), with 5 quadrate lobes with dissected pubescent margins, pale green; *glands* 5, transversely elliptic, 0.7–1.5 mm broad, contiguous, spreading, yellow-green, outer margins flat and entire; *stamens* with pubescent pedicels, bracteoles filiform and pubescent; *ovary* ± spherical, not obviously 3-angled, glabrous, raised on very short pedicel to 0.5 mm long; styles 0.5–0.7 mm long, divided nearly to base, spreading widely. *Capsule* 3.5–4.5 mm diam., almost spherical and not obviously 3-lobed, glabrous but not shiny, pale green sometimes suffused with red, sessile to exserted to 1 mm.

Euphorbia ephedroides is easily distinguished from the other members of this section by the ± parallel-sided larger leaves, the lack of any stipules at the bases of the leaves and (in subsp. *ephedroides*) also by the neat, parallel-branching within each shrub. The manner in which the younger segments of the branches become gradually thicker towards their apices (a feature somewhat reminiscent of plants of *Ephedra*, making this species very aptly named) and the blackish colour of the stem and lower parts of the branches at the base of the plant are also distinctive.

Three subspecies are recognized and may be separated as follows:

1. Plant of lowly habit with mature specimens rarely more than 0.1 m tall, branches drooping and often spreading horizontally on ground..subsp. **imminuta**
1. Plant an erect shrub usually 0.2–0.6 (1) m tall, branches erect...2.
2. Branches rising ± parallel in shrub and not interwoven, upper parts ± 2 mm thick and reaching 5 mm thick below, capsule sessile..subsp. **ephedroides**
2. Branches ascending and somewhat interwoven in older plants, upper parts 1 mm thick and rarely more than 3 mm thick below, capsule exserted ± 1 mm from cyathium when mature...subsp. **gamsbergensis**

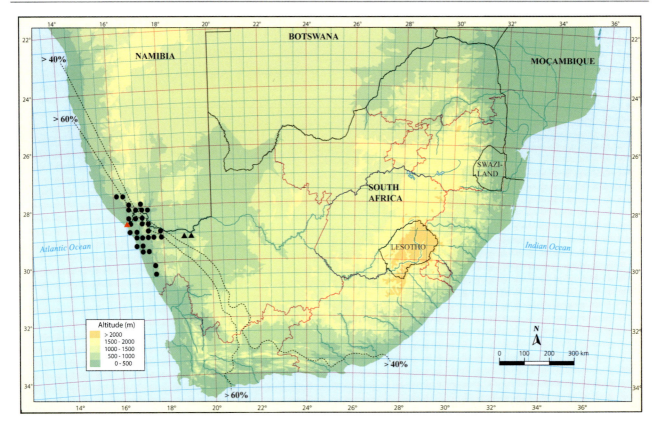

Fig. 3.20. Distribution of *Euphorbia ephedroides* (● = ssp. *ephedroides*, ▲ = ssp. *imminuta*, ▲ = ssp. *gamsbergensis*) (© PVB).

Euphorbia ephedroides subsp. **ephedroides**
Euphorbia ephedroides var. *debilis* L.C.Leach, *S. African J. Bot.* 56: 73 (1990). *Tirucalia ephedroides* var. *debilis* (L.C.Leach) P.V.Heath, *Calyx* 5: 90 (1996). Type: Namibia, north of Rosh Pinah, *Leach & Brunton 15893* (NBG, holo.; K, MO, PRE, iso.).

Plant an erect shrub usually 0.2–0.6 (1) m tall. *Branches* erect and ± parallel, 2–5 mm thick and becoming blackish below and ± 2 mm thick near tips. *Capsule* sessile.

Fig. 3.21. *Euphorbia ephedroides* ssp. *ephedroides*, ± 1 m diam., in firm loam among very short scrub (*Cheiridopsis denticulata* on right), foot of Kourkamma Mountain, Namaqualand, South Africa, 6 Jul. 2016 (© PVB).

3.2 Sect. Articulofruticosae

Fig. 3.22. *Euphorbia ephedroides* ssp. *ephedroides*, among quartz-gravel on gentle slopes, north of Eksteenfontein, Namaqualand, South Africa, 6 Sep. 1988 (© PVB).

Distribution & Habitat

Euphorbia ephedroides subsp. *ephedroides* is found in south-western Namibia and north-western South Africa. In Namibia it is known in the part receiving winter-rainfall, from north of the Aurus Mountains near the coast to near Rosh Pinah and eastwards to Marinkas Quellen (south of Ai-Ais). In South Africa it is more widely distributed, from Alexander Bay to Soebatsfontein and Wallekraal in the coastal plain and neighbouring hills, as well as from Eksteenfontein and the Rosyntjie Mountain to Jakkalswater and Steinkopf.

Subsp. *ephedroides* is generally a common plant. Mostly it grows in flats and on low hills and only much more rarely on steep, rocky slopes. Plants usually occur in large and prolific colonies and, locally, they often dominate the vegetation, being considerably taller than most of the short, succulent members of the Aizoaceae that grow together with them. While it may occur in quite hard ground, in many places (especially on the coastal plain of Namaqualand), subsp. *ephedroides* grows in the loamy, softer and deeper soils on and around old '*heuweltjies*' and is not found in the harder, stony ground away from these spots. Nevertheless, in the coastal plain it only occurs in the *hardeveld* and is never found in the softer sands of the *sandveld* closer to the coast.

Diagnostic Features & Relationships

Most specimens of subsp. *ephedroides* are around half a metre tall, with the branches mostly parallel within the shrubs and lending them a very neat appearance. Above, the branches are pliable and generally a slightly dull grey-green while lower down they become more rigid and grey-brown, though remaining smooth and almost shiny. The bases of the longer branches

Fig. 3.23. *Euphorbia ephedroides* ssp. *ephedroides*, steep rocky south-facing slopes, ± 0.8 m tall with blackish lower branches clearly visible, *PVB 10028*, west of Gamkab River, Namibia, 12 Jul. 2005 (© PVB).

and the base of the plant as a whole have a slightly coarser bark, which often assumes a distinctive blackish hue. This is only for a short distance above the ground and the general colour of the shrubs is grey-green. The branches are distinctly segmented at the nodes and one often finds that while still green, each segment becomes gradually thicker towards its apex. Once the colour changes to grey-brown, the segments are more distinctly thickened at their apices beneath the next branches. Leaves are present briefly when the branches grow and, while they are mostly small (at around 5–8 mm long), they can reach a length of 20 mm on young plants or on well sheltered specimens.

Flowering in subsp. *ephedroides* generally takes place between June and September, beginning with a solitary cyathium that terminates the branch. Other cyathia usually develop in the axils of the bracts subtending this terminal cyathium. By the repeated development of a succession of cyathia in the axils of bracts subtending earlier cyathia, a

Fig. 3.24. *Euphorbia ephedroides* ssp. *ephedroides*, male plant with cyathia, *PVB 11420*, near Kommagas, Namaqualand, South Africa, 12 Jul. 2009 (© PVB).

Fig. 3.25. *Euphorbia ephedroides* ssp. *ephedroides*, female plant with cyathia and developing capsules, near Naroegas, Namaqualand, South Africa, 12 Jul. 2007 (© PVB).

3.2 Sect. Articulofruticosae

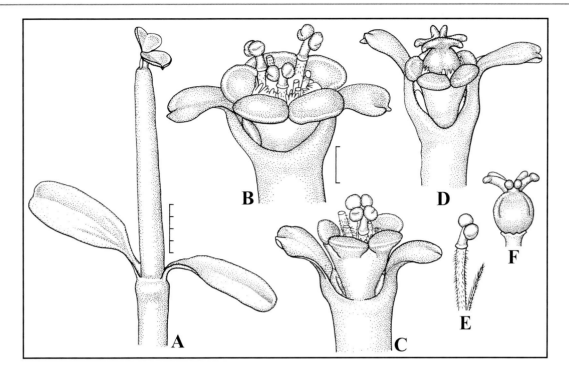

Fig. 3.26. *Euphorbia ephedroides* ssp. *ephedroides*. **A**, tip of branch with leaf-rudiments (scale 4 mm). **B**, **C**, male cyathium from side (scale 1 mm, as for **C–F**). **D**, female cyathium from side. **E**, anther and bracteole. **F**, female floret. Drawn from: *PVB 3205*, Aurus Mountains, west of Witpütz, Namibia (© PVB).

branch may give rise to large numbers of cyathia. This may result in a complex, dichotomously branched synflorescence, up to 50 mm long or more. The cyathia are very insignificant, both in colour and size and are considerably exceeded in length by their subtending bracts. The somewhat larger capsules are almost spherical but are also inconspicuously coloured.

History

Subsp *ephedroides* was first collected in September 1830 by J.F. Drège and his brother Carl during their expedition of June 1830 to January 1831 from the Cape to the Orange River. Their collection was made at 'Goedemanskraal', which lies somewhere between the Koperberg (near the present-day Springbok) and the town of Lekkersing. White et al. (1941) had little to say about it. Leach and Williamson (1990) added many previously unknown details and described two new varieties. One of these is recognised here (as subsp. *imminuta*) but the other (var. *debilis*) has been found to be impossible to separate reliably from the typical subspecies and is not recognised.

Euphorbia ephedroides subsp. **gamsbergensis** Bruyns, *Haseltonia* 25: 47 (2018). Type: South Africa, southern slopes of Pellaberg, *Bruyns 13516* (BOL, holo.; E, iso.).

Plant 0.2–0.4 m tall. *Branches* ascending and somewhat interwoven in older specimens, ± 3 mm thick below and upper parts 1 mm thick. *Capsule* exserted ± 1 mm from cyathium when mature.

Euphorbia ephedroides subsp. *gamsbergensis* is known from two of the large inselbergs of north-western Bushmanland, the Pellaberg and Gamsberg. These lie some 100 km east of other known localities for *E. ephedroides*. Plants occur in large numbers on the fairly steep southern slopes of these mountains, between altitudes of around 800 to 1000 m. Here they grow among schistose and quartzitic rubble, with a sparse cover that includes scattered shrubs of *Euphorbia avasmontana*, *E. gregaria*, *Montinia* and *Zygophyllum*.

Subsp. *gamsbergensis* differs from the other two subspecies by its generally more slender branches (most of the upper parts are around 1 mm thick, lower down rarely more than 3 mm thick) which are interwoven into a densely mat-like bush that is heavily grazed back to the lower, thicker branches during dry periods. The other difference is that the capsule is slightly exserted from the dried-out cyathium as it matures.

This subspecies has only become known relatively recently, with collections by E. van Jaarsveld on Pellaberg in March 1985 and T. Anderson and Van Heerden on Gamsberg in May 1999.

Fig. 3.27. *Euphorbia ephedroides* ssp. *gamsbergensis*, *PVB 13516*, Pellaberg, South Africa, Sep. 2019 (© PVB).

Fig. 3.28. *Euphorbia ephedroides* ssp. *gamsbergensis*, *PVB 13516*, Pellaberg, South Africa, Sep. 2019 (© PVB).

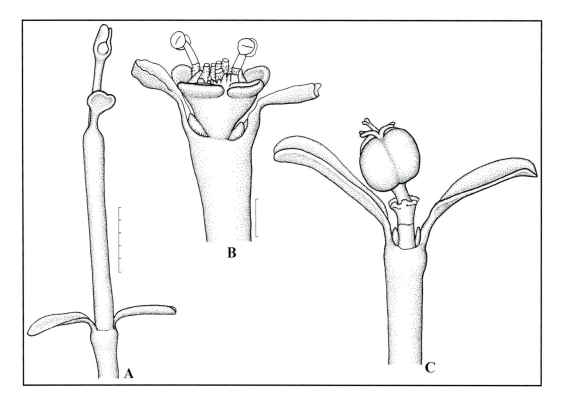

Fig. 3.29. *Euphorbia ephedroides* ssp. *gamsbergensis*. **A**, apical portion of branch (scale 5 mm). **B**, side view of male cyathium (scale 1 mm, as for **C**). **C**, side view of mature capsule. Drawn from: *PVB 13516*, Pellaberg, west of Pofadder, South Africa (© PVB).

Euphorbia ephedroides subsp. **imminuta** (L.C.Leach & G.Will.) Bruyns, *Haseltonia* 25: 47 (2018).
Euphorbia ephedroides var. *imminuta* L.C.Leach & G.Will., *S. African J. Bot.* 56: 72 (1990). *Tirucalia ephedroides* var. *imminuta* (L.C.Leach & G.Will.) P.V.Heath, *Calyx* 5: 90 (1996). Type: South Africa, Cape, Alexander Bay, *Williamson 3652* (NBG, holo.; K, PRE, iso.).

Plant 0.02–0.1 m tall. *Branches* drooping and often nearly prostrate.

Euphorbia ephedroides subsp. *imminuta* is only known on the southern side of the Orange River from near its mouth at Alexander Bay to near Beauvallon. It occurs among low outcrops of limestone with pebbles of quartz and with small accumulations of windblown sand. This area supports a wealth of other small succulents, including *E. mauritanica*, *E. stapelioides* and much reduced forms of *E. caput-medusae* (Fig. 2.375) and *E. rhombifolia* (Fig. 3.90).

This subspecies is distinguished from subsp. *ephedroides* by its small stature, with the stems frequently prostrate or drooping towards the ground. Larger plants may form low, almost hemispherical shrublets that may reach 100 mm tall and 200 mm in diameter.

Subsp. *imminuta* was first collected in July 1937 by R.A. Dyer.

Fig. 3.30. *Euphorbia ephedroides* ssp. *imminuta*, larger plant male and ± 20 cm diam., among stones and windblown sand, Kortdoringberg, east of Alexander Bay, South Africa, 13 Jul. 2014 (© PVB).

Fig. 3.31. *Euphorbia ephedroides* ssp. *imminuta*, male plant ± 12 cm diam., Kortdoringberg, east of Alexander Bay, South Africa, 13 Jul. 2014 (© PVB).

Fig. 3.32. *Euphorbia ephedroides* ssp. *imminuta*, male plant with cyathia, Kortdoringberg, east of Alexander Bay, South Africa, 13 Jul. 2014 (© PVB).

Euphorbia exilis L.C.Leach, *S. African J. Bot.* 56: 76 (1990). Type: South Africa, Cape, Aties, May 1984, *Leach & Bayer 17129* (NBG, holo.).

Euphorbia glandularis L.C.Leach & G.Will., *S. African J. Bot.* 56: 75 (1990). Type: South Africa, Cape, Klipfontein, near Steinkopf, Feb. 1984, *Leach & Hilton-Taylor 17019* (NBG, holo.; K, PRE, iso.).

Unisexual (occasionally bisexual) spineless and glabrous free-standing succulent shrub (rarely growing inside other shrubs) 0.2–0.6 (1) m tall, branching and rebranching mainly near base from similar stem to 0.3 m tall with thickened woody rootstock up to 0.3 m long bearing fibrous roots. *Branches* erect, repeatedly rebranching, cylindrical, 50–500 × 1–3 mm (sometimes to 5 mm thick near base), without distinct tubercles, smooth, soft and succulent when young and becoming woody at base only, dark green; *leaf-rudiments* on new growth towards apices of branches, opposite, (0.5) 1–2 × 0.5–1 mm, ascending and adpressed to branch then spreading towards tip, fleeting, spathulate, glabrous except sometimes for few hairs along midrib above, flat above, margins entire, obtuse, tapering below into slightly swollen base, subtended on either side at base by often inconspicuous ellipsoidal brown slightly flattened slightly rugulose stipule ± 0.3 mm diam. *Synflorescences* terminal on branches, glabrous, each of 1 unisexual (rarely bisexual) sessile cyathium subtended by 2 obovate to spathulate green to red-green bracts very like leaf-rudiments 1–2 × 0.5 mm in axils of which 2 more solitary cyathia develop on short peduncles 1–11 mm long; *cyathia* cupular, minutely pubescent towards base otherwise glabrous, 2.5–3 mm broad (± 1.5 mm long below insertion of glands), with 5 quadrate lobes with dissected pubescent margins, green; *glands* 5, transversely elliptic, 1–1.3 mm broad, spreading, yellow-green, outer margins flat and entire; *stamens* with glabrous pedicels, bracteoles filiform and pubescent; *ovary* ellipsoidal, obtusely 3-angled, glabrous, raised on short pedicel 0.7–1.5 mm long; styles ± 0.5 mm long, divided nearly to base, spreading. *Capsule* 3.5–4.5 mm diam., obtusely 3-lobed, glabrous and shiny, green with red lines down sutures and edges or wholly red above, slightly exserted from cyathium on short erect pedicel.

Distribution & Habitat
Exclusively a species of Namaqualand, *E. exilis* occurs from near Klawer through the Knersvlakte to around Bitterfontein and then more or less continuously northwards to the high areas between Nigramoep and Steinkopf. There are somewhat disjunct populations around Eksteenfontein in the Richtersveld and in the hills southwest of Loeriesfontein, a little further east of its main region of occurrence.

Euphorbia exilis is found in stony to hard, loamy ground, often among other bushes and sometimes growing inside them. It has often been observed to be more common around small, occasionally wet water-courses in valleys between hills, while it is much less plentiful away from these patches and rarely occurs on the slopes of the hills themselves.

Diagnostic Features & Relationships
Euphorbia exilis forms fairly dense shrubs which are usually between 0.2 and 0.5 m tall and arise from a quite stout, somewhat woody rootstock. These shrubs are made up of many slender branches, which are generally strictly erect and rebranch mainly near their bases. The branches are dark green and smooth, with quite long internodes and, unusually for *Euphorbia*, they are brittle and break easily if bent. Towards their apices they bear tiny, almost spathulate leaf-rudiments which are pressed to the surface of the branch in their lower half and spread slightly above, with two small, round, flattened brown stipules alongside their bases. These stipules are quite easily visible in live material but become harder to see in dried specimens. Plants of *E. exilis* are often heavily grazed and then seem to resprout from the thicker bases of the branches and from the stout rootstock. There also seems to be considerable die-back of the branches in years of poor rainfall and many of the older plants have a lot of dead branches among the living ones.

3.2 Sect. Articulofruticosae

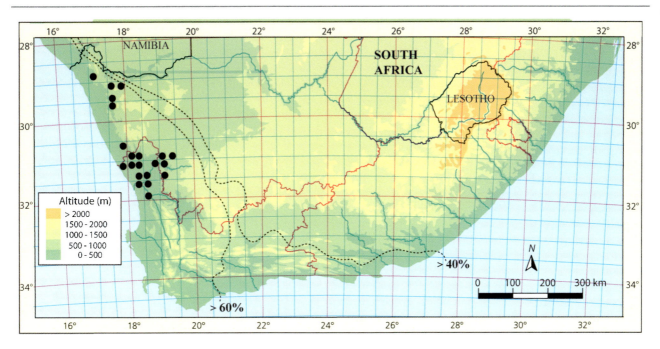

Fig. 3.33. Distribution of *Euphorbia exilis* (© PVB).

Fig. 3.34. *Euphorbia exilis*, typical densely and finely branched shrub (with many dead branches among the others) ± 30 cm tall, *PVB 10006*, south of Bitterfontein, South Africa, 10 Jul. 2005 (© PVB).

The small cyathia are produced mainly between June and August at the tips of the newer branches but are quite inconspicuous, with small yellow glands. The capsules, however, are more striking. They are a bright and shiny green, usually with red lines down the grooves between the locules and along the edges. Between Steinkopf to Nigramoep, they were observed to have a red blush above with dark green in the sutures and pale green below but still contrasted noticeably against the stems.

The branches of *E. exilis* are so slender that it cannot be confused with any other species, except occasionally for some slender forms of *E. rhombifolia* and for *E. tenax*, where the branches are also generally very slender. However, the leaf-rudiments are always distinctly spathulate in *E. exilis* and do not have the distinctive horns at their bases that are present in the other two. Stipules are absent in *E. rhombifolia* and *E. tenax*. In *E. rhombifolia* the branches are very variable in colour but are never the dark green of *E. exilis* and in *E. tenax* the branches are always grey-green and may be somewhat roughened.

History

Euphorbia exilis was recorded many times before it was eventually described by L.C. Leach in 1990. The first record was made by Carl Zeyher near Bitterfontein, probably in 1829 or 1830. Harry Bolus made a specimen in August 1883 near Steinkopf and Rudolf Schlechter collected it in July

Fig. 3.35. *Euphorbia exilis*, ± 0.8 m broad, Nuwerus, Namaqualand, South Africa, 12 Jul. 2006 (© PVB).

Fig. 3.36. *Euphorbia exilis*, male plant flowering, Nuwerus, Namaqualand, South Africa, 12 Jul. 2006 (© PVB).

1896 near the Sout River on the Knersvlakte and again in September 1897 at Steinkopf. These earlier collections were variously referred to as *E. ephedroides*, *E. perpera*, *E. rudolfii* and *E. spartaria*. Leach (1990) described two names for this species. One was based on material from the Knersvlakte and the other, *E. glandularis*, was based on specimens from around Steinkopf but I have shown (Bruyns 1992b) that only one species is involved.

3.2 Sect. Articulofruticosae

Fig. 3.37. *Euphorbia exilis*, female plant with shiny capsules (here green) in different stages, *PVB 3225*, north of Hol River siding, Namaqualand, South Africa, 13 Aug. 1988 (© PVB).

Fig. 3.38. *Euphorbia exilis*, female plant with cyathia and red capsules, *PVB 6151*, Katdoringvlei, west of Kotzesrus, Namaqualand, South Africa, 8 Sep. 1994 (© PVB).

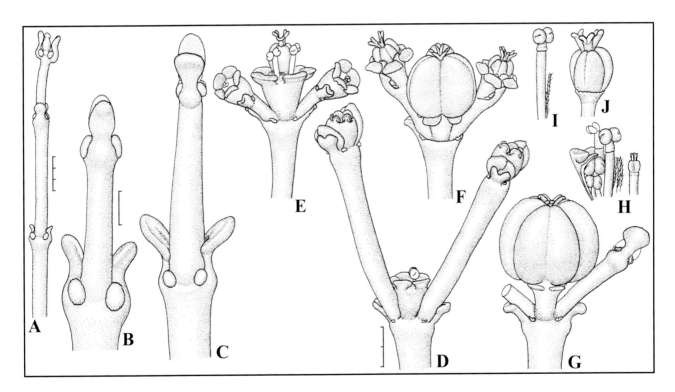

Fig. 3.39. *Euphorbia exilis*. **A–C**, tip of branch with leaf-rudiments (scale 3 mm for **A**, 1 mm for **B** as for **C**, **H–J**). **D–G**, cymes in different stages of development (scale 2 mm for **D**, as for **E–G**). **H**, anthers with rudimentary female. **I**, anther and bracteole. **J**, female floret. Drawn from: **A**, **B**, **D**, **G**, *PVB 3214*, west of Bulletrap, Namaqualand, South Africa. **C**, **H**, *PVB 3835*, north of Groen River, Namaqualand, South Africa. **E**, **F**, **I**, **J**, *PVB 3225*, north of Hol River siding, Namaqualand, South Africa (© PVB).

Euphorbia gentilis N.E.Br., *Fl. Cap.* 5 (2): 289 (1915). *Tirucalia gentilis* (N.E.Br.) P.V.Heath, *Calyx* 5: 90 (1996). Lectotype (Bruyns 2012): South Africa, Cape, Vanrhynsdorp Div., hills near Zout River, 500', 14 Jul. 1896, *Schlechter 8136* (BOL; BM, COI, GRA, HBG, K, PRE, S, iso.).

Euphorbia vaalputsiana L.C.Leach, *S. African J. Bot.* 54: 534 (1988). Type: South Africa, Cape, Vaalputs, near Gamoep, 28 Oct. 1984, *Leach & Perry 17232* (NBG, holo.; K, MO, PRE, iso.).

Euphorbia gentilis subsp. *tanquana* L.C.Leach, *S. African J. Bot.* 54: 538 (1988). Type: South Africa, Cape, near turnoff to Skitterykloof, *Leach & Perry 17247a* (NBG, holo.; K, M, MO, PRE, iso.).

Unisexual spineless glabrous succulent shrublet 50–150 mm tall branching often densely just above ground from short firm subterranean stem rising from apex of often slightly swollen irregularly cylindrical rootstock up to 100 × 5–15 mm bearing fibrous roots. *Branches* ascending to erect, often rebranching, irregularly cylindrical and tapering somewhat to apex and distinctly constricted at base, 5–150 × 3–6 mm, with indistinct tubercles surrounded by slight grooves on young growth, soft and succulent when young and remaining so with age, green often longitudinally banded with purple, smooth to distinctly papillate; *leaf-rudiments* towards apices of branches on slightly raised rounded tubercles, initially opposite (later sometimes shifted to appear alternating), 0.5–3 × 1–2 mm, ascending, spathulate, slightly fleshy, minutely papillate, flat above, obtuse, tapering into short petiole 0.5–1 mm long, estipulate. *Synflorescences* terminal or clustered around tips of branches, each of 1–3 unisexual (rarely bisexual) cyathia on short peduncle 1–2 mm long each subtended by 2 obovate to nearly circular often apically toothed sparsely pubescent bracts 1–2 × 1 mm resembling leaves; *cyathia* cupular to slightly urceolate, glabrous to sparsely pubescent, 2–3 mm broad (± 2 mm long below insertion of glands), with 5 obovate lobes with deeply dissected margins, green sometimes suffused with red; *glands* 5, transversely elliptic, 1–1.5 mm broad, spreading, yellow-green, outer margins flat and entire; *stamens* with glabrous pedicels, bracteoles filiform and pubescent towards tips; *ovary* globose, shortly pubescent, slightly raised then becoming exserted on decurved pedicel 3–4 mm long; styles 0.7–2 mm long, divided to lower third, widely spreading. *Capsule* 3–3.5 mm diam., obtusely 3-lobed, sparsely and shortly coarsely pubescent, green suffused with red, exserted from cyathium on decurved and later erect pedicel 3–6 mm long.

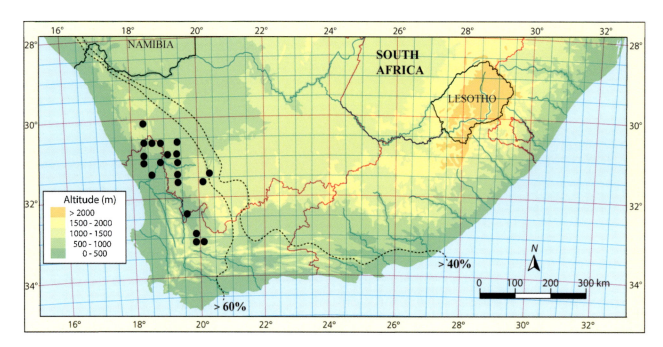

Fig. 3.40. Distribution of *Euphorbia gentilis* (© PVB).

Distribution & Habitat

Euphorbia gentilis is distributed towards the eastern margin of the winter-rainfall region of South Africa from just south of Gamoep in Bushmanland to the eastern side of the Knersvlakte, where it occurs from Quaggaskop to near Loeriesfontein. It has been recorded a few times from between Loeriesfontein and south-east of Calvinia and it appears again more regularly on the western and southern Tanqua Karoo from near Skitterykloof to some 40 km east of Karoopoort towards Sutherland.

Euphorbia gentilis is usually found in flat or gently sloping, often somewhat calcareous loam derived from gneiss or shales or on calcretes, sometimes with some windblown, red sand covering the surface (as on parts of the Knersvlakte). It is generally of humble habit and tends to grow inside other short shrubs (usually not exceeding 0.3 m tall) and more rarely grows in the open, unprotected by other bushes. Consequently plants are generally inconspicuous and it may be more widely distributed than the records show.

Fig. 3.41. *Euphorbia gentilis*, female plant with brightly coloured branches, Flaminkvlakte, Namaqualand, South Africa, 4 Aug. 2012 (© PVB).

Fig. 3.42. *Euphorbia gentilis*, male plant with dull branches and many cyathia, Flaminkvlakte, Namaqualand, South Africa, 4 Aug. 2012 (© PVB).

Diagnostic Features & Relationships

Euphorbia gentilis forms tiny shrubs which arise from the apex of an often somewhat swollen (sometimes slender as in Fig. 3.45) but rarely deeply seated tap-root. Branching takes place above the surface of the ground. Although the shrublets formed rarely exceed 0.1 m in diameter, they are variable in size and richness of branching, depending on the amount of shelter and water received. In some exposed plants, there is an occasional tendency for the branches to bend to face northwards. The branches are comparatively stout (relative to their length) and are always soft and succulent, tapering slightly towards their tips but without any tendency to dry out and become spiky at their tips. Their surface is often quite prominently banded with purple on grey-green (with a grey-green patch around and on each tubercle and the rest of the internode a purplish colour –

Fig. 3.43. *Euphorbia gentilis*, robust and boldly marked male plant, Donkiedam, west of Loeriesfontein, South Africa, 12 Jul. 2007 (© PVB).

Fig. 3.44. *Euphorbia gentilis*, robust male plant ± 15 cm diam., Sekretarspan, east of Karoopoort, Tanqua Karoo, South Africa, 17 Jun. 2006 (© PVB).

3.2 Sect. Articulofruticosae

Fig. 3.45. *Euphorbia gentilis,* excavated plants showing very slender tap-root found at this locality, *PVB 10013*, south of Grootriet, Namaqualand, South Africa, 10 Jul. 2005 (© PVB).

though this is variable in intensity). The surface of the branches is smooth to papillate within populations. Tiny, almost circular, fleshy leaf-rudiments are present near the tips of young growth. Each leaflet is seated on a tubercle which merges gradually with the surface of the branch as the branch ages.

Flowering takes place during the winter, mainly in July and August. In years of good rainfall the tips of the branches become covered with the small yellowish cyathia. Each cyathium arises on a very short peduncle and, although initially solitary, further cyathia may develop in the axils of the bracts just below it. The cyathia seem to be variable in shape, with more urceolate ones found in the north and slightly shallower, more conical ones found further south.

Euphorbia gentilis is closely related to and not always easy to distinguish from *E. juttae* and the differences between them are discussed under *E. juttae*.

Fig. 3.46. *Euphorbia gentilis,* male cyathia, *PVB 10101*, Bokkraal, east of Garies, South Africa, 22 Jul. 2005 (© PVB).

Fig. 3.47. *Euphorbia gentilis*, female plant with many capsules, *PVB 11145*, near Platbakkies, Bushmanland, South Africa, 20 Oct. 2008 (© PVB).

History

Euphorbia gentilis was first gathered by Carl Zeyher (under the number *Zeyher 1531*) near Bitterfontein, probably during 1829 (Gunn and Codd 1981). This specimen was quoted by N.E. Brown under *E. gentilis* but Boissier considered it to represent *E. spicata* (= *E. muricata*). Leach (1988b) wrote extensively about *E. gentilis* and described a closely related new species, *E. vaalputsiana*, as well as a new subspecies of *E. gentilis*. However, further collecting has indicated that these are both part of a broader concept of *E. gentilis*.

According to Leach (1988b), *E. vaalputsiana* was distinguished from both *E. gentilis* and *E. juttae* by its 'velvety'-papillate branches, while the others were 'glabrous to the eye (actually minutely asperulous)'. As Leach mentioned, they are all minutely papillate but longer papillae are not restricted to plants from south of Gamoep and are typical of most material from between there and Loeriesfontein, with many populations on the eastern Knersvlakte also containing plants with both obviously papillate and almost glabrous branches.

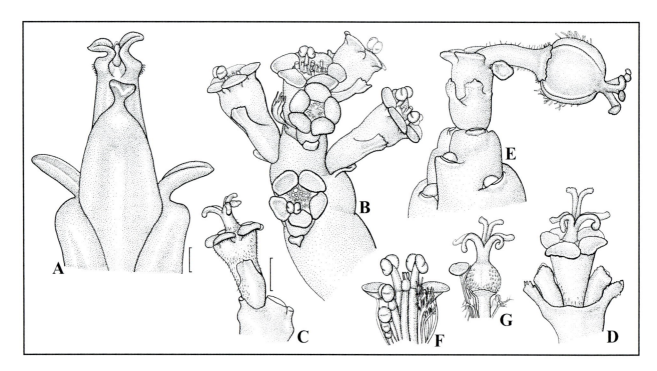

Fig. 3.48. *Euphorbia gentilis*. **A**, tip of branch with leaf-rudiments (scale 1 mm, as for **B**). **B**, tip of flowering branch on male plant. **C–E**, female cyathium at different stages (scale 1 mm, as for **D–G**). **F**, dissected male cyathium. **G**, part of dissected female cyathium. Drawn from: **A**, **B**, **D–G**, *PVB 4708*, Stofkraal, SE of Garies, South Africa. **C**, *PVB 4707*, Kliprand, South Africa (© PVB).

3.2 Sect. Articulofruticosae

Euphorbia giessii L.C.Leach, *Dinteria* 16: 27 (1982). *Tirucalia giessii* (L.C.Leach) P.V.Heath, *Calyx* 5: 90 (1996). Type: Namibia, 18 km east of Henties Bay, Dec. 1976, *Giess 14809 (sub Leach 15940)* (PRE, holo.; M, WIND, iso.).

Unisexual spineless glabrous succulent ± spherical shrub 0.3–0.8 m tall branching often densely above ground from short subterranean stem at apex of wiry or slightly fleshy taproot 100–200 × 5–20 mm bearing fibrous roots. *Branches* erect to spreading, often rebranching, cylindrical, 50–400 × 3–8 mm, ± without tubercles, soft and succulent when young becoming rigid with age and occasionally drying off towards tips, grey-green to purple-green, smooth; *leaf-rudiments* towards apices of branches on slightly raised rounded tubercles, initially opposite (later sometimes shifted to appear alternating), 2–7 × 1–3 mm, ascending and spreading, caducous, ovate-deltate, slightly fleshy, glabrous, flat to slightly channelled above, acute, with short petiole 0.5–1 mm long, estipulate. *Synflorescences* in terminal cymes, of 1 to many unisexual sessile cyathia each subtended by 2 obovate often apically toothed glabrous to sparsely pubescent bracts 1–2 × 1 mm resembling leaves in axils of which further cyathia develop on short branches, these cyathia again subtended by bracts from whose axils further cyathia develop in the same way (repetition and branching most frequent in males); *cyathia* conical-cupular in females to slightly urceolate in males, usually sparsely pubescent, 1.5–2.5 mm broad and almost twice as broad in males as females (± 1.5 mm long below insertion of glands), with 5 obovate pubescent lobes with deeply dissected margins, pale green rarely suffused with red; *glands* 5, transversely elliptic to nearly circular, 0.8–1 mm broad, ascending, yellow-green, outer margins entire; *stamens* with glabrous pedicels, bracteoles filiform and pubescent; *ovary* globosely 3-lobed, sparsely pubescent, raised on pedicel ± 1 mm long; styles 0.8–1.0 mm long, divided to near base, slightly to widely spreading. *Capsule* 3–4.5 mm diam., obtusely 3-lobed, sparsely to densely pubescent, green suffused with red, ± sessile ontop of cyathium on erect pedicel ± 1 mm long.

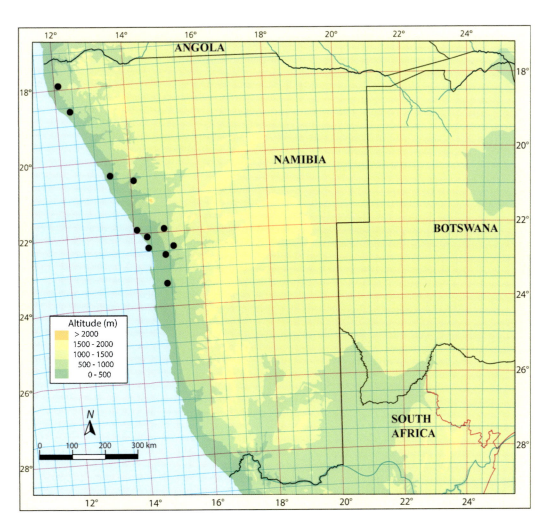

Fig. 3.49. Distribution of *Euphorbia giessii* (© PVB).

Distribution & Habitat

Found in Namibia only, *E. giessii* is confined to the Namib Desert from near Swakopmund to some 80 km south of the mouth of the Cunene River.

In some spots *E. giessii* is a common plant, usually in gravelly, flattish places with a very scanty cover of shrubs and isolated tufts of grass or in patches of gravelly sand between larger rocks. It grows in areas of remarkable aridity, receiving less than 100 mm of rain annually, probably supplemented by moisture from fog from the sea.

Diagnostic Features & Relationships

Plants of *E. giessii* form shrubs usually around 0.5 m tall and very variably shaped, from dense and hemispherical to

Fig. 3.50. *Euphorbia giessii*, on gravelly plains, large plant ± 0.8 m tall, *PVB 12835*, near Rössing, Namibia, 20 Dec. 2014 (© PVB).

Fig. 3.51. *Euphorbia giessii*, ± 0.3 m tall, *PVB 12835*, near Rössing, Namibia, 20 Dec. 2014 (© PVB).

taller and spherical or roughly inverted conical. They are also very variably branched, sometimes densely with large numbers of branches and sometimes only sparsely branched. Young plants often have distinctly weeping branches and more conspicuous leaf-rudiments but these disappear in older specimens, where the branches are erect and the leaves are much smaller.

Florally this species is typical of the section, with small and insignificant cyathia that are slightly narrower in the females than the males. Initially the cyathia are terminal on

3.2 Sect. Articulofruticosae

Fig. 3.52. *Euphorbia giessii*, among boulders east of Wlotzkabaken, Namibia, 21 Dec. 2014 (© PVB).

the branches but with successive branching from the axils of their bracts, a quite complexly branched structure may develop and this is more pronounced in males than in females. As it matures, the ovary is pushed slightly out of the cyathium so as to protrude just beyond the cyathial lobes. It remains like this so that the usually noticeably pubescent capsules are seated ontop of the remains of the cyathium and are not exserted further as they mature.

Leach (1982) compared *E. giessii* to *E. negromontana*, another species endemic to the Namib Desert, though at that time only known in its northernmost part around the town of Namibe in southern Angola. The two differ by the considerably shorter bracts around the cyathia in *E. negromontana* (where they are around half the length of those in *E. giessii*) and the rounder ± glabrous cyathia (narrower and more cylindrical in *E. giessii* and also more pubescent). Both *E. giessii* and *E. negromontana* were included in the complete sampling of this section in Bruyns et al. (2011) and were found to be very closely related sisters, bearing out Leach's opinion that they were 'most closely related'. From the widespread *E. spartaria*, *E. giessii* differs by the pubescent capsules. In habit they also differ: in *E. giessii* the branches are

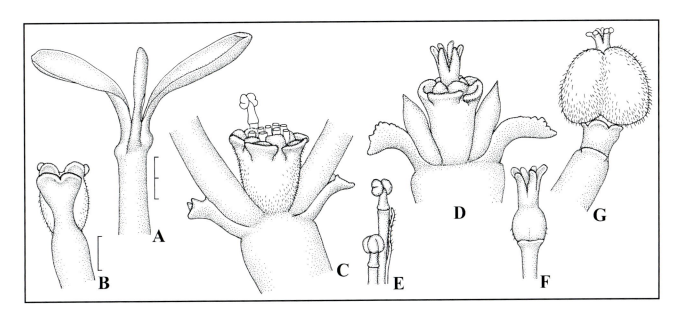

Fig. 3.53. *Euphorbia giessii*. **A**, tip of branch with leaf-rudiments (scale 2 mm). **B**, tip of flowering branch on male plant (scale 1 mm, as for C–G). **C**, male cyathium. **D**, female cyathium. **E**, anthers and bracteole. **F**, female floret. **G**, capsule. Drawn from: *PVB 12835*, near Rössing, Namibia (© PVB).

more succulent and the lower branches spread more horizontally, while in *E. spartaria* the branches are more slender and are all more erect.

History

The first record of *Euphorbia giessii* is a collection of Richard H.W. Seydel, made in January 1957 (*Seydel 888*, K, PRE). It is hard to explain why this species went unnoticed by Dinter, except for the fact that it does not occur near the old railway-line from Swakopmund to Windhoek, along which most of the early collecting took place.

Euphorbia herrei A.C.White, R.A.Dyer & B.Sloane, *Succ. Euphorb.* 2: 962 (1941). *Tirucalia herrei* (A.C.White, R.A.Dyer & B.Sloane) P.V.Heath, *Calyx* 5: 90 (1996). Type: South Africa, Cape, near Swartwater, Oct. 1930, *Herre sub PRE 46025* (PRE, holo.).

Unisexual or bisexual dwarf spineless glabrous succulent 50–150 mm tall branching extensively from small subglobose stem with 2–4 nodes and arising from slender wiry rootstock bearing fibrous roots. *Branches* erect to spreading and rebranching extensively, strongly constricted at base and irregularly rebranched into subglobose irregularly 4- or 5-angled segments, tapering to apex, 5–50 (150) × 3–10 mm and usually with 2–4 nodes only, without distinct tubercles, smooth to finely papillate, soft and succulent when young and becoming rigid with age, grey-green or brown-green to reddish on new growth; *leaf-rudiments* towards apices of branches on very slightly raised tubercles, opposite or nearly opposite, 1.5–2 × 0.5–1.5 mm, spreading and slightly recurved, caducous, ovate with often conspicuously toothed margins, glabrous, obtuse to acute, with short petiole 0.5–1 mm long, estipulate. *Synflorescences* terminal on branches, each of 1–5 unisexual or bisexual cyathia on very short peduncle ± 1 mm long each subtended by 2 obovate leaf-like bracts 1–2 × 1–1.5 mm; *cyathia* cupular, minutely pubescent, 2–2.5 mm broad (± 1.5 mm long below insertion of glands), with 5 obovate lobes with deeply dissected margins, green; *glands* 5, transversely elliptic, 0.7–1.2 mm broad, ascending to spreading, yellow-green, outer margins flat and entire; *stamens* with glabrous pedicels, bracteoles filiform and pubescent; *ovary* globose, glabrous to sparsely pubescent, exserted on eventually decurved pedicel 3–4 mm long; styles ± 1 mm long, divided to below middle, spreading, sometimes horizontally. *Capsule* 2.5–3 mm diam., slightly longer than broad, obtusely 3-lobed, glabrous, green, exserted on decurved and later erect pedicel 3.5–4.5 mm long.

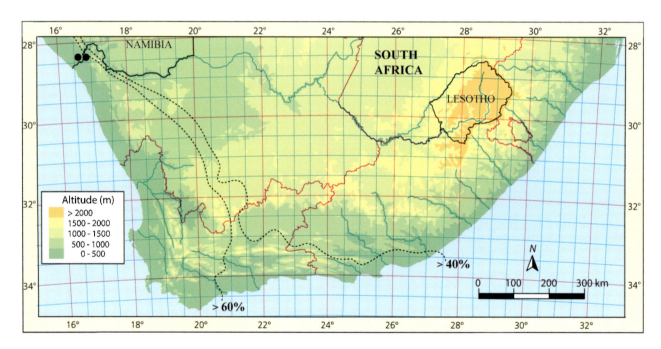

Fig. 3.54. Distribution of *Euphorbia herrei* (© PVB).

Distribution & Habitat

Euphorbia herrei is found only along the hyper-arid portion of the Orange River Valley that runs in a roughly north-south direction between Sendelings Drift and Grootderm. It has been recorded on both sides of the river, i.e. both in Namibia and in South Africa, at altitudes of between 100 and 250 m.

In these localities, it grows mostly in the open on flat patches in hard, loamy, shallow ground among pebbles, often on soils derived from grey to black limestones.

3.2 Sect. Articulofruticosae

Fig. 3.55. *Euphorbia herrei*, plump after rain, *PVB 3943*, Bloeddrift, east of Alexander Bay, South Africa, 16 Jul. 1989 (© PVB).

Fig. 3.56. *Euphorbia herrei*, in dry state among alluvial pebbles, *PVB 3943*, Bloeddrift, east of Alexander Bay, South Africa, 8 Jul. 2007 (© PVB).

Diagnostic Features & Relationships

In habitat, plants of *Euphorbia herrei* form a minute shrublet usually only 40 to 60 mm tall (a little larger if sheltered or in deeper ground) and they may become considerably larger and more luxuriant in cultivation, where they may survive for many years. Beneath the ground, the stem tapers off into a slender, wiry tap-root and the stem itself is a short, plump body consisting of two to four nodes and internodes. In the axils of the upper nodes, similar plump branchlets arise, consisting also of two to four nodes, with the lowest internode usually the fattest and those above becoming more slender. After these the branch ceases to grow and growth continues from further branchlets near its apex. With time much rebranching takes place and older plants may develop into a small, tight cluster. Each softly fleshy branch is then often only slightly longer than it is

Fig. 3.57. *Euphorbia herrei*, two plants removed to show the slender rootstock, *PVB 3943*, Bloeddrift, east of Alexander Bay, South Africa, 16 Jul. 1989 (© PVB).

thick and, with the deep articulations between it and the next, the appearance of small sausages strung together is created. Towards its apex each branchlet bears generally only two to three pairs of small, rapidly deciduous leaf-rudiments, which are recurved and often have a toothy margin.

As is typical for this section, the cyathia are small and inconspicuous, but they are not always unisexual. At an early stage the capsule is exserted from the cyathium on a relatively long, somewhat decurved pedicel, which becomes erect as the capsule matures (as seen in Fig. 1.51). The plants are sometimes self-fertile.

While *E. herrei* is closely related to species such as *E. rhombifolia* and *E. stapelioides*, it is easily separated from them by the unusually and irregularly swollen, clearly angled branchlets. Differences are clear also in the leaf-rudiments, where those of *E. herrei* lack the lateral teeth at the base that are present in the others. The capsule is also exserted further from the cyathium than is common in the other species.

History

Euphorbia herrei was first collected by Hans Herre in October 1930 near Swartwater, near Beesbank along the southern bank of the Orange River in South Africa. At the time, Herre (1936) believed that he had re-discovered the little-known *E. stapelioides*, which he believed had not been collected since its discovery by the brothers Drège in 1830. Once it was realised that this was incorrect, these plants were named after their discoverer by White et al. (1941). The first record from Namibia was made only in 1993.

Fig. 3.58. *Euphorbia herrei*, bisexual cyathia, *PVB 3943*, Bloeddrift, east of Alexander Bay, South Africa, 14 Oct. 2018 (© PVB).

3.2 Sect. Articulofruticosae

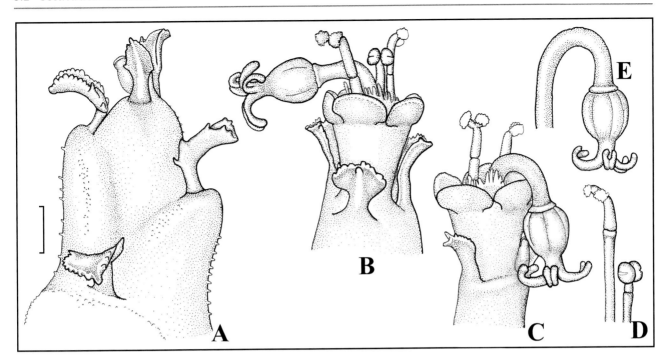

Fig. 3.59. *Euphorbia herrei*. **A**, side view of tip of branch with leaf-rudiments (scale 1 mm, as for **B–E**). **B, C**, side view of cyathium. **D**, anthers. **E**, female floret. Drawn from: *PVB 3943*, Bloeddrift, east of Alexander Bay, South Africa (© PVB).

Euphorbia juttae Dinter, *Neue Pflanzen Deutsch-SWA's*: 30 (1914). *Tirucalia juttae* (Dinter) P.V.Heath, *Calyx* 5: 90 (1996), as '*judithae*'. Lectotype (Leach 1988b): Namibia, Garub, 900 m, 9 Jan. 1910, *Dinter 1047* (SAM; NY, iso.).

Euphorbia siliciicola Dinter, *Neue Pflanzen Deutsch-SWA's*: 31 (1914). Lectotype (Leach 1988b): Namibia, Büllsport, 5 Apr. 1911, *Dinter 2132* (SAM).

Euphorbia aequoris N.E.Br., *Fl. Cap.* 5 (2): 279 (1915). *Tirucalia aequoris* (N.E.Br.) P.V.Heath, *Calyx* 5: 87 (1996). Lectotype (Bruyns 2012): South Africa, Cape, Middelburg div., Schoombie, Feb. 1897, *Trollip (sub SAM 20091)* (SAM; K, iso.).

Unisexual spineless glabrous succulent shrublet 30–220 mm tall branching sparingly to densely above ground from short firm subterranean stem rising from apex of wiry taproot or slightly swollen irregularly cylindrical tuber up to 100 × 5–15 (25) mm bearing fibrous roots. *Branches* prostrate to erect, often rebranching, cylindrical and tapering somewhat to apex and sometimes constricted at base, 5–150 × 3–6 mm, with indistinct tubercles surrounded by slight grooves on young growth, soft and succulent when young and remaining so with age, grey-green to dark green sometimes lined with purple or red, smooth; *leaf-rudiments* towards apices of branches on slightly raised rounded tubercles, initially opposite (later sometimes shifted to appear alternating), 1–12 × 1–4 mm, ascending to spreading, spathulate, slightly fleshy, glabrous, flat to slightly channelled above, obtuse, tapering into short petiole 0.5–1 mm long, estipulate. *Synflorescences* terminal, each of 1–3 unisexual (rarely bisexual) sessile cyathia each subtended by 2 obovate to elliptic often apically toothed glabrous to sparsely pubescent bracts 1–2 × 1 mm resembling leaves; *cyathia* cupular to slightly urceolate, glabrous, 1.5–2 mm broad (1–2 mm long below insertion of glands), with 5 obovate lobes with deeply dissected margins, green sometimes suffused with red; *glands* 4–5, transversely elliptic, 1–1.5 mm broad, spreading, yellow-green, outer margins flat and entire; *stamens* with glabrous or sparsely pubescent pedicels, bracteoles filiform and pubescent towards tips; *ovary* globose, glabrous, slightly raised then becoming exserted on decurved pedicel 3–4 mm long; styles 0.7–2 mm long, divided to between middle and lower third, widely spreading. *Capsule* 3.5–4 mm diam., obtusely 3-lobed, glabrous, green suffused with red, exserted on decurved and later erect pedicel 2–5 mm long.

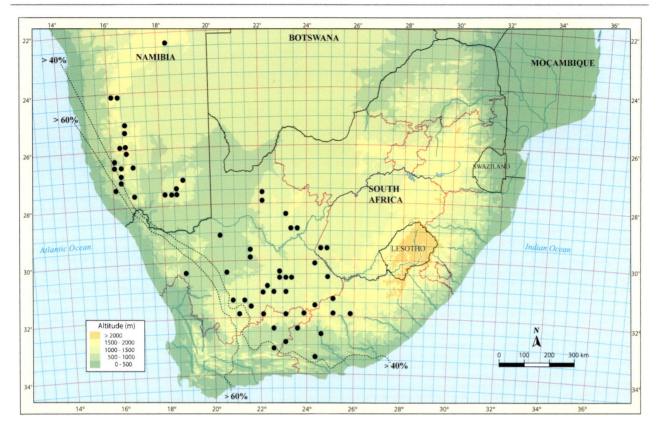

Fig. 3.60. Distribution of *Euphorbia juttae* (© PVB).

Fig. 3.61. *Euphorbia juttae*, female plant ± 6 cm long, pressed to ground and facing north, in granitic gravel, *PVB 10079*, near Aus, Namibia, 17 Jul. 2005 (© PVB).

Distribution & Habitat

Euphorbia juttae is a widely, though rather sparsely distributed species that has been collected in Namibia at altitudes of up to 1900 m, from the Naukluft Mountains to a little north of Rosh Pinah. It is also well-documented from the Great and Little Karas Mountains and there is an isolated record of Dinter's from near Gobabis. In South Africa it is known from Kakamas, Kenhardt and Kliprand to the Free State near Lückhoff and south across the Great Karoo to Fraserburg, Kendrew and Steytlerville.

Fig. 3.62. *Euphorbia juttae*, male plant ± 5 cm long, nearly prostrate and facing north, among calcrete chips on edge of pan, *PVB 3474*, Pearson's Hunt, north of Upington, South Africa, 9 Jan. 1989 (© PVB).

Fig. 3.63. *Euphorbia juttae*, female plant, nearly erect, among calcrete chips on edge of pan, *PVB 3474*, Pearson's Hunt, north of Upington, South Africa, 20 Dec. 2012 (© PVB).

Fig. 3.64. *Euphorbia juttae*, female plant, ± erect, *PVB 11353*, plateau of Great Karas Mountains, north of Grünau, Namibia, 5 Apr. 2009 (© PVB).

Around Aus in Namibia it is may be found in patches of quartz-gravel, but typically it grows inside low bushes, often in calcareous patches, where the vegetation is very short but the soil is fairly deep.

Diagnostic Features & Relationships
Euphorbia juttae is a lowly plant that varies considerably in size and vigour across its large distribution. On the gravel-plains around Aus they are very short, with the stem and few branches remaining close to the ground and bent strongly towards the north or west. The branches in these plants are comparatively thick, swollen and noticeably jointed with quite conspicuous tubercles and the whole plant usually arises from a slender, wiry tap-root, though this may be swollen also. Outside of such habitats it generally forms a small, often very diffuse, few- and slender-branched shrublet arising from a slender, slightly woody stem which descends into the ground and eventually swells into a slightly tuberous structure from which finer roots spread out.

When rains bring on new growth, the young branches bear small to quite prominent leaves that taper into a short petiole, with the whole structure on a small, raised, cushion-like tubercle. The size of the leaves seems to depend on the time of the year in which rain is received, with the smallest on plants receiving rain in winter.

Flowering takes place over most of the year after rain has stimulated growth and the tiny cyathia are produced singly at the tips of the branches, sometimes with two more cyathia arising later in the axils of the bracts around the first one. Although the plants are usually unisexual, they are often bisexual as well, with some cyathia producing viable florets

Fig. 3.65. *Euphorbia juttae*, female plants erect and ± 10 cm tall, extracted from growing inside other shrublets and with fairly stout tubers, *PVB 13066*, Snyderskraal, north of Murraysburg, South Africa, 1 Nov. 2015 (© PVB).

Fig. 3.66. *Euphorbia juttae*, plants erect ± 15 cm tall, removed from growing inside other shrublets, with more slender tubers, *PVB 11021*, Gansegat, NW of Colesberg, South Africa, 30 Apr. 2008 (© PVB).

Fig. 3.67. *Euphorbia juttae*, leafy plant with capsules, erect one about to dehisce, *PVB 10079*, near Aus, Namibia, 30 Nov. 2012 (© PVB).

of both sexes. The male florets are small and inconspicuous but the female floret becomes more prominent after fertilization and then the capsule is exserted and usually dependent on a very slender pedicel.

Euphorbia juttae and *E. gentilis* are very similar in many respects and are hard to separate reliably. Leach (1988b) distinguished the two by the fact that the leaves are 'elliptic or oblong-elliptic, petiolate' and the cyathia have four glands in *E. juttae*, while the leaves are 'rudimentary, sessile, ovate or almost circular' and the cyathia have five glands in *E. gentilis*. The sketches here of the two species show that the differences in the leaves cannot be used to separate them. Furthermore, the number of glands on the cyathium is variable in *E. juttae*, even in the area from which the type plants came. The two differ primarily in the much denser branching of *E. gentilis*, with its production of many cyathia around the tips of the branches, rather than the sparse branching and much smaller groups of one to three cyathia found in *E. juttae*. Plants of both species may produce subterranean tubers (also occasionally present in *E. rhombifolia*), while on others the taproot tapers off quickly into slender fibrous rootlets. Leach (1988b) also mentioned the habit of *E. juttae*, where the branches bend strongly to the north or west, sometimes becoming almost prostrate and the manner in which the tubercles may be particularly prominent in some material from around Aus. However, he pointed out that *E. juttae* was variable in both these features and that plants with more erect and more cylindrical stems had been referred to as *E. siliciicola*, but were now considered to fall within the range of *E. juttae*.

Fig. 3.68. *Euphorbia juttae*, leafless plant with capsules, *PVB 13066*, Snyderskraal, north of Murraysburg, South Africa, 1 Nov. 2015 (© PVB).

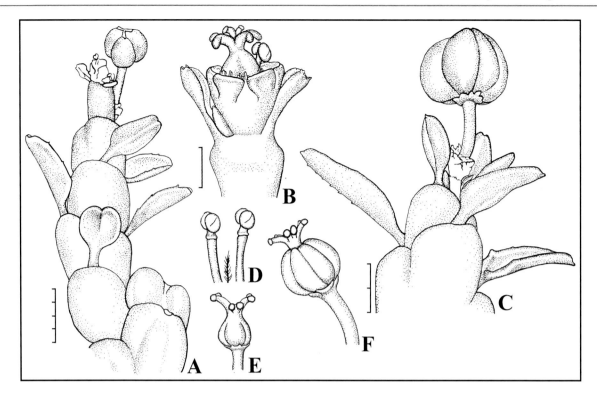

Fig. 3.69. *Euphorbia juttae*. Short plants. **A**, tip of branch with leaf-rudiments (scale 4 mm). **B**, side view of bisexual cyathium (scale 1 mm, as for **D–F**). **C**, tip of fruiting branch (scale 2 mm). **D**, anthers and bracteole. **E, F**, female floret. Drawn from: *PVB 3243*, Anib, near Pockenbank, south of Aus, Namibia (© PVB).

History

Plants described as *Euphorbia aequoris* were first collected between Colesburg and Hanover by Harry Bolus in 1871. Those described as *E. juttae* were first collected early in January 1910 by Jutta Dinter at Garub, near Aus.

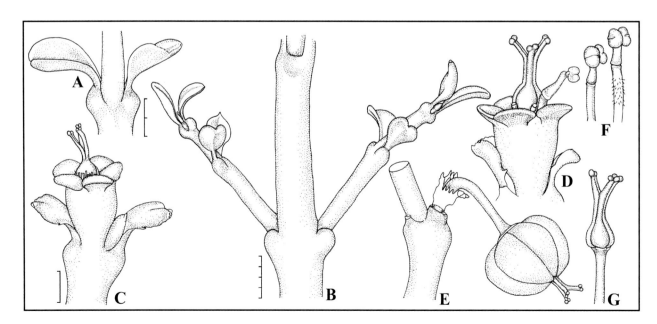

Fig. 3.70. *Euphorbia juttae*. Taller plants ('*aequoris*'). **A, B**, near tip of branch with leaf-rudiments (scale 2 mm for **A**, as for **E**; scale 4 mm for **B**). **C, D**, side view of female/bisexual cyathium (scale 1 mm, as for **D, F, G**). **E**, part of fruiting branch. **F**, anthers. **G**, female floret. Drawn from: *PVB 3137*, Rietbron, South Africa (© PVB).

Leach (1988b) discussed *E. juttae* in detail but considered that *E. aequoris*, although closely related, was 'sufficiently distinct in vegetative characters and habit alone for it to be disregarded' in those discussions. He did not say in what way it was so distinct but one may speculate that this distinctiveness lay in the much more robust plants formed by *E. aequoris*, the longer and more slender stem and branches, with more widely spaced and less prominent tubercles and a longer rootstock, as well as the lack of the peculiar habit in *E. juttae*, where the plant bends over to the north or west. Nevertheless, among the material that he cited under *E. juttae* were two specimens from near Olifantshoek and Kenhardt respectively that are rather more typical of *E. aequoris* than of *E. juttae*. While many specimens of *E. aequoris* are unmistakeable (especially those from the Great Karoo and the drier parts of the Eastern Cape), those from the Northern Cape and calcareous pans on the southern edge of the Kalahari are not clearly referable to either species and have been variably placed under both names in some herbaria. Some of these from exposed spots even exhibit a similar, almost prostrate habit to *E. juttae* and have shorter stems and branches with more prominent tubercles, while more protected plants are erect, more slender and are then more typical of *E. aequoris*. I have found no clear distinctions between the two species and so I have placed *E. aequoris* in synonymy.

Euphorbia muricata Thunb., *Prodr. Fl. Cap.* 2: 86 (1800). *Tirucalia muricata* (Thunb.) P.V.Heath, *Calyx* 5: 91 (1996). Type: South Africa, Cape, between the Olifants River and Bockland, fl. October, *Thunberg* (UPS-THUNB, holo.; drawing and fragment at K, iso.).

Euphorbia spicata E.Mey. ex Boiss. in DC., *Prodr.* 15 (2): 97 (1862). *Tirucalia spicata* (E.Mey. ex Boiss.) P.V.Heath, *Calyx* 5: 93 (1996). Lectotype (Bruyns 2012): South Africa, 31 Aug. 1830, *Drège 2946* (K; S, iso.).

Euphorbia aspericaulis Pax, *Bot. Jahrb. Syst.* 28: 26 (1899). *Tirucalia aspericaulis* (Pax) P.V.Heath, *Calyx* 5: 88 (1996). Type: South Africa, Cape, Hantam Mtns, Calvinia distr., Feb. 1869, *ex Dr Meyer* (B, missing; drawing and fragment at K, iso.).

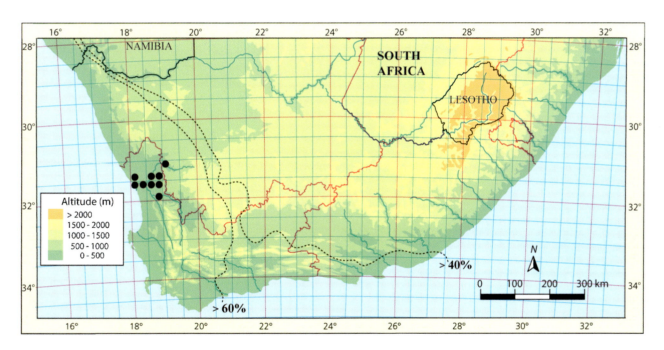

Fig. 3.71. Distribution of *Euphorbia muricata* (© PVB).

Unisexual (occasionally bisexual) spineless glabrous low spreading succulent shrublet 30–150 × 30–500 mm, branching sparsely to densely above ground from short subterranean stem rising from apex of slender wiry taproot 100–200 × 3–8 mm bearing fibrous roots. *Branches* erect to usually spreading close to ground, often rebranching, cylindrical or somewhat flattened just below nodes, 50–150 × 3–10 mm, with indistinct swollen tubercle beneath each leaf-rudiment, soft and succulent, grey-green to blue-green and often suffused with purple between nodes, coarsely asperulous with raised irregular wart-like papillae; *leaf-rudiments* towards apices of branches on slightly raised rounded tubercles, initially opposite (later sometimes shifted to appear alternating), 1–2 × 1–2 mm, ascending and later spreading, often persistent as small black scales, hastate-deltate with two spreading hornlets at bases, slightly fleshy, minutely puberulous, flat to slightly channelled above, acute, with very short petiole < 0.5 mm long, estipulate. *Synflorescences* in terminal cymes, of 1 to many unisexual sessile cyathia each subtended by 2 ovate to elliptic often apically and marginally toothed glabrous or puberulous bracts 0.7–1 × 1 mm resembling leaves (but lacking basal hornlets) in axils of which further cyathia develop on short branches, these cyathia again subtended by bracts from whose axils process repeated (repetition most frequent in males); *cyathia* conical-cupular in females to slightly urceolate in males, glabrous to puberulous, 1.5–2 mm broad and slightly broader in males than females (0.8–2 mm long below insertion of glands), with 5 obovate lobes with deeply dissected margins, pale grey-green sometimes suffused with red or wholly red; *glands* 5, nearly circular, 0.8–1.2 mm broad, erect, yellow to grey-green suffused with red, outer margins flat and entire; *stamens* with glabrous pedicels, bracteoles filiform and pubescent; *ovary* globose, glabrous to sparsely pubescent, raised and partly exserted on pedicel ± 0.8 mm long; styles 0.7–1 mm long, divided to near base, widely spreading. *Capsule* 2–3 mm diam., obtusely 3-lobed, glabrous to sparsely pubescent, dull grey-green suffused with red, exserted slightly from cyathium on erect pedicel ± 1.5 mm long.

Distribution & Habitat

Euphorbia muricata is a characteristic species of the Knersvlakte, where it is widespread. Records exist from near Komkans to Strandfontein (near the mouth of the Olifants River) in the west and eastwards to the base of the escarpment between Nieuwoudville and Klawer. A few collections from around Calvinia are also included here under *E. muricata* (though they may be considered to be a scabrid form of *E. rhombifolia*).

Typically *E. muricata* grows in firm, reddish brown loam with a scanty cover of short, succulent shrubs. More rarely it occurs in patches of quartz-gravel among the dense concentrations of small, clump-forming succulent Aizoaceae that characterise these habitats. Plants are often found in the open but may also shelter within another shrub. They are usually rather scattered and rarely seem to develop into dense populations.

Diagnostic Features & Relationships

In some areas, plants of *E. muricata* remain quite small, with relatively few branches arising from a slender tap-root. In other places, especially where they occur in deeper and looser loam, the plants may reach 0.3–0.5 m in diameter, forming a low, spreading shub. The branches are typically short and relatively thick, rapidly tapering off towards their tips into parts bearing flowers that soon dry off and persist as grey-black twigs. Some branches are erect, but mostly they spread out quite close to the ground. Their surface is grey-green, but this is greatly modified by suffusion with purple in parts exposed to the sun. Consequently, the undersides of the branches are grey-green, while above they are purple, with a V-shaped grey-green mark at each node, delineating the beginning of the next node above (though in some plants these marks are absent and the upper surface is uniformly purplish grey-green). The surface of the branches is mostly noticeably roughened by irregular, wart-like papillae (see Fig. 1.28 A, B), some of which coalesce into longer ridges

Fig. 3.72. *Euphorbia muricata*, in firm loam, *PVB 6026*, road to Kalkgat, NE of Vanrhynsdorp, South Africa, 10 Jul. 2005 (in a dry year) (© PVB).

3.2 Sect. Articulofruticosae

Fig. 3.73. *Euphorbia muricata*, ± 10 cm tall with attractively marked branches, *PVB 6026*, road to Kalkgat, NE of Vanrhynsdorp, South Africa, 10 Jul. 2005 (dry year) (© PVB).

Fig. 3.74. *Euphorbia muricata*, in reddish loam with *Malephora*, Strandfontein, west of Vanrhynsdorp, South Africa, 28 Sep. 2013 (© PVB).

running down within the node. Similar, though finer, papillae are also found on the peduncles subtending the cyathia. The nodes are quite clearly defined, widening noticeably and also becoming slightly flattened towards their apices, where a small leaf-rudiment develops at the tip of each rounded tubercle. There are usually slight ridges in the sides of the tubercle beneath the leaf-rudiment, running down towards the next node, but vanishing before it. Though small, the leaf-rudiments in *E. muricata* are very characteristic and are identical to those in *E. rhombifolia*, persisting, as in that species, as tiny, recurved, blackish structures with small, lateral horns at their bases.

The small cyathia of *E. muricata* are borne at the tips of the branches. They also arise on the tips of further short branchlets that develop in the axils of the pair of bracts below the first cyathium. The males are a little broader and longer than the females and are usually also a bit more noticeable because of the yellow anthers and pollen. The

Fig. 3.75. *Euphorbia muricata*, large male plant ± 30 cm broad in firm red sand, west of Hol River siding, South Africa, 9 Sep. 1994 (© PVB).

Fig. 3.76. *Euphorbia muricata*, male plant flowering (branches much smoother), Strandfontein, west of Vanrhynsdorp, South Africa, 28 Sep. 2013 (© PVB).

Fig. 3.77. *Euphorbia muricata*, female plant with capsules (branches clearly papillate), Quaggaskop, north of Vanrhynsdorp, South Africa, 24 Sep. 2011 (© PVB).

small capsules usually have short, white hairs scattered over their surface, which is otherwise a dull grey-green faintly suffused with red.

Euphorbia muricata is a close relative of *E. rhombifolia* and differs from it by its thicker, much shorter branches generally held close to the ground, most of which have a scabrid surface that is easily detected by touch (though in some populations occasional plants with smooth branches are found). The tubercles on which the leaf-rudiments are raised are more conspicuous and consequently the branches are more clearly angled at the nodes than in *E. rhombifolia*. The cyathia are also smaller, with more erect glands and are not produced in such complex structures as one may encounter in *E. rhombifolia*.

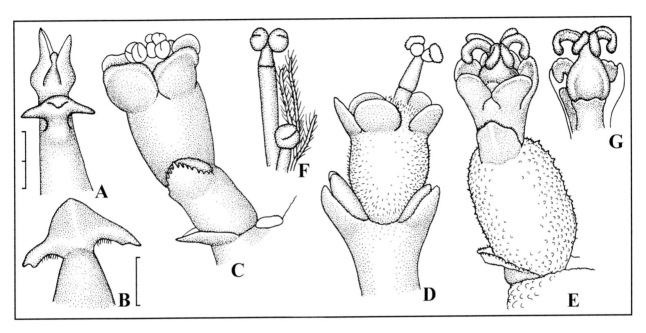

Fig. 3.78. *Euphorbia muricata*. **A, B**, tip of branch with leaf-rudiments (scale 2 mm for **A**; 1 mm for **B** as for **C–G**). **C, D**, male cyathium. **E**, female cyathium. **F**, anthers and bracteoles. **G**, female floret in dissected cyathium. Drawn from: **A, B, D**, *PVB*, north of Hol River siding, Namaqualand, South Africa. **C, E–G**, *PVB*, Quaggaskop, north of Vanrhynsdorp, Namaqualand, South Africa (© PVB).

History

Euphorbia muricata was collected by C.P. Thunberg in flower around or just after 30 October 1774 during his third and last journey at the Cape, made in the company of Francis Masson (Gunn and Codd 1981). Apart from saying it was an unarmed shrub with angled, rough branches and terminal flowers, he gave no information about this collection in the original publication and these details were first supplied in Thunberg (1823: 405), where a slightly more detailed description also appeared. Boissier (1862) placed *E. muricata* as a synonym of *E. brachiata* and commented on the inappropriateness of the name 'muricata', as the 'branches certainly of the type specimen were only scabrid'. Brown (1915) resurrected the name *Euphorbia muricata* but still had not seen much material of it, as he cited only the type collection and a specimen of Pearson, collected at Aties (about 15 km south of Vanrhynsdorp) in December 1910. White et al. (1941) appeared to be unfamiliar with it.

3 Euphorbia subg. Chamaesyce

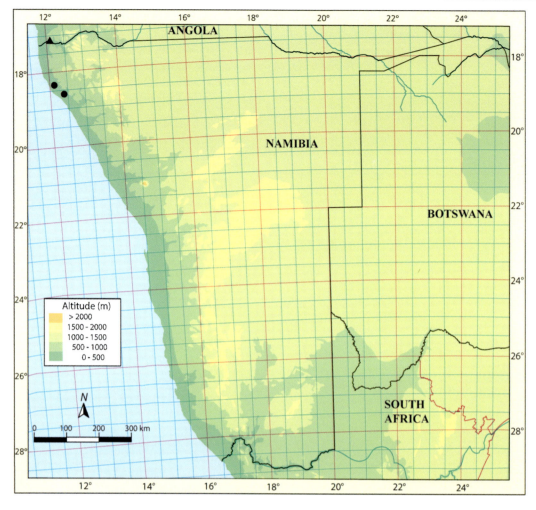

Fig. 3.79. Distribution of *Euphorbia negromontana* in southern Africa (▲ = ssp. *negromontana*, ● = ssp. *rimireptans*) (© PVB).

Fig. 3.80. *Euphorbia negromontana* ssp. *negromontana*, hemispherical plant ± 0.4 m diam., *PVB 13294*, east of Namibe, Angola, Dec. 2016 (© PVB).

Euphorbia negromontana N.E.Br., *Fl. Trop. Afr.* 6 (1): 557 (1911). *Tirucalia negromontana* (N.E.Br.) P.V.Heath, *Calyx* 5: 91 (1996). Lectotype (Leach 1974): Angola, Namibe distr., Serra de Montes Negros, 10 Aug. 1859, *Welwitsch 632* (BM; G, K, LISU, iso.).

Euphorbia conformis N.E.Br., *Fl. Trop. Afr.* 6 (1): 601 (1912). Type: Angola, Namibe distr., Serra de Montes Negros, 10 Aug. 1859, *Welwitsch 631* (LISU, holo.).

Euphorbia fragiliramulosa L.C.Leach, *Bol. Soc. Brot.*, sér. 2, 44: 201 (1970). Type: Angola, Namibe distr., 16 km from Moçamedes towards Dois Irmaos, 24 Jan. 1956, *Mendes 1374* (LISC, holo.; BM, LUA, iso.).

Euphorbia negromontana is found in southern Angola, within 40 km of the coast, where it is known from Namibe southwards to along the Cunene River, east of Foz do Cunene. In Angola, plants occur in very arid situations and are mainly found in the pale calcareous ground that forms the coastal plain, but also occur on shales and granitic gravel along the Giraul River, east of Namibe.

In Namibia, similar plants are found in the Skeleton Coast National Park, some 150 km south of the mouth of the Cunene River and these were recently described as *E. rimireptans*. This new species was compared in great detail with *E. giessii*, but it was not considered how it differs from the little-known Angolan *E. negromontana*. Particular emphasis was placed on the unusual prostrate and sprawling habit of *E. rimireptans*, which occasionally is even pendulous over rocks. In Angola, plants of *E. negromontana* generally consist of spreading to erect branches (so that the plant forms a hemispherical shrub), but they may sprawl on the ground on occasion, in which case the whole plant may be almost flat. The Namibian collections differ from Angolan material of *E. negromontana* in that the cyathia are sparsely pubescent outside (those in Angolan *E. negromontana* are glabrous to sparsely pubescent outside), while the bracts surrounding the cyathia are ± 1.5 mm long in Namibian collections and ± 1 mm long in Angolan plants. In Namibian plants the ovary is glabrous to sparsely pubescent while in Angolan *E. negromontana* it is densely pubescent. Although the capsule was mentioned as being only 2 mm diam. in Leach (1970b), it was clearly stated on this occasion that these capsules were immature and mature capsules are also ± 3.5 mm diam. The pubescence of the capsules provides the only clear-cut difference between them. Consequently, this taxon is recognized here as a subspecies of *E. negromontana*. The two subspecies may be separated as follows:

1. Ovary glabrous to sparsely pubescent, plants generally with all branches prostrate..............................subsp. **rimireptans**
1. Ovary densely pubescent, plants generally with some to most of the branches ascending to erect..
....subsp. **negromontana**

Fig. 3.81. *Euphorbia negromontana* ssp. *negromontana*, partly prostrate plant ± 0.3 m diam., *PVB 13294*, east of Namibe, Angola, Dec. 2016 (© PVB).

Fig. 3.82. *Euphorbia negromontana* ssp. *negromontana*, female plant in flower, *PVB 13286*, south of Namibe, Angola, Dec. 2016 (© PVB).

Euphorbia negromontana subsp. **rimireptans** (Swanepoel, R.Becker & Möller) Bruyns, **comb. et stat. nov.**, *Euphorbia rimireptans* Swanepoel et al., *Phytotaxa* 414: 166 (2019). Type: Namibia, Skeleton Coast National Park, between Okau and Sarusas, 20.5 km east of Cape Fria, 198 m, 21 Feb. 2019, *Swanepoel & Becker 358* (WIND, holo.; PRE, PRU, iso.).

Unisexual spineless and glabrous prostrate or sprawling succulent shrublet 0.05–0.3 (0.5) m in diam., branching and rebranching from similar stem to 0.1 m long with slender slightly fleshy rootstock bearing fibrous roots. *Branches* spreading, repeatedly rebranching, cylindrical, 20–300 × 2–6 mm, without distinct tubercles, smooth, soft and succulent, reddish green becoming grey with age towards base; *leaf-rudiments* opposite, 2–3 × ± 2 mm, ascending and spreading towards tips, fleeting, ± lanceolate, glabrous, acute or obtuse, without stipules. *Synflorescences* terminal on branches, glabrous, each of 1–3 (–10) unisexual crowded cyathia on short peduncles 1–3 × ± 2.5 mm long each subtended by 2 obovate entire bracts ± 1.5 × 1.5 mm resembling leaves in axils of which further cyathia develop on short branches; *cyathia* urceolate in males to ± cupular in females, sparsely pubescent towards base, 1.5–2.5 mm broad and ± 1.5 × as broad in males as females, with 5 obovate lobes with dissected margins, yellow-green; *glands* 5, transversely elliptic, 0.8–1.5 mm broad, spreading, yellow-green, outer margins flat and entire; *stamens* with glabrous pedicels, bracteoles filiform and pubescent towards tips; *ovary* globose, glabrous to sparsely pubescent, raised on pedicel ± 0.5 mm long; styles 0.7–1.2 mm long, divided to above base, widely spreading. *Capsule* 3.5–4 mm diam., obtusely 3-lobed, glabrous to sparsely pubescent, green suffused with red, exserted slightly from cyathium on erect pedicel ± 1.2 mm long.

Distribution & Habitat

Subsp. *rimireptans* is known from the coastal region of the Skeleton Coast of the Namib Desert. This hyper-arid region receives 15–25 mm of rain annually (Jacobsen 1988) and also significant moisture from fog off the sea (Jacobsen 1987). Here subsp. *rimireptans* occurs on low hillocks among pieces of black, basaltic rock partly covered with fine, wind-blown sand.

Fig. 3.83. *Euphorbia negromontana* ssp. *rimireptans*, ± 0.3 m broad, *Moss & Jacobsen 308*, among black rocks and fine wind-blown sand, near Sarusas, Skeleton Coast, Namibia, 18 Apr. 1985 (© N. Jacobsen).

Diagnostic Features & Relationships

Subsp. *rimireptans* forms clusters of branches which may reach 0.5 m in diameter. The sprawling branches are usually yellowish green and cylindrical, with tiny leaves on new growth.

Flowering has been noted between February and June. As is often the case in this section, the male cyathia are larger than the females and they have here a distinctive urceolate shape (found also in subsp. *negromontana* in Angola, but not in *E. giessii*). The female cyathia tend to be more cup-shaped and, as in subsp. *negromontana*, there is a short pedicel under the ovary, which is about half the length of the pedicel in *E. giessii*.

History

Euphorbia negromontana was discovered by F. Welwitsch in August 1859, east of Namibe in southern Angola. Subsp. *rimireptans* was first observed in 1982 by M.A.N. Müller & R. Loutit and it was again found nearby in April 1985 by H. Moss and N. H. G. Jacobsen, during a herpetological expedition to the Skeleton Coast. Some of the photographs taken by Jacobsen in 1985 are used here to illustrate it.

Fig. 3.84. *Euphorbia negromontana* ssp. *rimireptans*, male plant with cyathia, *Moss & Jacobsen 308*, near Sarusas, Skeleton Coast, Namibia, 18 Apr. 1985 (© N. Jacobsen).

Euphorbia rhombifolia Boiss., *Cent. Euphorb.*: 19 (1860). *Tirucalia rhombifolia* (Boiss.) P.V.Heath, *Calyx* 5: 92 (1996). Lectotype (Bruyns 2012): South Africa, Cape, arid places on southern Karoo, *Drège 8217* (G; K, S, W, iso.).

Arthrothamnus densiflorus Klotzsch & Garcke, *Abh. Königl. Akad. Wiss. Berlin* 1859: 62 (1860). Lectotype (Bruyns 2012): South Africa, Cape, Karoo near Olifants River, Oudtshoorn distr., Jan. 1820, *Mund & Maire* (K).

Euphorbia brachiata (E.Mey. ex Klotzsch & Garcke) Boiss. in DC., *Prodr.* 15 (2): 74 (1862). *Arthrothamnus brachiatus* E.Mey. ex Klotzsch & Garcke, *Abh. Königl. Akad. Wiss. Berlin* 1859: 62 (1860). *Tirucalia brachiata* (E.Mey. ex Klotzsch & Garcke) P.V.Heath, *Calyx* 5: 88 (1996). Lectotype (Bruyns 2012): South Africa, Cape, near Ebenezer, *Drège 2948* (K; S, iso.).

Euphorbia decussata E.Mey. ex Boiss. in DC., *Prodr.* 15 (2): 74 (1862), *nom illegit.*, non Salisb. (1796).

Euphorbia angrae N.E.Br., *Fl. Cap.* 5 (2): 279 (1915). *Tirucalia angrae* (N.E.Br.) P.V.Heath, *Calyx* 5: 87 (1996). Lectotype (Bruyns 2012): Namibia, Lüderitz (Angra Pequeña), 18 Jan. 1907, *Galpin & Pearson 7549* (K; PRE, SAM, iso.).

Euphorbia chersina N.E.Br., *Fl. Cap.* 5 (2): 274 (1915). *Tirucalia chersina* (N.E.Br.) P.V.Heath, *Calyx* 5: 89 (1996). Type: Namibia, Lüderitz (Angra Pequeña), 18 Jan. 1907, *Galpin & Pearson 7584* (K, holo.; PRE, iso.).

Euphorbia amarifontana N.E.Br., *Fl. Cap.* 5 (2): 275 (1915). *Tirucalia amarifontana* (N.E.Br.) P.V.Heath, *Calyx* 5: 87 (1996). Lectotype (Bruyns 2012): South Africa, Cape, near Springbokkuil River, Bitterfontein, *Zeyher 1534* (K; BOL, SAM, iso.).

Euphorbia caterviflora N.E.Br., *Fl. Cap.* 5 (2): 286 (1915). *Tirucalia caterviflora* (N.E.Br.) P.V.Heath, *Calyx* 5: 88 (1996). Lectotype (Bruyns 2012): South Africa, Cape, Nieweveld, Beaufort West, *Drège 8218* (K; BM, G, MO, S, W, iso.).

Euphorbia hastisquama N.E.Br., *Fl. Cap.* 5 (2): 288 (1915). Lectotype (Bruyns 2012) South Africa, Cape, fields by the Swartkops River, *Zeyher 1099* (BOL; K, iso.).

Euphorbia indecora N.E.Br., *Fl. Cap.* 5 (2): 274 (1915). *Tirucalia indecora* (N.E.Br.) P.V.Heath, *Calyx* 5: 90 (1996). Type: South Africa, Cape, between Dabenoris and Houms Drift, 11 Jan. 1909, *Pearson 3387* (K, holo.; K, iso.).

Euphorbia mundii N.E.Br., *Fl. Cap.* 5 (2): 287 (1915). Lectotype (Bruyns 2012): South Africa, Cape, Montagu, 1 Jan. 1903, *Marloth 2805* (K; PRE, iso.).

Euphorbia perpera N.E.Br., *Fl. Cap.* 5 (2): 277 (1915). *Tirucalia perpera* (N.E.Br.) P.V.Heath, *Calyx* 5: 92 (1996). Type: South Africa, Cape, along Orange River, between Verleptpram and its mouth, *Drège* (K, holo.).

Euphorbia rudolfii N.E.Br., *Fl. Cap.* 5 (2): 276 (1915). *Tirucalia rudolfii* (N.E.Br.) P.V.Heath, *Calyx* 5: 92 (1996). Lectotype (Bruyns 2012): South Africa, Cape, Vanrhynsdorp div., Bitterfontein, Sept. 1897, *Schlechter 11047* (K; BR, COI, GRA, K, L-2 sheets, PRE, S, iso.).

Euphorbia bayeri L.C.Leach, *S. African J. Bot.* 54: 539 (1988). Type: South Africa, Cape, 2 km west of Mossel Bay, 11 Sept. 1985, *Bayer 4875* (NBG, holo.; K, MO, PRE, iso.).

Euphorbia einensis G.Will., *Haseltonia* 10: 57 (2004). Type: Namibia, southern Schakalberg, 70 km NE of Oranjemund, *Williamson 5143* (BOL, holo.).

Euphorbia einensis var. *anemoarenicola* G.Will., *Haseltonia* 10: 62 (2004). Type: South Africa, Kortdoorn, *Williamson 5985* (BOL, holo.).

Unisexual (occasionally bisexual) spineless glabrous succulent shrublet to hemispherical shrub 30–600 mm tall branching often densely above ground from short subterranean stem rising from apex of wiry or slightly tuberous taproot 100–200 (300) × 5–25 mm bearing fibrous roots. *Branches* erect to spreading, often rebranching, cylindrical or often somewhat flattened just below nodes, 50–200 × 3–8 mm, ± without tubercles, soft and succulent when young becoming rigid with age and occasionally drying off and sharp towards tips, yellow- to grey-green to dark blue-green and often purple or red between nodes, smooth to slightly asperulous; *leaf-rudiments* towards api-

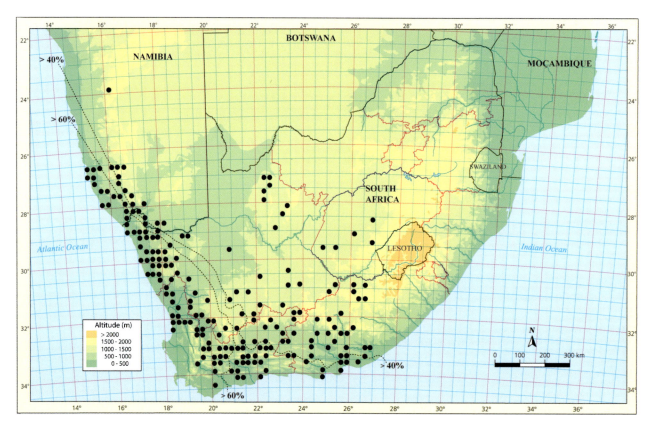

Fig. 3.85. Distribution of *Euphorbia rhombifolia* (© PVB).

Fig. 3.86. *Euphorbia rhombifolia*, male plant ± 20 cm diam., branches grey and relatively slender, Nieuwoudville, South Africa, 24 Sep. 2011 (© PVB).

Fig. 3.87. *Euphorbia rhombifolia*, male plant ± 15 cm diam., heavily grazed, Tafelberg, SW of Fraserburg, South Africa, 30 Sep. 2012 (© PVB).

ces of branches on slightly raised rounded tubercles, initially opposite (later sometimes shifted to appear alternating), 1–2 × 1–2 mm, ascending and later spreading, often persistent as small black scales, hastate-deltate with two spreading hornlets at base, slightly fleshy, minutely puberulous, flat to slightly channelled above, acute, with short petiole 0.5–1 mm long, estipulate. *Synflorescences* in terminal cymes, of 1 to many unisexual (rarely bisexual) sessile cyathia each subtended by 2 obovate to elliptic often apically toothed glabrous to sparsely pubescent bracts 1–2 × 1 mm resembling leaves in axils of which further cyathia develop on short branches, these cyathia again subtended by bracts from whose axils further cyathia develop in the same way (repetition most frequent in males); *cyathia* conical-cupular in females to slightly urceolate in males and bisexuals, glabrous, 1.5–2.5 mm broad and almost twice as broad in males as females (1–2 mm long below insertion of glands), with 5 obovate lobes with deeply dissected margins, yellow-green sometimes suffused with red; *glands* 5, transversely elliptic, 1–1.5 mm broad, spreading, yellow-green, outer margins flat and entire; *stamens* with glabrous pedicels, bracteoles filiform and pubescent towards tips; *ovary* globose, glabrous, raised on pedicel ± 1 mm long; styles 0.7–1.5 mm long, divided to near base, widely spreading. *Capsule* 3–4 mm diam., obtusely 3-lobed, glabrous, green suffused with red, ± sessile to exserted slightly from cyathium on erect pedicel ± 1.5 mm long.

Fig. 3.88. *Euphorbia rhombifolia*, plant ± 50 cm diam., gravelly area at base of hills, Holgat River, South Africa, 8 Jul. 2013 (© PVB).

Distribution & Habitat

Euphorbia rhombifolia is widely distributed over southern Namibia and South Africa. It is especially prevalent in the parts of south-western Namibia and the western portion of South Africa which receive rainfall in winter, particularly in the karroid areas of the Cape Floristic Region and the Succulent Karoo Region. It is of more sporadic occurrence further afield in the parts of South Africa receiving rain in summer, where it is known from Griqualand West across the Great Karoo to the drier parts of the Eastern Cape and the Free State.

Generally *E. rhombifolia* is a common element of semi-arid vegetation in Namibia and South Africa. Some of its larger forms may become dominant in certain patches in the manner of *E. mauritanica*. This is especially the case with the very substantial shrubs often called '*E. chersina*' that are a common feature in the winter-rainfall 'desert' of the Richtersveld and south-western Namibia, but it is also encountered in the substantial, shrubby forms of *E. rhombifolia* that occur on the Tanqua Karoo. These large and dense populations may develop on stony slopes or on hard loam in flat, gravelly places and are much more rarely found in looser sands. The ubiquitous, less robust to very small, more scattered plants of *E. rhombifolia* (often referred to as '*E. caterviflora*' or '*E. mundii*') are typical of flat, gravelly or stony spots with shallow soil (often overlaying outcrops of rock) and low, scanty vegetation that are often surrounded by deeper soils and denser vegetation (or more steeply sloping areas). This is typically how *E. rhombifolia* occurs over most of the Karoo and especially in the areas receiving rainfall in summer.

Diagnostic Features & Relationships

As one would expect from its wide distribution and broad diversity of habitat, *E. rhombifolia* is vegetatively variable. Beneath the surface of the soil the stem continues into a tap-root which may be either slender or swollen and fleshy and somewhat tuberous. Both swollen and not-at-all fleshy tap-roots may be found within the same population. In habitat, plants are often found ripped out by animals (presumably porcupines) that seem to relish the tap-root. The larger forms of *E. rhombifolia* generally take the shape of a hemispherical shrub which may be densely or sparsely branched. Smaller forms are considerably more variable in shape (and often much influenced by predation), from dense, low-growing plantlets (only rarely slightly spiky) to very diffuse, sparsely branched shrubs. The appearance of trichotomous branching is common, where almost every node, especially towards the tips of the main branches, gives rise to two further branches from the axils of the opposite leaf-rudiments. The colour of the plants also varies considerably. Robust plants in the arid north-west are often a striking yellow-green (occasionally red) to bright green. However, even here blue-green plants will be found and blue-green is the more typical colour of *E. rhombifolia* over most of its distribution. In smaller plants the branches are often suffused with purple, with a blue-green band around each node. The surface of the branches is smooth to slightly scabrid. Especially tiny plants are often encountered in the hyper-arid southern Namib Desert, close to the coast between Lüderitz and Alexander Bay, where they may consist of only a few short branches held close to the ground. These usually occur together with others of much larger proportions, with which they share the same colour and texture of the branches and the same flowers and they are clearly much reduced forms of the same species (though very small ones were previously distinguished as '*E. angrae*' and '*E. einensis*').

Characteristic of *E. rhombifolia* are the small, swollen leaf-rudiments produced in opposite pairs on the tips of young growth. They are roughly triangular in outline with a hornlet on either side at the base (and further teeth may be present along their margins). These leaflets are green or reddish green for a short while, after which they dry out and often persist as small black husks in which their basal hornlets can still usually be distinguished.

Fig. 3.89. *Euphorbia rhombifolia,* plant ± 50 cm diam., branches paler and more slender, foot of Kourkamma Mountain, near Komaggas, South Africa, 6 Jul. 2016 (© PVB).

Fig. 3.90. *Euphorbia rhombifolia*, ± 12 cm diam., particularly small form among stones and windblown sand, Lüderitz, Namibia, 11 Jul. 2014 (© PVB).

Flowering takes place between June and November in the areas receiving winter-rainfall and later in the parts with summer-rainfall, depending on the occurrence of rain. Lots of cyathia are produced at the tips of the branches in variously dense clusters, with successive cyathia developing from the axils of the bracts, sometimes sufficiently swiftly that the tips of the branches are covered with cyathia. The cyathia are very small, yellow-green in the males to green in the females. Usually the males are produced in larger numbers and their yellowish colour makes them a bit more conspicuous than the females. Usually in the male plants the peduncles branch more extensively as well. Cyathia bearing fertile female florets may sometimes have some male florets in them too and bear a slightly protruding ovary that is, on fertilization, exserted a little further on a short, erect pedicel.

Fig. 3.91. *Euphorbia rhombifolia*, male ± 20 cm diam., somewhat larger than previous, Lüderitz, Namibia, 11 Jul. 2014 (© PVB).

Fig. 3.92. *Euphorbia rhombifolia*, small male plant, heavily grazed but flowering, Tafelberg, SW of Fraserburg, South Africa, 30 Sep. 2012 (© PVB).

Fig. 3.93. *Euphorbia rhombifolia*, branch of more laxly branched male plant, Ouberg Pass, east of Montagu, South Africa (© PVB).

Fig. 3.94. *Euphorbia rhombifolia*, capsules, Wyke siding, NE of Laingsburg, South Africa, 27 Oct. 2012 (© PVB).

Fig. 3.95. *Euphorbia rhombifolia*, small rhizomatous form described as *E. bayeri*, in shallow soil among sandstone outcrops, *PVB 9976*, Mossel Bay, South Africa, 28 May 2005 (© PVB).

With its great variability, *E. rhombifolia* is not easy to characterize. It will often be found that several forms of it occur together, differing in the colour of the plants and their size and shape. If the spot is searched properly and widely, the extremes will usually be found to be connected by many intermediates. Thus, for example, the particularly robust and imposing forms known as '*E. chersina*' appear distinctive. However, they are only distinguished by colour and thickness of the branches from other less robust plants nearby, usually in slightly different ecological circumstances. Another example is provided by '*E. bayeri*', a very small-stemmed, somewhat rhizomatous variant (as in Fig. 3.95) from shallow soils on sandstones near Mossel Bay (and limestones further west). At the type locality it is distinctive. However, along the banks of the Little Brak River nearby, this distinctive small form gradually merges with larger ones that grow in deeper soils and on shales, which become impossible to separate from typical *E. rhombifolia*. Another distinctive form was described from limestone outcrops on the farm Namuskluft in SW Namibia as *Euphorbia lavrani* (see Fig. 3.96). It is also regarded here as a very local ecotype of the widespread *E. rhombifolia*.

Euphorbia rhombifolia differs from *E. spartaria* by the generally smaller habit (with stouter branches and a distinctly hemispherical shape when it is as large as *E. spartaria* often is) and by the slender, narrowly elliptic leaf-rudiments in *E. spartaria* which never bear the two hornlets at their bases that are typical of *E. rhombifolia*. In both *E. tenax* and *E. muricata* these lateral hornlets are present. *Euphorbia tenax* is separated from *E. rhombifolia* by its more slender

Fig. 3.96. *Euphorbia rhombifolia*, small densely-clustered form described as *E. lavrani*, in shallow soil in crevices among limestone outcrops, *PVB*, Namuskluft, Namibia, 26 Aug, 2003 (© PVB).

branches with their invariably rough surface. In *E. muricata* the surface of the branches is rougher than in *E. tenax* and the plant is small, with a few rather randomly directed, quite stout and longitudinally grooved branches. Florally all these species are difficult to distinguish.

History

With its very wide distribution, *E. rhombifolia* was naturally discovered as soon as collecting activities began in the interior of South Africa. The oldest collection I have seen is an undated one collected by Francis Masson before March 1795 ('Promont. B. Spei Mr Masson' at BM). J.L.L. Mund and L. Maire, two collectors sent to the Cape in 1816 from the Berlin Museum, made another early collection in January 1820.

Euphorbia rhombifolia has the longest list of synonyms of all species of *Euphorbia* in Southern Africa, which is testimony to its extraordinary variability and the frequency with which it has been collected over the last two hundred years. N.E. Brown particularly provided many names for these different collections without knowing whether distinct species were involved. Marloth (on *Marloth 12357* at PRE) commented that there was no real difference between *E. chersina* and *E. amarifontana*, suggesting that he might have had doubts about the distinctness of some of the many 'species' recognised by Brown. White et al. (1941) did not address this problem and only with the Cape Plants and Succulent Karoo projects of SANBI (Bruyns 2000, 2013) were the first attempts made to assess the number of species involved in this complex. This also involved determining the large amount of material of this species that has been collected, in a consistent manner across all the herbaria.

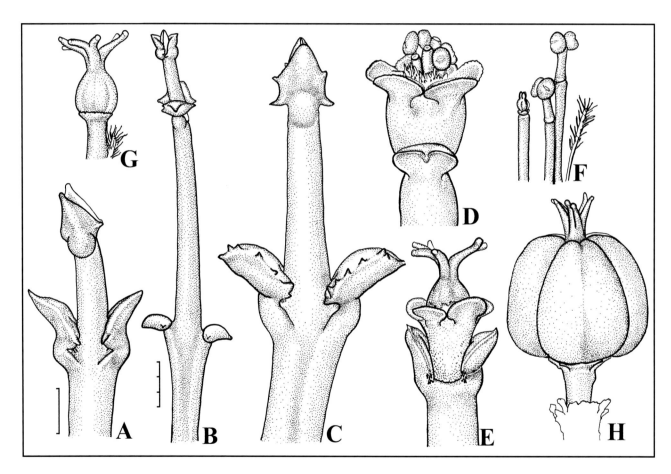

Fig. 3.97. *Euphorbia rhombifolia*. **A–C**, tip of young branch with leaf-rudiments (scale of **A** 1 mm, as for **C–H**; for **B** 3 mm). **D**, side view of male cyathium. **E**, female cyathium from side. **F**, anthers, bracteole and rudimentary female floret from male cyathium. **G**, female floret. **H**, capsule. Drawn from: *PVB*, Warmbad, south of Calitzdorp, South Africa (© PVB).

3.2 Sect. Articulofruticosae

Euphorbia spartaria N.E.Br., *Fl. Trop. Afr.* 6 (1): 558 (1911). *Tirucalia spartaria* (N.E.Br.) P.V.Heath, *Calyx* 5: 93 (1996). Type: Namibia, Hoffnung, Feb. 1907, *Galpin & Pearson 7560* (K, holo.; PRE, SAM, iso.).

Euphorbia racemosa E.Mey. ex Boiss. in DC., *Prodr.* 15 (2): 75 (1862), *nom. illegit.*, *non* Tausch ex Rchb. (1832). Lectotype (Bruyns 2012): South Africa, Cape, near Hamerkuil, *Drège* (MO; S, numbered 8204, iso.).

Euphorbia rhombifolia var. *laxa* N.E.Br., *Fl. Cap.* 5 (2): 285 (1915). Lectotype (Bruyns 2012): South Africa, Cape, among rocks along Chichaba River between Komgha and Kei Mouth, Aug. 1891, 1000', *Flanagan 838* (GRA; PRE, SAM, iso.).

Euphorbia rhombifolia var. *triceps* N.E.Br., *Fl. Cap.* 5 (2): 285 (1915). Lectotype (Bruyns 2012): South Africa, Cape, Queenstown distr., mountains near Imbasa River, 1860, *Cooper 318* (K; BOL, TCD, W, iso.).

Euphorbia cibdela N.E.Br., *Fl. Cap.* 5 (2): 275 (1915). *Tirucalia cibdela* (N.E.Br.) P.V.Heath, *Calyx* 5: 89 (1996). Type: Namibia, on hills at Schakalskuppe, 4900–5600', 18 Jan. 1909, *Pearson 4428* (K, holo.; BOL, LD, SAM, iso.).

Euphorbia rectirama N.E.Br., *Fl. Cap.* 5 (2): 283 (1915). *Tirucalia rectirama* (N.E.Br.) P.V.Heath, *Calyx* 5: 92 (1996). Lectotype (Bruyns 2012): South Africa, Cape, Klipfontein, Griqualand West, 29 Dec. 1812, *Burchell 2633* (K).

Unisexual (occasionally bisexual) spineless glabrous succulent shrub 0.1–1 m tall branching from apex of slightly swollen taproot 100–300 × 5–25 mm bearing fibrous roots. *Branches* almost all erect (occasionally slightly spreading), only sparsely rebranched, cylindrical or often somewhat flattened just below nodes, 50–600 × 1.5–6 mm, without tuber-

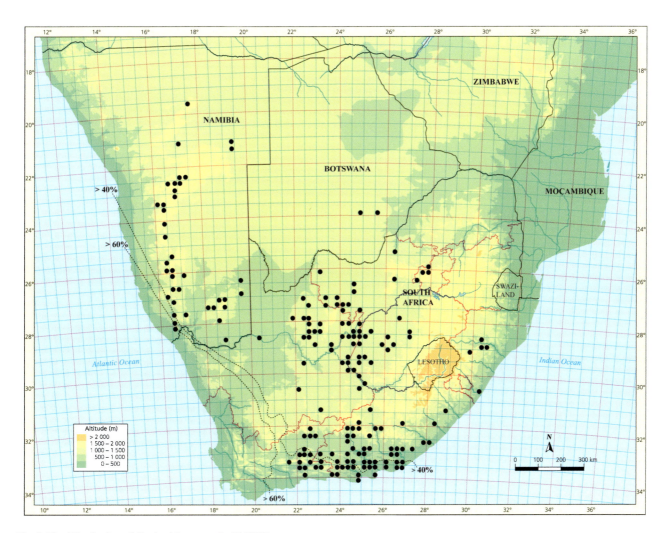

Fig. 3.98. Distribution of *Euphorbia spartaria* (© PVB).

cles, soft and succulent when young becoming slightly rigid with age, grey-green, smooth to slightly asperulous; *leaf-rudiments* towards apices of branches on very slightly raised rounded tubercles, initially opposite (later sometimes shifted to appear alternating), 1–6 × 0.5–1 mm, ascending and slightly spreading towards tips, soon deciduous, narrowly spathulate to almost linear, slightly fleshy, minutely puberulous towards base and towards midrib above, often channelled above, obtuse to minutely apiculate, tapering gradually into short petiole 0.5–1 mm long, estipulate. *Synflorescences* in terminal cymes, of 1 to many unisexual (sometimes bisexual) sessile cyathia each subtended by 2 obovate to elliptic obtuse sparsely pubescent (around margins and midrib otherwise glabrous) shortly spathulate bracts ± 1 × 1 mm in axils of which further cyathia develop on further branches with long internodes, these cyathia again subtended by bracts from whose axils process repeated (repetition most frequent in males); *cyathia* conical-cupular in females to slightly urceolate in males and bisexuals, glabrous, 2–3.5 mm broad (± 2 mm long below insertion of glands), with 5 obovate lobes with deeply dissected margins, yellow-green sometimes suffused with red; *glands* 5, transversely elliptic, 1–1.5 mm broad, ascending, yellow-green, outer margins erect and entire; *stamens* with glabrous pedicels, bracteoles filiform and pubescent towards tips; *ovary* globose, glabrous, raised on pedicel ± 1.5 mm long; styles 0.7–1.5 mm long, divided at most to middle, widely spreading. *Capsule* 3–4 mm diam., obtusely 3-lobed, glabrous, green suffused with red, ± sessile ontop of cyathium on short erect pedicel ± 1.5 mm long.

Distribution & Habitat

Euphorbia spartaria is second only to *E. mauritanica* in its ubiquitousness and is known in Botswana, Namibia and South Africa. In Botswana it was known (as *E. rectirama*) in the south of the country, near Tsabong and Kweneng (Carter and Leach 2001; Hargreaves 1992a) but it is actually much more widespread and has now also been recorded close to the centre of the country, north of Molepolole. In Namibia it is found from the mountains around Otavi via Windhoek to the mountains around Rosh Pinah and to the Great Karas Mountains, from where its distribution continues in South Africa, generally east of the winter-rainfall area, through Griqualand West to Oudtshoorn and Grahamstown as well as the Free State. There are also a few records from Kwazulu-Natal, mainly along the upper reaches of the Tugela River.

Euphorbia spartaria generally occurs in semi-arid places among grasses and scattered trees on stony to very rocky slopes or in deep sand. Sometimes, as in the mountains around Rosh Pinah and between Oudtshoorn and Willowmore, it is found in succulent scrub. Occasionally, *E. spartaria* may be found in exceedingly dry areas, such as near the Augrabies National Park and in the hills east of Aus, where it grows among rocks on otherwise almost bare slopes (in the former place near to *E. avasmontana*).

Fig. 3.99. *Euphorbia spartaria*, shrub ± 1 m tall in deep sand among trees, *PVB 12299*, north of Molepolole, Botswana, 23 Dec. 2012 (© PVB).

Fig. 3.100. *Euphorbia spartaria*, shrub ± 40 cm tall on outcrop of quartz, ± 10 km south of Windhoek and within 30 km of the type-locality, Namibia, 28 Dec. 2018 (© PVB).

Diagnostic Features & Relationships

Plants of *E. spartaria* may become very substantial, often reaching 1 m in height and sometimes as much in breadth. Each specimen arises from a thick tap-root, to which (in some areas) the whole plant may be eaten back occasionally and from which it then sprouts again. Fairly dense rebranching takes place, mainly from the lower parts of the branches so that distinctively brush-shaped shrubs usually develop. In the arid area west of the Augrabies National Park somewhat more rigid shrubs with an almost spherical shape (brought about by more dense rebranching) and a bright green colour are found (as in Fig. 3.101), but this is unusual, only seen otherwise in this species in the dry area east of Aus, where the plants are less obviously spherical and are a more bluish green. Generally the branches are long and very slender, often not more than 3 mm thick for most of their length and only 2 mm thick towards their tips. On young growth one finds the distinctive leaf-rudiments, which are generally pressed against the branch towards their bases and spread away from the branch towards their tips. They are very slender, slightly channelled on the inner face and become progressively narrower towards their bases, which then swell into a slight tubercle on the branch, with the branch slightly flattened below each pair of leaves.

Flowering takes place between October and April, with many small cyathia produced at the tips of the branches and others developing on slender peduncles from the axils of the bracts surrounding the first, terminal cyathium. Flowering does not make a show since the branches are rather diffuse and the cyathia are scattered over the top of the plant, without forming dense clusters. The male (and occasional bisexual) cyathia are urceolate, while those that are purely female are more conical below the glands. In the female, the ovary is partly exserted and matures just outside the cyathium, though the pedicel is usually not visible beneath it (except by dissection).

With its large distribution, it is not surprising that *E. spartaria* has several synonyms, but most specimens have been placed either under the name *E. rectirama* or under *E. spartaria* (with somewhat fewer erroneously placed under one of the various synonyms of *E. rhombifolia*). White et al. (1941) were doubtful of the distinctness of *E. rectirama* from *E. spartaria*, though they separated it from *E. spartaria* by its larger size and the shorter persistence of the bracts. Generally plants from South Africa have been determined as *E. rectirama* while those from Namibia were referred to *E. spartaria*. However, I have been unable to find any reliable features on which to distinguish them and so both are included now under a single name, *E. spartaria*.

Euphorbia spartaria and *E. rhombifolia* are most readily distinguished when the plants are growing, by the hornlets at the base of the leaves in *E. rhombifolia*, where the leaves are also shorter and thicker and the young shoots often somewhat stouter. *Euphorbia spartaria* is a considerably taller

Fig. 3.101. *Euphorbia spartaria,* shrub ± 80 cm tall on stony arid quartz-strewn slopes with *E. gregaria*, east of Vredesvallei, Riemvasmaak, South Africa, 9 Apr. 2009 (© PVB).

Fig. 3.102. *Euphorbia spartaria,* shrub ± 1 m tall among shrubs of *E. radyeri*, Jansenville, South Africa, 3 Sep. 2015 (© PVB).

Fig. 3.103. *Euphorbia spartaria,* shrub ± 60 cm tall among succulent vegetation, Klaarstroom, South Africa, 28 May 2005 (© PVB).

3.2 Sect. Articulofruticosae

Fig. 3.104. *Euphorbia spartaria,* female plant with cyathia and capsules, Klaarstroom, South Africa, 28 May 2005 (© PVB).

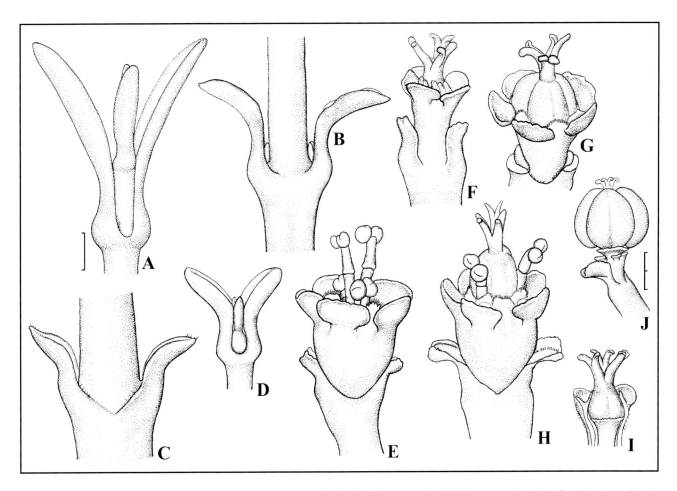

Fig. 3.105. *Euphorbia spartaria.* **A–D**, tip of young branch with leaf-rudiments (scale of **A** 1 mm, as for **B–I**). **E**, side view of male cyathium. **F, G**, female cyathium from side at different stages. **H**, side view of bisexual cyathium. **I**, female floret from dissected cyathium. **J**, capsule (scale 2 mm). Drawn from: **A**, *PVB 2923*, Huis River Pass, west of Calitzdorp, South Africa. **B**, *PVB 3778*, Kleinwaterval, NE of Laingsburg, South Africa. **C, F, I**, *PVB 3916*, Sebrafontein, NE of Rosh Pinah, Namibia. **D, J**, *PVB 5625*, SE of Windhoek, Namibia. **E, G, H**, *PVB 2900*, Tierberg, east of Prince Albert, South Africa (© PVB).

plant with much more slender branches which are also always smooth (their surface is sometimes roughened in *E. rhombifolia*). Florally the two are not easy to separate, though they are known to occur together in some areas without appearing to interbreed.

History

The first record of *E. spartaria* appears to be that of William Burchell, which was made late in November 1812 near the source of the Kuruman River. Material that he collected in December 1812 became the type of the name *Euphorbia rectirama*. In Namibia it was first collected near Windhoek by Dinter in January 1899 and then was gathered repeatedly on the farm Hoffnung, also near Windhoek, first by Galpin and Pearson in February 1907 and again by Dinter in August 1909. These were described by N.E. Brown as *E. spartaria*. An early record of the species from the Mooi River Valley in 'Weenen Country', Natal is that of Peter C. Sutherland, made in October 1858 (K).

Leach and Williamson (1990) briefly reviewed the case of *E. spartaria*. They concluded that it was a Namibian endemic and that none of the material from South Africa placed under it by N.E. Brown (1915: 281) belonged there. Since Brown had included such specimens as *Schlechter 8146* and *11381*, that belong to *E. exilis*, they were essentially correct in their conclusion. However, they did not investigate its relationships to *E. rectirama* and, once this name is brought into the picture, *E. spartaria* is found to encompass that as well. White et al. (1941) did not add anything to the knowledge of these species and appear to have been unfamiliar with both *E. rectirama* and *E. spartaria*. Marloth placed his collection *13120* (PRE), from Huis River Pass near Calitzdorp under *E. spartaria* and was therefore the first to indicate that *E. spartaria* was probably far more widely distributed than most authors had considered before.

Euphorbia spinea N.E.Br., *Fl. Cap.* 5 (2): 272 (1915). *Tirucalia spinea* (N.E.Br.) P.V.Heath, *Calyx* 5: 93 (1996). Lectotype (Bruyns 2012): Namibia, among rocks near Dabegabis, *Pearson 4380* (K; BOL, iso.).

Unisexual (occasionally bisexual) rigid spiny glabrous succulent ± hemispherical low shrublet 60–300 × 15–400 mm, branching densely above ground from short subterranean stem rising from apex of wiry taproot 100–200 × 5–10 mm, bearing fibrous roots. *Branches* briefly erect then spreading, densely rebranching with ultimate branchlets often slightly downcurved and becoming spike-like at tips, cylindrical or often somewhat flattened just below nodes, 50–200 × 1–8 mm, ± without tubercles, soft and succulent when young becoming rapidly rigid and drying off to form a sharp spike at tips, grey-green to dark blue-green and often purple or red between nodes on upper sides, smooth; *leaf-rudiments* towards apices of branches on very slightly raised rounded tubercles, initially opposite (later sometimes shifted to appear alternating), 1–2 × 1–2 mm, ascending and later spreading, often persistent as small black scales, hastate-deltate with two spreading hornlets at base, slightly fleshy, minutely puberulous, flat to slightly channelled above, acute, with short petiole 0.5–1 mm long, estipulate. *Synflorescences* in terminal cymes, of 1 to many unisexual (rarely bisexual) sessile cyathia each subtended by 2 obovate to elliptic often apically toothed glabrous to sparsely pubescent bracts 1–2 × 1 mm resembling leaves in axils of which further cyathia develop on short branches, these cyathia again subtended by bracts from whose axils process repeated (repetition most frequent in males); *cyathia* conical-cupular in females to slightly urceolate in males and bisexuals, glabrous, 1.5–2.5 mm broad (1–2 mm long below insertion of glands), with 5 obovate lobes with deeply dissected margins, yellow-green sometimes suffused with red; *glands* 5, transversely elliptic, 1–1.5 mm broad, spreading, yellow-green, outer margins flat and entire; *stamens* with glabrous pedicels, bracteoles filiform and pubescent towards tips; *ovary* globose, glabrous, raised on pedicel ± 1 mm long; styles 0.7–1.5 mm long, divided to near base, widely spreading. *Capsule* 3–4 mm diam., obtusely 3-lobed, glabrous, green suffused with red, exserted from cyathium on decurved and later erect pedicel 3–5 mm long.

Distribution & Habitat

Euphorbia spinea is found in south-eastern Namibia from west of Grünau to near Onseepkans. In South Africa it is recorded mainly along the valley of the Orange River, from near Vioolsdrift to Pofadder and eastwards to Prieska, as well as near Griquatown.

Euphorbia spinea usually grows in gently sloping areas covered by pale quartzitic gravels and scattered stones derived from gneiss. Where it occurs, the rainfall is low (mostly not exceeding 125 mm per year and often only averaging around 60 mm per annum) and the vegetation is very scanty. Nevertheless, shrubby succulents are frequently well represented, with species like *Aloe claviflora*, *A. dichotoma*, *A. hereroensis*, *Cynanchum pearsonianum*, *C. viminale*, *Euphorbia gariepina*, *E. gregaria*, *E. crassipes*, *Ceropegia gordonii*, *Kleinia longiflora* and species of *Monsonia* quite often present. In these associations there is a marked preponderance of plants with pencil-shaped branches.

3.2 Sect. Articulofruticosae

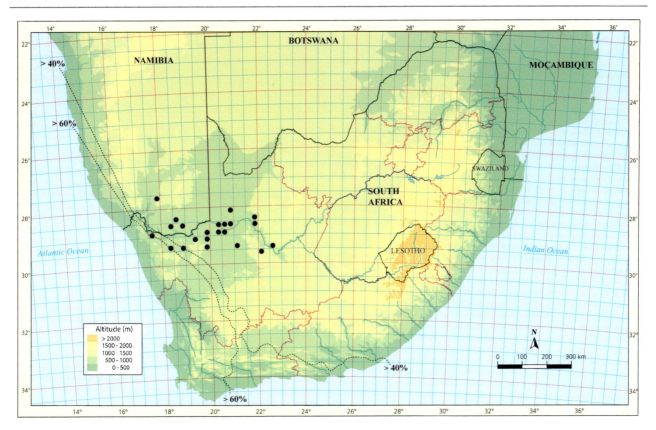

Fig. 3.106. Distribution of *Euphorbia spinea* (© PVB).

Fig. 3.107. *Euphorbia spinea*, pale low dense and spiky shrublet ± 20 cm diam., among quartz-gravel, *PVB 13499*, just west of Alheit, near Kakamas, South Africa, 23 May 2018 (© PVB).

Diagnostic Features & Relationships

Euphorbia spinea forms small, dense shrubs that rarely exceed 0.2 m tall and are frequently less than 0.1 m tall, though they are often somewhat broader than they are tall. Arising from a slightly swollen tap-root, they usually consist of a few thicker branches which produce a fairly dense network of opposite pairs of branchlets towards their tips. These ultimate branchlets usually remain short (often only 10–30 mm long), with spreading and slightly downward pointed tips that soon dry out to form very sharp, rigid, but quite fine spikes. The succulent bark on these spikes gradually separates into blackish rings which eventually fall off, leaving a paler surface on the spike. This network of fine spikes partly protects the plant from predation, though many of them are still heavily eaten, especially when conditions are very dry. The succulent bark on the branches usually has a banded appearance, with purple on the parts between the nodes and a somewhat irregular, narrower, pale blue-green band around each node. However, the purple colour only develops on the upper surfaces and those parts that are sheltered from the sun remain blue-green.

One may suspect that *E. spinea* is an ecotype of *E. rhombifolia* that is restricted to the kind of habitats described above. However, it is distinctive and easily recognized over a wide area and has thus been retained at the level of species. It differs from *E. rhombifolia* by the manner in which the

Fig. 3.108. *Euphorbia spinea*, *PVB 13499*, just west of Alheit, near Kakamas, South Africa, 23 May 2018 (© PVB).

Fig. 3.109. *Euphorbia spinea*, tips of branches drying into hard spikes, *PVB 13499*, just west of Alheit, near Kakamas, South Africa, 23 May 2018 (© PVB).

ultimate branchlets develop (more or less without exception) into short, sharp spikes with a slightly decurved habit. The capsule is also further exserted from the cyathium than in *E. rhombifolia* and in this respect it is reminiscent of *E. gentilis* and others.

Euphorbia spinea bears a strong, superficial resemblance to *E. lignosa* and White et al. (1941) placed it next to *E. lignosa*. However, the opposite leaflets, the much smaller unisexual cyathia and the tiny capsules will immediately dispel any confusion between the two.

Fig. 3.110. *Euphorbia spinea*, male shrublet with brightly mottled branches, gravelly plains at foot of mountain, *PVB 11132*, Naip, west of Pofadder, South Africa, 19 Oct. 2008 (© PVB).

Fig. 3.111. *Euphorbia spinea*, more brightly coloured female shrublet with capsules, *PVB 11132*, Naip, west of Pofadder, South Africa, 19 Oct. 2008 (© PVB).

History

When N.E. Brown described *E. spinea*, he included several collections, but not all of these represented the same species. Consequently a lectotype was selected, which belongs to the species considered here. This was collected by H.H.W. Pearson in south-eastern Namibia in January 1909. Since N.E. Brown's time, several species continued to be included under this name so that, for example, Figure 130 in White et al. (1941) is of a robust, slightly spiky plant of *E. rhombifolia* rather than *E. spinea* and the position in many southern African herbaria was similarly muddled.

Euphorbia stapelioides Boiss., *Cent. Euphorb.*: 26 (1860). *Tirucalia stapelioides* (Boiss.) P.V.Heath, *Calyx* 5: 93 (1996). Type: South Africa, Cape, at the mouth of the Gariep (Orange), 4 Oct. 1830, *Drège 8199* (P, holo.; MEL, S, W, iso.).

Euphorbia lumbricalis L.C.Leach, *S. African J. Bot.* 52: 369 (1986). Type: South Africa, Cape, north of Koekenaap, 10 May 1984, *Leach & Bayer 17123* (NBG, holo.; K, MO, PRE, iso.).

Unisexual dwarf spineless glabrous succulent branching extensively underground to spread diffusely over and area of 0.2–0.5 (1) m from apex of swollen ± cylindrical tuberous rootstock up to 100 × 10–20 mm bearing fibrous roots. *Branches* spreading underground as rhizomes (often developing subsidiary tubers) then erect above surface, often rebranching beneath ground, cylindrical and slender underground becoming thicker above surface where irregularly cylindrical and tapering somewhat to apex, above-ground portions 5–50 (to 200 mm long if sheltered) × 3–10 mm, sometimes branching above ground, without distinct tubercles, soft and succulent when young and remaining so with age, grey-green, smooth; *leaf-rudiments* towards apices of branches on very slightly raised tubercles, initially opposite (later sometimes shifted to appear alternating), 0.5–1.5 × 1–2.5 mm, ascending and later spreading, often persistent as small black scales, hastate-deltate with two spreading hornlets at base, minutely puberulous, with slight medial groove above, acute, with short petiole 1 mm long, estipulate. *Synflorescences* terminal on branches, each of 1–5 unisexual cyathia on short peduncle 1–8 mm long each subtended by 2 obovate bracts 1–2 × 1 mm; *cyathia* cupular, minutely pubescent, 2–3 mm broad (± 1.5 mm long below insertion of glands), with 5 obovate lobes with deeply dissected margins, green; *glands* 5, transversely elliptic, 0.7–1.2 mm broad, ascending to spreading, yellow-green, outer margins flat and entire; *stamens* with glabrous pedicels, bracteoles filiform and pubescent; *ovary* globose, glabrous to sparsely pubescent, slightly exserted on short pedicel 0.5 mm long; styles 0.6–1 mm long, divided nearly to base, spreading, sometimes horizontally. *Capsule* 2.5–3 mm diam., obtusely 3-lobed, glabrous to sparsely pubescent, green, exserted from cyathium on short sometimes slightly decurved and later erect pedicel ± 1 mm long.

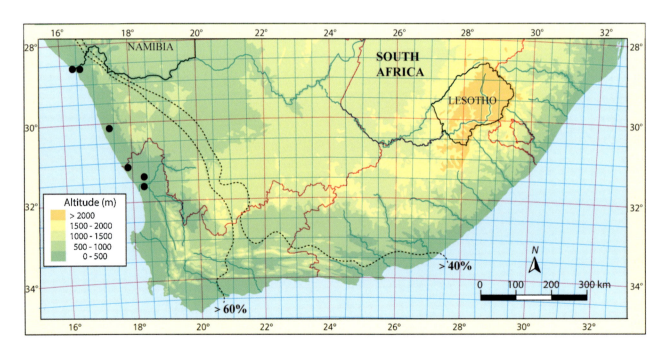

Fig. 3.112. Distribution of *Euphorbia stapelioides* (© PVB).

3.2 Sect. Articulofruticosae

Fig. 3.113. *Euphorbia stapelioides*, with gravel and windblown sand, most branches projecting ± 1.5 cm from ground, *PVB 3947*, foot of Kortdoringberg, east of Alexander Bay, South Africa, 13 Jul. 2014 (© PVB).

Distribution & Habitat

Euphorbia stapelioides is known along the banks of the Orange River near its mouth, from Grootderm to Alexander Bay and from three widely scattered localities further south in Namaqualand: around Riethuis, south-west of Kotzesrus and north of Koekenaap, a little north of the Olifants River. All known localities are within 20 km of the sea.

Euphorbia stapelioides grows in gravelly or firm, loamy ground among scattered, short bushes or between occasional, larger stones in almost completely flat areas. Around the mouth of the Orange River it occurs among low outcrops of limestone with a fair amount of wind-blown sand and some quartz-gravel, which may partly cover the above-ground portions of the plants and often blasts the sur-

Fig. 3.114. *Euphorbia stapelioides*, most branches projecting ± 1.5 cm from ground but no new growth, *PVB 3947*, foot of Kortdoringberg, east of Alexander Bay, South Africa, 13 Jul. 1997 (© PVB).

faces of these small branches as well. Here it occurs with a wealth of other succulents, including *E. caput-medusae*, *E. ephedroides*, *E. mauritanica* and *E. rhombifolia*. At the other known localities it mostly grows together with *E. celata*, also among many other succulents.

Diagnostic Features & Relationships
The plant in *E. stapelioides* consists of a tuber of somewhat variable shape that is also located at very variable depths in the soil, though it is rarely found deeper than about 20 cm beneath the surface and even more rarely it is right at the surface. From the tuber, slender branches rise to the surface and often spread laterally as well by slender horizontal rhizomes before emerging from the ground (as seen in Fig. 3.115). These rhizomes often bear roots and subsidiary tubers may develop on them as well. Consequently, a single specimen may spread over anything between 10 and 50 cm of the surface of the ground (occasionally covering sparsely an even larger area) and the central tuber may be hard to locate. Above the ground the plant consists of many short, cylindrical, grey-green branches. In the driest area where it occurs (namely around the mouth of the Orange River) the branches are between 5 and 10 mm thick and often project for only 10 to 20 mm from the ground, so that most of the plant is underground.

Fig. 3.115. *Euphorbia stapelioides,* plant with some sand blown away to reveal rhizomatous branches, here ± 8 cm long, *PVB 3947*, foot of Kortdoringberg, east of Alexander Bay, South Africa, 13 Jul. 2014 (© PVB).

Further to the south the above-ground portions are longer, more slender and more cylindrical, though in some plants the branches are also short. However, in cultivation, plants from near the mouth of the Orange River become much more luxuriant and slender, with their rhizomatousness greatly reduced and they become identical to those from further south. The tiny, opposite pairs of leaf-rudiments are only briefly present on young branches and have two small spreading horns at their bases. After drying out, they may persist as small blackish scales.

The very small, terminal, unisexual cyathia are produced at the tips of young branches, usually towards the beginning of the growing season, often between April and June but they may appear as late as September.

A typical member of sect. *Articulofruticosae*, *E. stapelioides* is almost the smallest member of the section, where it is rivalled in size only by *E. herrei*. It is the only species with a tuber and a strongly rhizomatous habit, though this habit appears again in *E. rhombifolia* in the population described as *E. bayeri*. It shares with *E. rhombifolia* the small spreading horns at the bases of the leaf-rudiments and the two differ mainly vegetatively.

When describing *E. lumbricalis*, Leach compared it to *E. brachiata*. However, it is to *E. stapelioides* that it is closest and here it is regarded as a synonym of *E. stapelioides*. The southern populations differ from those at the mouth of the Orange River by the longer above-ground branches which are more slender and more rounded (as in Figs. 3.118 and 3.119), the styles are shorter and more closely pressed to the top of the ovary and the ovary is very finely pubescent, where it is usually glabrous in plants from the Orange River Mouth. As remarked above, the vegetative differences disappear in cultivation and even Herre remarked on one of his early collections of this plant that specimens in habitat looked just like plants of *E. stapelioides* after a few years in cultivation.

History
Euphorbia stapelioides was discovered by J.F. Drège and his brother Carl in October 1830 near the mouth of the Orange River. It was recollected for the first time in June 1928 by F.C. Kolbe and again in 1930 by P. Ross Frames. These and later collections by Dyer and others established the identity

3.2 Sect. Articulofruticosae

Fig. 3.116. *Euphorbia stapelioides,* female plant flowering in cultivation, *PVB 3947,* foot of Kortdoringberg, east of Alexander Bay, South Africa, 13 Nov. 2007 (© PVB).

Fig. 3.117. *Euphorbia stapelioides,* capsule, in cultivation, *PVB 3947,* foot of Kortdoringberg, east of Alexander Bay, South Africa, 18 Oct. 2015 (© PVB).

Fig. 3.118. *Euphorbia stapelioides,* growing in quartzitic and gneissic gravel, most branches projecting ± 2 cm from ground, *PVB 4598,* Riethuis, NE of Hondeklip Bay, South Africa, 4 Jul. 1991 (© PVB).

Fig. 3.119. *Euphorbia stapelioides*, growing in reddish sand, most branches projecting up to ± 6 cm from ground, *PVB 5159*, west of Kotzesrus, South Africa, 9 Sep. 1994 (© PVB).

of the species. Herre (1936) believed that he had rediscovered this species, but actually he had found another superficially similar one which was later described as *E. herrei*. The relationships of *E. stapelioides* led to much speculation: Boissier (1860) believed it was related to *E. hamata* and *E. clavarioides*; N.E. Brown (1915) found that it was close to *E. gentilis*, *E. karroensis* (= *E. burmanni*) and *E. spicata* (= *E. rhombifolia*) while Herre (1936) felt it was an ally of *E. mauritanica*.

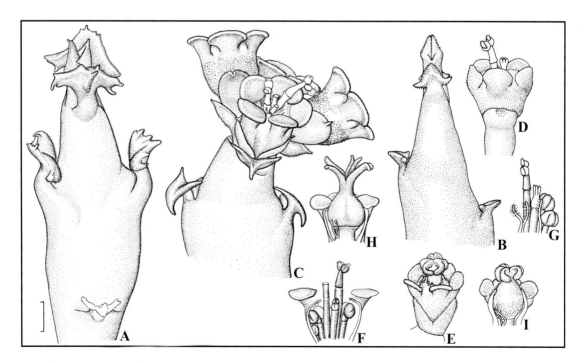

Fig. 3.120. *Euphorbia stapelioides*. **A**, **B**, tip of young branch with leaf-rudiments (scale 1 mm, as for **B–I**). **C**, side view of cyme of male cyathia. **D**, side view of male cyathium. **E**, side view of female cyathium. **F**, **G**, anthers, bracteoles and sterile female in male cyathium. **H**, **I**, female floret in dissected cyathium. Drawn from: **A**, **C**, **F**, **H**, *PVB 3947*, foot of Kortdoringberg, east of Alexander Bay, South Africa. **B**, **E**, **I**, *PVB 1083*, north of Koekenaap, near Vredendal, South Africa. **D**, **G**, *PVB 4598*, Riethuis, NE of Hondeklip Bay, South Africa (© PVB).

The material described by Leach as *E. lumbricalis* was collected at its type locality first by myself in June 1975. Although Leach declared that this 'species' was only known from this locality (Leach 1986c), this was not the first collection of such plants, for they were gathered near Riethuis by H. Herre already in 1940 (BOL records) and near Koekenaap in May 1949, also by Herre (PRE records). On these specimens, Herre had commented on their remarkable resemblance to *E. stapelioides*.

Euphorbia suffulta Bruyns, *S. African J. Bot.* 56: 129 (1990). Type: South Africa, Cape, Tierberg, Prince Albert distr., 6 Dec. 1987, *Bruyns 2902* (BOL, holo.; K, PRE, iso.).

Unisexual creeping to scandent spineless glabrous succulent undershrub leaning on stems of surrounding bush for support (rarely a small free-standing bushlet) branching and rebranching extensively from similar stem to 0.3 m tall with slender woody rootstock bearing fibrous roots. *Branches* spreading to erect, repeatedly rebranching, cylindrical (indistinctly 4-angled when young), 50 mm –1.3 m × 3–5 mm, without distinct tubercles, papillate, soft and succulent when young and remaining so with age, dark green suffused with purple when young becoming pale grey-green with age (with waxy. slightly flaking covering); *leaf-rudiments* towards apices of branches, opposite, 1–2 × 1 mm, ascending and spreading towards tip, fleeting, spathulate, minutely puberulous, slightly channelled above by upwardly folded somewhat toothed margins, obtuse, tapering below into very short petiole 1 mm long, subtended on either side at base by conspicuous ellipsoidal dark brown tuberculate stipule 1–1.5 mm diam. *Synflorescences* terminal on branches, papillate, each of 1–3 unisexual cyathia on short peduncle 1–8 mm long and each subtended by 2 obovate bracts very like leaf-rudiments 1–2 × 1 mm (many more cyathia sometimes produced by repeated development from axils of bracts); *cyathia* conical, minutely pubescent, 1.5–2 mm broad (± 1 mm long below insertion of glands), with 5 obovate lobes with deeply dissected margins, green; *glands* 5, transversely elliptic, 1 mm broad, ascending to spreading, yellow-green, outer margins flat and entire; *stamens* with pubescent pedicels, otherwise glabrous, bracteoles filiform and pubescent; *ovary* globose, pubescent, raised and slightly exserted on erect pedicel 0.5 mm long; styles ± 0.6 mm long, divided nearly to base, horizontally spreading. *Capsule* 2.5–3 mm diam., obtusely 3-lobed, minutely pubescent, dull green, sessile.

Distribution & Habitat

Euphorbia suffulta is fairly widespread on the western side of the Great Karoo, where it is known from Merweville to near Rietbron and southwards towards Prince Albert and Klaarstroom.

This species grows on the slopes of and between low hills on shales and tillites of the Beaufort, Dwyka and Ecca series among shrubs of *Rhigozum obovatum* and some succulents (often including a wide selection of other species of *Euphorbia*). Plants rarely grow in the open and usually shel-

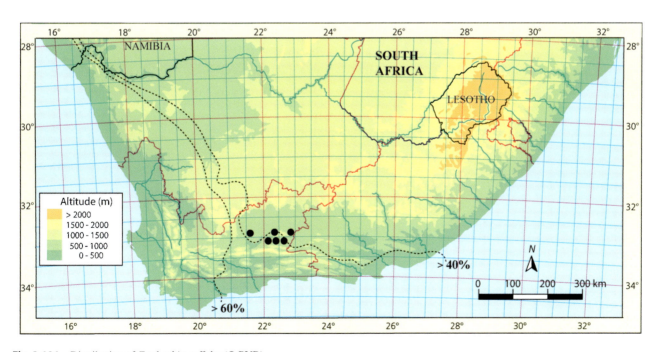

Fig. 3.121. Distribution of *Euphorbia suffulta* (© PVB).

Fig. 3.122. *Euphorbia suffulta*, three small plants removed to show habit and slender rootstock, *PVB 2902*, Tierberg, east of Prince Albert, South Africa, 26 Mar. 1989 (© PVB).

ter in other shrubs, creeping and clambering among their branches, both for support and protection, in a similar manner to various species of *Ceropegia*.

Diagnostic Features & Relationships

Euphorbia suffulta forms small, diffuse clumps (very occasionally large bushes are found and these tend to develop only where they have been protected from grazing) with nearly cylindrical, but fairly slender, dark green, softly pliable and finely papillate, succulent branches. Opposite pairs of small, spathulate leaf-rudiments are present on the younger branches and each leaflet is subtended by a pair of relatively prominent, swollen stipules which are initially pale yellow, then red and soon turn dark brown, almost black. The stipules remain on the branches after the leaf-rudiment has fallen off.

In *E. suffulta* flowering takes place over much of the summer months but generally only happens if and when some rain is received. The unisexual cyathia are small and inconspicuous. The first ones on each branch are borne terminally but much rebranching may take place from the axils of the bracts beneath them.

Euphorbia suffulta is a typical member of sect. *Articulofruticosae* but is more insignificant even than most other members of this group. Its relationships within this section are still unclear. Its habit of growing best inside other shrubs is occasionally also found in *E. rhombifolia* but these two are easily separated by the rather more clearly angled, more glabrous branches in *E. rhombifolia* and the lack of the conspicuous stipules of *E. suffulta*. Other species with such noticeable stipules are *E. burmanni* and *E. exilis*.

Fig. 3.123. *Euphorbia suffulta*, male plant with cyathia and glandular stipules, *PVB 8182*, Rondawel, east of Prince Albert, South Africa, 1 Nov. 2007 (© PVB).

3.2 Sect. Articulofruticosae

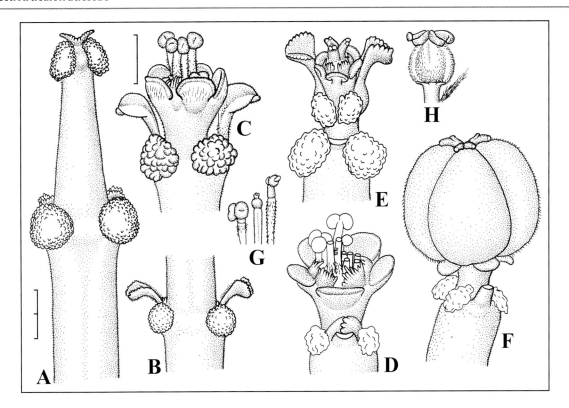

Fig. 3.124. *Euphorbia suffulta*. **A, B**, part of young branch with leaf-rudiments and stipules (scale 2 mm, as for **B**). **C, D**, side view of male cyathium (scale 1 mm, as for **D–H**). **E**, female cyathium from side. **F**, capsule. **G**, anthers and sterile female from male cyathium. **H**, female floret. Drawn from: *PVB 2902*, Tierberg, east of Prince Albert, South Africa (© PVB).

With its very slender, brittle branches and minute stipules, *E. exilis* cannot be confused with *E. suffulta* but *E. burmanni* is more similar. From *E. burmanni*, *E. suffulta* is separated by the fact that *E. burmanni* is mostly a robust, free-standing shrub (though this is not always true, especially in the Eastern Cape, but it never has the creeping-climbing habit of *E. suffulta*). *Euphorbia burmanni* has smooth as opposed to papillate branches and the stipules (though similar in size and colour) are smooth rather than tuberculate. Furthermore, the styles are more erect and not horizontally spreading ontop of the ovary as in *E. suffulta* and the capsule is slightly exserted as it matures in *E. burmanni*.

History

I first observed *Euphorbia suffulta* in April 1987 near Prince Albert and the type was collected in December of the same year. The species does not appear to have been recorded before these collections were made.

Euphorbia tenax Burch., *Trav. S. Africa* 1: 219 (1822). *Tirucalia tenax* (Burch.) P.V.Heath, *Calyx* 5: 93 (1996). Type: South Africa, Cape, Hangklip, near Ongeluks River, Ceres div., 17 July 1811, *Burchell 1219* (K, holo.).

Euphorbia arceuthobioides Boiss., *Cent. Euphorb.*: 20 (1860). *Tirucalia arceothobioides* (Boiss.) P.V.Heath, *Calyx* 5: 87 (1996). Type: South Africa, Cape, 70.10, *Ecklon & Zeyher, Euphorb. 76*, (*Ecklon 1312*) (G, holo.; W, iso.).

Arthrothamnus ecklonii Klotzsch & Garcke, *Abh. Königl. Akad. Wiss. Berlin* 1859: 63 (1860). Lectotype (Bruyns 2012): South Africa, Cape, *Ecklon & Zeyher, Euphorb. 24*, (*Ecklon 1871*) (W; MEL, iso.).

Arthrothamnus scopiformis Klotzsch & Garcke, *Abh. Königl. Akad. Wiss. Berlin* 1859: 63 (1860). Type: South Africa, Cape, *Bergius* (missing).

Euphorbia rhombifolia var. *cymosa* (Klotsch & Garcke) N.E.Br., *Fl. Cap.* 5 (2): 285 (1915). *Arthrothamnus cymosus* Klotzsch & Garcke, *Abh. Königl. Akad. Wiss. Berlin*

1859: 63 (1860). *Tirucalia cymosa* (Klotzsch & Garcke) P.V.Heath, *Calyx* 5: 92 (1996). Lectotype (Bruyns 2012): South Africa, Cape, *Ecklon & Zeyher, Euphorb. 24* (W; MEL, iso.).

Euphorbia serpiformis Boiss. in DC., *Prodr.* 15 (2): 75 (1862). Lectotype (Bruyns 2012): South Africa, Cape, Berg River Valley, *Zeyher 1535* (BOL; K, S, SAM, W, WU, Z, iso.).

Euphorbia mixta N.E.Br., *Fl. Cap.* 5 (2): 585 (1925). *Tirucalia mixta* (N.E.Br.) P.V.Heath, *Calyx* 5: 91 (1996). *Euphorbia arrecta* N.E.Br., *Fl. Cap.* 5 (2): 283 (1915), *nom. illegit.*, *non* N.E.Br. (1914). Type: South Africa, Cape, Berg River Valley, *Zeyher 1535* (K, holo.; BOL, S, SAM, W, WU, Z, iso.).

Unisexual spineless and glabrous succulent shrub sometimes growing inside other shrubs, 0.2–0.8 m tall, branching and rebranching extensively around similar stem, all arising from somewhat swollen or woody rootstock bearing fibrous roots. *Branches* ascending to erect, repeatedly rebranching, cylindrical to slightly angled, 30–500 × 2–5 mm, without distinct tubercles, roughened with low transparent ridges to smooth, soft and succulent, dark to light grey-green sometimes becoming grey and woody with age towards base; *leaf-rudiments* on young growth towards apices of branches on very slightly raised tubercles, initially opposite (later sometimes shifted to appear alternating), 2–6 × 1–3 mm, ascending and later spreading, fleeting, hastate-deltate to almost spathulate with two spreading hornlets at base, puberulous to glabrous, slightly channelled above by upwardly folded margins, obtuse, tapering below into short petiole 1–2 mm long, estipulate. *Synflorescences* terminal on branches, each of 1 unisexual cyathium on peduncle 1–3 mm long closely subtended by 2 obovate to spathulate bracts very like leaf-rudiments (often with basal or near apical hornlets) 1–2 × 1–3 mm in axils of which further branches 3–30 mm long with further terminal cyathia (and further branchlets with cyathia) may arise; *cyathia* conical in females to sometimes slightly urceolate in males, finely pubescent around bases, 2–3.5 mm broad and almost twice as broad in males as females (± 1 mm long below insertion of glands), with 5 obovate lobes with deeply dissected margins, greyish green; *glands* 5, transversely elliptic, 0.7–1.5 mm broad, spreading, yellow-green, outer margins slightly recurved and entire; *stamens* with glabrous pedicels, bracteoles filiform and pubescent; *ovary* globose, glabrous, raised on pedicel ± 0.7 mm long; styles ± 1 mm long, divided to near base, often horizontally spreading. *Capsule* 3–4 mm diam., obtusely 3-lobed, sparsely pubescent to glabrous, dull green, slightly exserted from cyathium on short erect pedicel ± 1.5 mm long.

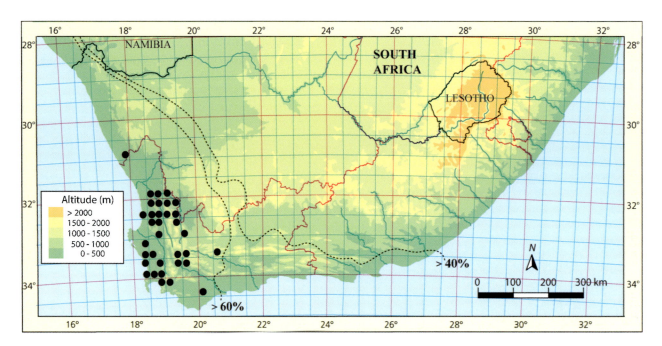

Fig. 3.125. Distribution of *Euphorbia tenax* (© PVB).

3.2 Sect. Articulofruticosae

Fig. 3.126. *Euphorbia tenax*, slender shrub ± 50 cm tall in white sand with restios and other *fynbos*, *PVB 10890*, Olof Bergh Pass, Redelinghuis, South Africa, 30 Sep. 2007 (© PVB).

Distribution & Habitat

Euphorbia tenax is mainly a species of the Cape Floristic Region, to which it is mostly confined. Within the Cape Floristic Region, it is known from near Nieuwoudtville, Clanwilliam and near the coast at Verlorevlei southwards to around Paarl and Cape Town and is also widely known in the Cedarberg and southwards to Worcester and near Elgin. Outside the Cape Floristic Region, it has been found in Namaqualand, near the coast to the west of Kotzesrus, west of Bitterfontein.

Within the Cape Floristic Region, where the rainfall is relatively high, *E. tenax* is found in a wide variety of habitats. These range from nutrient-poor, white, acidic sands inhabited mainly by Restionaceae (as in Fig. 3.126) and other *fynbos* elements to dry slopes and outcrops of shales and granites with a scanty cover of grass, other small succulents and many geophytes. The latter habitat is exemplified by the slopes of Signal Hill at Cape Town and the hills around Mamre and Darling (as in Figs. 3.127 and 3.128). In Namaqualand it is also found among restios, other *fynbos* elements and *renoster* on deep, pale sand.

Diagnostic Features & Relationships

A typical plant of *Euphorbia tenax* arises from a somewhat swollen, elongated rootstock, which may begin anything up to 0.1 m below the surface of the soil (depending on the looseness of the ground). Occasionally the plants are eaten off right down to this rootstock, but they readily resprout from it. Depending on habitat, they vary greatly in form. Those typical of sandy habitats among restios are very slender, diffuse, sparsely branched and almost reed-like with long internodes and these can be hard to distinguish from the

Fig. 3.127. *Euphorbia tenax*, small grey shrub ± 15 cm tall in recently burnt area, Signal Hill, Cape Town, South Africa, 13 Jul. 2015 (© PVB).

Fig. 3.128. *Euphorbia tenax*, male shrub ± 30 cm tall covered with cyathia, Signal Hill, Cape Town, South Africa, Sep. 1996 (© PVB).

surrounding vegetation. When they are in exposed, stony habitats, they usually form small, fairly dense and rounded shrublets. The branches are also of variable thickness (generally with the slightly stouter ones on exposed places) but they are often scabrid from small raised circular to longitudinal warts scattered over the surface. Their dark grey-green colour makes the whole plant quite inconspicuous, especially when conditions are dry. Tiny leaf-rudiments are borne on young shoots and these too are very variable in shape, though a constant feature is the pair of hornlets at their bases.

Flowering in *E. tenax* is an inauspicious event. Many of the small and inconspicuous cyathia are produced at the tips of the branches during the rainy and growing season between June and September. Additional cyathia usually develop from the axils of the bracts surrounding the earlier cyathia and may form a fairly richly branched structure. The female cyathia may be as small as half the diameter of the males and this, combined with their greener colour, makes them much less conspicuous than even the males. The males are more substantial and have more obviously yellowish glands.

Euphorbia tenax is closely related to and may be difficult to distinguish from *E. muricata* and *E. rhombifolia*. The main differences lie in the much more diffuse habit of *E. tenax*, with more slender, reed-like branches which never tend to dry out towards their tips.

Fig. 3.129. *Euphorbia tenax*, female plant with cyathia and capsules, *PVB 10890*, Olof Bergh Pass, Redelinghuis, South Africa, 30 Sep. 2007 (© PVB).

3.2 Sect. Articulofruticosae

Fig. 3.130. *Euphorbia tenax*, male plant with cyathia, Signal Hill, Cape Town, South Africa, Sep. 1996 (© PVB).

History

As it is found around Cape Town, it is not surprising that *Euphorbia tenax* was known early on in the botanical exploration of South Africa and there is a plate in Burman (1738) which is referable to this species. This plate was made somewhat earlier and is Folio 192 in the volume of the *Codex witsenii* known as *Icones plantarum et animalium* (Macnae and Davidson 1969).

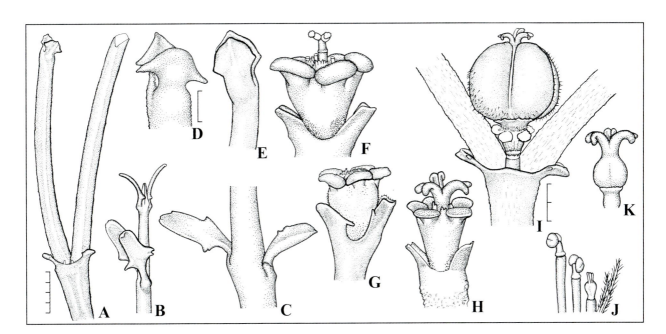

Fig. 3.131. *Euphorbia tenax*. **A–C**, tip of young branch with leaf-rudiments (scale of **A** 4 mm, as for **B**, **C**). **D**, **E**, bracts in inflorescence (scale 1 mm, as for **E–K**). **F**, **G**, side view of male cyathium. **H**, female cyathium from side. **I**, capsule. **J**, anthers, bracteoles and rudimentary female floret from male cyathium. **K**, female floret. Drawn from: **A**, **D–K**, *PVB*, Signal Hill, Cape Town, South Africa. **B**, **C**, *PVB*, Elandsvlei, west of Piketberg, South Africa (© PVB).

In Bruyns (2000) and Bruyns et al. (2006) this species was referred to as *E. arceuthobioides*. However, the name *Euphorbia tenax* of Burchell was published forty years before Boissier's *E. arceuthobioides*. Consequently it has precedence and is the correct name for these plants. Burchell collected the specimens on which he based *E. tenax* near Ceres in July 1811.

Boissier gave this species the rather formidable name *E. arceuthobioides*. This name refers to its resemblance to *Arceuthobium*, which is a genus of mistletoe (belonging to the Viscaceae) that is found widely in the northern hemisphere.

Euphorbia verruculosa N.E.Br., *Fl. Cap.* 5 (2): 585 (1925). *Tirucalia verruculosa* (N.E.Br.) P.V.Heath, *Calyx* 5: 93 (1996). Type: Namibia, Lüderitz (Angra Pequeña), 10 miles from coast, Nov. 1908, *Marloth 4639* (PRE, holo.; K, iso.).

Unisexual dwarf spineless glabrous succulent branching from apex of somewhat swollen, ± woody, divided rootstock 20–100 × 8–20 mm bearing fibrous roots. *Branches* usually spreading to rarely erect above surface, often rebranching extensively above and close to ground and branches deeply articulated, older branches cylindrical and grey, younger shoots more irregularly cylindrical and more prominently knobbly-tuberculate, 10–75 (rarely to 150 mm if sheltered) × 4–10 mm, soft and succulent when young and becoming more rigid with age but brittle at joints, grey-green, surface convoluted-warty; *leaf-rudiments* towards apices of branches on slightly raised tubercles, initially opposite (later often shifted to appear alternating), 0.5–1 × 1–2 mm, ascending and later slightly spreading, persistent briefly as small black scales, hastate-deltate with 2–5 spreading hornlets at base and along margins, minutely puberulous around base, without medial groove above, obtuse to mucronate, without obvious petiole or stipules. *Synflorescences* terminal on branches, each of 1–5 unisexual cyathia on short peduncle 1–3 mm long each subtended by 2 ovate reddish bracts ± 0.8 × 1 mm; *cyathia* cupular, minutely pubescent behind glands, 1.5–2.5 mm broad (± 1 mm long below insertion of glands), with 5 minute reddish lobes with deeply dissected margins, green suffused with red to bright red; *glands* 5, transversely elliptic, 0.5–0.8 mm broad, ascending to spreading, dull green suffused with red to red, outer margins flat and entire; *stamens* with glabrous pedicels, bracteoles filiform and pubescent; *ovary* globose, pubescent with adpressed white hairs, becoming slightly exserted on short pedicel 0.5–1.5 mm long; styles 1 mm long, divided nearly to base, spreading, sometimes horizontally. *Capsule* 1.5–2 mm diam., obtusely 3-lobed, pubescent with white hairs, dull green, exserted from cyathium on short slightly decurved later erect pedicel 1–2 mm long.

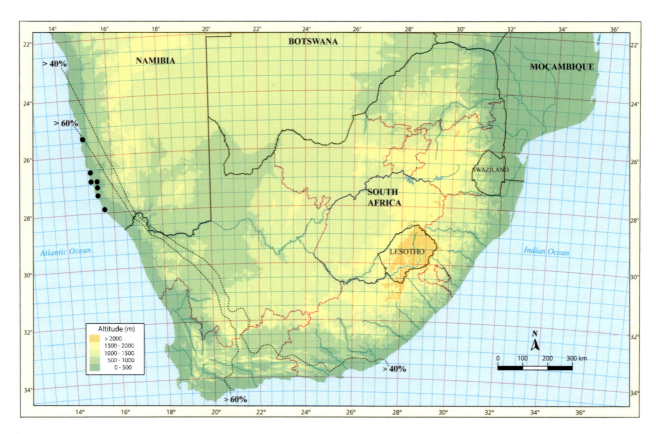

Fig. 3.132. Distribution of *Euphorbia verruculosa* (© PVB).

3.2 Sect. Articulofruticosae

Distribution & Habitat

Euphorbia verruculosa occurs in Namibia, in the coastal parts of the southern Namib Desert, where it is known from near Chameis Bay (north of Oranjemund and about 170 km south of Lüderitz) to Lüderitz. There is also a single collection from Oyster Cliffs (actually given as 'Osterklippen, zwischen Lüderitz und Walvis Bay' but assumed to be Oyster Cliffs), some 150 km north of Lüderitz and the species probably occurs between these places as well.

Euphorbia verruculosa is found from very close to the coast (some localities are within one kilometre of the sea) to 16 km from the coast. This is a hyper-arid region and at Lüderitz the average annual rainfall is below 20 mm and usually falls in winter. Despite the great aridity and the barren appearance of the immediate surroundings of Lüderitz, Dinter (1923: 40) recorded over 80 species of plants there. Most of these are succulents, with such genera as *Crassula*, *Pelargonium*, *Monsonia*, *Tylecodon*, various succulent Aizoaceae (including *Conophytum*, *Fenestraria* and *Lithops*) and even the ubiquitous *Euphorbia rhombifolia* is present. *Euphorbia verruculosa* grows on low outcrops of dark grey gneiss and schist, often in the open between stones, with the rootstock tightly wedged among the rocks.

This is one of several small and localized species belonging to sect. *Articulofruticosae* that are associated with the hyper-arid Namib Desert. These are *E. herrei*, *E. stapelioides* and *E. verruculosa* from the southern Namib and *E. negromontana* from much further north.

Fig. 3.133. *Euphorbia verruculosa*, relatively dry state in year of poor rains, plant very coral-like (± 80 mm diam.) and with a few orange lichens growing on some of the branches, *PVB 12542*, Lüderitz, Namibia, 6 wJul. 2013 (© PVB).

Diagnostic Features & Relationships

Brown (1915) mentioned that *E. verruculosa* was 'apparently creeping extensively underground' and must have believed that the plant pressed by Marloth was part of a much larger, rhizomatous individual. This was quoted by White et al. (1941). However, the plants are not rhizomatous at all and each one consists of a single cluster of branches arising from the apex of a thickened, often quite rigid rootstock that is usually somewhat divided beneath the surface and whose thickened parts are rarely more than 120 mm long. Each rootstock gives rise to an often quite dense cluster of knobbly and clearly articulated branches that remain close to the ground and rarely exceed 75 mm long. Older branches become somewhat whitish (probably from blasting

Fig. 3.134. *Euphorbia verruculosa*, large plant ± 100 mm diam., plump after some rain, among rocks and windblown sand, *PVB 12542*, Lüderitz, Namibia, 11 Jul. 2014 (© PVB).

Fig. 3.135. *Euphorbia verruculosa*, male plant with tiny cyathia, *PVB 12542*, Lüderitz, Namibia, 11 Jul. 2014 (© PVB).

by wind-blown sand) but the younger ones are greyish green and very young growth is briefly bright green (as in Fig. 3.137). The surface of the branches is deeply convoluted (Fig. 1.28 C, D) and somewhat translucent (this dries to a finely wrinkled, whitish, papery consistency, according to Brown 1915), with quite prominent tubercles and a thick covering of wax. Initially opposite, the tubercles are often shifted away from this position so as to appear to be alternating. Each tubercle bears a small, thick leaf-rudiment that is usually considerably shorter than broad and bears two to several lateral hornlets towards its base. Each leaf-rudiment also has a small cluster of hairs on the surface next to the stem near its base.

3.2 Sect. Articulofruticosae

Fig. 3.136. *Euphorbia verruculosa*, male plant with few cyathia, *PVB 12542*, Lüderitz, Namibia, 11 Jul. 2014 (© PVB).

Fig. 3.137. *Euphorbia verruculosa*, female plant flowering and with reddish new growth, *PVB 12542*, Lüderitz, Namibia, 11 Jul. 2014 (© PVB).

Flowering seems to take place during winter when some rain is received but probably occurs after occasional rainfall-events at other times too. The cyathia are borne on very short peduncles and are also particularly small, with the females about half the diameter of the males. Each is subtended by two tiny reddish bracts and the cyathia may be wholly reddish too. Others are dull green suffused with red and that is also the usual colour of the glands. Male florets are short, with bright yellow anthers and the female florets are raised on a short pedicel which later elongates to push the small, pubescent capsule out of the cyathium.

Fig. 3.138. *Euphorbia verruculosa*, male cyathia, *PVB 12542*, Lüderitz, Namibia, 11 Jul. 2014 (© PVB).

History

Euphorbia verruculosa was discovered by Rudolf Marloth in November 1908, with another collection apparently made in May 1910 by Paul Range. Dinter (1923: 43) also recorded seeing it near Lüderitz in 1922, but he believed that he had found *E. stapelioides* and Range's material was also initially identified in Berlin as *E. stapelioides*. Nevertheless, the 'warty' surface that Dinter mentioned makes it clear that he had encountered *E. verruculosa* rather than *E. stapelioides*, which does not occur as far north as Lüderitz and does not have verruculose branches.

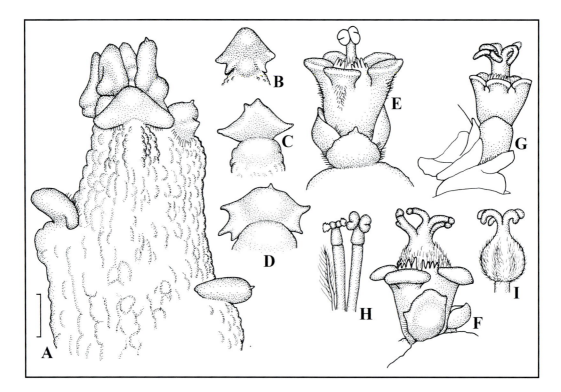

Fig. 3.139. *Euphorbia verruculosa*. **A**, tip of young branch with leaf-rudiments (scale 1 mm, as for **B–I**). **B–D**, leaf-rudiments. **E**, side view of male cyathium. **F, G**, female cyathium from side. **H**, anthers and bracteole. **I**, female floret. Drawn from: *PVB 12542*, Lüderitz, Namibia (© PVB).

3.3 Sect. Espinosae

Euphorbia sect. **Espinosae** Pax & K.Hoffm. in Engl., *Veg. Erde* 9, 3 (2): 149 (1921). *Euphorbia* subsect. *Espinosae* (Pax & K.Hoffm.) Pax & K.Hoffm. in Engl., *Nat. Pflanzenfam.*, ed. 2, 19c: 213 (1931). Type: *E. espinosa* Pax.

Bisexual woody shrubs to small trees often with shiny, peeling papery bark on trunk, not spiny, sometimes with swollen roots. *Branches* slender, grey to brown, occasionally drying out towards tips but not forming spikes. *Leaves* ephemeral, to 60 mm long, alternate, elliptical to obovate, petiolate, with small glandular stipules. *Synflorescence* a solitary bisexual cyathium terminating short shoots, surrounded by bracts that are scale-like to slightly shorter and broader than leaves, glands 5, entire, yellow-green, without appendages. *Capsule* ≥ 5 mm diam., exserted on erect to decurved and later erect pedicel. *Seeds* smooth, attractively mottled with brown, with cap-like caruncle.

This section contains two species that are widely distributed in Africa south of the equator. In southern Africa *E. guerichiana* is particularly associated with the edges of the Namib Desert, but *E. espinosa* is of very scattered occurrence in moister (though locally dry) areas of the tropics. Both species occur in the northern parts of South Africa, from near Thabazimbi in the north-west to the Kruger National Park, but they have never been recorded growing together. Since they flower at different times, hybrids between them are unlikely.

Key to the species of sect. *Espinosae*

Plant flowering when leafless, leaves elliptical with petiole 5–12 mm long..**E. espinosa**
Plant flowering with leaves, leaves narrowly oblong to lanceolate with petiole 1–3 mm long..........................**E. guerichiana**

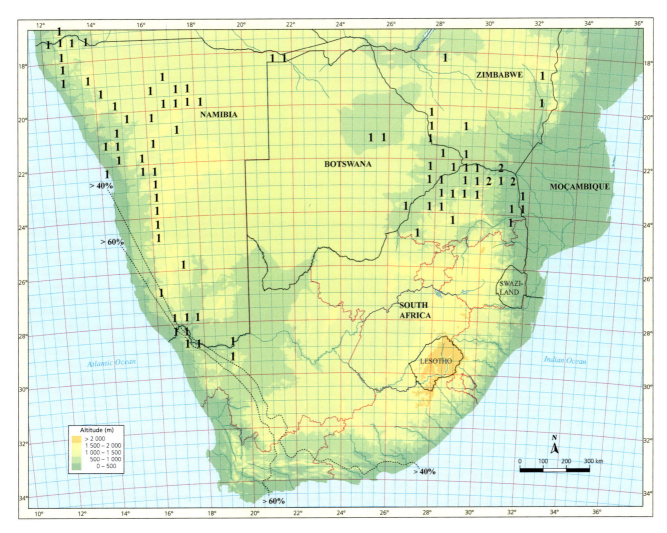

Fig. 3.140. Distribution of *Euphorbia* sect. *Espinosae* in southern Africa (showing number of species per half-degree square) (© PVB).

Euphorbia espinosa Pax, *Bot. Jahrb. Syst.* 19: 120 (1894). Lectotype (Bruyns 2012): Tanzania, without precise locality, 1885/86, *G.A.Fischer 285* (K).

Euphorbia gynophora Pax, *Bot. Jahrb. Syst.* 34: 374 (1904). Lectotype (Bruyns 2012): Tanzania, Pare Mountains, between Kisuani and Madji-ya-juu, 700 m, 13 Oct. 1902, *Engler 1579* (drawing at K).

Euphorbia nodosa N.E.Br., *Fl. Trop. Afr.* 6 (1): 548 (1911). Lectotype (designated here): Angola, along Rio Cavaco, east of Benguella, 7 July 1905, *Gossweiler 1695* (BM; LISU, iso.).

Bisexual spineless glabrous woody shrub 0.5–5 (7) m tall, branching from base or with slender trunk 20–110 mm thick with shiny grey-brown bark peeling in papery strips near base, with several large swollen spreading roots bearing fibrous roots. *Branches* erect to spreading, often rebranching, cylindrical, younger branches 1.5–3 mm thick, ± without tubercles, somewhat pliable but not succulent, grey-brown, glabrous (finely pubescent); *leaves* mainly on young growth on very slightly raised tubercles, alternate, 20–45 (60) × 15–30 mm, ascending to spreading, deciduous, elliptic, glabrous with sparsely pubescent margins (finely pubescent), ± flat, green (paler below than above), obtuse to apiculate, with slender petiole 5–12 mm long, with small brown glandular stipules. *Synflorescences* almost sessile, in axils of fallen leaves with 1 terminal usually bisexual cyathium closely subtended by 2–8 adpressed scale-like reddish bracts 0.5–2 × 0.5–1.5 mm; *cyathia* conical-cupular, ± glabrous, 4–5 mm broad, with 5 nearly quadrate reddish lobes with denticulate pubescent margins, green to suffused with red near base; *glands* 5, transversely elliptic, 2–2.5 mm broad, spreading, contiguous, green, outer margins flat and entire or finely emarginate; *stamens* with glabrous pedicels, bracteoles filiform with pubescent tips, with yellow or reddish anthers; *ovary* 3-lobed, glabrous, dull reddish green, raised on glabrous pedicel 1 mm long soon elongating to 4 mm or more to push ovary out of cyathium; styles 1.3–2 mm long, divided for most of length, widely spreading. *Capsule* 6–10 mm diam., obtusely 3-lobed, glabrous, greyish green with reddish blush near base, exserted from cyathium on decurved and later erect pedicel 6–12 mm long.

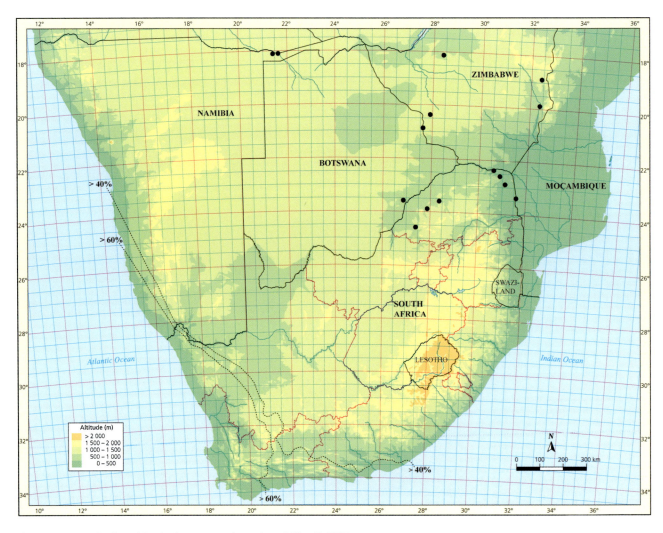

Fig. 3.141. Distribution of *Euphorbia espinosa* in southern Africa (© PVB).

Distribution & Habitat

Euphorbia espinosa is widespread in south tropical Africa, where it is known in Angola, Botswana, southern Kenya, Malawi, Moçambique, northern Namibia, Tanzania and Zimbabwe. It has been scantily recorded also in the northern parts of South Africa, from near Thabazimbi to Punda Milia in the Kruger National Park. In Namibia it is only known in the relatively moist Caprivi Strip in the north-east on rocky places near the Okavango River.

Plants usually grow on fairly densely vegetated, stony, flat or slightly sloping ground and they are often difficult to distinguish from the surrounding shrubs. They sometimes occur with a wealth of other succulents, including *Aloe*, various stapeliads and other species of *Euphorbia*.

Fig. 3.142. *Euphorbia espinosa*, plant around 0.75 m tall, on conglomerates with boulders and bush, *PVB 12052a*, south of Marken, South Africa, 2 Nov. 2011 (© PVB).

Fig. 3.143. *Euphorbia espinosa*, excavated plant showing large swollen roots, here 3 to 5 cm thick, *PVB 12052a*, south of Marken, South Africa, 2 Nov. 2011 (© PVB).

Diagnostic Features & Relationships

Euphorbia espinosa forms shrubs that may reach 3 m tall, but are often 0.5 to 1.5 m in height and sparingly branched from the base. The plant arises from several much swollen roots which spread out just beneath the surface of the soil. Its stem and thicker branches are covered with a shiny, grey-brown bark that peels off in strips towards the base of larger specimens. Conspicuous, deciduous leaves are produced in summer towards the tips of the branches. They have a fresh green colour with fairly prominent veining.

Flowering of *E. espinosa* takes place from May to October, when the plants are bare of leaves. Many cyathia are produced on the slender, younger shoots, one in each leaf-axil. Since the leaves are no longer there, the flowering plants are very inconspicuous indeed among the many other leafless plants and

Fig. 3.144. *Euphorbia espinosa*, shiny, slightly peeling bark at base of plant and showing leaves, *PVB 12052a*, south of Marken, South Africa, 2 Nov. 2011 (© PVB).

Fig. 3.145. *Euphorbia espinosa*, shiny peeling bark on small tree around 2 m tall, *PVB 13328a*, east of Benguela, Angola, 26 Dec. 2016 (© PVB).

Fig. 3.146. *Euphorbia espinosa*, flowering plant with green cyathial glands, *PVB 12052a*, south of Marken, South Africa, 9 Jun. 2019 (© PVB).

3.3 Sect. Espinosae

dead twigs where they grow. In addition, the cyathia are small and green (though often reddish below) and this does not make them more easily seen. The ovary is initially hidden in the cyathium but, if fertilized, it is soon pushed out and becomes rather loosely pendulous on a slender pedicel, though the capsule becomes erect again just before dehiscing. The ovary was said to be 'subtended by an obvious 3-lobed perianth' (Carter 1988; Carter and Leach 2001) but Brown (1911–12) mentioned that it was 'very rudimentary' and it was not observed in the material illustrated here (see Fig. 3.149 C, G).

Euphorbia espinosa is easily separated from *E. guerichiana*, to which it is closely related, by the differently shaped and broader, bright green leaves with a considerably longer petiole. They also flower at very different times, *E. guerichiana* during the growing season and *E. espinosa* when the plant is leafless. In both, the ovary is initially included in the cyathium and it is later pushed out of the cyathium, but in *E. espinosa* the pedicel is decurved and may reach 12 mm long, while it is not more than 5 mm long and is only slightly curved or even erect in *E. guerichiana*.

Fig. 3.147. *Euphorbia espinosa,* flowering plant with brownish green cyathial glands and reddish exserted capsules, *PVB 12052a*, south of Marken, South Africa, 9 Jun. 2019 (© PVB).

Fig. 3.148. *Euphorbia espinosa,* flowering plant with fully developed capsule ± 10 mm in diam., *PVB 12052a*, south of Marken, South Africa, 9 Jun. 2019 (© PVB).

History

Euphorbia espinosa was described from leafless twigs collected by Gustav Adolf Fischer in 1885 or 1886 at an unknown locality in Tanzania (Pax 1894). This material was sent on loan from Berlin to Kew in 1910 and on this occasion a drawing of the whole specimen was made and a single cyathium was taken off and kept at Kew. Today this is all that remains of the type, as the rest was destroyed in Berlin during World War 2. The synonym *E. nodosa* was described from three Angolan collections made by Gossweiler.

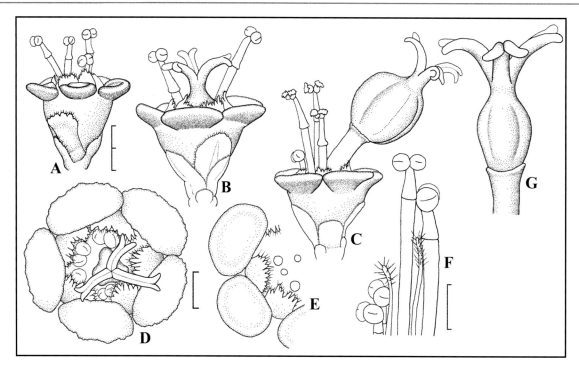

Fig. 3.149. *Euphorbia espinosa*. **A**, male cyathium (scale 2 mm, as for **B–C**). **B**, bisexual cyathium. **C**, bisexual cyathium with ovary exserted after fertilization. **D**, face view of cyathium with convex glands (scale 1 mm, as for **E**). **E**, face view of part of cyathium with concave glands. **F**, anthers and bracteoles (scale 1 mm, as for **G**). **G**, female floret. Drawn from: *PVB 12052a*, south of Marken, South Africa (© PVB).

Euphorbia espinosa was not recorded from South Africa until L.E. Codd and R.A.Dyer collected it in the northern part of the Kruger National Park in November 1948. In Namibia the first record was made in the Caprivi Strip of the extreme north-east by B. de Winter and W. Marais in February 1956.

Euphorbia guerichiana Pax, *Bot. Jahrb. Syst.* 19: 143 (1894). Type: Namibia, rocks south of Khorixas, 14 Nov. 1888, *Gürich 73* (missing). Neotype (Bruyns 2012): Namibia, Ababes, banks of Tsondap River, 30 Dec. 1915, *Pearson 9119* (BOL; J, iso.).

Euphorbia commiphoroides Dinter, *Deut. Südw. Afrik.*: 90 (1909). Neotype (Bruyns 2012): Namibia, Tsumeb distr., Auros, 10 Feb. 1925, *Dinter 5596* (BOL; SAM, iso.).

Euphorbia frutescens N.E.Br., *Fl. Cap.* 5 (2): 270 (1915). Type: Namibia, lower mountain slopes of Aus, 3000', Jan. 1909, *Pearson 4714* (K, holo.; BOL, SAM, iso.).

Bisexual (occasionally unisexual) spineless glabrous woody untidy shrub branching close to the base to small tree 1–6 m tall branching above slender trunk 70–300 mm thick with shiny yellow-brown bark peeling in papery strips, with slightly fleshy taproot bearing fibrous roots. *Branches* erect to spreading, often rebranching, cylindrical, younger branches 1.5–3 mm thick, ± without tubercles, somewhat pliable but not succulent becoming more rigid with age, dark brown, smooth; *leaves* on young and older growth on very slightly raised tubercles, alternate, 3–15 (35) × 1.3–5 (15) mm, ascending to spreading, deciduous, narrowly oblong to lanceolate, very slightly fleshy, minutely puberulous, flat to slightly channelled above, glaucous grey-green, obtuse or acute, with short petiole 1–3 mm long, estipulate. *Synflorescences* on short minutely puberulous axillary shoots 2–10 mm long developing in axils of fallen leaves with 1 (2) terminal usually bisexual sessile cyathia each subtended by 2 or more obovate bracts shorter and narrower than leaves; *cyathia* conical-cupular, glabrous to sparsely puberulous, 4–5 mm broad (1.5–2 mm long below insertion of glands), with 5 obovate lobes with deeply dissected margins, pale green; *glands* 5, transversely elliptic, 1.5–2 mm broad, spreading, contiguous to slightly separated, yellow-green later becoming bright yellow or reddish, outer margins flat and entire, inner margins often somewhat raised; *stamens* with glabrous pedicels, bracteoles filiform and pubescent; *ovary* 3-lobed, glabrous, raised on pedicel ± 2 mm long; styles ± 2 mm long, divided for two thirds of length, widely spreading. *Capsule* 6–10 mm diam., obtusely 3-lobed, glabrous, grey-green with red blush on sunny side, exserted 3–5 mm from cyathium on erect to slightly curved pedicel.

Distribution & Habitat

Euphorbia guerichiana is widely distributed, as Fig. 3.150 shows, in two disjunct areas. One of these lies mainly along the edges of the escarpment on the eastern margin of the Namib Desert. Here it occurs along the valley of the Orange River in South Africa from Hells Kloof and the Rosyntjie Mountain to near Pofadder, then northwards through Namibia to the Cunene River and further northwards into Angola. In Namibia its distribution is more or less continuous into the mountains between Otavi and Grootfontein, well east of the Namib. Disjunct from and some 500 km east of these records, there are also collections in an area from north-eastern Botswana and the northern part of the former Transvaal in South Africa from near Thabazimbi and the

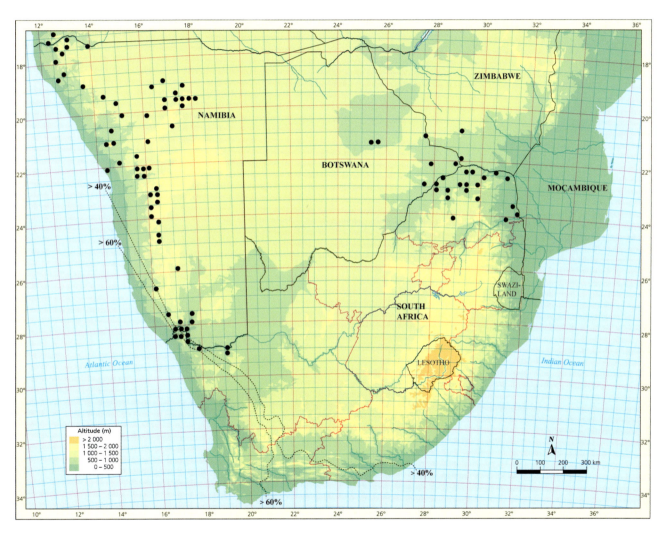

Fig. 3.150. Distribution of *Euphorbia guerichiana* in southern Africa (© PVB).

Blouberg into the Kruger National Park. It is also known in the adjacent part of southern Zimbabwe.

Typically, *Euphorbia guerichiana* is a plant of dry, rocky habitats where it grows in open woodland among other, usually deciduous trees (often also with their bark peeling in paper-like strips) with its roots wedged tightly among the rocks. However, in north-eastern Botswana it occurs in flat areas among short shrubs.

Diagnostic Features & Relationships

The plant in *E. guerichiana* is of variable appearance. In the valley of the Orange River and northwards to around Aus (Namibia) it usually forms an untidy shrub (as in Fig. 3.154), often not more than 1.5 m tall and quite densely branched from near the base, so that the characteristic peeling yellow-brown bark is only visible (if visible at all) on the bases of the thickest parts of the rather indistinct

Fig. 3.151. *Euphorbia guerichiana*, trees to 5 m tall on sparsely wooded slope (with *mopane*) overlooking the Namib Desert and Hartmann Mountains, south of Kunene River, Namibia, 27 Dec. 2014 (© PVB).

Fig. 3.152. *Euphorbia guerichiana*, leafy tree ± 4 m tall on wooded rocky slope (with brown shrubs of *Myrothamnus* at its base), Otjihipa, Kaokoveld, Namibia, 23 Dec. 1999 (© PVB).

Fig. 3.153. *Euphorbia guerichiana*, richly branched tree ± 2 m tall in bed of Swakop River, south of Karibib, Namibia, 20 Dec. 2018, with Cornelia Klak (© PVB).

stem. North of Aus the plants develop a slender trunk 1–2 m long and 10 cm or more thick and are small, relatively sparsely branched trees. The leaves are also smaller, somewhat thicker and more glaucous in the south, but they are generally ephemeral, dropping off as soon as conditions become dry.

Flowering takes place from August (in the parts receiving winter-rainfall) until January and on to March in areas with summer-rainfall. Cyathia develop on short and leafy branchlets that arise from the axils of leaf-scars, mainly on the younger and more slender branches. Each branchlet bears a single cyathium at its tip (though occasionally a sec-

Fig. 3.154. *Euphorbia guerichiana*, shrub ± 2 m tall on arid gneissic slopes, *PVB 10027*, west of Gamkab River, Namibia, 12 Jul. 2005 (© PVB).

Fig. 3.155. *Euphorbia guerichiana*, shiny peeling bark on trunk of tree, south of Kunene River, Namibia, 27 Dec. 2014 (© PVB).

ond may develop too) subtended by a few bracts that are smaller than but similar to the leaves. The yellowish glands make the cyathia clearly visible against the somewhat glaucous leaves, though they are widely separated and so they do not alter the colour of the plant.

Although very similar in habit and appearance to *E. currorii* (and sometimes growing with or near it on the edges of the Namib Desert of northern Namibia and southern Angola), this similarity is a parallel development. Analysis of DNA-data showed clearly that *E. guerichiana* is closely related to *E. espinosa* and belongs to subg. *Chamaesyce*, while *E. currorii* belongs to subg. *Athymalus* (Steinmann and Porter 2002; Bruyns et al. 2006). *Euphorbia currorii* differs by the clusters of cyathia at the tips of the branches and the triangular fruit (also 3-angled in *E. guerichiana* but with plump, swollen locules). The seeds of *E. currorii* also lack a caruncle, while a cap-like caruncle is present in *E. guerichiana* (see Fig. 1.57 F). The differences between *E. espinosa* and *E. guerichiana* are discussed under *E. espinosa*.

Fig. 3.156. *Euphorbia guerichiana,* male cyathia, *PVB 12849,* east of Orupembe, Namibia, 25 Dec. 2014 (© PVB).

Fig. 3.157. *Euphorbia guerichiana,* cyathia and capsule, *PVB 10027,* west of Gamkab River, Namibia, 12 Jul. 2005 (© PVB).

Fig. 3.158. *Euphorbia guerichiana,* capsules, *PVB 12849,* east of Orupembe, Namibia, 25 Dec. 2014 (© PVB).

3.4 Sect. Frondosae

History

Euphorbia guerichiana was discovered by the geologist Georg Gürich near Khorixas in Damaraland, Namibia in November 1888. It was also collected by Dinter in January 1899 south of Windhoek in Namibia. White et al. (1941) maintained *E. frutescens* as distinct from *E. guerichiana*, but they expressed considerable doubt that the two were different and Meyer (1967) reduced *E. frutescens* to synonymy. Material from Botswana and further east in South Africa and Zimbabwe has been included in *E. guerichiana* by most authors (e.g. White et al. 1941; Carter & Leach 2001).

According to records at PRE, this seems to have followed from the identification as *E. guerichiana* of material that was first collected around the western base of the Soutpansberg near Soutpan in 1932. Unpublished results from analysis of molecular data suggest that this material (i.e. from Botswana, northern South Africa and Zimbabwe) is not synonymous with the typical *E. guerichiana* from Angola, Namibia and western South Africa. Plants from this eastern part appear to lack the conspicuous yellowish peeling bark of those from the west, but other morphological differences are not known at present. This requires further investigation.

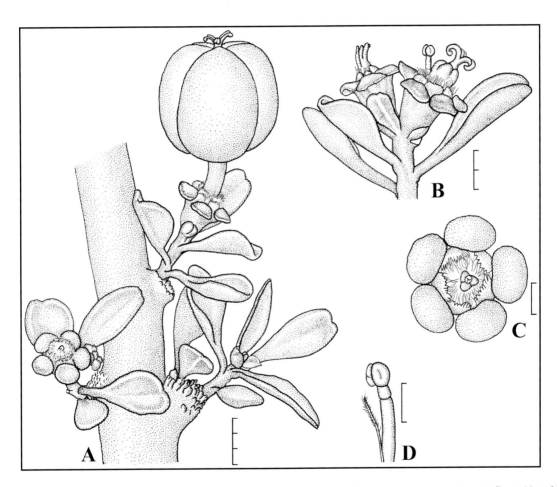

Fig. 3.159. *Euphorbia guerichiana*. **A**, branch with cyathium and capsule (scale 3 mm). **B**, pair of cyathia (scale 2 mm). **C**, cyathium from above (scale 1 mm). **D**, anther and bracteole (scale 1 mm). Drawn from: *PVB 3297*, foot of Oemsberg, north of Eksteenfontein, South Africa (© PVB).

3.4 Sect. Frondosae

Euphorbia sect. **Frondosae** Bruyns, *Taxon* 55: 416 (2006). Type: *Euphorbia goetzei* Pax (= *Euphorbia transvaalensis* Schltr.).

Bisexual non-spiny perennial semisucculent to succulent shrubs sometimes somewhat woody near base, sometimes with swollen roots. *Branches* terete and semisucculent to succulent, green to brownish green, with scattered to densely distributed tubercles bearing leaves. *Leaves* deciduous, alternate to opposite, prominent, elliptical to obovate, petiolate, sometimes with small glandular stipules. *Synflorescence* a 3- to 5-branched terminal umbel, surrounded by bracts similar to leaves in size and shape or ± circular; glands 4 (5), bilobed or emarginate to entire, yellow-green, without appendages. *Capsule* ≥ 8 mm diam., exserted on decurved and later erect pedicel. *Seeds* wrinkled to prominently tuberculate, without caruncle (in southern African species).

Of the six species in this section (Yang et al. 2012, but including *E. goetzei* in *E. transvaalensis*) only two are found in southern Africa. Of these, *E. leistneri* is endemic to Namibia while *E. transvaalensis* is widespread in locally dry areas of the tropics.

Key to the species of sect. *Frondosae* in southern Africa

1. Branches 15–25 mm thick, tubercles on branches 5–15 mm long..E. leistneri
1. Branches 3–6 mm thick, tubercles on branches 1–2 mm long...E. transvaalensis

Euphorbia leistneri R.Archer, *S. African J. Bot.* 64: 258 (1998). Type: Namibia, east of Epupa Falls, Jul. 1976, *Leistner et al. 264* (PRE, holo.; B, K, WIND, iso.).

Bisexual spineless glabrous succulent shrub 0.3–1 m tall, unbranched to branching in upper half from erect, grey-green to reddish tuberculate stem 20–70 mm thick bearing fibrous roots from base. *Branches* ascending and sometimes rebranching, 30–250 × 15–25 mm, without angles, slightly constricted into segments, smooth, pale green with longitudinal red stripes; *tubercles* ± spiralling along stem and branches but not arranged into angles, conical, spreading, 5–15 mm long; *leaves* on tips of new tubercles towards tips of branches, 50–90 × 10–30 mm, spreading, deciduous, obovate, glaucous, with entire smooth margins, midrib prominent beneath, on petiole 5–15 mm long (petiole persisting as tubercle), with minute brown stipules up to 0.2 mm long on either side at base. *Synflorescences* 1 to several per branch near apex in axils of tubercles, each on peduncle 50–120 × 2–4 mm with 2–4-rays at apex, rays

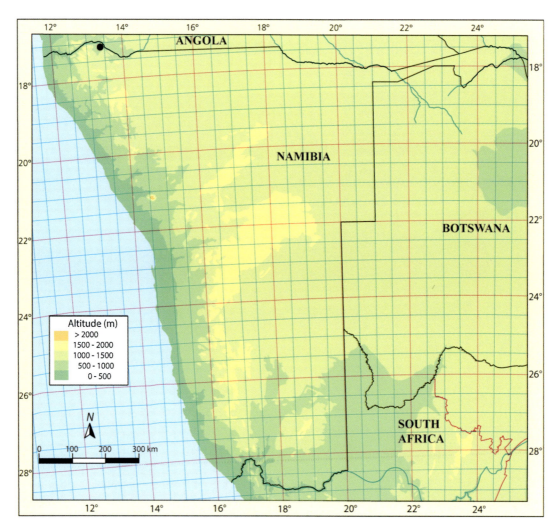

Fig. 3.160. Distribution of *Euphorbia leistneri* (© PVB).

usually simple or 2-forked and 10–60 × (2–4 mm), peduncle and each ray and fork terminated by a single bisexual cyathium subtended by 2–4 almost circular sessile to shortly petiolate bracts 8–20 × 8–12 mm (peduncle and rays with few smaller bracts scattered along their length); *cyathia* conical-cupular, glabrous, 4–5 mm broad (2 mm long below insertion of glands), with 5 obovate lobes with toothed margins, yellow-green; *glands* 4–5, transversely rectangular, well separated, ± 1.5 × 1.7–2.5 mm, pale green, spreading, inner margins not raised, outer margins shallowly to deeply 2-toothed, upper surface rugulose; *stamens* with glabrous pedicels, anthers yellow, bracteoles partitioning groups of males, with ciliate tips, glabrous; *ovary* globose, glabrous, green, exserted on erect and later decurved pedicel 3–5 mm long; styles 2–2.5 mm long, slightly spreading, branched to near middle. *Capsule* 8–10 mm diam., deeply 3-angled, glabrous, green, exserted on decurved and later erect pedicel 3–6 mm long.

Distribution & Habitat

Euphorbia leistneri is known in a small area east of Epupa Falls near the Cunene River in the northernmost part of Namibia. Here it occurs in flattish ground in the shelter of *mopane* trees on calcareous, stony soil, with *Cissus cactiformis*, *Cyphostemma uter*, *Euphorbia monteiroi*, *E. subsalsa* subsp. *fluvialis* and *Ceropegia floriparva* (*Hoodia parviflora*).

Diagnostic Features & Relationships

Euphorbia leistneri is an unusual-looking plant, with an irregularly knobbly stem and branches, each of which bears a cluster of prominent leaves around its apex during the summer months. The plants are initially single-stemmed but they branch with time, eventually forming a roughly spherical shrub. Most of the branches are around 20 mm thick, remaining pale green with occasional red streaks. The surface of the branches is covered quite densely with tubercles. Initially each of these bears a leaf and, once the leaves have fallen off,

Fig. 3.161. *Euphorbia leistneri*, plant about 0.6 m tall, under trees among pieces of calcrete, *PVB 5598*, east of Epupa Falls, Namibia, 24 Feb. 1993 (© PVB).

Fig. 3.162. *Euphorbia leistneri*, plant about 0.4 m tall cultivated in the stock-inspector, Mr Venter's garden in Opuwa, showing the very tuberculate branches, Namibia, 10 Jan. 1990 (© PVB).

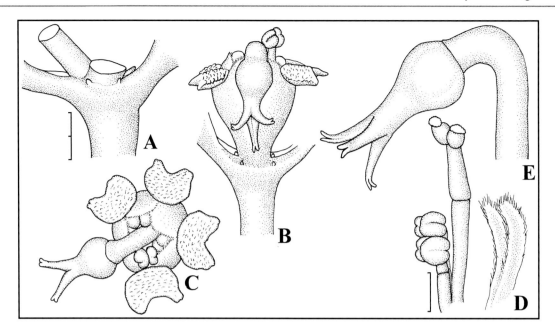

Fig. 3.163. *Euphorbia leistneri*. **A**, bases of leaves with minute stipules (scale 2 mm, as for **B**, **C**). **B**, cyathium from side. **C**, cyathium from above. **D**, anthers and bracteole (scale 1 mm, as for **E**). **E**, female floret. Drawn from: *PVB 5598*, east of Epupa Falls, Namibia (© PVB).

the projecting, blunt tubercles give rise to the characteristic knobbliness of the stem and branches. Towards the base of the plant the tubercles gradually become less prominent so that the stem is nearly smooth there.

Flowering takes place during January and February, with several slender peduncles arising in the axils of tubercles among the uppermost leaves. At the tip of the peduncle a rosette of between two and four leaf-like bracts subtends the comparatively small terminal cyathium. Rays may develop from the axils of the bracts just beneath the cyathium and each of these bears another terminal cyathium. The cyathium has four or five glands, each with two broad teeth separated by a deep indentation on the outer edge and a considerably exserted female floret.

When published, *E. leistneri* (Archer 1998) was considered to be most closely related to *E. transvaalensis* on account of their similar floral structures and leaves and this was corroborated in Bruyns et al. (2006, 2011) with data from DNA. It differs from *E. transvaalensis* primarily in the lack of swollen, fleshy roots, the much stouter stem and stouter branches, which are much more densely and prominently tuberculate and in the manner in which the leaves are clustered near the tips of the stem and branches. Florally the two species are very similar.

History

Euphorbia leistneri was first recorded by O.A. Leistner, E.G.H. Oliver, P.J. Vorster and P. Steenkamp in 1976. It was also known to Petrus Ignatius Venter (18 Aug. 1925 –), the stock inspector for the Kaokoveld, based in Opuwo from February 1982 to March 1990 and I observed it growing on his rockery (Fig. 3.162) in Opuwo in January 1990, alongside what was later described as *Orbea maculata* subsp. *kaokoensis* (= *Ceropegia rangeana* subsp. *kaokoensis*).

Euphorbia transvaalensis Schltr., *J. Bot.* 34: 394 (1896). *Tirucalia transvaalensis* (Schltr.) P.V.Heath, *Calyx* 5: 93 (1996). Lectotype (Bruyns 2012): South Africa, Transvaal, near Edwin Bray Battery, shady kloofs in Kaap River Valley, Barberton, 2000', fl. Nov. 1890, *Galpin 1198* (GRA; K, NH, SAM, Z, iso.).

Euphorbia galpinii Pax, *Bull. Herb. Boiss.* 6: 742 (1898). Lectotype (Bruyns 2012): South Africa, Transvaal, near Edwin Bray Battery, shady kloofs in Kaap River Valley, Barberton, 2000', fl. Nov. 1890, *Galpin 1198* (SAM; GRA, K, NH, Z, iso.).

Euphorbia ciliolata Pax, *Bull. Herb. Boiss.* 6: 743 (1898). Lectotype (Bruyns 2012): Angola, Sierra Chella and Gambos, 900–1100 m, *Antunes & Dekindt 781* (BR; COI, LISC, Z, iso.).

Euphorbia goetzei Pax, *Bot. Jahrb. Syst.* 28: 420 (1900). Type: Tanzania, Iringa distr., Ruaha River, *Goetze 450* (B, holo., destroyed; K, iso.).

Bisexual spineless glabrous (occasionally thinly pubescent beneath on leaves and petioles) slightly succulent shrub 0.3–1.5 m tall, branching from base and above from erect,

3.4 Sect. Frondosae

pale green later pale brown and slightly woody sparsely and only slightly tuberculate stem 3–10 mm thick, with many slightly swollen roots spreading from base bearing fibrous roots. *Branches* ascending and usually forked, 30–400 × 3–6 mm, cylindrical and without angles, slightly swollen at nodes, smooth, pale green becoming pale brown with age; *tubercles* very sparsely arranged along stem and branches, very indistinct, slightly spreading, 1–2 mm long; *leaves* alternate to occasionally opposite and usually forming whorl-like cluster at tips of branches, 30–110 × 15–50 mm, spreading, deciduous, oblong-lanceolate to elliptic, slightly glaucous green (paler below than above), with entire smooth margins (sometimes with few hairs near base and on petiole), midrib prominent beneath, on petiole 5–30 mm long, with minute grey-brown stipules up to 1 mm long on either side at base. *Synflorescences* in terminal pedunculate cymes on some branches, each on peduncle 20–200 mm long and 1–2 mm thick with 2–4-rays at apex, rays usually simple or 2-forked and 10–60 × 2–4 mm, peduncle and each ray and fork terminated by a single ± sessile bisexual cyathium subtended by 2–4 almost circular to lanceolate, sessile to shortly petiolate bracts 8–20 × 6–12 mm; *cyathia* conical-cupular, glabrous, 4–5.5 mm broad (2 mm long below insertion of glands), with 5 obovate lobes with pubescent margins, pale green; *glands* 4–5, irregularly transversely rectangular, well separated, ± 1.5 × 1.7–3 mm, dark green to yellowish green, spreading, inner margins not raised, outer margins shallowly to deeply 2- to 3-toothed, upper surface rugulose-pitted; *stamens* with glabrous pedicels, anthers yellow, bracteoles short and slender, pilose; *ovary* 3-angled, glabrous, green, erect then gradually exserted on somewhat decurved pedicel 3–5 mm long; styles 1.5–2.5 mm long, slightly spreading, branched to around middle. *Capsule* 8–9 mm diam., very obtusely 3-angled, glabrous, green, exserted on decurved and later erect pedicel 4–10 mm long.

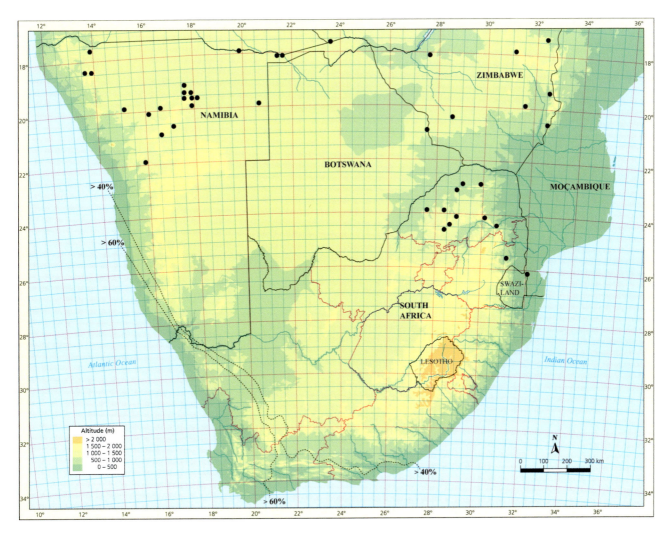

Fig. 3.164. Distribution of *Euphorbia transvaalensis* in southern Africa (© PVB).

Distribution & Habitat

Euphorbia transvaalensis is widespread across southern Africa from northern Namibia to the northern parts of the former Transvaal in South Africa, with only a few records in Botswana. It has also been recorded in Swaziland. In tropical Africa it is known in Angola, Moçambique, Zambia, Zimbabwe and northwards to Kenya, southern Ethiopia and southern Somalia (where it is usually known as *E. goetzei*).

Generally a plant of open forest on the lower slopes of hills, *E. transvaalensis* often grows in shallow soils in leaf-litter on rocks or in deeper ground in crevices between these rocks. It has also been observed under trees on raised ground on and around termite-mounds.

Fig. 3.165. *Euphorbia transvaalensis*, diffuse shrub ± 30 cm diam., among rocks under trees, *PVB 12089*, south of Ellisras, South Africa, 28 Dec. 2011 (© PVB).

Diagnostic Features & Relationships

Euphorbia transvaalensis usually forms shrubs around 0.6 m in diameter and slightly less in height. In Namibia they are more densely branched into round-topped shrubs, while they are rather more diffusely branched elsewhere. The stem and branches are slightly succulent and soft and pliable, only becoming harder and slightly woody towards their bases. Some of the roots may be slightly fleshy (as in Fig. 3.167).

Fig. 3.166. *Euphorbia transvaalensis*, round-topped shrub ± 40 cm tall among dolomitic rocks and short grasses under trees, *PVB 10346*, north of Gauss, Otavi Mountains, Namibia, 2 Jan. 2006 (© PVB).

3.4 Sect. Frondosae

Fig. 3.167. *Euphorbia transvaalensis*, young plants ± 15 cm tall with swollen rootstock, *PVB 12089*, south of Ellisras, South Africa, 28 Dec. 2011 (© PVB).

On young plants, leaves are scattered along the branches, while on older plants they are mainly produced near the tips of the branches in small, whorl-like clusters of between three and ten. They persist for most of the growing season but drop off in autumn, leaving small, scar-tipped tubercles along the branches. The leaves are conspicuous, mostly elliptic and not at all succulent.

Flowering takes place from November onwards to February, once growth has resumed after the leafless, winter period. Cyathia are borne on short branches or rays and each ray is terminated by a single cyathium which is closely subtended by a whorl of usually three, leaf-like bracts. The cyathium is inconspicuous, since it and the cyathial glands have more or less the same colour as the surrounding bracts (which may be slightly more yellowish than the leaves) It gives off a strong, sweet scent that makes up for its inconspicuousness.

History

Euphorbia transvaalensis was described from material collected by Ernest E. Galpin near Barberton in 1890 and this appears to be the earliest collection of this widely distributed plant. Dinter discovered it in Namibia, where he made the first record in December 1908.

White et al. (1941) included the widespread *E. goetzei* Pax under *E. transvaalensis*. This follwied Brown (1911–1912) and is probably a sensible arrangement, but one that was not followed in recent accounts of the genus for East Africa (Carter 1988), north-east Africa (Holmes 1993; Gilbert 1995) and south tropical Africa (Carter and Leach 2001). In these accounts *E. goetzei* is separated by being

Fig. 3.168. *Euphorbia transvaalensis*, *PVB 12089*, with obtuse bracts, south of Ellisras, South Africa, 28 Dec. 2011 (© PVB).

Fig. 3.169. *Euphorbia transvaalensis, PVB 12089*, with acute bracts, south of Ellisras, South Africa, 28 Dec. 2011 (© PVB).

Fig. 3.170. *Euphorbia transvaalensis,* with capsules, *PVB 10346*, north of Gauss, Otavi Mountains, Namibia, 2 Jan. 2006 (© PVB).

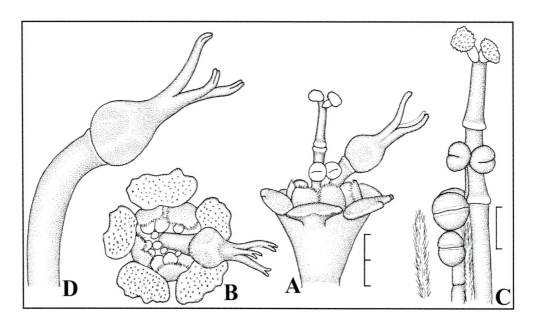

Fig. 3.171. *Euphorbia transvaalensis*. **A**, side view of cyathium (scale 2 mm, as for **B**). **B**, cyathium from above. **C**, anthers and bracteoles (scale 1 mm, as for **D**). **D**, female floret. Drawn from: *PVB 12089*, south of Ellisras, South Africa (© PVB).

slightly taller (reaching 3 m high), with the leaves sparsely pubescent underneath. Since a sparse pubescence is sometimes present on the leaves in *E. transvaalensis* too, this feature cannot be used to separate them. Plants in East Africa are often also around 0.4–0.6 m tall, so *E. goetzei* is only occasionally taller than *E. transvaalensis*. There appears to be no reliable way to separate them and they are considered once more to be conspecific.

3.5 Sect. Gueinziae

Euphorbia sect. **Gueinziae** Riina, *Taxon* 61: 780 (2012). Type (and only species): *Euphorbia gueinzii* Boiss.

Euphorbia gueinzii was found by Yang et al. (2012) to belong to *Euphorbia* subg. *Chamaesyce*, where it proved to be closely related to a group of perennial, shrubby, even sometimes somewhat succulent species endemic to Brazil in South America. Consequently it was placed in a section of its own. It is not closely allied to any other southern African species. In the past (e.g. White et al. 1941) it was believed to be related to other geophytic species of the summer-rainfall areas such as *E. pseudotuberosa* and *E. trichadenia*, but these are now known to belong to subg. *Athymalus* (Peirson et al. 2013). The monotypic sect. *Gueinziae* is endemic to South Africa and Swaziland.

Euphorbia gueinzii Boiss. in DC., *Prodr.* 15 (2): 71 (1862). Lectotype (Bruyns 2012): South Africa, at Natal Bay, *Gueinzius* (G; W, iso.).
Euphorbia gueinzii var. *albovillosa* (Pax) N.E.Br., *Fl. Cap.* 5 (2): 252 (1915). *Euphorbia albovillosa* Pax, *Bot. Jahrb. Syst.* 34: 373 (1904). Lectotype (Bruyns 2012): South Africa, Natal, Inchanga, 1180 m, 16 Sept. 1893, *Schlechter 3245* (BOL; GRA, K, PRE, iso.).

Usually unisexual spineless geophyte with often branched irregular subterranean woody tuberous stem bearing fibrous roots, giving rise to annual herbaceous branches above ground forming cluster 40–100 × 50–200 mm. *Branches* 1–10, ascending to erect above ground, deciduous, simple to branched, 50–150 × 1–4 mm, glabrous to densely pubescent, greyish green; *tubercles* absent; *spines* absent; *leaves* alternating, becoming opposite just below synflorescences, 6–30 × 2–10 mm, ascending, deciduous, lanceolate to ovate-lanceolate, acute, grey-green, glabrous to pubescent especially on midrib, shortly petiolate (to 2 mm) to ± sessile, usually somewhat longitudinally folded upwards, with small round stipules. *Synflorescences* few, usually solitary in fork of branch or terminal in small groups, on grey-green glabrous to pubescent peduncle 1–10 mm long sometimes with 2–4 small leaf-like bracts beneath terminal cyathium and sometimes with 2 more cyathia developing on short peduncles from their axils, glabrous to pubescent; bracts sessile, lanceolate, acute to obtuse, glabrous or pubescent, 4–10 × 1–3 mm; *cyathia* cupular, glabrous or pubescent, 2–6 mm broad (± as broad in male as in female), with 5 lobes with deeply toothed margins, grey-green; *glands* 5, cuneate to transversely elliptic, 0.6–2.5 mm broad, usually ascending, separated to nearly contiguous, grey-green to yellow-green, outer margins entire to minutely crenulate, stamens with glabrous pedicels, bracteoles filiform and apically pubescent; *ovary* obscurely 3-angled, pubescent to glabrous, raised on finely pubescent pedicel ± 1 mm long and soon elongating to push capsule out of cyathium; styles 1–2 mm long, branched and spreading from near base, glabrous. *Capsule* 4–6 mm diam. (slightly longer than broad), obscurely 3-angled, glabrous to pubescent, grey-green, exserted on decurved and later erect finely pubescent pedicel 2–10 mm long. *Seeds* without caruncle.

Distribution & Habitat

Euphorbia gueinzii is widespread along the eastern flank of southern Africa in the wetter parts that receive rainfall in summer. It is known from Swaziland and from the eastern parts of the former Transvaal southwards into Natal, the eastern side of the Free State and to near Kokstad in the Eastern Cape. Although not recorded in Lesotho, it may occur there too (Hargreaves 1992c).

Euphorbia gueinzii mostly grows in frequently burnt grasslands. Habitats include shallow soils overlying submerged rocks on slopes or in deeper, often rocky ground in flat areas.

Diagnostic Features & Relationships

An inconspicuous species, *Euphorbia gueinzii* resembles many other small herbs of its grassland-habitats, with slender grey-green leaves in small clusters close to the ground. They emerge quickly after the first rains and are most easily seen before the grass grows strongly. Flowering takes place mainly in August and September, with the fairly small and insignificantly coloured cyathia in small numbers at the tips of some of the branches and in the forks between others. The whole plant is usually covered with fine hairs. Its hairiness is variable and, while hairs may be densely distributed along the branches and the peduncles, on the leaves they may be confined to the midrib and near the base.

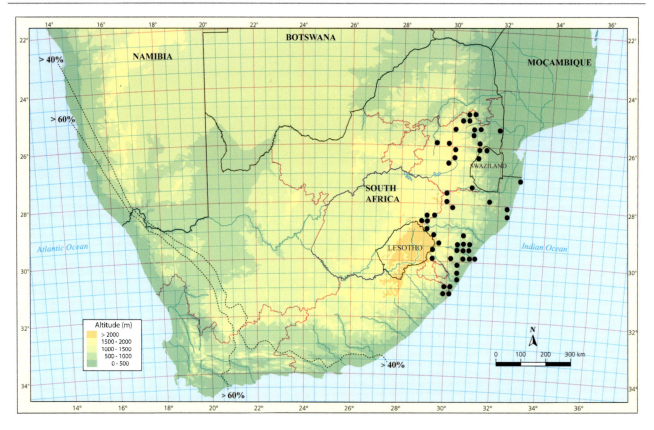

Fig. 3.172. Distribution of *Euphorbia gueinzii* (© PVB).

Fig. 3.173. *Euphorbia gueinzii*, branches ± 8 cm tall, in bare patch between grasses, *PVB 13094*, near Underberg, Drakensberg, South Africa, 27 Dec. 2015 (© PVB).

Fig. 3.174. *Euphorbia gueinzii*, female plant with bright green foliage of the first growth in the new growing season, Beacon Hill, Umtamvuna Reserve, near Port Edward, South Africa, 26 Aug. 2017 (© Graham Grieve).

Fig. 3.175. *Euphorbia gueinzii*, female plant with very glaucous foliage and capsules on short pedicels, *PVB 13094*, near Underberg, Drakensberg, South Africa, 27 Dec. 2015 (© PVB).

Fig. 3.176. *Euphorbia gueinzii*, capsules pubescent and on much longer pedicels, *PVB 13094*, near Underberg, Drakensberg, South Africa, 27 Dec. 2015 (© PVB).

Fig. 3.177. *Euphorbia gueinzii*, capsules glabrous (the left hand one just starting to become erect to dehisce), *PVB 13094*, near Underberg, Drakensberg, South Africa, 27 Dec. 2015 (© PVB).

There was some uncertainty whether plants of *E. gueinzii* are unisexual or not. Brown (1915) declared it to be 'dioecious' i.e. with male and female on separate plants, but Hargreaves (1992c) claimed to have observed cultivated plants with both male and female florets in some cyathia, though he never observed the species in habitat nor did he cultivate it while he lived in Lesotho. Plants seen in preparation for this account have always possessed cyathia of only one sex, so Brown was almost certainly correct, though this feature could occasionally be variable.

History

This somewhat cryptic species was first observed in October 1858 by Peter C. Sutherland and again in April 1861 by M.J. McKen and W.T. Gerrard. Wilhelm Gueinzius collected it sometime before 1862 as well. It was also collected by the ever-observant Rudolf Schlechter in August and September 1893 in the same area where Gueinzius found it (though Schlechter mixed some other similar-looking species with this collection, according to N.E. Brown 1915).

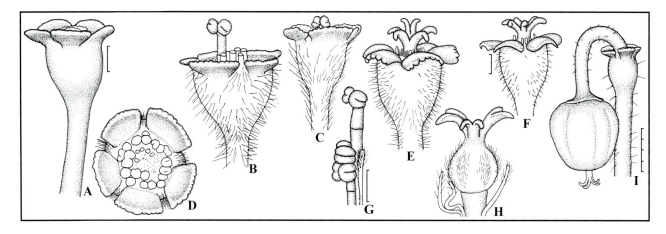

Fig. 3.178. *Euphorbia gueinzii*. **A–C**, male cyathium (scale 1 mm, as for **B–E**). **D**, male cyathium from above. **E, F**, female cyathium from side (scale for **F** 1 mm). **G**, anthers and bracteole (scale 1 mm, as for **H**). **H**, female floret. **I**, capsule (scale 4 mm). Drawn from: **A**, **I**, *PVB 13094*, near Underberg, Drakensberg, South Africa. **B**, **D–H**, *K.W. Grieve 1723*, Umtamvuna Reserve, near Port Edward, South Africa. **C**, *K.W. Grieve 2140*, Umtamvuna Reserve, near Port Edward, South Africa (© PVB).

3.6 Sect. Tenellae

Euphorbia sect. **Tenellae** Pax & K.Hoffm. in Engl., *Veg. Erde* 9, 3 (2): 147 (1921). Type: *Euphorbia glaucella* Pax (= *Euphorbia glanduligera* Pax).
Euphorbia subsect. *Capensis* Boiss. in DC., *Prodr.* 15 (2): 66 (1862). Type: *Euphorbia phylloclada* Boiss.

Bisexual annual herbs (sometimes persisting for a few years), with short soon aborting stem and branches arising in axils of few leaves on stem. *Branches* prostrate to ascending. *Leaves* to 30 mm long, alternating on stem, opposite on branches, linear-lanceolate to orbicular-ovate, petiolate on stem, usually sessile on branches, sometimes with slender stipules. *Synflorescences* many, each a solitary bisexual cyathium terminating branchlets, surrounded by leaf-like sessile bracts, glands 4, entire to emarginate, green, sometimes with small pale petaloid appendages along outer margins. *Capsule* 2–3 mm diam., exserted on decurved and later erect pedicel. *Seeds* roughened, ± 4-angled, with small cap-like caruncle.

This section contains the species *E. claytonioides*, *E. glanduligera*, *E. macra*, *E. parifolia* and *E. phylloclada* and is found only in Angola, Botswana, South Africa and Zimbabwe. In southern Africa it is especially associated with the Namib Desert. *Euphorbia claytonioides* is endemic to the Angolan part of the Namib, while *E. macra* and *E. parifolia* occur in moister parts of southern Angola east of the most arid parts of the desert.

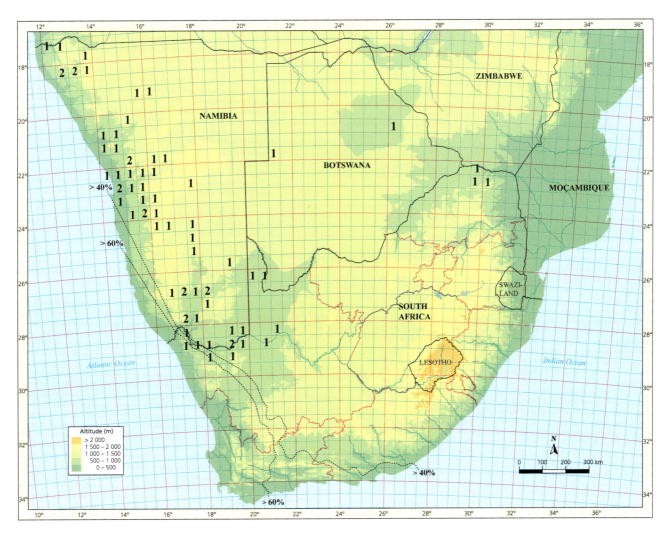

Fig. 3.179. Distribution of *Euphorbia* sect. *Tenellae* in southern Africa (showing number of species per half-degree square) (© PVB).

Key to the species of sect. *Tenellae* in southern Africa

Leaves mostly broader than long (usually larger ones at least 10 mm broad), branches prostrate and with dense clusters of leaf-like bracts at their tips broader than leaves, leaves on branches with a conspicuous white patch at their base, cyathial glands with small pale petaloid appendages on outer margins, capsule glabrous..**E. phylloclada**
Leaves longer than broad (usually largest ones < 5 mm broad), branches ascending and with small clusters of leaf-like bracts at their tips, leaves on branches without white patch, cyathial glands without petaloid appendages, capsule with hairs......
....**E. glanduligera**

Euphorbia glanduligera Pax, *Bot. Jahrb. Syst.* 19: 142 (1894). *Chamaesyce glanduligera* (Pax) Koutnik, *S. African J. Bot.* 3: 263 (1984). Type: Namibia, bei Nawas am Swakop, bei Salem, 12 Dec. 1888, *Gürich 3* (missing; sketch of type at K). Neotype (Bruyns 2012): Namibia, Naukluft Mtns, between Ababes and Homnus, *Pearson 9106* (BOL).

Euphorbia pfeilii Pax., *Bot. Jahrb. Syst.* 23: 534 (1897). Type: Namibia, Stolzenfels, Rietfontein, 1890/1891, *Pfeil 91* (missing).

Euphorbia glaucella Pax., *Bull. Herb. Boiss.* 6: 737 (1898). Lectotype (Bruyns 2012): Namibia, Okahandja, Mar. 1883, *Köpfner 68* (Z).

Euphorbia anomala Pax, *Bull. Herb. Boiss.*, Ser. 2, 8: 636 (1908), *nom. illegit.*, non Boissier (1862).

Euphorbia kwebensis N.E.Br., *Bull. Misc. Inform.* 1909: 137 (1909). Lectotype (Bruyns 2012): Botswana, Kwebe Hills, 3300', 7 Jan. 1897, *Lugard 143* (K).

Bisexual spineless glabrous annual or short-lived perennial with slender taproot bearing fibrous roots, giving rise to short pale green stem (usually 10–30 mm long) with ascending branches forming diffuse clump 50–300 mm broad. *Branches* 1–8, simple to repeatedly rebranched (sometimes in whorls) in larger plants, 25–150 × 1–2 mm, pale grey-green; *tubercles* absent; *leaves* alternating on stem, opposite on branches, 5–20 × 5–9 mm, spreading, linear-lanceolate-obovate on stem, linear-lanceolate to orbiculate on branches, obtuse to acute, grey-green, tapering into petiole 2–20 mm long, often with slender stipules. *Synflorescences* many, cyathia solitary at tips of branchlets, sessile, with 2 leaf-like bracts beneath cyathium and with 2 further sessile cyathia developing in their axils; *cyathia* narrowly urceolate, glabrous, ± 1 mm broad, with 5 minute pubescent lobes with deeply toothed margins, grey-green; *glands* 4, transversely elliptic, ± 0.5 mm broad, spreading, usually separated, dark green, outer margins entire, stamens with glabrous pedicels, bracteoles filiform and glabrous, anthers often orange; *ovary* obscurely 3-angled, sparsely hairy, raised on glabrous pedicel ± 0.5 mm long and soon elongating to push capsule out of cyathium; styles ± 0.5 mm long, branched from near base, glabrous. *Capsule* 2–3 mm diam. (longer than broad), obscurely 3-angled, with short thick hairs, dark red-green, exserted on decurved and later erect pedicel ± 2 mm long.

Distribution & Habitat

Euphorbia glanduligera is best-known from Namibia, where it occurs along the whole length of the Namib Desert, extending into southern Angola and also eastwards along the Orange River to Pofadder and beyond. Much more scattered records are known from Botswana and the northern parts of the former Transvaal.

Usually *E. glanduligera* is found in stony ground, often with little other surrounding vegetation. Plants are often rather scattered, though they may become common if good rains are received.

Diagnostic Features & Relationships

Euphorbia glanduligera is often a very small herb, close to but not pressed to the ground, with slightly ascending branches. Nevertheless, the whole plant may be as large as 0.3 m in diameter but this does not make them conspicuous as it is very diffuse, with slender branches and small, fairly slender leaves. Flowering is equally inconspicuous and even the capsules are small and insignificant.

Although the two have a very different appearance, *Euphorbia glanduligera* is closely related to *E. phylloclada*, a fact that only became apparent once information from DNA-data became available. In *E. glanduligera* the leaves are not densely clustered along the branches and the internodes are much longer and more slender. The leaves are also much narrower and smaller and have an altogether different shape.

3.6 Sect. Tenellae

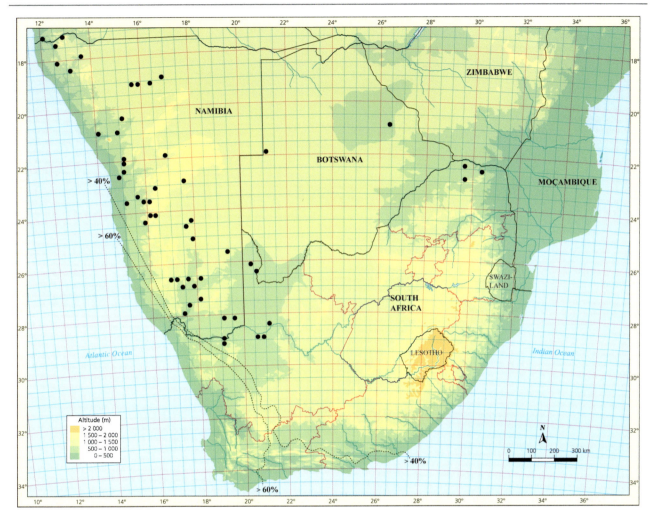

Fig. 3.180. Distribution of *Euphorbia glanduligera* in southern Africa (© PVB).

Fig. 3.181. *Euphorbia glanduligera,* stony plains at base of hill, plant ± 12 cm broad, *PVB 13564*, Namib Desert south of Solitaire, Namibia, 18 Dec. 2018 (© PVB).

Fig. 3.182. *Euphorbia glanduligera,* plant ± 10 cm broad, among stones on slope near bank of Orange River, *PVB 13525*, near Pella, South Africa, 29 May 2018 (© PVB).

Fig. 3.183. *Euphorbia glanduligera*, cyathia and capsules, *PVB 13525*, banks of Orange River near Pella, South Africa, 29 May 2018 (© PVB).

History

Euphorbia glanduligera was collected several times in the last 20 years of the nineteenth century, with the first being that of Köpfner, north of Windhoek in Namibia. Most of these early collections were given different names, all of which are now regarded as belonging to the same species.

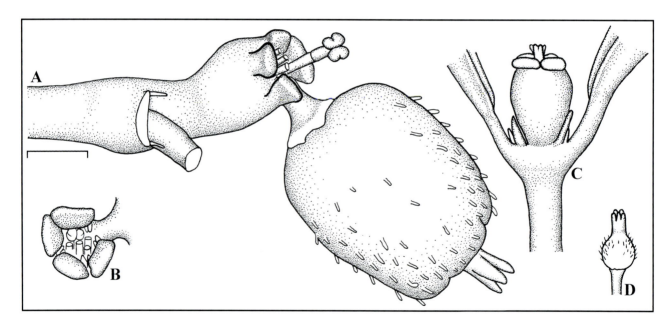

Fig. 3.184. *Euphorbia glanduligera*. **A**, side view of cyathium, also showing small stipules alongside base of cut-off bract (scale 1 mm, as for **B–D**). **B**, cyathium from above. **C**, female cyathium. **D**, female floret. Drawn from: *PVB 13525*, banks of Orange River near Pella, South Africa (© PVB).

3.6 Sect. Tenellae

Euphorbia phylloclada Boiss. in DC., *Prodr.* 15 (2): 66 (1862). Lectotype (Bruyns 2012): South Africa, between Verleptpram and mouth of Gariep, Sept., *Drège 238* (S).
Euphorbia hereroensis Pax, *Bot. Jahrb. Syst.* 10: 35 (1888). Type: Namibia, Hereroland, Hykamkab, 300 m, May 1886, *Marloth 1190* (missing).

Bisexual spineless glabrous annual or short-lived perennial with slender taproot bearing fibrous roots, giving rise to short purple to pale green stem (usually 10–30 mm long) with ± prostrate branches forming clump 50–300 mm broad. *Branches* 1–10, simple to repeatedly rebranched in larger plants, 25–150 × 1–2 mm, pale grey-green; *tubercles* absent; *leaves* alternating on stem, opposite on branches, 5–30 × 5–20 mm, spreading, spathulate-obovate on stem, broadly cordate-ovate to orbicular-ovate on branches, obtuse to acute, grey-green with large whitish patch towards base and usually with reddish margins, tapering into petiole 2–20 mm long on stem, sessile on branches and often imbricate and crowded towards tips, sometimes with slender stipules. *Synflorescences* many, cyathia solitary at tip of branchlets in very crowded flowering parts, on short peduncle <1 mm long surrounded by 2 leaf-like bracts beneath cyathium and with 2 more cyathia developing on short peduncles from their axils; bracts leaf-like, sessile and imbricate; *cyathia* narrowly urceolate, glabrous, ± 2 mm broad, with 5 minute pubescent lobes with deeply toothed margins, grey-green suffused with red; *glands* 4, transversely elliptic, 0.6–1 mm broad, spreading, usually separated, green, outer margins entire and with ascending white to pinkish slightly crenulate-margined petaloid appendage, stamens with glabrous pedicels, bracteoles filiform and glabrous, anthers

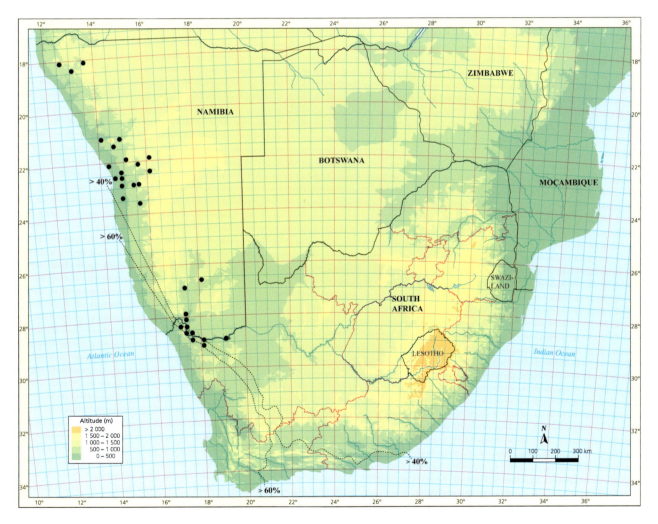

Fig. 3.185. Distribution of *Euphorbia phylloclada* (© PVB).

often orange; *ovary* obscurely 3-angled, glabrous, raised on glabrous pedicel ± 1 mm long and soon elongating to push capsule out of cyathium; styles ± 0.7 mm long, branched from near base, glabrous. *Capsule* 2–3 mm diam. (slightly longer than broad), obscurely 3-angled, glabrous, grey-green suffused with red, exserted on decurved and later erect pedicel 4–5 mm long.

Distribution & Habitat

Euphorbia phylloclada is almost entirely associated with the Namib Desert, from Orupembe in the northern part of Namibia to Goodhouse and Pella along the Orange River. There is a considerable gap between Walvis Bay and the Orange River, where this species is not known in the parts of the Namib receiving rainfall in winter. In many of the places where it occurs, less than 50 mm of rain is received annually.

Euphorbia phylloclada usually grows in loose, alluvial gravel in dry watercourses with many other annuals and often becomes extremely common after rare events of good rainfall.

Fig. 3.186. *Euphorbia phylloclada*, plant ± 15 cm diam., growing in dried-up watercourse, *PVB 10082*, Chamaites, Namibia, 18 Jul. 2005 (© PVB).

Fig. 3.187. *Euphorbia phylloclada*, barren stony area on banks of Orange River, *PVB 13524*, north of Pella, South Africa, 29 May 2018 (© PVB).

Diagnostic Features & Relationships

After good rains, plants of *E. phylloclada* form clumps of prostrate branches up to 0.3 m diameter but, if rains are less generous, most plants are only 6 to 10 cm in diameter and with smaller leaves. The whole plant consists of a short taproot which continues above-ground into a short stem 10–20 mm long with a few alternating, petiolate and usually somewhat spathulate leaves, terminated by a small knob. This knob-like structure is probably a rudimentary cyathium so that the whole plant is reduced to a series of synflorescences arising around it. Growth continues from prostrate branches in the axils of the leaves and on these branches the leaf-like bracts are sessile, usually clasping the stem slightly and with a white patch towards their bases.

Flowering takes place within a month of rains having fallen and many of the small cyathia are produced around the perimeter of the plant, with the capsules ripening soon after.

While *Euphorbia phylloclada* may occur together with the closely related *E. glanduligera,* it is most similar to *E. claytonioides* from the vicinity of Namibe in southern Angola. *Euphorbia claytonioides* is a larger plant (often 0.3 m in diameter or more) in which the branches are some-

Fig. 3.188. *Euphorbia phylloclada,* most of the plant consists of sessile bracts and cymes, two leaves which are petiolate are visible in the centre, *PVB 13524*, north of Pella, South Africa, 29 May 2018 (© PVB).

Fig. 3.189. *Euphorbia phylloclada,* here four of the spathulate and petiolate leaves are visible, while the other leaf-like structures are bracts, *PVB 13524*, north of Pella, South Africa, 29 May 2018 (© PVB).

Fig. 3.190. *Euphorbia phylloclada*, again two petiolate leaves visible on left, otherwise only sessile bracts and inflorescences can be seen, *PVB 13524*, north of Pella, South Africa, 29 May 2018 (© PVB).

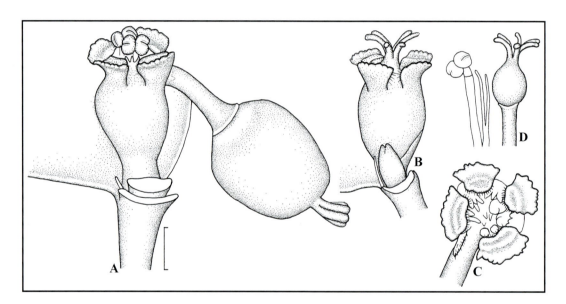

Fig. 3.191. *Euphorbia phylloclada*. **A**, side view of cyathium, with nearly mature capsule and with small linear stipule at base of one cut-off bract (scale 1 mm, as for **B–D**). **B**, young cyathium from side. **C**, cyathium from above. **D**, anther, bracteole and female floret (scale 1 mm). Drawn from: *PVB 13524*, north of Pella, South Africa (© PVB).

what ascending (and not pressed to the ground), it has a broader cyathium and hairy capsules (as does *E. glanduligera*). It also does not have the white patches on the leaves of the branches that are typical of *E. phylloclada*. Stipules were said to be absent in *E. phylloclada* (Brown 1915) and present (as fine linear structures) in *E. glanduligera*, but the same fine structures are occasionally present in *E. phylloclada* too (as seen in Fig. 3.191 A).

History

Euphorbia phylloclada is another of the species discovered by J.F. Drège and his brother Carl in the middle of September 1830 along the Orange River near the place they called Verleptpram, which is situated near the confluence of the Fish and Orange Rivers.

Euphorbia subg. *Esula*

Euphorbia subg. **Esula** Pers., *Syn. Pl.* 2: 14 (1806). *Esula* (Pers.) Haw., *Syn. Pl. Succ.*: 153 (1812). *Tithymalus* sect. *Esula* (Pers.) Prokh., *Sist. Obzor Moloch. Sr. Azil*: 166 (1933). Type: *Euphorbia esula* L. *Esula* (Pers.) Haw. (see under sect. *Esula*).

Bisexual, geophytic herbs or spineless succulent shrubs with slender cylindrical branches. *Leaves* alternate, ovate to lanceolate, fleeting in succulents, not arising on raised tubercles, estipulate. *Synflorescences* terminal and ± sessile on branches surrounded by whorl of bracts in whose axils rays terminated by a further cyathium develop forming umbel (sometimes further rays developing around these cyathia), bracts often differently shaped from leaves. *Cyathia* bisexual or unisexual (then males larger than females and with more glands); glands 4–8 (10), elliptic and flat with entire margins or crescent-shaped with two lateral horns on outer margins, without petaloid appendages; male florets with glabrous or pubescent pedicels. *Capsule* obtusely 3-lobed, smooth, glabrous, 3–8 mm diam., exserted, dehiscing explosively. *Seeds* 3 per capsule, ellipsoidal to oblong, smooth to slightly rugulose, carunculate.

Of the 21 sections recognized in subg. *Esula* by Riina et al. (2013), only two are represented in southern Africa, namely sect. *Aphyllis* and sect. *Esula*, with a total of only 11 species. All are perennials. The nine members of sect. *Esula* are slender herbs, sometimes with a geophytic habit or perennating by means of rhizomes. The three members of sect. *Aphyllis* are shrubby succulents with ephemeral leaves (Table 4.1) and green photosynthetic branches. The extraordinary variability of members of subg. *Esula* was remarked on by Croizat (1945) and considerable variation is found in southern Africa in both *E. mauritanica* and in many of the members of sect. *Esula*.

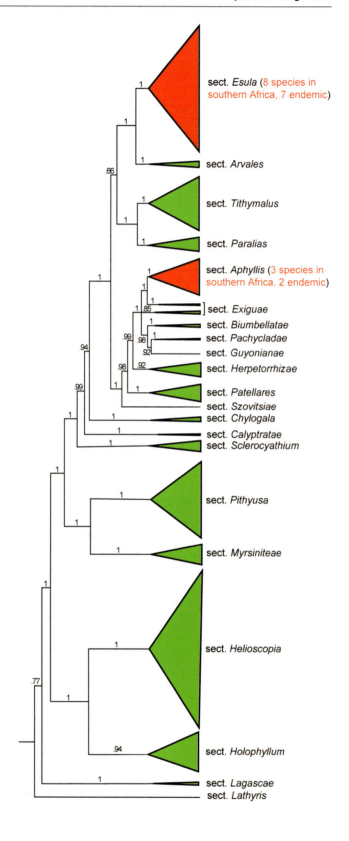

Fig. 4.1. Relationships between the members of subg. *Esula*. The sections represented in southern Africa are red (those not represented are green). The length of the vertical side of each of the triangles is proportional to the number of species in that section (adapted from Riina et al. 2013).

Table 4.1. The sections and subsections of *Euphorbia* subg. *Esula* occurring naturally in southern Africa.

Section	Sub-section	Namibia (endemic)	South Africa (endemic)	Southern Africa (endemic)	Elsewhere (total)	Annuals	Perennial, non-succulent herbs	Woody shrub to tree	Succulents	Geophytes
Aphyllis		3 (0)	2 (0)	3 (2)	21 (23)	0	0	0	3	0
	Africanae	3 (0)	2 (0)	3 (2)	10 (12)	0	0	0	3	0
Esula		1 (1)	7 (4)	8 (7)	89 (96)	0	5	0	0	3

Key to the sections of subg. *Esula* in southern Africa:

Plant a succulent with branches fleshy and persistent with rapidly caducous leaves, cyathial glands ± elliptical and without appendages on outer margins, male florets with glabrous pedicels...................................sect. **Aphyllis** (subsect. **Africanae**)
Plant not succulent, with slender branches sometimes becoming slightly woody, sometimes geophytic with ephemeral above-ground parts but leaves not caducous, cyathial glands crescent-shaped with horn-like appendages on outer edges, male florets with pubescent pedicels..sect. **Esula**

4.1 Sect. Aphyllis

Euphorbia sect. **Aphyllis** Webb & Berthel., *Hist. Nat. Iles Canaries* 2 (3): 253 (1847). Type: *Euphorbia aphylla* Brouss. ex Willd.

Of the two subsections recognized in Riina et al. (2013) only subsect. *Africanae* occurs in southern Africa. Here there are two widely distributed, endemic species. The third species is much more localized in northern Namibia and is also found in Angola.

4.1.1 Subsect. Africanae

Euphorbia subsect. **Africanae** Molero & Barres, *Taxon* 62: 338 (2013). Type: *Euphorbia mauritanica* L.

Bisexual succulent sometimes rhizomatous and often densely branched shrub. *Branches* cylindrical to fusiform, green to grey-green. *Leaves* ephemeral, to 25 mm long, alternate, lanceolate, sessile, estipulate. *Synflorescences* terminal, with central ± sessile often male cyathium with 5–9 glands surrounded by bracts slightly shorter and broader than leaves from whose axils 3–8 short rays arise each terminated by bisexual cyathium with 4 or 5 glands, glands entire, without appendages. *Capsule* ≥ 5 mm diam., exserted on initially decurved and later erect pedicel. *Seeds* often faintly rugulose, 3–4 × 2–3 mm, grey to brown, with cap-like apical caruncle.

This subsection contains 12 species that are widely distributed across the drier (but not desertic) parts of Africa. Three of them occur in southern Africa and two are endemic. Here they are widely distributed on the western side of the sub-continent, mainly along the edge of the Namib Desert and also both in and around the western parts of South Africa with a winter-rainfall regime.

Key to the species of subsect. *Africanae* in southern Africa.

1. Low shrub often broader than tall and neither hemispherical nor spherical in outline, branches erect and usually arising slightly beneath ground (without any central stem visible above ground), mainly simple above ground (sometimes rebranching near apex only around positions of old terminal cyathia), ± fusiform........................**E. stolonifera**
1. Shrub roughly hemispherical to spherical in outline with branches arising from short central stem above ground, branches rebranching frequently from their bases to near tips, branches cylindrical and not fusiform..........................**2**.
2. Roots close to surface giving rise to adventitious plantlets (later shrublets) for some distance around parent-shrub, bracts surrounding lateral cyathia with eciliate margins, central terminal cyathium usually with 5 glands, lateral cyathia with (3) 4 glands..**E. berotica**
2. Roots not giving rise to plantlets and plant forming solitary shrub, bracts surrounding lateral cyathia with ciliate margins, central terminal cyathium usually with 8 glands, lateral cyathia with (4) 5 glands....................................**E. mauritanica**

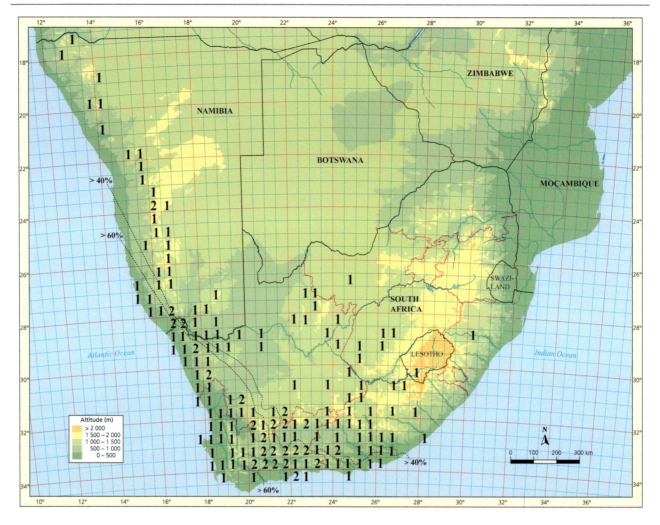

Fig. 4.2. Distribution of *Euphorbia* subsect. *Africanae* in southern Africa (showing number of species per half-degree square) (© PVB).

Euphorbia berotica N.E.Br., *Fl. Trop. Afr.* 6 (1): 600 (1912). *Tirucalia berotica* (N.E.Br.) P.V.Heath, *Calyx* 5: 88 (1996). Type: Angola, Moçamedes distr., foot of Sierra Negros, behind Boca do Rio Bero, July 1859, *Welwitsch 633* (BM, holo.; LISU, iso.).

Bisexual spineless glabrous much-branched succulent spherical to irregularly shaped shrub 0.3–1.5 × 0.3–1.5 m, branching from base and from above from small central trunk with fibrous roots, readily forming new plants from thicker shallow roots. *Branches* ascending to erect, simple or repeatedly rebranched along length sometimes in whorls, cylindrical, 0.15–1 m × 3–6 mm, without tubercles, smooth, yellowish grey-green; *leaves* towards apex of branches on young growth, alternate, 8–15 (30) × 2–3 mm, ascending, fleeting, lanceolate to linear-lanceolate, glabrous, grey-green, slightly channelled above, acute, sessile but narrowing slightly towards base, estipulate. *Synflorescences* terminal or on axillary shoots near tips of branches, glabrous, consisting of very shortly pedunculate (peduncle to 3 mm long) deciduous male cyathium (sometimes bisexual) surrounded by 2–5 simple terete rays 4–10 × 1.5–2 mm, rays subtended by 2–5 deciduous concave grey-green bracts shorter and broader than leaves and each tipped with bisexual cyathium on short peduncle, secondary bisexual cyathium surrounded by 2 (3) sessile grey-green bracts shorter and broader than leaves (2.5–8.5 × 2.0–4.5 mm) with eciliate margins; *cyathia* shallowly conical, glabrous, 5.5–7 mm broad (± 2 mm long below insertion of glands), with 5 lobes with finely and irregularly dissected margins, yellow; *glands* 5 (–8) in central cyathium, (3) 4 (5) in laterals, transversely broadly elliptic to nearly circular, 2–3 mm broad, spreading and usually well separated, yellow, concave and slightly pitted above, outer margins entire or crenulate; *stamens* with glabrous pedicels, bracteoles filiform and pubescent towards apices; *ovary* obtusely 3-lobed, glabrous, initially included on pedicel 1.5–3 mm long then

4.1 Sect. Aphyllis

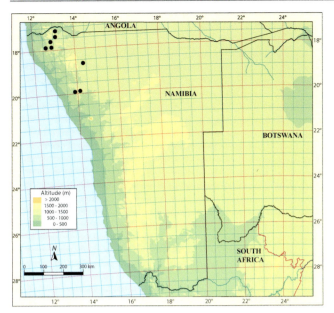

Fig. 4.3. Distribution of *Euphorbia berotica* in southern Africa (© PVB).

soon exserted and spreading horizontally on pedicel 4–5 mm long; styles 2–3 mm long, branched to at least two thirds. *Capsule* 5–6 mm diam., obtusely 3-lobed, glabrous, exserted from cyathium on gradually erect pedicel ± 7 mm long.

Distribution & Habitat

Euphorbia berotica is known in the western part of southern Angola (mainly in the gently sloping hills north-east of Namibe) and in north-western Namibia. In Namibia it is found from Grootberg Pass (west of Kamanjab) and near Orupembe to the higher parts of the Baynes Mountains close to the Cunene River. Most of the known localities are between 1000 and 1650 m, usually above the escarpment or near its edge, unlike in Angola where it occurs closer to the sea at altitudes of 200–500 m.

Euphorbia berotica grows in very varied habitats, from steep rocky slopes with a few trees to flat, gravelly areas among scattered trees and various other succulents. Populations usually consist of many plants and are relatively isolated from other similar groups.

Fig. 4.4. *Euphorbia berotica*, at base of hills among scattered *mopane*-trees, *PVB 12846*, east of Orupembe, Namibia, 25 Dec. 2014 (© PVB).

Diagnostic Features & Relationships

Euphorbia berotica forms round to irregularly-shaped, often densely and randomly branched shrubs. Some of the thicker, superficial roots are able to give rise to new plants and large plants may be surrounded by several much smaller ones that arise from its own roots. While the parent is still alive, these offshoots rarely achieve the same size as it. Near the base, the stem may become woody and covered with a grey bark, but the branches remain pliable and succulent and have a faintly yellowish, grey-green colour. Slender leaves are present in the growing season but are soon lost.

Flowering generally takes place in the height of summer (January to March) and many clusters of cyathia develop near the tips of the branches, in good years giving the plant a yellow tinge from the distance. Many of these terminate the branches, but they may also be borne on short shoots near their tips (especially in older branches which have already borne terminal cyathia). Each cluster consists of a central, sometimes bisexual cyathium with five glands surrounded, on short rays, by several bisexual cyathia each usually with four glands (in Angolan material the central cyathium has between 6 and 8 glands). The glands on the cyathia are almost circular (occasionally slightly radially shorter

Fig. 4.5. *Euphorbia berotica*, smaller plants developing from rhizomes on larger ones, *PVB 12846*, east of Orupembe, Namibia, 25 Dec. 2014 (© PVB).

Fig. 4.6. *Euphorbia berotica*, rhizomes giving rise to young plants, *PVB 12846*, east of Orupembe, Namibia, 25 Dec. 2014 (© PVB).

than broad) and widely spaced out around the cyathium with the developing capsule usually dependent in the gap provided by the missing gland (if there are four glands).

Both vegetatively and florally *E. berotica* is obviously closely related to *E. mauritanica* and the tropical *E. gossypina*. Results from molecular data placed it closest to *E. mauritanica*, *E. orthoclada* and *E. stolonifera* (Riina et al. 2013), though this was not strongly supported. It differs from *E. mauritanica* and *E. stolonifera* by the suckering habit and also by the different numbers of glands (usually 5 as opposed to 8 on the central male cyathium and 4 rather than 5 on the lateral bisexual cyathia). In their shape, the glands may look different, being often somewhat shorter radially than broad in *E. stolonifera* and in *E. mauritanica*, but these features are variable and are not reliable. *Euphorbia berotica* also has smaller and more rounded seeds than *E. mauritanica* and in *E. mauritanica* the apical caruncle on the seed is also usually much larger (Leach 1975a).

History

Discovered near the mouth of the Bero River near Namibe, Angola in July 1859 by Welwitsch, *E. berotica* was only incompletely described from a very poor sterile specimen and was considered 'imperfectly known' by N.E. Brown. Nevertheless, the unusual colour of the branches and their suckering habit make them unmistakable. It was collected again in southern Angola in September 1967 by Leach and Cannell to the north-east and north of Namibe (Leach 1975a). In Namibia it was first recorded by De Winter and Leistner in April 1957.

4.1 Sect. Aphyllis

Fig. 4.7. *Euphorbia berotica*, cymes with central bisexual cyathium, *PVB 13442*, west of Opuwa, Namibia, 23 Mar. 2017 (© PVB).

Fig. 4.8. *Euphorbia berotica*, cymes with central male cyathium, 5 km towards Sao Nicolau, Angola, 19 Dec. 2016 (© PVB).

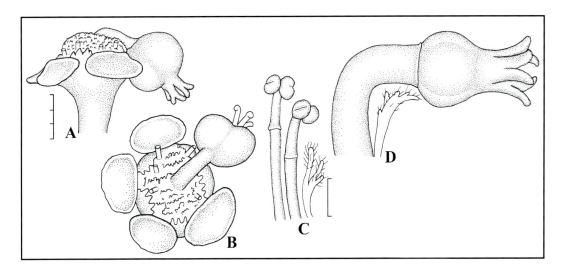

Fig. 4.9. *Euphorbia berotica*. **A**, bisexual cyathium from side (scale 3 mm, as for **B**). **B**, bisexual cyathium from above. **C**, anthers and bracteole (scale 1 mm, as for **D**). **D**, female floret with bracteole. Drawn from: *PVB 13442*, west of Opuwa, Namibia (© PVB).

Euphorbia mauritanica L., *Sp. Pl.* 1: 452 (1753). *Tithymalus mauritanicus* (L.) Haw., *Syn. Pl. Succ.*: 139 (1812). *Tirucalia mauritanica* (L.) P.V.Heath, *Calyx* 5: 91 (1996). Lectotype (Croizat 1945): Dill., *Hort. Eltham.* 2: 384, t. 289, f. 373 (1732).

Tithymalus zeyheri Klotzsch & Garcke, *Abh. Königl. Akad. Wiss. Berlin* 1859: 71 (1860). Lectotype (designated here): South Africa, Cape, *Ecklon & Zeyher 26* (MEL).

Tithymalus brachypus Klotzsch & Garcke, *Abh. Königl. Akad. Wiss. Berlin* 1859: 74 (1860). Type: South Africa, Cape, *Bergius* (missing).

Euphorbia phymatoclada Boiss., *Cent. Euphorb.*: 24 (1860). Type: South Africa, Namaqualand, rocky hills at Ebenaeser, *Drège 2943* (P 00576083, holo.). [the lectotype designated in Bruyns 2012 is *E. burmanni*. This does not fit Boissier's description, since Boissier mentioned caruncles on the seeds. Furthermore, there is a specimen at P that is likely to have been the one examined by Boissier and therefore is probably the holotype. It belongs to *E. mauritanica*].

Euphorbia melanosticta E.Mey. ex Boiss. in DC., *Prodr.* 15 (2): 95 (1862). Lectotype (designated here): South Africa, towards Goedemanskraal, 2500', *Drège 2945* (P 00576084; P 00576085, K, MO, iso.).

Euphorbia hydnorae E.Mey. ex Boiss. in DC., *Prodr.* 15 (2): 95 (1862). Type: South Africa, between Kaus and Doornpoort, *Drège 2943* (P 00576082, holo.; S, iso.). [the lectotype of Bruyns 2012 is set aside as a probable holotype has been found at P].

Euphorbia mauritanica var. *namaquensis* N.E.Br., *Fl. Cap.* 5 (2): 292 (1915). Lectotype (Bruyns 2012): South Africa, Pofadder distr., Groot Rosynbos, 9 Jan. 1909, *Pearson 3845* (K; BOL, NBG, Z, iso.).

Euphorbia sarcostemmatoides Dinter, *Feddes Repert. Spec. Nov. Regni Veg.* 17: 304 (1921). Lectotype (Bruyns 2012): Namibia, (Tsamkubis ?) Klein Aub, 7 Apr. 1911, *Dinter 2149* (SAM).

Euphorbia paxiana Dinter, *Feddes Repert. Spec. Nov. Regni Veg.* 17: 265 (1921). *Tirucalia paxiana* (Dinter) P.V.Heath, *Calyx* 5: 92 (1996). Type: Namibia, Klein Aub, am schwarzem Kam Rivier im Bastardland, *Dinter 2652* (SAM, holo.).

Euphorbia mauritanica var. *foetens* Dinter ex A.C.White, R.A.Dyer & B.Sloane, *Succ. Euphorb.* 2: 961 (1941). Type: Namibia, 8 km east of Pomona, 14 June 1929, *Dinter 6418* (PRE, holo.; BOL, BM, HBG, K, M, NBG, S, SAM, iso.).

Euphorbia mauritanica var. *minor* A.C.White, R.A.Dyer & B.Sloane, *Succ. Euphorb.* 2: 961 (1941). Type: South Africa, Cape, 30 miles north of Laingsburg, Aug. 1939, *Dyer 4105* (PRE, holo.; K, iso.).

Euphorbia mauritanica var. *lignosa* A.C.White, R.A.Dyer & B.Sloane, *Succ. Euphorb.* 2: 961 (1941). Type: Namibia, Namib near Lüderitzbucht, Nov. 1908, *Marloth 4638* (PRE, holo.; PRE, iso.).

Euphorbia mauritanica var. *corallothamnus* Dinter ex A.C.White, R.A.Dyer & B.Sloane, *Succ. Euphorb.* 2: 961 (1941). Type: Namibia, dunes near Buchuberge, 1 July 1929, *Dinter 6467* (PRE, holo.; BOL, BM, HBG, K, LD, M, NBG, S, SAM, iso.).

Bisexual spineless glabrous much-branched hemispherical to spherical succulent shrub 0.3–1.5 × 0.3–1.5 m branching from base and from above from small central trunk with fibrous roots. *Branches* ascending to erect, simple or repeatedly rebranched along length, cylindrical, 0.15–1 m × 3–8 mm, without tubercles, smooth, grey-green to bright green or yellow-green sometimes suffused with red; *leaves* towards apex of branches on young growth, alternate, 8–15 × 3–5 mm, ascending, fleeting, lanceolate, glabrous, glaucous to yellow-green, somewhat channelled above by upwardly folded margins, acute, sessile but narrowing towards base, estipulate. *Synflorescences* terminal, glabrous, consisting of very shortly pedunculate (peduncle to 3 mm long) deciduous male cyathium (rarely absent or bisexual) surrounded by (2-) 5–8 simple terete rays 10–20 mm long, rays subtended by 2–5 deciduous concave yellow-green bracts shorter and broader than leaves with glabrous or cilate margins and each tipped with a bisexual cyathium, secondary bisexual cyathium surrounded by 2–4 sessile pale green to reddish or yellow bracts shorter and broader than leaves (4–8 × 3–6 mm) with ciliate margins; *cyathia* shallowly conical, glabrous, 6–14 mm broad (± 2 mm long below insertion of glands), with 5–8 lobes with finely to deeply dissected margins, green to yellow; *glands* 5–8 in central cyathium, (4) 5 in laterals, elliptic (often radially elliptic on bisexual cyathia) to circular or cuneate, 2–3 mm broad, spreading and usually slightly separated, yellow or yellow-green turning orange, concave on central cyathium to flat on surrounding cyathia and ± smooth above, outer margins entire or notched; *stamens* with glabrous pedicels, bracteoles filiform and pubescent towards apices; *ovary* globose, glabrous, initially included on pedicel 1.5–3 mm long then becoming exserted and pendulous on decurved pedicel 4–6 mm long; styles 2–3 mm long, branched in upper half to two thirds. *Capsule* 5–8 mm diam., obtusely 3-lobed, glabrous, exserted from cyathium on gradually erect pedicel (3) 4–7 mm long.

Distribution & Habitat

Euphorbia mauritanica is a very widespread species. In Namibia it has been recorded from near Twyfelfontein in Damaraland and Klein Spitzkop in the Namib Desert over much of the south to the Orange River, though here it occurs mainly in the west and around the Karas Mountains. In South Africa it is ubiquitous. In the winter-rainfall region, it occurs from some of the driest parts of Namaqualand to locally dry spots on the Cape Peninsula (where it still occurs around Camps Bay in Cape Town). In the summer-rainfall region it is more sporadic, but it is still widespread, from Gordonia and Griqualand West across the Karoo to Natal and the Eastern Cape.

Plants occur on flats in deep, hard loam to looser sandier soils and are equally often encountered on the lower slopes of rocky hills. On the western side of South Africa (especially in the winter-rainfall region) they frequently dominate the vegetation locally to form stands of hundreds of individuals which are visible from a long way off. These populations are often dotted randomly and very diffusely across the landscape.

4.1 Sect. *Aphyllis*

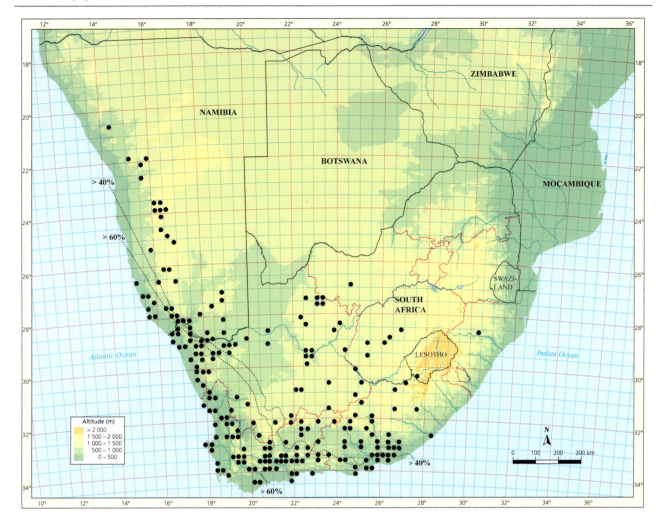

Fig. 4.10. Distribution of *Euphorbia mauritanica* (© PVB).

Fig. 4.11. *Euphorbia mauritanica*, typical yellow-green colour ('*geelbos*'), Witteberg road, west of Laingsburg, South Africa, 9 Aug. 2013 (© PVB).

Fig. 4.12. *Euphorbia mauritanica*, less floriferous, thicker-branched form typical of Richtersveld and SW Namibia, south of Eksteenfontein, South Africa, 1 Sep. 2013 (© PVB).

Fig. 4.13. *Euphorbia mauritanica*, rudimentary leaves, Katdoringvlei, west of Kotzesrus, Namaqualand, South Africa, 8 Sep. 1994 (© PVB).

Fig. 4.14. *Euphorbia mauritanica*, rudimentary leaves and darker young growth of northern form, Numeis, north of Rosh Pinah, Namibia, 5 Jul. 1997 (© PVB).

Diagnostic Features & Relationships

In the winter-rainfall parts of the former Cape Province, colonies of robust bushes of *E. mauritanica* are a common and distinctive sight. Here individuals are usually bright yellow-green, almost spherical shrubs which often exceed 1 m in height and are densely branched. Their colour is rather different from that of the surrounding vegetation, even when that is green after good rains, but especially when it is brown and dry during the summer. Many of these populations have a few individuals too whose branches are reddish green. Outside the winter-rainfall region, *E. mauritanica* is not always so conspicuous, especially since it frequently grows with and has a similar colour and shape to such succulents as *Kleinia longiflora*, *Cynanchum pearsonianum* and *C. viminale*. It also exhibits greater vegetative variation: both large and stout-branched plants (that are usually bright green) and some very slender-branched, often much smaller and more delicate forms (that are usually slightly grey-green) occurring, sometimes even together.

Fig. 4.15. *Euphorbia mauritanica,* typical arrangement of cyathia, Klawer, Namaqualand, South Africa, 10 Jul. 2005 (© PVB).

When growing during the moist season, the branches bear quite obvious but narrow and sharp-tipped, alternating leaves towards their tips. These leaves do not have a petiole but become more slender near their base. They rapidly fall off to leave a very narrow, slightly sickle-shaped scar on the branch.

In the winter-rainfall region in years of good rains, flowering takes place between mid-June and September. Outside of this region flowering is dependent on rains and usually takes place between spring and early summer (though it may occur as late as April). Many of the branches around the top of the plant bear cyathia and the flowering sequence on each branch starts with a large, terminal, male cyathium with eight glands that is surrounded by five bracts that are a bit shorter and broader than the leaves and are often also slightly yellowish. In the axil of each of these bracts a further, smaller bisexual cyathium develops (sometimes also in the axils of lower bracts too) on a peduncle up to 20 mm long so that the original cyathium becomes surrounded by these smaller ones. The central cyathium rapidly withers and usually falls off by the time the outer ones are mature, leaving only a small scar in the centre. The outer cyathia are often not quite as brightly yellow (though their glands also turn orange later) and have smaller glands that are usually radially rather than transversely elliptic. In these cyathia initially the female floret is just contained within the cyathium on a quite long pedicel but as the males mature the female become exserted and somewhat pendulous on a much longer pedicel. As the capsules ripen their pedicels become erect once more.

Among the southern African 'pencil-stemmed' species, *E. mauritanica* and *E. stolonifera* are distinguished by their alternating, stipule-free leaves and the structure of the synflorescence. As in many species the synflorescence is termi-

Fig. 4.16. *Euphorbia mauritanica,* capsules held erect just before dehiscing, Vrolikheid, Robertson, South Africa, 29 Oct. 2017 (© PVB).

nal on the branches but here it consists of an almost sessile cyathium that terminates the shoot, surrounded by several others on rays of varying length. *Euphorbia mauritanica* and *E. stolonifera* are very similar in that the central cyathium is functionally male (having only a rudimentary female floret) and has eight glands around its edge, while the other cyathia are bisexual and have five glands. These two species may be found growing close to each other over a large portion of the Karoo and flower at the same time. They may be distinguished by the rhizomatous habit of *E. stolonifera*. Here the branches arise from a slightly submerged base of the plant and from rhizomes and they usually remain more or less unbranched, whereas in *E. mauritanica* the plant branches and rebranches extensively above a small basal trunk. The branches of *E. stolonifera* are also characteristically somewhat narrower towards their bases and tips and are blue-green, while in *E. mauritanica* they are of even thickness and mostly yellow-green.

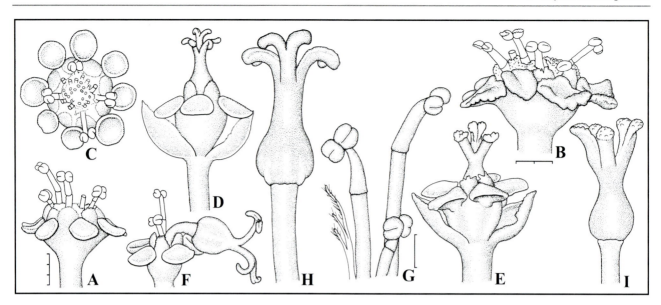

Fig. 4.17. *Euphorbia mauritanica*. **A**, **B**, central male cyathium from side (scale of **A** 3 mm, as for **C**, **D**, **F**; scale of **B** 2 mm, as for **E**). **C**, central male cyathium from above. **D**, **E**, **F**, side view of bisexual cyathium at different stages. **G**, anthers and bracteole (scale 1 mm, as for **H**, **I**). **H**, **I**, female floret. Drawn from: **A**, **C**, **D**, **F–H**, *PVB*, Yserfontein, north of Cape Town, South Africa. **B**, **E**, **I**, *PVB*, Buffelsklip, east of Oudtshoorn, South Africa (© PVB).

Fig. 4.18. *Euphorbia mauritanica*, *PVB 10569*, Rietbron, South Africa, 1 Oct. 2006 (var. '*minor*') (© PVB).

Furthermore, in *E. mauritanica* the cyathia-bearing rays around the central terminal cyathium fall off after flowering and fruiting, whereas in *E. stolonifera* they persist and form a small rosette of branchlets around the tips of some of the branches. Since both may flower at the same time it is remarkable that no hybrids are ever observed.

White et al. (1941) recognized *E. mauritanica* var. *minor* (mainly from the eastern Klein Karoo, southern and central Great Karoo). Sometimes this is distinctive with slender branches only 2–4 mm thick gathered into delicate and neat shrubs usually under 30 cm tall, with a grey-green colour. As with *E. stolonifera*, this small form also often grows together with larger plants of 'var. *mauritanica*' and they may flower at the same time, but without hybridizing or intergrading. Florally nothing has been found to distinguish it, although occasionally the central cyathium has been observed to be bisexual with five glands and produces a capsule. In other areas these two cannot readily be distinguished and herbarium material is difficult to assign to either variety with certainty, so the name has been abandoned here.

Fig. 4.19. *Euphorbia mauritanica*, cymes with central male cyathium with around eight glands, *PVB 10569*, Rietbron, South Africa, 1 Oct. 2006 (© PVB).

Fig. 4.20. *Euphorbia mauritanica*, cymes with central bisexual cyathium with pendulous capsules and only five or six glands, *PVB 10569*, Rietbron, South Africa, 1 Oct. 2006 (© PVB).

History

Dillenius (1732) referred to the present species as '*Tithymalus aphyllus mauritaniae*' and clearly believed that it came from NW Africa. This led to Linnaeus' adoption of the rather inappropriate and confusing name *Euphorbia mauritanica*. In his Encyclopédie, Lamarck (1786: 418) repeated that it was the 'Euphorbe de Mauritanie' but, in the later Supplement to the same work, Poiret (1812: 610) corrected this to 'Cape of Good Hope' (i.e. around Cape Town, South Africa).

However, *E. mauritanica* was in cultivation well before Dillenius published his illustration. A figure of it was made by Jan Moninckx between 1686 and 1690 from material introduced before 1689 and grown in the Hortus Medicus in Amsterdam. This figure is t. 32 in the Moninckx Atlas Volume 1, compiled in 1690 (Wijnands 1983). *Euphorbia mauritanica* was collected on 21 September 1685 near the present-day Vredendal by Simon van der Stel's Expedition to the Copper Mountains. A painting of this collection, ostensibly by Heinrich Claudius, is Folio 25 in the *Codex Witsenii* (Wilson et al. 2002; also Folio 337 in the volume of the *Codex witsenii* known as *Icones plantarum et animalium*, Macnae and Davidson 1969). This figure appeared again in Petiver (1709–11: t. 90, fig. 3) and Burman (1738: t. 6, fig. 2). Since it is common around Cape Town, it is highly probable that it had been observed even before this.

When Croizat (1945) selected Dillenius' figure 373 as the lectotype of *E. mauritanica*, he asserted that it 'is next to impossible to deny' that it belonged to what he called *E. obtusifolia* (now *E. lamarckii*), a species from the Canary Islands and NW Africa. However, it is not obvious what led him to this assertion. Dillenius' figure is, as Croizat said, very clear. In it, one sees the manner in which many branches are slightly swollen above their bases (while they swell into their join in *E. lamarckii*), the leaves taper to sharp tips and are around 3 times as long as the branches are thick (much longer and obtuse in *E. lamarckii*). The relatively large size of the cyathia relative to the leaves (with breadth about a third of the length of a fully-grown leaf), the short simple rays bearing them, with narrow leaf-like bracts and the usually eight-glanded central cyathium in the cyme are all very different to those of *E. lamarckii* (Kunkel 1978).

Euphorbia stolonifera Marloth ex A.C.White, R.A.Dyer & B.Sloane, *Succ. Euphorb.* 2: 961 (1941). *Tirucalia stolonifera* (Marloth ex A.C.White, R.A.Dyer & B.Sloane) P.V.Heath, *Calyx* 5: 93 (1996). Type: South Africa, Cape, near Matjiesfontein and Dwars in die Weg, 900 m, Oct. 1920, *Marloth 9836* (PRE, holo.; PRE, iso.).

Bisexual spineless glabrous succulent shrub (often broader than tall), branching extensively but mainly diffusely at base (usually beneath surface of ground) from soon indistinguishable stem, 0.15–0.5 (1) × 0.15–0.5 m in diameter, often with horizontally spreading tough and somewhat woody rhizomes around base rising to surface some distance from main plant, with slender woody rootstock and fibrous roots. *Branches* erect, mainly simple or sometimes rebranching near apex only around position of old cyathia, fleshy and cylindrical-fusiform (usually thickest around middle and tapering to base and tip), below-ground portion tough and slightly woody changing to fleshy more or less at surface of ground, 0.1–0.5 (1)) m × 5–15 mm, without distinct tubercles, smooth, soft and fleshy above ground when young and remaining so with age, grey-green sometimes suffused with red; *leaf-rudiments* towards apex of young branches, alternating, 4–10 × 2–6 mm, ascending, fleeting, ovate to ovate-lanceolate, glaucous, glabrous, slightly concave above, with entire eciliate sometimes red margins, acute, sessile, estipulate. *Synflorescences* termi-

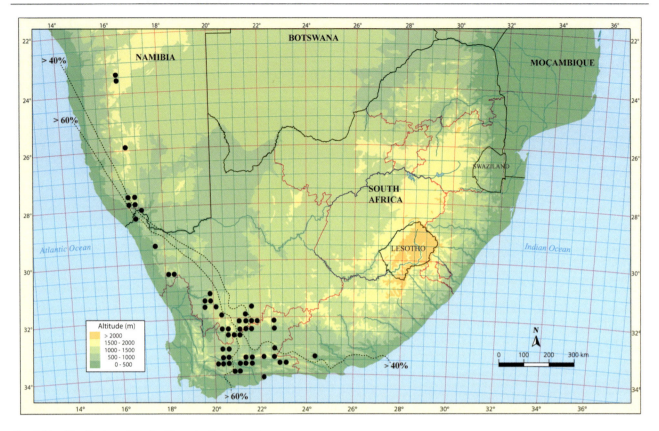

Fig. 4.21. Distribution of *Euphorbia stolonifera* (© PVB).

nal, smooth, with central male or bisexual cyathium on short often slightly angled peduncle 1–5 mm long surrounded by 3–8 simple short often slightly angled rays 5–10 mm long each subtended by leaf-like bract (sometimes already fallen off when cyathia mature) and each tipped with a bisexual to female cyathium surrounded by 1–3 rapidly caducous sometimes reddish leaf-like bracts 2–5 × 2–4 mm with cilate margins; *cyathia* bowl-shaped to very shallowly conical, glabrous, 5–8 mm broad (8–10 mm in central cyathium, 1–2.5 mm long below insertion of glands), with 5 obovate lobes with ciliate margins, green to yellow; *glands* 5–10 in central cyathium (sometimes on one plant), 4–5 in laterals, ± square to elliptic, 2–3 mm broad, spreading, yellow to olive-green, outer margins flat and entire; *stamens* with glabrous pedicels, bracteoles filiform and sparsely pubescent; *ovary* globose, glabrous, exserted to partly included on erect to decurved pedicel 1.5–6 mm long; styles 1.5–2 mm long, divided nearly to base, slightly spreading. *Capsule* 5–6 mm diam., obtusely 3-lobed, glabrous, green, sessile to exserted on decurved and later erect pedicel to 5 mm long.

Distribution & Habitat

Euphorbia stolonifera is widely distributed between southern Namibia and the southern parts of the former Cape Province, mainly around the margins of the winter-rainfall region. It is well-known in the southern part of its distribution, from Calvinia to Matjiesfontein and to Prince Albert, as well as on the Klein Karoo but it occurs over a much wider area, where it has been poorly recorded generally. Outliers occur on the southern coastal plain in the dry area around Mossel Bay. It also occurs in Namaqualand in some of the higher areas such as the Khamiesberg and on the Vandersterrberg of the Richtersveld. There are similar-looking plants in the mountains of south-western Namibia which are also included here under this species. Plants from around Nauchas, south-west of Windhoek in central Namibia are also included here, though the identification is not certain in this case.

Euphorbia stolonifera is always found on stony, often gentle slopes (most commonly on shales, but sometimes gneiss and quartzites of the Nama-series) among and often

4.1 Sect. *Aphyllis*

Fig. 4.22. *Euphorbia stolonifera*, large plant ± 1 m broad, *PVB 12675*, SE of Fraserburg, South Africa, 29 Mar. 2014 (© PVB).

partly inside low bushes, though some of the more robust forms grow entirely in the open.

Diagnostic Features & Relationships

In *E. stolonifera* the plant often has a rhizomatous habit, with rhizomes spreading around it so that some branches may emerge from the ground at a short distance from the main plant. However, not all populations of this species are so obviously rhizomatous and this is especially true of the robust plants found on the Roggeveld near Sutherland (and further east to near Fraserburg) and those from the Richtersveld and the neighbouring parts of Namibia, where the rhizomatous habit is often absent. It is wholly absent in plants from around Nauchas, Namibia. In typically rhizomatous specimens of *E. stolonifera* the branches spread from the plant with a slender, somewhat rigid and slightly woody, brown, underground portion which soon rises to the surface, after which it becomes erect and fleshy. Most of the branches are simple and remain like this, but small branchlets may develop on older branches near their tips. These branchlets appear if the branch had been nibbled off apically or they sometimes develop where flowering previously took place. In *E. stolonifera* the branches are usually distinctly fusiform, that is thicker around their middle and tapering to both ends. Since they are mainly simple, erect and more or less parallel, larger plants have an almost cylindrical shape. In the dry season a small cluster of leaf-rudiments forms a distinct and tight cone around the apical bud, protecting it from drying out. With the onset of rains, these leaf-rudiments spread out to become fairly conspicuous and persist down the branch for a short distance before falling off. The branches remain

Fig. 4.23. *Euphorbia stolonifera*, ± 30 cm diam., *PVB 9491*, south of Sebrafontein, east of Rosh Pinah, Namibia, 16 Jul. 2005 (© PVB).

Fig. 4.24. *Euphorbia stolonifera*, plant ± 0.4 m broad flowering richly, *PVB 10603*, SE of Fraserburg, South Africa, 14 Oct. 2006 (© PVB).

Fig. 4.25. *Euphorbia stolonifera*, cyme with central male cyathium, *PVB 9491*, south of Sebrafontein, east of Rosh Pinah, Namibia, 11 May 2014 (© PVB).

Fig. 4.26. *Euphorbia stolonifera*, cymes in female stage (females well exserted) and central cyathium absent, *PVB 11009*, Buffelsklip, east of Oudtshoorn, South Africa, 15 Aug. 2010.

somewhat pliable, not becoming rigid with age and they die off and are replaced with new branches regularly. They always have a grey-green colour.

As in most members of subg. *Esula* the cyathia in *E. stolonifera* are produced at the apices of the branches. A single, usually functionally male cyathium develops first terminally. This cyathium is surrounded by several bracts that are indistinguishable from the leaves (though they often fall off before the cyathium matures). In plants from southern Namibia the female floret in this cyathium may develop as well, but it is usually rudimentary. This terminal cyathium usually has eight glands (but sometimes only 5 or 6, and sometimes terminal cyathia on the same plant have five glands while others have eight). In the axils of the bracts, short, often slightly angled rays develop and each of these ends in a cyathium. These lateral cyathia are smaller than the terminal one and typically have five glands. They mature after the central one has finished flowering and often after it has even dropped off. Mostly these lateral cyathia are bisexual but, in some plants from southern Namibia, they are female only, without traces even of rudimentary males. The pedicel of the female floret is very variable in length and may be included inside the lobes or may project from the cyathium and may be recurved over the glands. The cyathia are usually very sweetly scented.

4.1 Sect. Aphyllis

Fig. 4.27. *Euphorbia stolonifera*, cymes in female stage (females almost sessile) with central cyathium fallen off, *PVB 10540*, road to Matjiesvlei, west of Calitzdorp, South Africa, 29 Sep. 2006 (© PVB).

Euphorbia stolonifera and *E. mauritanica* are similar, with their fleshy green photosynthetic branches and with their sessile, alternating leaf-rudiments that leave small, sickle-shaped scars on falling off. In their manner of flowering and in the details of the cyathia, they cannot be separated easily and it is only vegetatively that they can be distinguished. The main differences between them lie in the frequent subterranean spreading via short, rigid and woody rhizomes in some of the branches in *E. stolonifera* and the manner in which its quite strictly erect branches hardly rebranch at all except sometimes around their apices. Consequently, plants of *E. stolonifera* do not form the nearly spherical bushes that one often finds in *E. mauritanica*, where the branches arise from a distinct, small trunk (the base of the stem just above the ground), they rebranch extensively along their length and have a much more outwardly spreading habit. The yellow-green colour that one often (though not always) finds in *E. mauritanica*, is never present in *E. stolonifera*.

History

Euphorbia stolonifera was known for a long time before it was validly published. Although the name is associated with Rudolf Marloth and he first collected it in the Komsberg Pass and near Matjiesfontein before 1918 (the type was collected in October 1920), he was not the first to observe it. The first recorded collection was made in November 1898 at Matjiesfontein by Peter MacOwan and it was also collected in December 1908 south-west of Calvinia by H.H.W. Pearson. It was mentioned by R.H. Compton (1929) in his informal account of the plants of the Karoo as one of the few 'rhizomatous and stoloniferous plants' in the Karoo and again (Compton 1931: 298) in his account of the Flora of the Whitehill district. Here he mentioned that it was 'a widely distributed species, confused in herbaria with *E. mauritanica*, but entirely distinct in the wild'. N.E. Brown saw the specimen *Marloth 5745* (PRE) in January 1918 and annotated it as '*Euphorbia mauritanica*'.

As White et al. (1941) pointed out, the name '*stolonifera*' is not really appropriate. The growth of this species is not 'stoloniferous', but is more accurately described by the terms 'rhizomatous' or 'soboliferous'. The name *Euphorbia stolonifera* was validated by White et al. (1941), who used Marloth's name, somewhat modified his unpublished description and provided the Latin diagnosis.

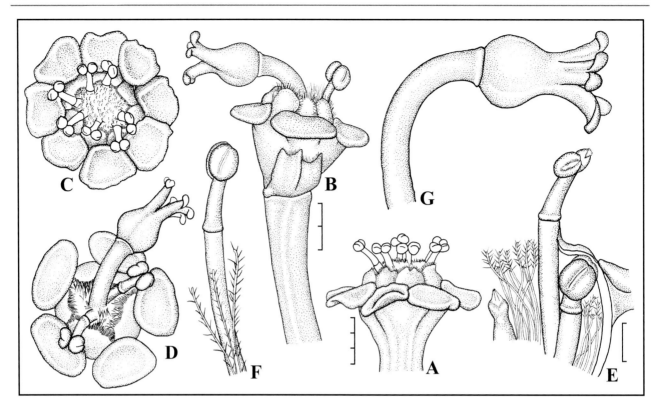

Fig. 4.28. *Euphorbia stolonifera,* South African material. **A**, side view of central male cyathium (scale 3 mm, as for **C**). **B**, side view of lateral bisexual cyathium (scale 2 mm, as for **D**). **C** face view of central male cyathium. **D**, face view of lateral bisexual cyathium. **E**, **F**, anthers, bracteoles and rudimentary female in central cyathium (scale 1 mm, as for **F**, **G**). **G**, female floret. Drawn from: *PVB,* just south of Sutherland, South Africa (© PVB).

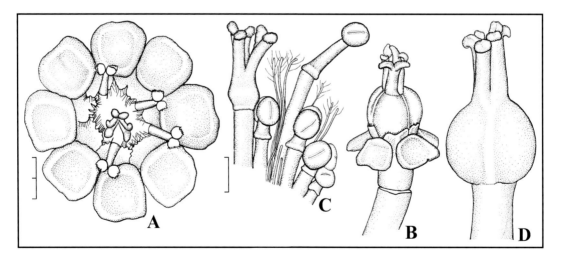

Fig. 4.29. *Euphorbia stolonifera,* Namibian material. **A**, face view of central male cyathium (scale 2 mm, as for **B**). **B**, side view of lateral (female) cyathium. **C**, anthers, bracteoles and rudimentary female in central cyathium (scale 1 mm, as for **D**). **D**, female floret. Drawn from: *PVB 3918*, Sebrafontein, NE of Rosh Pinah, Namibia (© PVB).

4.2 Sect. Esula

Euphorbia sect. **Esula** (Pers.) Dumort., *Fl. Belg.*: 87 (1827). *Esula* (Pers.) Haw., *Syn. Pl. Succ.*: 153 (1812). Type: *E. esula* L.

Bisexual non-succulent sometimes geophytic herbs to small shrubs, often densely branching. *Branches* slender, cylindrical, grey-green often suffused with red and becoming woody lower down. *Leaves* prominent and often densely crowded, alternate, linear to obovate, sessile to shortly petiolate, estipulate. *Synflorescences* terminal, with central often shortly pedunculate usually bisexual cyathium with 5 glands surrounded by 3–8 bisexual cyathia with 4 or 5 glands terminating rays, each arising in axil of prominent leaf-like but usually much broader bracts surrounding terminal cyathium, sometimes further rays arising from axils of bracts around lateral cyathia, glands crescent-shaped with two spreading horn-like appendages on outer margins. *Capsule* 3–5 mm diam., exserted on initially decurved and later erect pedicel. *Seeds* smooth, ± 2 × 1.5 mm, grey to nearly black, with cap-like apical caruncle.

The eight members of this section found in southern Africa (out of a total of 96 species) form a small group of closely related species (Riina et al. 2013: Fig. 5D). It is possible that *E. citrina* from the Chimanimani Mountains is also closely related. The following species are now included here: *E. corneliae*, *E. dumosoides*, *E. genistoides*, *E. kraussiana*, *E. minuscula*, *E. natalensis*, *E. sclerophylla* and *E. striata*. Of these, *E. natalensis* occurs in Moçambique and Zimbabwe (where it was recently redescribed as *E. crebrifolia*), while the other seven species are endemic to southern Africa. Only *E. corneliae* occurs in Namibia, in the south-western corner of the country. The group is most common on the moister, eastern side of the subcontinent, becoming rarer towards the west. Some of the species (such as *E. genistoides* and *E. sclerophylla*) are difficult to distinguish from one another. Several of them exhibit the tendency to die back into the rootstock in the dry season, some with a real geophytic habit, others with a system of underground rhizomes. Such perennating rhizomes have been observed in many members of

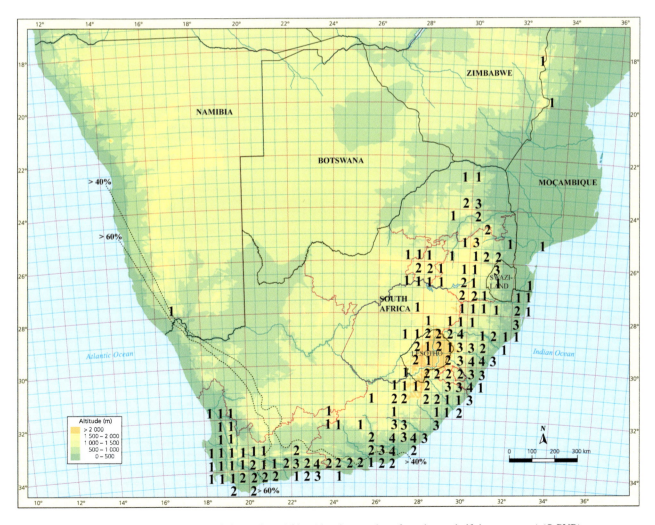

Fig. 4.30. Distribution of *Euphorbia* sect. *Esula* in southern Africa (showing number of species per half-degree square) (© PVB).

sect. *Esula* in the northern hemisphere (Croizat 1945), where they sometimes enable the plants to survive very low temperatures.

Key to the species of sect. *Esula* in southern Africa.

1. Plant very small and low-growing, 10–20 × 20–80 mm, consisting of clusters of leaves on short branches spreading close to ground, cyathial glands red..**E. minuscula**
1. Plant not as above, with ± erect branches > 50 mm long, cyathial glands yellow or yellow-green......................................2.
2. Leaves densely crowded, often somewhat deflexed above bases, usually cordate towards bases, often abruptly narrowed near tips..**E. natalensis**
2. Leaves usually not densely crowded, spreading or ascending, tapering towards bases sometimes into short petiole, also tapering towards tips...3.
3. Plants without perennial or swollen rootstock, but only with short fibrous taproot beneath soil..4.
3. Plants with perennial often slightly swollen rootstock or series of small slender tubers (to which plant may die back in dry season) from which fibrous roots arise...5.
4. Cyathial glands ± elliptic and without horns on outer edges, styles 2–3 mm long, leaves 15–100 × 4–15 mm..**E. kraussiana**
4. Cyathial glands crescent-shaped and distinctly horned on outer edges, styles 1–1.5 mm long, leaves 5–30 × 2–5 mm..**E. dumosoides**
5. Leaves few and distant on branch, erect or ascending, usually > 25 mm long, plant with 1 to few branches not rebranching above base..**E. striata**
5. Leaves numerous and sometimes crowded, ascending-spreading to spreading widely, usually < 20 mm long, plant with many branches which often rebranch..6.
6. Cyathia 5–6 mm broad, bracts beneath cyathium not expanded, linear to elliptic and similar to leaves..**E. corneliae**
6. Cyathia 2.5–4 mm broad, bracts beneath cyathium much expanded (shorter and broader than leaves) so as to be ± as long as broad..7.
7. Leaves broadest below middle, with sharply pointed tips, styles ± sessile on ovary......................................**E. sclerophylla**
7. Leaves not broadest below middle, with acute sometimes recurved tips, styles with short united part above ovary..**E. genistoides**

Euphorbia corneliae Bruyns, *Haseltonia* 25: 48 (2018). Type: Namibia, Rosh Pinah distr., Nasapberg, Aub, 1600 m, 5 July 1997, *Bruyns 7207* (BOL, holo.).

Bisexual spineless glabrous sparingly to densely branched non-succulent herb 0.15–0.3 × 0.1–0.2 m, branching from top of slightly swollen subterranean rootstock 0.1–0.3 m × 4–10 mm bearing fibrous roots. *Branches* ascending to erect, simple or sparingly rebranched, cylindrical, 0.1–0.3 m × 1–2 mm, without tubercles, slightly roughened from old leaf-bases, grey- to red-green becoming brown near base, often dying back to rootstock during dry periods; *leaves* present throughout branches or becoming absent in lower parts, numerous but not crowded, alternate, 8–20 × 1–3 mm, ascending, persistent for several seasons or shed during dry periods, grey-green and often with pale margins, linear-lanceolate, glabrous, flat to slightly channelled above, acute, finely apiculate, narrowing towards base into very short petiole 1–2 mm long, estipulate. *Synflorescences* terminal, glabrous, consisting of very shortly pedunculate (peduncle ± 0.5 mm long, glabrous) bisexual cyathium surrounded by 3–5 usually simple terete almost filiform rays 6–30 mm long, rays subtended by whorl of 2–3 slightly glaucous bracts like leaves (linear to elliptic, 6–30 × 2–6 mm) and each tipped with a bisexual cyathium, secondary cyathia sessile and surrounded by 2 sessile linear bracts often broader than leaves (6–12 × 2–4 mm); *cyathia* cupular, glabrous, 5–6 mm broad (± 2 mm long below insertion of glands), with 5 lobes with slightly toothed margins, grey-green; *glands* 4, thickly crescent-shaped, ± 2.5 mm broad, spreading and contiguous, yellow-green, ± flat and slightly pitted above, outer margins produced into two spreading horns at their ends and hollowed out between them (giving crescent-shape), inner margins slightly raised; *stamens* with sparsely and finely pubescent pedicels,

4.2 Sect. *Esula*

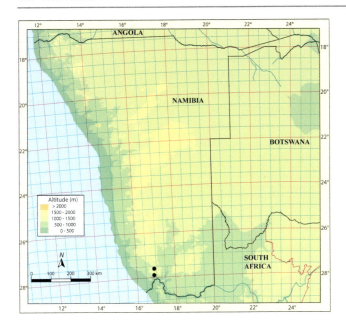

Fig. 4.31. Distribution of *Euphorbia corneliae* (© PVB).

Fig. 4.32. *Euphorbia corneliae*, ± 15 cm tall between rocks, *PVB 7207*, NE of Witpütz, Namibia, 12 Jul. 2014 (© PVB).

bracteoles few, finely filiform, pubescent; *ovary* globose and slightly 3-angled, glabrous, soon slightly exserted on pedicel 2 mm long; styles 1–1.5 mm long, branched to near base (tips again bifid). *Capsule* not seen.

Distribution & Habitat

Euphorbia corneliae is known from the flattish summits of a few of the higher peaks in south-western Namibia between Witpütz and Namuskluft, north and east of Rosh Pinah. It occurs in heavy, loamy soil in very rocky ground among a fairly uniform covering of short shrubs (many of them succulent), not exceeding 0.5 m in height.

This is an area receiving under 150 mm of rain per year, mainly in winter and is by far the driest where any member of this group is found.

Diagnostic Features & Relationships

A slender herb that arises from a slightly thickened, often etiolated perennial rootstock, *E. corneliae* is almost certainly geophytic in habitat, dying back into this rootstock in the

drier times, for older specimens have many old, short dead branches attached to the underground stem. In seasons of good rains, the slender branches reach 30 cm or more, with fine leaves sparingly spread out along most of their length and these persist all year in cultivation.

Flowering takes place in September and October, after which the plants begin to die back and lose some of their leaves. The cyathia are bright yellow-green and have always been observed with only four glands.

Euphorbia corneliae is closely allied to *E. genistoides*, from which it differs by the considerably larger cyathia (around twice as broad across the top as those of *E. genistoides* also with a broader involucre beneath the glands) with broader glands, longer styles and longer mature anthers. A further difference lies in the narrower bracts beneath the cyathia, which are the same shape as the leaves and not as broadly expanded as in *E. genistoides* (where they are shorter and broader than the leaves).

History

Euphorbia corneliae was first observed in July 1989 on the upper slopes of some of the high peaks near the boundary of the farms Namuskluft and Sebrafontein, east of Rosh Pinah in south-western Namibia. A second area

Fig. 4.33. *Euphorbia corneliae*, plant excavated to show slender tubers and rhizomatous habit, *PVB 7207*, NE of Witpütz, Namibia, 12 Jul. 2014 (© PVB).

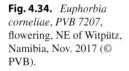

Fig. 4.34. *Euphorbia corneliae*, *PVB 7207*, flowering, NE of Witpütz, Namibia, Nov. 2017 (© PVB).

4.2 Sect. *Esula*

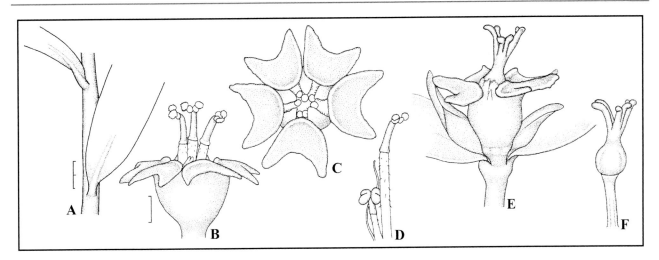

Fig. 4.35. *Euphorbia corneliae*. **A**, portion of branch with bases of leaves (scale 2 mm). **B**, side view of central male cyathium (scale 1 mm, as for C–F). **C**, face view of central male cyathium. **D**, male florets. **E**, side view of lateral bisexual cyathium with receptive female floret. **F**, female floret. Drawn from: *Bruyns 7207*, NE of Witpütz, Namibia (© PVB).

slightly further north was found in July 1997 near the farm Aub. These localities are more than 400 km from the nearest member of this group (*E. genistoides*) and extend its distribution into much drier areas than previously known. It was named for Cornelia Klak.

Euphorbia dumosoides Bruyns, *Phytotaxa* 423: 95 (2019). *Euphorbia dumosa* E.Mey. ex Boiss. in DC., *Prodr.* 15 (2): 168 (1862), *nom. illegit.*, *non* A. Rich. (1850). Lectotype (Bruyns 2019): South Africa, Pondoland, near the Umsikaba River, *Drège 4619* (BM; K, 2 sheets, MO, TCD, iso.).

Bisexual spineless glabrous sparingly branched non-succulent herb 0.15–0.70 m tall and same in breadth, with sometimes slightly woody stem 2–5 mm thick, with slender taproot bearing fibrous roots. *Branches* ascending to erect, simple or rebranched often in umbel-like whorls, persistent, cylindrical, 0.1–0.5 m × 1–2 mm, without tubercles, grey-green to brown; *leaves* becoming absent in lower parts of branches, numerous, alternate, 4–30 × 1.5–5 mm, spreading or ascending, persistent for several seasons, slightly glaucous, linear to obovate-oblong tapering towards base into short petiole to 1 mm long or ± sessile, glabrous to finely pubescent, upper surface flat, obtuse to tapering to acute tip, apiculate, estipulate. *Synflorescences* in lax terminal umbels, 15–40 mm diam., glabrous, consisting of ± sessile bisexual cyathium surrounded by 1–5 simple terete rays 3–15 mm long, rays subtended by whorl of 2–5 slightly glaucous or yellowish bracts shorter than and broader than leaves (broadly ovate to semi-circular, 2–10 × 3–10 mm) and each tipped with a bisexual cyathium, secondary cyathia ± sessile and surrounded by 2–3 sessile broad and semi-circular bracts (3–5 × 5–8 mm); *cyathia* narrowly cupular, glabrous, 3–4 mm broad (± 2 mm long below insertion of glands), with 5 lobes with finely toothed margins, grey-green; *glands* 4 or 5, crescent-shaped, 1–2 mm broad, spreading and contiguous to slightly separated, yellow-green, ± flat and slightly pitted above, outer margins produced into two spreading horns at their ends and flat or hollowed out between them (giving crescent-shape), inner margins slightly raised; *stamens* with finely pubescent pedicels, bracteoles few, finely filiform, pubescent; *ovary* globose and strongly 3-angled, glabrous, erect and slightly exserted on pedicel ± 2.5 mm long and soon pendulous on decurved pedicel 3–5 mm long; styles 1–1.5 mm long, branched and spreading to near base (tips again bifid). *Capsule* 3–4 mm diam., obtusely 3-lobed, glabrous, exserted from cyathium on gradually erect pedicel 3–5 mm long.

Distribution & Habitat

Euphorbia dumosoides is known only in Kwazulu-Natal, where it is of scattered occurrence from Hlabisa to Durban, as well as the Port Edward area and in the north-eastern corner of the Eastern Cape in 'Pondoland'. It mostly grows at low altitudes, but it has been recorded up to 1500 m on occasion. It is usually found in shady, moist spots under trees, sometimes in shallow soils among rocks and it is often extremely common where it occurs.

Diagnostic Features & Relationships

Although *E. dumosoides* somewhat resembles *E. natalensis*, the two differ vegetatively in several ways. Their rootstocks are different: in *E. dumosoides* there is a fine and short tap-

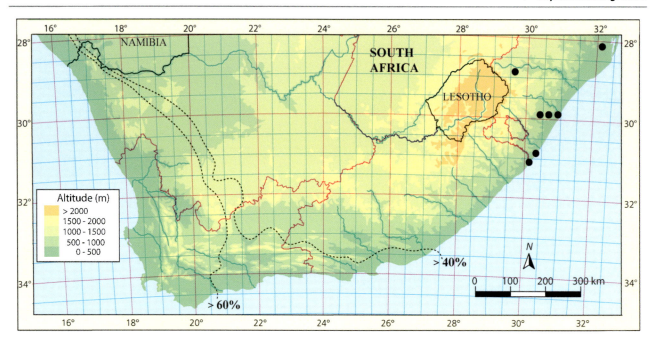

Fig. 4.36. Distribution of *Euphorbia dumosoides* (© PVB).

Fig. 4.37. *Euphorbia dumosoides*. Grassy habitat, *K.W. Grieve 1855*, Braemar, near Port Edward, South Africa (© Kate Grieve).

Fig. 4.38. *Euphorbia dumosoides*. Among grasses and rocks, *K.W. Grieve 1855*, Braemar, near Port Edward, South Africa (© Kate Grieve).

4.2 Sect. *Esula*

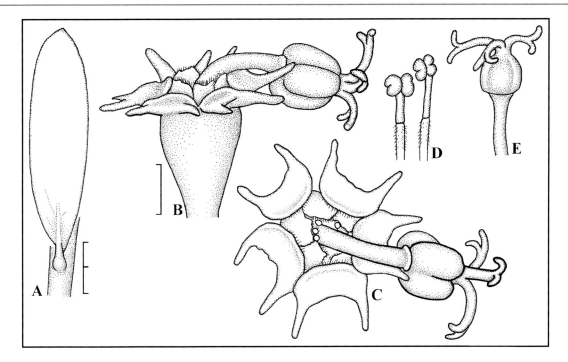

Fig. 4.39. *Euphorbia dumosoides*. **A**, portion of branch with leaf (scale 2 mm). **B**, cyathium from side (scale 1 mm, as for **C–E**). **C**, cyathium from above. **D**, anthers. **E**, female floret. Drawn from: *K.W. Grieve 2428*, Umtamvuna Reserve, near Port Edward, South Africa (© PVB).

root from which fine fibrous roots arise with no evidence of the ability to resprout from the rootstock (which is common in *E. natalensis*). In *E. dumosoides* the leaves are also flat (lacking the revolute margins of *E. natalensis*) and they taper towards their bases, where they are not cordate as in *E. natalensis*. Florally, *E. dumosoides* cannot be separated from *E. natalensis*, but the cyathia are fewer and not organized into such large heads.

History

Euphorbia dumosoides was first collected by J.F. Drège and his brother Carl and it remains known from relatively few gatherings. The present illustrations are all from a colony found by Graham and Kate Grieve near the Umtamvuna River of the Port Edward district in the south of Kwazulu-Natal.

Euphorbia genistoides P.J.Bergius, *Descriptiones Plantarum ex Capite Bonae Spei*: 146 (1767). *Tithymalus genistoides* (P.J.Bergius) Klotzsch & Garcke, *Abh. Königl. Akad. Wiss. Berlin* 1859: 97 (1860). *Galarhoeus genistoides* (P.J.Bergius) Haw., *Syn. Pl. Succ.*: 144 (1812). Type: South Africa, Cape of Good Hope, *Auge (Grubb)* (SBT, holo.).

Euphorbia ericoides Lam., *Encycl.* 2 (2): 430 (1788). Type: South Africa, Cape of Good Hope, *Sonnerat* (P-LAM, holo.; K, fragment, iso.).

Tithymalus revolutus Klotzsch & Garcke, *Abh. Königl. Akad. Wiss. Berlin* 1859: 99 (1860). Type: South Africa. Cape of Good Hope, *Ecklon & Zeyher 2* (missing).

Euphorbia genistoides var. *puberula* N.E.Br., *Fl. Cap.* 5 (2): 264 (1915). Lectotype (Bruyns 2012): South Africa, Cape, Lion Mountain, *Wolley-Dod 3104* (K; BOL, iso.).

Euphorbia genistoides var. *corifolia* (Lam.) N.E.Br., *Fl. Cap.* 5 (2): 264 (1915). *Euphorbia corifolia* Lam., *Encycl.* 2 (2): 431 (1788). Type: South Africa, Cape of Good Hope, *Sonnerat* (P-LAM, holo.; K, iso.).

Euphorbia genistoides var. *leiocarpa* Boiss., in DC, *Prodr.* 15 (2): 168 (1862). Lectotype (designated here): South Africa, Cape, in monte Drakenstein, *Drège* (S; TCD (labelled 8193), iso.).

Euphorbia genistoides var. *major* Boiss., in DC, *Prodr.* 15 (2): 168 (1862). Lectotype (designated here): South Africa, Cape, in monte leonis, *Drège 8192* (BM; HBG, TCD, iso.).

Euphorbia erythrina Link, *Enum. Pl. Hort. Berol.* 2: 12 (1822). *Tithymalus erythrinus* (Link) Klotzsch & Garcke, *Abh. Königl. Akad. Wiss. Berlin* 1859: 91 (1860). Type: South Africa, Cape of Good Hope, *Bergius* (missing). Neotype (Bruyns 2012): South Africa. Cape, Paarl Mountain, *Drège 2197* (K; K, iso.).

Euphorbia erythrina var. *meyeri* N.E.Br., *Fl. Cap.* 5 (2): 262 (1915). *Euphorbia meyeri* Boiss., *Cent. Euphorb.*: 35 (1860), *nom. illegit.*, *non* Steud. (1840). Lectotype (Bruyns 2012): South Africa. Cape, Paarl Mountain, *Drège 2197* (K; K, iso.).

Tithymalus apiculatus Klotzsch & Garcke, *Abh. Königl. Akad. Wiss. Berlin* 1859: 94 (1860). Lectotype (Bruyns 2012): South Africa. Cape, *Mund & Maire* (K).

Tithymalus confertus Klotzsch & Garcke, *Abh. Königl. Akad. Wiss. Berlin* 1859: 94 (1860). Lectotype (Bruyns 2012): South Africa. Cape, *Mund & Maire* (K).

Euphorbia erythrina var. *burchellii* Boiss. in DC., *Prodr.* 15 (2): 169 (1862). Type: South Africa, near Cape Town, 3 Jan. 1811, *Burchell 458* (G-DC, holo.; K, iso.).

Euphorbia foliosa (Klotzsch & Garcke) N.E.Br., *Fl. Cap.* 5 (2): 262 (1915). *Tithymalus foliosus* Klotzsch & Garcke, *Abh. Königl. Akad. Wiss. Berlin* 1859: 67 (1860). Lectotype (Bruyns 2012): South Africa, Cape Flats, near Cape Town, *Ecklon & Zeyher 12* (K; K, SAM, iso.).

Euphorbia artifolia N.E.Br., *Fl. Cap.* 5 (2): 263 (1915). Type: South Africa, Milkwoodfontein, Riversdale div., ± 600', 7 Oct. 1897, *Galpin 4562* (K, holo.; PRE, iso.).

Bisexual spineless glabrous sparingly (to densely) branched non-succulent herb to shrub 0.15–0.3 (0.6) × 0.15–0.3 m, branching from top of subterranean perennial rootstock bearing fibrous roots and sometimes spreading by underground rhizomes. *Branches* ascending to erect, simple or sparingly (densely) rebranched along length, cylindrical, 0.1–0.3 m × 1–2 mm, without tubercles, slightly roughened from old leaf-bases, grey-green to red-green becoming brown near base; *leaves* present throughout branches or becoming absent in lower parts, numerous and often crowded, alternate, 4–20 × 1–3 mm, ascending, persistent for several seasons, slightly glaucous and often with reddish margins, linear-lanceolate to linear-oblong, glabrous, flat to slightly channelled above and occasionally with the margins recurved, with acute to obtuse sometimes recurved tips, apiculate, narrowing towards base into very short petiole 1–2 mm long, estipulate. *Synflorescences* terminal, glabrous, consisting of very shortly pedunculate (peduncle 1–2 mm long, slightly pubescent) bisexual cyathium surrounded by 3–8 simple to rebranched terete rays 6–50 mm long, rays subtended by whorl of 3–5 slightly glaucous slightly yellowish bracts shorter than and broader than leaves (oblong, ovate or rhomboid, 4–10 × 2–10 mm) and each tipped with a bisexual cyathium, secondary cyathia very shortly pedunculate and surrounded by 2–4 sessile broadly rhomboid bracts much shorter and broader than leaves (4–8 × 5–12 mm) with further rays developing in their axils; *cyathia* cupular, glabrous, 3–4 mm broad (± 1.5 mm long below insertion of glands), with 5 lobes with finely toothed margins, grey-green; *glands* 4–5 (5 on central cyathium, 4 on laterals), crescent-shaped, ± 1.5 mm broad, spreading and contiguous to slightly separated, green with reddish horns, ± flat and slightly pitted above, outer margins produced into two spreading horns at their ends and hollowed out between them (giving crescent-shape), inner margins slightly raised; *stamens* with finely pubescent pedicels, bracteoles few, finely filiform, pubescent; *ovary* globose and strongly 3-angled, glabrous, soon exserted and pendulous on decurved pedicel 4–5 mm long; styles 1–1.2 mm long, branched to near base (tips again bifid). *Capsule* 3–4 mm diam., obtusely 3-lobed, glabrous, exserted from cyathium on gradually erect pedicel 4–6 mm long.

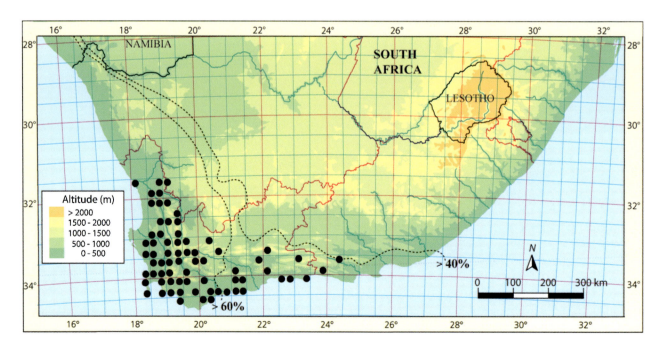

Fig. 4.40. Distribution of *Euphorbia genistoides* (© PVB).

Fig. 4.41. *Euphorbia genistoides*, open ground alongside gravel road, exposed plant with reddish branches, Paarl Mountain, South Africa, 29 Aug. 2014 (© PVB).

Fig. 4.42. *Euphorbia genistoides*, 'foliosa', recently burnt, sandy area, *PVB 13750*, Arniston, South Africa, 18 Sep. 2019 (© PVB).

Fig. 4.43. *Euphorbia genistoides*, rhizomatous habit in sand, *PVB 13750*, Arniston, South Africa, 18 Sep. 2019 (© PVB).

Distribution & Habitat

Euphorbia genistoides is found on the western flank of South Africa, extending into the southern Cape. It is recorded from south of Nieuwoudville and the Gifberg to the Cape Peninsula, with more sporadic occurrence to the east, into the Long Kloof and Baviaanskloof on the western margin of the Eastern Cape.

Plants usually occur on the lower slopes of hills and mountains on heavy loam derived from shales of the Table Mountain Series, often close to the underlying granites and below the sandstones, sometimes on soils of granitic origin as on Paarl Mountain or on coastal limestones (as around Still Bay), more rarely on steep slopes on shales covered with dense, scrubby *renoster*-veld (Worcester-Robertson Karoo). The associated vegetation consists of scattered to dense grasses and small restios, some shrublets of *Gnidia*, *Struthiola* and *Muraltia* and isolated shrubs of *Montinia* and *renoster*. After this vegetation is burnt, *E. genistoides* may sometimes appear in large numbers.

Diagnostic Features & Relationships

Euphorbia genistoides forms small colonies of inconspicuous plants. Often these are separate plants, but they are sometimes connected underground by slender perennial rhizomes. They die off into the perennial rootstock during the dry season in some areas, but are perennial above-ground in other, moister places. If they die back during the dry summers, branches usually appear again from the rootstock by

Fig. 4.44. *Euphorbia genistoides*, red branches and yellowish bracts, Still Bay, South Africa, 6 Sep. 2014 (© PVB).

Fig. 4.45. *Euphorbia genistoides,* small bracts beneath cyathium the same colour as leaves, Paarl Mountain, South Africa, 29 Aug. 2014 (© PVB).

May or June. In disturbed spots or in recently burnt areas, where the other vegetation has been cleared away, they may become relatively conspicuous and may even tend to dominance in small areas. The branches bear slender distinctly bluish-green leaves (often with fine red margins) by which one can recognize them from the other surrounding shrublets.

Flowering begins in late July and early August and continues until October. The bracts around the central cyathium may be somewhat yellow towards their bases which gives a paler patch at the apex of the branch just before flowering commences. Within a few days of its maturing the central cyathium is overtopped by the rays from axils of bracts around it and their cyathia also develop quickly. Rays have much shorter and broader bracts surrounding their cyathium than the bracts around the central cyathium (which are just slightly broader and slightly shorter than the leaves). The central cyathium has five glands and most of those on rays have only four, but all are bisexual. The glands are crescent-shaped, with the horns of the crescent produced by two spreading outgrowths of the ends of the outer margins of the gland, with a deep bay between these horns. When the styles mature the ovary usually already slightly protrudes from among the lobes and it is soon pushed further out on a decurved pedicel.

Euphorbia genistoides is said to differ from *E. erythrina* by the revolute margins of the leaves, which is especially a feature of plants from further inland (such as around Ceres). *Euphorbia foliosa* is supposed to be an altogether smaller plant with a denser arrangement of shorter and broader leaves on the branches. Both the shape and the size of the leaves is variable within populations, especially depending on how

4.2 Sect. *Esula*

Fig. 4.46. *Euphorbia genistoides*, broad and more brightly yellow bracts beneath cyathium, *PVB 13750*, Arniston, South Africa, 18 Sep. 2019 (© PVB).

Fig. 4.47. *Euphorbia genistoides*, larger bracts beneath cyathium with developing rays, Tokai, Cape Town, South Africa, 17 Aug. 2014 (© PVB).

Fig. 4.48. *Euphorbia genistoides*, developing capsules, some bracts suffused with yellow, Still Bay, South Africa, 6 Sep. 2014 (© PVB).

exposed the plants are. The degree of exposure also greatly affects the size of the plants. So, in some spots (as on the slopes of Paarl Mountain) one finds plants assignable to *E. foliosa* and *E. genistoides* growing together and connected by many intermediates. In *E. foliosa* the rays are also shorter so that the synflorescence is denser and this seems to be fairly typical of drier, often calcareous habitats (especially such as those of the coastal limestones from Bredasdorp to Riversdale). Plants with exceedingly narrow leaves have usually been attributed to *E. ericoides*, but these are also connected by many intermediates to the others with broader leaves. Florally all are the same with similar cyathia with crescent-shaped glands. *Euphorbia genistoides* differs from *E. natalensis* by the less dense arrangement of the leaves (though much denser than in *E. sclerophylla* and *E. striata*) and the manner in which the leaves taper towards their bases. From *E. striata* it differs by the denser arrangement of the more spreading leaves and by the much narrower bracts under the first cyathium. In *E. sclerophylla* the leaves are more spaced out along the branches, they are differently shaped (usually broader below the middle and with a particularly noticeable sharp apical point) and the styles are sessile above the ovary.

History

Euphorbia genistoides was the first of the herbaceous species to be described from the Cape and was based on material collected by Johann Andreas Auge, possibly as early as 1750. It was among Auge's specimens that were bought in Cape Town in 1764 by Michael Grubb, the director of the Swedish East India Company, who was at the time on a return voyage from China. Grubb passed these dried specimens on to P.J. Bergius in Stockholm and Bergius (1767) described sev-

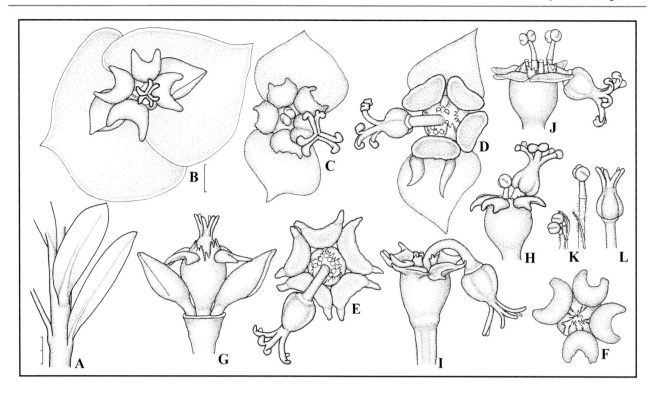

Fig. 4.49. *Euphorbia genistoides*. **A**, portion of branch with bases of leaves (scale 2 mm). **B–F**, cyathium from above (scale 1 mm, as for **C–L**). **G–J**, side view of cyathium. **K**, anthers and bracteoles. **L**, female floret. Drawn from: **A**, Rhodes' Memorial, Cape Town, South Africa. **B, E, G, I, K, L**, slopes of Constantiaberg, Cape Town, South Africa. **C, H**, Still Bay, near Riversdale, South Africa. **D, F, J**, Paarl Mountain, South Africa (© PVB).

eral of them in an early account of the Cape Flora (Gunn and Codd 1981). As it is common around Cape Town, it was often collected, for example by Sonnerat and Thunberg around 1770 and somewhat later by Burchell in 1814 and 1815.

Euphorbia kraussiana Bernh., *Flora* 28: 87 (1845). Type: South Africa, Natal, forest margins near Pietermaritzburg, 1 Sept. 1839, 2000-2500', *Krauss 256* (MO, holo.; BM, K-2 sheets, TCD, TUB, iso.).
Tithymalus truncatus Klotzsch & Garcke, *Abh. Königl. Akad. Wiss. Berlin* 1859: 75 (1860). Type: South Africa, Cape, *Krebs* (missing).
Tithymalus meyeri Klotzsch & Garcke, *Abh. Königl. Akad. Wiss. Berlin* 1859: 75 (1860). Lectotype (Bruyns 2012): South Africa, Cape, *Ecklon & Zeyher Euphorb. 13* (Z; MEL, SAM, TCD, iso.).
Euphorbia kraussiana var. *erubescens* (E.Mey. ex Boiss.) N.E.Br., *Fl. Cap.* 5 (2): 268 (1915), *nom. illegit. Euphorbia erubescens* E.Mey. ex Boiss. in DC., *Prodr.* 15 (2): 116 (1862), *nom. illegit.*, *non* Boiss. (1849). Lectotype (Bruyns 2012): South Africa, Natal, between Umzimkulu & Umkomaas, Apr., *Drège* (S; BM, iso.).

Bisexual spineless glabrous much-branched perennial shrub 0.3–3 × 0.3–2 m, branching well above base from slender pliable and later slightly woody stem with fibrous roots. *Branches* often somewhat whorled, ascending to erect, simple or rebranched (also often in whorls) along length, cylindrical, 0.15–1 m × 1–3 mm, without tubercles, slightly roughened from old leaf-bases, grey-green to red-green; *leaves* absent in lower parts of branches and stem, alternate and whorled at base of umbels, 15–120 × 4–15 mm, spreading to slightly pendulous, persistent for several seasons, slightly glaucous, linear to oblanceolate, glabrous, flat to slightly channelled above, acute to obtuse, apiculate, tapering gradually towards base into short petiole 2–5 mm long, estipulate. *Synflorescences* terminal, glabrous, consisting of very shortly pedunculate (peduncle 1–2 mm long) bisexual cyathium surrounded by 5–8 simple to rebranched terete rays 15–100 mm long, rays subtended by whorl of 2–5 slightly glaucous bracts shorter and narrower than leaves (8–30 × 3–10 mm) and each tipped with a bisexual cyathium, secondary cyathia shortly pedunculate and surrounded by 2–4 sessile lanceolate to deltate bracts with cordate bases much shorter and often broader than leaves (4–15 (25) × 3–15 (22) mm); *cyathia* slenderly cupular, glabrous, 2–3 mm broad (1.5–2 mm long below insertion of glands), with 5 lobes with pilose margins, pale grey-green; *glands* 4–5, transversely elliptic to nearly rectangular, ± 1 mm broad, spreading and usually slightly separated, pale yellow-green, ± flat and slightly pitted above, outer margins entire or slightly notched, inner margins slightly raised; *stamens*

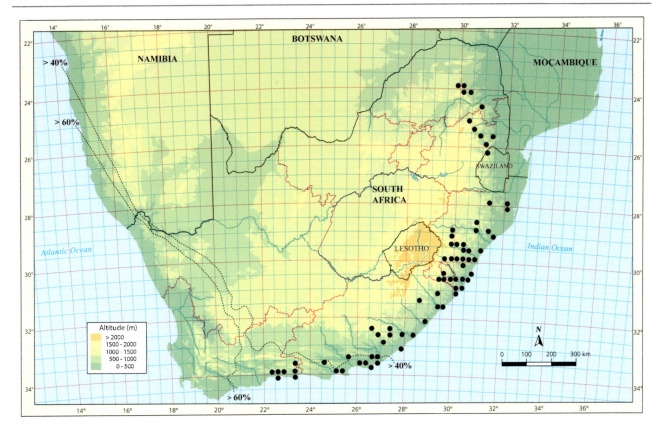

Fig. 4.50. Distribution of *Euphorbia kraussiana* (© PVB).

with glabrous or finely pubescent pedicels, bracteoles few, finely filiform and pubescent near tips; *ovary* globose and faintly 3-angled near base, glabrous, erect and partly exserted on pedicel ± 1.5–2 mm long soon further exserted and pendulous on decurved pedicel 4–5 mm long; styles 2–3 mm long, branched in upper half to two thirds (tips again bifid). *Capsule* 4–5 mm diam., obtusely 3-lobed, glabrous, exserted from cyathium on gradually erect pedicel 4–6 mm long.

Distribution & Habitat

Euphorbia kraussiana is widely distributed down the eastern side of South Africa. Somewhat isolated from the rest are the records of it from the valleys between mountains in the Tzaneen area and around Barberton. In Natal it is widespread in coastal to lower montane areas from Ngome to the Umtamvuna River, continuing to the vicinity of George in the narrow coastal plain of the southern part of the former Cape Province. It was not listed by Compton (1976) for Swaziland, but one collection has been made since his account, from the north-west of the country.

Euphorbia kraussiana usually grows in partial shade in coastal bush or on the edges of forests. Sometimes it occurs in deeper shade under trees in forests, where plants are usually smaller and more diffuse than more exposed specimens.

Diagnostic Features & Relationships

Euphorbia kraussiana forms shrubs which, because of their diffuse habit, slender branches and pale leaves, are usually not conspicuous. Nevertheless, for a southern African species of *Euphorbia*, the leaves are relatively prominent (often between 40 and 50 mm long and at least 4 mm broad), quite loosely clustered near the tips of the branches, mostly slightly oblanceolate and tapering gradually towards their bases into a short petiole. The plant sometimes dies back to the rootstock and is able to regenerate from it.

Cyathia are produced over most of the summer. The bracts (usually there are five of them) subtending the first, terminal cyathium on a branch are often a bit shorter than and slightly narrower than the leaves. Rays sprouting from the axils of these bracts (and around the base of the primary cyathium) bear further, secondary cyathia that are identical to the primary cyathium but the bracts subtending these secondary cyathia (and any further cyathia on the next series of rays) are usually entirely differently shaped – short and broad, more or less deltate and with a cordate base. The cyathia are very tiny, elongated and are the same colour as the leaves except for the faintly yellow, entire, spreading glands.

Fig. 4.51. *Euphorbia kraussiana*, flowering branch, *PVB 12248*, Zuurberg, west of Grahamstown, South Africa, 14 Dec. 2018 (© PVB).

Fig. 4.52. *Euphorbia kraussiana*, tip of flowering branch, *PVB 12248*, Zuurberg, west of Grahamstown, South Africa, 7 May 2014 (© PVB).

Fig. 4.53. *Euphorbia kraussiana*, tip of flowering branch showing narrow cyathia and long styles, *PVB 12248*, Zuurberg, west of Grahamstown, South Africa, 8 Mar. 2016 (© PVB).

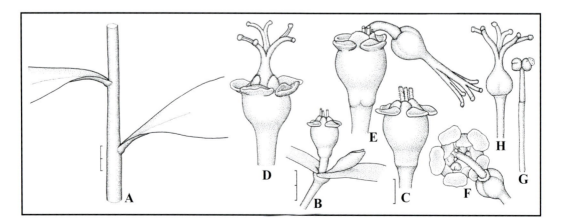

Fig. 4.54. *Euphorbia kraussiana*. **A**, portion of branch with bases of leaves (scale 2 mm). **B**, **C**, male cyathium from side (scale for **B** 2 mm, for **C** 1 mm as for **D–H**). **D**, **E**, side view of bisexual cyathium at different stages. **F**, bisexual cyathium from above. **G**, anther. **H**, female floret. Drawn from: *PVB 12248*, Zuurberg, west of Grahamstown, South Africa (© PVB).

Although its habit mostly makes *Euphorbia kraussiana* easily distinguishable from the other species in the group, the petioles are longer than in any of the others, while the slender cyathia with entire (not horned) glands and long styles also make it one of the easier species to recognise reliably.

History
Euphorbia kraussiana was first collected in September 1813 by William Burchell along the banks of the Great Fish River, with specimens gathered by J.F. Drège on 5 November 1829 as well. Most of these early collections were informally referred to as *Euphorbia erubescens* but, as N.E. Brown (1915) mentioned, this is not distinguishable from *E. kraussiana*, which is the earliest valid name. *Euphorbia kraussiana* was described by Bernhardi from plants collected by C. Ferdinand F. Krauss in September 1839.

Euphorbia minuscula Bruyns, *Phytotaxa* 423: 94 (2019). Type: South Africa, Cape, Kathoek, 50 m, 21 Aug. 1980, *Burgers 2493* (NBG, holo.).

Bisexual spineless glabrous sparingly branched non-succulent geophytic herb 10–50 × 15–120 mm with 1–10 (50) deciduous branches arising annually from top of slightly thicker wiry brown subterranean perennial stem 10–50 × 0.5–5 mm joined to ± spherical tuber 5–20 mm diam. bearing fibrous roots. *Branches* ascending to prostrate, simple or sparingly branched, cylindrical, 15–50 × ± 0.5 mm, without tubercles, usually reddish; *leaves* many, alternate, 2.5–10 × 1.5–5 mm, spreading, deciduous, glaucous green above and paler green or reddish below, elliptical to obovate, flat to margins slightly recurved, acute, narrowing towards base into short petiole 0.25–1 mm long, estipulate. *Synflorescences* terminal, glabrous, consisting of 1 shortly pedunculate (peduncle ± 0.5 mm long and sparsely pubescent) male cyathium surrounded by 1–4 mostly simple terete rays 1.5–5 mm long, rays subtended by whorl of 3 glaucous bracts very similar to leaves, each ray tipped with shortly pedunculate bisexual cyathium surrounded by 2 (–4) sessile broadly lanceolate apiculate bracts 2–5 × 1.5–4 mm sometimes with reddish margins (these bracts sometimes with further rays in their axils); *cyathia* cupular, glabrous, 2–2.5 mm broad (± 1 mm long below insertion of glands), with 5 lobes with toothed pubescent margins, grey-green; *glands* 4–5, ± crescent-shaped, ± 1 mm broad, spreading and contiguous, red, ± flat and slightly pitted above, outer margins produced into two spreading pale yellow horns at their ends and hollowed out between them (giving crescent-shape); *stamens* with finely pubescent pedicels, bracteoles few, filiform, sparsely pubescent; *ovary* globose and strongly 3-angled, glabrous, initially erect on pedicel ± 1 mm long and soon exserted and pendulous on decurved pedicel 1.5–3 mm long; styles ± 1 mm long, branched to near base and spreading just above ovary. *Capsule* ± 3 mm diam., obtusely 3-lobed, glabrous, exserted from cyathium on gradually erect pedicel 1.5–3 mm long. Seeds smooth, ± 1.2 × 1 mm, with small brown caruncle.

Distribution & Habitat
Euphorbia minuscula is confined to the southern coastal plain of South Africa between Riviersonderend and Bredasdorp, where it is known from four collections and found between altitudes of 50 and 300 m.

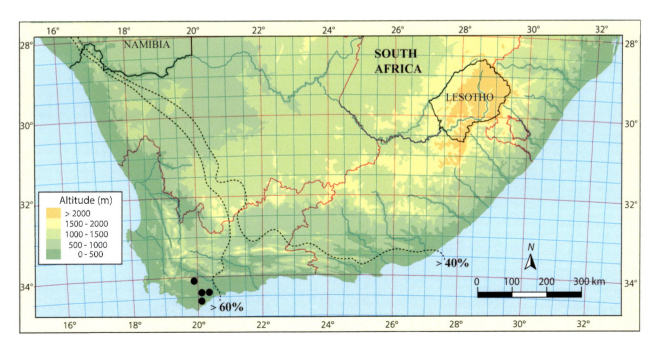

Fig. 4.55. Distribution of *Euphorbia minuscula* (© PVB).

Fig. 4.56. *Euphorbia minuscula,* greyish green plant ± 4 cm diam., among pieces of limestone, *PVB 13717*, east of Bredasdorp, South Africa, 20 Apr. 2019 (© PVB).

Fig. 4.57. *Euphorbia minuscula,* three plants excavated to show short branches, underground stems and neat tubers (scale in cm), *PVB 13717*, east of Bredasdorp, South Africa (© PVB).

The type collection was made on dry slopes on shale. Close to Bredasdorp it is common on a plateau of coastal limestone, where it grows in shallow pockets of black soil in crevices in the rock with a wealth of other succulents, including *E. caput-medusae*, *E. pseudoglobosa*, a small, rhizomatous form of *E. rhombifolia* ('*bayeri*') and a *Trichodiadema*, as well as *fynbos*-elements such as restios and *Acmademia* of the Rutaceae.

Diagnostic Features & Relationships

With small reddish or glaucous leaves in a cluster close to the ground, *E. minuscula* looks like one of the lowly, annual members of sect. *Anisophyllum*. Nevertheless, it has alternating leaves and a geophytic habit that is unlike anything known among these species. The lack of stipules and the caruncle on the seeds places it clearly in sect. *Esula*. *Euphorbia minuscula* is a geophyte, with the above-ground branches disappearing entirely in the dry season, a phenomenon also known in *E. corneliae*, *E. genistoides* and *E. striata*. In these other geophytic species the underground parts consist of

4.2 Sect. *Esula*

Fig. 4.58. *Euphorbia minuscula*, flowering plant among *Haworthia* and *Trichodiadema*, *PVB 13717*, east of Bredasdorp, South Africa (© PVB).

Fig. 4.59. *Euphorbia minuscula*, flowering plant, *PVB 13717*, east of Bredasdorp, South Africa (© PVB).

Fig. 4.60. *Euphorbia minuscula*, plant flowering and beginning to bear capsules, among clumps of *Trichodiadema*, *PVB 13717*, east of Bredasdorp, South Africa (© PVB).

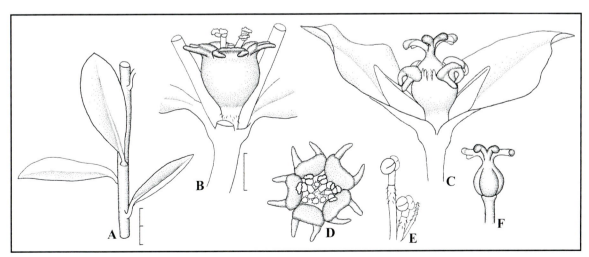

Fig. 4.61. *Euphorbia minuscula*. **A**, portion of branch with leaves (scale 2 mm). **B**, central male cyathium from side (scale 1 mm, as for **C**–**F**). **C**, bisexual cyathium from side with bracts. **D**, central male cyathium from above. **E**, anthers and bracteole. **F**, female floret. Drawn from: *PVB 13717*, NE of Bredasdorp, South Africa (© PVB).

slender stems and strings of slender tubers while in *E. minuscula* there is a single almost spherical tuber. *Euphorbia minuscula* is far smaller than any other members of sect. *Esula*, where the plants are mostly at least 10 cm tall and usually between 0.3 and 1.5 m tall.

Flowering in *E. minuscula* does not make the plants more conspicuous and the tiny cyathia are barely noticeable at the tips of the branches. In *Euphorbia minuscula* the bracts around the first terminal cyathium on a branch are more or less identical to the leaves and are not differently coloured either. Those bracts that subtend the secondary cyathia (those on rays arising around the first terminal cyathium) are differently shaped to the leaves, being both shorter and broader. Generally, the terminal cyathium is on a tiny sparsely pubescent peduncle, is male only and has five glands, while the secondary cyathia are subtended by a glabrous peduncle and are bisexual with four glands. The cyathial glands are also purple-red with pale yellow horns, while in all the other species they are yellow or yellow-green (though occasionally they may become suffused with red as they age in the others). Flowering takes place from May to September and the small cyathia have a strong sweet scent.

History

Discovered by C.A. Smith in June 1926, *E. minuscula* was observed again in August 1962 by H.C. Taylor at Riviersonderend. Another collection was made in August 1980 by Christiaan J. Burgers of the Cape Department of Nature Conservation, during his survey of the vegetation of the area that was later incorporated into the De Hoop Nature Reserve. After good rains in late summer 2019, it was found to be plentiful near Bredasdorp and from this material and the other earlier collections a description was drawn up and it was finally published in 2019.

Euphorbia natalensis Bernh., *Flora* 28: 86 (1845). Type: South Africa, Natal, base of Tafelberg, Aug. 1839, *Krauss 434* (MO, holo.; BM, FI, K, M, TCD, TUB, iso.).

Tithymalus epicyparissias Klotzsch & Garcke, *Abh. Königl. Akad. Wiss. Berlin* 1859: 88 (1860). Lectotype (Bruyns 2012): South Africa, Cape, *Drège* (HBG; MO, W-3 sheets, iso.).

Tithymalus involucratus Klotzsch & Garcke, *Abh. Königl. Akad. Wiss. Berlin* 1859: 91 (1860). Lectotype (Bruyns 2012): South Africa, *Drège* (HBG; MO, S, iso.).

Euphorbia epicyparissias E.Mey. ex Boiss. in DC., *Prodr.* 15 (2): 168 (1862). Lectotype (Bruyns 2012): South Africa, Transvaal, near Vaal River, *Burke* (K; MPU, iso.).

Euphorbia involucrata E.Mey. ex Boiss. in DC., *Prodr.* 15 (2): 168 (1862). Lectotype (Bruyns 2012): South Africa, near Phillipstown, *Ecklon & Zeyher n. 8* (HBG; MEL, S, SAM, iso.).

Euphorbia bachmannii Pax, *Bot. Jahrb. Syst.* 23: 535 (1897). Type: South Africa, Pondoland, end Oct. 1888, *Bachmann 755* (missing).

Euphorbia involucrata var. *megastegia* Boiss. in DC., *Prodr.* 15 (2): 168 (1862). Lectotype (designated here): South Africa, Cape, near Katberg, *Drège* (TCD).

Euphorbia epicyparissias var. *puberula* N.E.Br., *Fl. Cap.* 5 (2): 267 (1915). Type: South Africa, Cape, Kentani, 1200', 8 Oct. 1910, *Pegler 460* (K, holo.; BOL, SAM, iso.).

Euphorbia epicyparissias var. *wahlbergii* (Boiss.) N.E.Br., *Fl. Cap.* 5 (2): 267 (1915). *Euphorbia wahlbergii* Boiss. in DC., *Prodr.* 15 (2): 169 (1862). Lectotype (Bruyns 2012): South Africa, 1842, *Wahlberg* (S).

Euphorbia crebrifolia S.Carter, *Kew Bull.* 45: 334 (1990). Type: Zimbabwe, Inyanga distr., Inyangani Mountain, near source of Matendere River, 2220 m, 1 Aug. 1988, *Carter & Coates-Palgrave 2077* (K, holo.; EA, SRGH, iso.).

Bisexual spineless glabrous sparingly branched non-succulent herb 0.15–0.30 m tall and same in breadth to shrub reaching 1.5 × 1.3 m, branching from top of and above slightly thicker subterranean perennial rootstock bearing fibrous roots. *Branches* ascending to erect, simple or rebranched, persistent, cylindrical, 0.1–1.0 m × 1–5 mm, without tubercles but leaf-bases projecting slightly after leaf falls off, grey-green to red-green becoming brown near base; *leaves* present throughout branches or becoming absent in lower parts, numerous and often densely crowded, alternate, 4–30 × (0.5) 1–6 (8) mm, often deflexed and spreading or ascending towards tips, persistent for several seasons, slightly glaucous and often with reddish margins, linear to linear-oblong (ovate-lanceolate, rhomboid or deltate) and usually rounded or cordate (often wing-like) at base, glabrous to finely pubescent, margins usually revolute and upper surface convex, often with abruptly obtuse apex (acute) and usually mucronate, ± sessile, estipulate. *Synflorescences* terminal and often head-like, 15–80 mm diam., ± glabrous, consisting of very shortly pedunculate (peduncle 1–2 mm long, slightly pubescent) bisexual cyathium surrounded by 3–8 simple to rebranched terete rays 6–80 mm long, rays subtended by whorl of 3–5 slightly glaucous or yellowish to reddish bracts shorter than and broader than leaves (oblong, ovate or rhomboid,

4.2 Sect. *Esula*

4–10 × 3–12 mm) and each tipped with a bisexual cyathium, secondary cyathia very shortly pedunculate and surrounded by 2–4 sessile broadly rhomboid bracts much shorter and broader than leaves (4–10 × 5–15 mm) with further rays developing in their axils; *cyathia* cupular and slightly pentagonal, glabrous, 3–5 mm broad (± 2 mm long below insertion of glands), with 5 lobes with finely toothed margins, grey-green; *glands* 4 or 5, crescent-shaped to cuneate, 1–3 mm broad, spreading and contiguous to slightly separated, yellow-green, ± flat and slightly pitted above, outer margins produced into two spreading horns at their ends and flat or hollowed out between them (giving crescent-shape), inner margins slightly raised; *stamens* with finely pubescent pedicels, bracteoles few, finely filiform, pubescent; *ovary* globose and strongly 3-angled, glabrous, erect and slightly exserted on pedicel ± 2.5 mm long and soon pendulous on decurved pedicel 3–5 mm long; styles 1–1.5 mm long, branched and spreading to near base (tips again bifid). *Capsule* 3–4 mm diam., obtusely 3-lobed, glabrous, exserted from cyathium on gradually erect pedicel 3–6 mm long.

Distribution & Habitat

Euphorbia natalensis is the most widely distributed member of this group and occurs in Lesotho, Moçambique, South Africa, Swaziland and Zimbabwe. In South Africa it is found from the Soutpansberg to coastal Natal, it is of wide and frequent occurrence in the Drakensberg (often at high altitudes

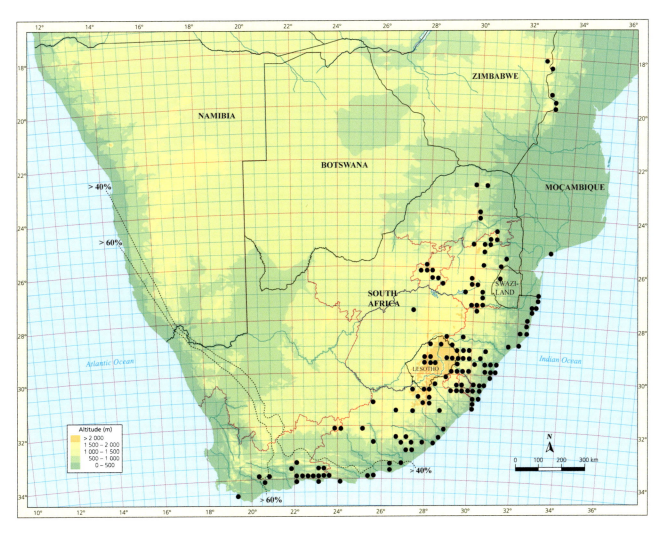

Fig. 4.62. Distribution of *Euphorbia natalensis* (© PVB).

Fig. 4.63. *Euphorbia natalensis*, ± 1 m tall, among dense grasses, foothills of Drakensberg, *PVB 13084*, Garden Castle, west of Underberg, South Africa, 23 Dec. 2015 (© PVB).

Fig. 4.64. *Euphorbia natalensis*, ± 0.8 m tall, *PVB 13084*, Garden Castle, west of Underberg, South Africa, 23 Dec. 2015 (© PVB).

in Lesotho) and southwards to around the higher mountains of the southern Cape to Swellendam (with an isolated occurrence near Gansbaai too). West of the escarpment, a few collections have been made in the former Transvaal around Krugersdorp (west of Johannesburg).

Mainly a plant of moist grasslands in summer-rainfall areas, *E. natalensis* may occasionally also occur on the margins of clumps of bush or among bushes around large boulders. It is often very common, forming diffuse to dense colonies that may be quite isolated from the next such colony, though plants in coastal grasslands (such as around Port Edward and in southern Moçambique) are often quite scattered.

Diagnostic Features & Relationships

Euphorbia natalensis is especially variable in the shape and size of the plant. In some areas (especially in sheltered spots alongside streams at 1000–1500 m along the eastern side of the Drakensberg) they form much-branched shrubs with slightly woody stems that may reach 1.5 m tall. On these shrubs the leaves may also be relatively large (to 25 × 5 mm) but small side-shoots may bear much smaller leaves (which are often only 12 × 1–2 mm). Many of the leaves are densely clustered on the newer parts of the branches, whose lower parts are bare. In these areas smaller plants (20–30 cm tall) are also found, often quite close to larger ones, but usually in more open and exposed, slightly drier, readily burnt grasslands and usually in considerable numbers. Among these smaller plants there is also much variation in the breadth of the leaves (1–5 mm broad) but they are mostly shorter (10–15 mm long). Coastal grasslands may have somewhat different-looking plants, often solitary and consisting of a single unbranched stem, bear-

Fig. 4.65. *Euphorbia natalensis*, shrubs ± 0.3 m tall, *PVB 13102*, 2600 m, between Roma and Semongkong, Lesotho, 29 Dec. 2015 (© PVB).

ing very slender downward-pointing leaves (1–2 mm broad).

While not geophytic like *E. striata*, *E. natalensis* is nevertheless able to die back or be burnt back and resprout from the rootstock, especially in smaller plants in grassland. Strings of slender tubers, such as one finds in *E. striata*, are not produced.

Despite its great vegetative variability, *E. natalensis* is characterized by the crowded groups of leaves with their revolute margins and their somewhat expanded, usually at least slightly (and often markedly) cordate bases.

Plants from the Drakensberg in Kwazulu-Natal with very narrow leaves have often been placed under *Euphorbia ericoides*. The type of *E. ericoides* is a plant without cordate leaf-bases from the 'Cape of Good Hope' (collected around 1770) and belongs to *E. genistoides*. Once it has been observed how variable the breadth of the leaves is within populations from the Drakensberg and how, sometimes, plants with generally broader leaves produce short branches with very narrow leaves, it becomes clear that these narrow-leaved plants in Kwazulu-Natal are variants of the typical *E. natalensis*. Since they also share the cordate bases of the leaves with *E. natalensis*, they cannot belong to *E. ericoides* (= *E. genistoides*).

The fairly recently described *E. crebrifolia* has also now been placed into synonymy here. This has all the typical features of *E. natalensis*, especially the densely crowded, downwardly inclined leaves with rounded bases. Both

Fig. 4.66. *Euphorbia natalensis*, single-stemmed herb ± 0.3 m tall in short coastal grassland, Umtamvuna Reserve, South Africa, 18 Dec. 2015 (© PVB).

Fig. 4.67. *Euphorbia natalensis*, densely clustered leaves at tips of branches and lightly yellow-tinted bracts, *PVB 13084*, Garden Castle, west of Underberg, South Africa, 23 Dec. 2015 (© PVB).

Fig. 4.68. *Euphorbia natalensis*, capsules from immature to ready to dehisce (one already gone), *PVB 13084*, Garden Castle, west of Underberg, South Africa, 23 Dec. 2015 (© PVB).

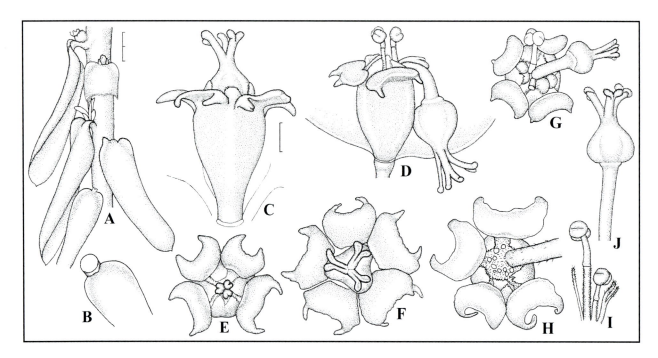

Fig. 4.69. *Euphorbia natalensis*. **A**, portion of branch with bases of leaves (scale 2 mm, as for **B**). **B**, base of leaf from above. **C, D**, cyathium from side at different stages (scale 1 mm, as for **D–H**). **E–H**, cyathium at different stages from above. **I**, anthers and bracteoles. **J**, female floret. Drawn from: **A–D, F, G, I, J**, *PVB 13084*, Garden Castle, west of Underberg, South Africa. **E**, *K.W. Grieve 1713*, Oribi, near Port Edward, South Africa. **H**, *K.W. Grieve 1858*, Ngele, near Kokstad, South Africa (© PVB).

cyathia and capsules are identical with those of *E. natalensis*.

History

Euphorbia natalensis was described by Johann J. Bernhardi from a collection of C. Ferdinand F. Krauss that was made in 1839. However, it had been discovered earlier than this, since it was collected several times between September 1813 and September 1814 by Burchell. There is also an even earlier collection 'Prom bon spei Thunberg' (at BM), which was named '*Euphorbia genistoides*' and '*Euphorbia erubescens* E. Meyer'. This was gathered by Thunberg probably in 1773, but without a precise locality being given.

Euphorbia sclerophylla Boiss., *Cent. Euphorb.*: 37 (1860). Lectotype (Bruyns 2012): South Africa, Cape, ad Grahamstown, Jul. 1829, *Ecklon & Zeyher n° 11* (G; LE, MO (only piece on right hand side), SAM, W, iso.).

Tithymalus multicaulis Klotzsch & Garcke, *Abh. Königl. Akad. Wiss. Berlin* 1859: 98 (1860). Lectotype (Bruyns 2019): South Africa, Cape of Good Hope, 1830, *Krebs 296* (G-DC; MO, iso.).

4.2 Sect. *Esula*

Euphorbia ovata (E.Mey. ex Klotzsch & Garcke) Boiss. in DC., *Prodr*. 15 (2): 167 (1862). *Tithymalus ovatus* E.Mey. ex Klotzsch & Garcke, *Abh. Königl. Akad. Wiss. Berlin* 1859: 97 (1860). Lectotype (Bruyns 2012): South Africa, Cape of Good Hope, *Drège* (LD; BM, MO, NY, REG, TUB, iso.).

Euphorbia sclerophylla var. *myrtifolia* E.Mey. ex Boiss. in DC., *Prodr*. 15 (2): 169 (1862). Type: South Africa, Albany Div., near Assegaaibosch, *Drège 3563* (P, holo.; HBG, K-2 sheets, iso.).

Euphorbia striata var. *brachyphylla* Boiss. in DC., *Prodr*. 15 (2): 170 (1862). Lectotype (Bruyns 2012): South Africa, Sterkstroom div., plains ontop of Katberg, *Drège* (K).

Euphorbia sclerophylla var. *puberula* N.E.Br., *Fl. Cap*. 5 (2): 260 (1915). Type: South Africa, Bathurst div., Rietfontein, between Kariega River and Port Alfred, 3 Oct. 1813, *Burchell 3961* (K, holo.).

Euphorbia muraltioides N.E.Br., *Fl. Cap*. 5 (2): 264 (1915). Lectotype (Bruyns 2012): South Africa, Albany div., Brookhuisens Valley, *MacOwan 642* (K; GRA, iso.).

Euphorbia ruscifolia (Boiss.) N.E.Br., *Fl. Cap*. 5 (2): 259 (1915). *Euphorbia sclerophylla* var. *ruscifolia* Boiss. in DC, *Prodr*. 15 (2): 169 (1862). Type: South Africa, between Kei and Gekau, *Drège 4621* (P 00576460, holo.)

Euphorbia albanica N.E.Br., *Fl. Cap*. 5 (2): 258 (1915). Type: South Africa, Albany div., Brookhuisens Poort, near Grahamstown, October, *MacOwan 657* (GRA, holo.; K, iso.)

Bisexual spineless mostly glabrous sparingly branched non-succulent geophytic herb 0.1–0.3 m tall branching annually from top of slightly thicker subterranean perennial rootstock with string of several small swollen tubers (10–30 × 3–15 mm) bearing fibrous roots. *Branches* erect, simple or sparingly branched, cylindrical, 0.1–0.3 m × 1–2 mm, without tubercles (often slightly roughened from old leaf-bases), grey-green or glaucous often becoming red near base; *leaves* present throughout or becoming absent in lower parts, alternate, 5–16 × 2–6 (10) mm, ascending often close to stem, deciduous, glaucous, lanceolate to ovate-lanceolate (ovate), puberulous to glabrous, flat to slightly channelled above, sharply acute and sometimes sharp-tipped, narrowing or rounded (cordate) towards base, sessile to very shortly petiolate, estipulate. *Synflorescences* terminal, glabrous, consisting of pedunculate (peduncle 1–2 mm long) usually bisexual cyathium surrounded by 3–5 simple to rebranched terete rays 12–50 mm long, rays subtended by whorl of 3–5 glaucous broadly ovate to transversely elliptic-ovate (cordate) acutely (almost sharp-) tipped bracts much shorter and broader than leaves 3–10 × 4–14 mm, each ray tipped with a ± sessile bisexual cyathium also surrounded by 2–4 sessile broadly ovate acutely-tipped bracts with further rays developing in their axils; *cyathia* cupular and slightly angled, glabrous, 2.5–4 mm broad (± 1.5 mm long below insertion of glands), with 5 lobes with finely toothed margins, grey-green; *glands* 4–5, ± crescent-shaped, 1.2–1.5 mm broad, spreading and ± contiguous, yellow-green, ± flat and slightly pitted above, outer margins produced into two spreading horns at their ends and hollowed out between them (giving crescent-

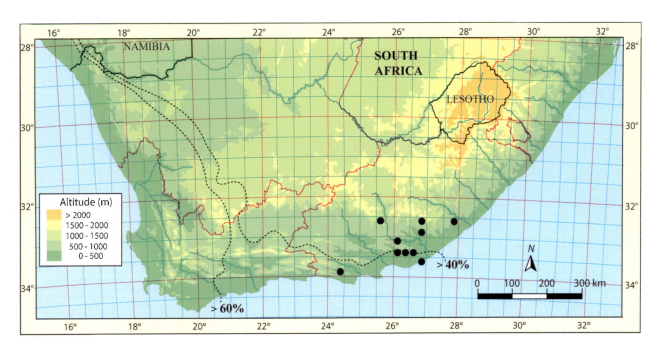

Fig. 4.70. Distribution of *Euphorbia sclerophylla* (© PVB).

Fig. 4.71. *Euphorbia sclerophylla*, cyathia among small sharp-tipped bracts, *PVB 11069*, Mt Marlow, SW of Cradock, South Africa, 23 Oct. 2018 (© PVB).

Fig. 4.72. *Euphorbia sclerophylla*, *PVB 11069*, Mt Marlow, SW of Cradock, South Africa, 23 Oct. 2018 (© PVB).

shape); *stamens* with finely pubescent pedicels, bracteoles few, finely filiform, sparsely pubescent; *ovary* globose and strongly 3-angled, glabrous, initially erect and soon exserted and pendulous on decurved pedicel 1.5–2 mm long; styles 0.6–1.2 mm long, branched to near base and spreading just above ovary. *Capsule* 3–4.5 mm diam., obtusely 3-lobed, glabrous, exserted from cyathium on gradually erect pedicel 1.5–2.5 mm long.

Distribution & Habitat

Euphorbia sclerophylla is mainly found in the Eastern Cape, where there are few, widely scattered records from near Maclear to Cradock and southards to Humansdorp, but with a concentration of collections in the sandstone hills around Grahamstown.

Euphorbia sclerophylla inhabits heavy loamy soils in short grasslands, sometimes associated with sandstones (as around Grahamstown) and sometimes with scattered dolerites overlaying shales (as for example near Cradock). Plants may become common, especially in shallow soils towards the edges of hilltops.

Diagnostic Features & Relationships

Euphorbia sclerophylla is a typical member of this group, with slender stems arising from a series of slender, irregularly shaped, brown tubers that may descend as far as 0.5 m into the ground and are connected by easily broken, thread-like brown stems. Above the ground the branches can be short (6–10 cm long) to much longer depending on the degree to which they are protected. They frequently arise in diffuse groups spread over a few square metres, perhaps where the activities of moles have broken up the underground rootstocks. Although the leaves may be fairly delicate, they often dry out to be quite rigid and even sharp-tipped and this is reflected in such names as *E. muraltioides* and *E. ruscifolia*. Mostly, the leaves taper roughly equally to the base and tip, with the tip usually sharply acute. Occasional collections have leaves that become broader towards their bases and unusually broad leaves are found on material described as *E. albanica* and *E. ovata* but these appear to be rare forms of the more typical slender-leaved plants.

Plants grow during the summer months, flowering at the tips of young growth in October and November and shedding seeds before mid-December, before the grass of their habitats becomes dense and too tall. After flowering the branches remain for the rest of the summer and some of them die back during winter. The cyathia are small and insignificant, differing only slightly in colour from the leaves by being yellow-green rather than greyish. They are produced in clusters of very variable breadth but usually there is at least one series of additional cyathia arising in the axils of bracts around the first one.

Euphorbia sclerophylla is very similar to *E. genistoides* in vegetative appearance, though the plants are often more slender and the leaves are less densely clustered along the stem. The leaves and bracts beneath the cyathia differ by their more slender and sharper tips which dry out to a somewhat rigid point. Flowering in *E. sclerophylla* takes place at the

4.2 Sect. *Esula*

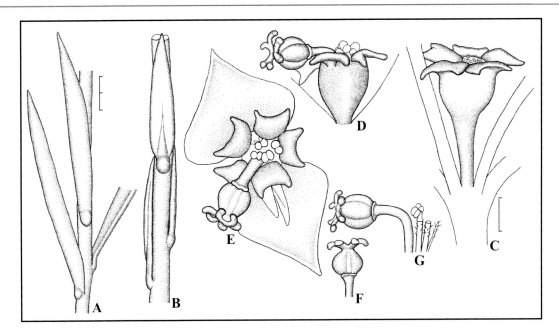

Fig. 4.73. *Euphorbia sclerophylla*. **A**, **B**, portion of branch with leaves (scale 2 mm, as for **B**). **C**, central male cyathium from side (scale 1 mm, as for **D–G**). **D**, bisexual cyathium from side. **E**, bisexual cyathium from above with bracts. **F**, female floret. **G**, female floret later with anthers and bracteole. Drawn from: *PVB 11069*, Mt Marlow, SW of Cradock, South Africa (© PVB).

beginning of the growing period, in *E. genistoides* at the end of it, but the cyathia are very similar, differing mainly in the shorter styles of *E. sclerophylla* (almost sessile ontop of the ovary) and the slightly angled involucre on a particularly short peduncle.

History

First collected by Burchell in August 1813 and soon afterwards by J.F. Drège around 1829, *E. sclerophylla* has never been particularly well-known and the position is much complicated by the description of the rather similar *E. albanica*, *E. myrtifolia* and *E. ruscifolia*, with *E. albanica* and *E. myrtifolia* even described from the plants collected around the same spot (Brookhuizen's Poort) south-west of Grahamstown.

Some specimens, such as *Bowker* (K) are very variable in the breadth of the leaves on the different branches gathered, some even looking rather like *E. striata*. However, the leaves are mostly denser on the branches than in *E. striata* (though in those with the broadest leaves they are the least densely spaced). The type of *E. albanica* has particularly broad and especially widely spaced leaves, with short broad bracts and long rays. *Drège 3561* (from the Katberg) has short stems, with rounded, quite broad leaves widely spaced along them. This was particularly distinctive and was called *E. ovata* by E. Meyer (and described as *Tithymalus ovatus* by Klotzsch & Garcke). Similarly, *Burchell 3961* also has these unusually broad leaves, but all these were considered by N.E. Brown to belong to *E. sclerophylla*. Some more recently gathered specimens, such as *Bayliss 8953* (GRA) with broader, sometimes round-tipped leaves also have an especially firm texture, but there are no other features to distinguish them and consequently they are also included here.

Euphorbia striata Thunb., *Prodr. Fl. Cap.* 2: 86 (1800). *Tithymalus striatus* (Thunb.) Klotzsch & Garcke, *Abh. Königl. Akad. Wiss. Berlin* 1859: 98 (1860). Type: South Africa, without locality, *Thunberg* (UPS-THUNB 11560, holo.).

Tithymalus capensis Klotzsch & Garcke, *Abh. Königl. Akad. Wiss. Berlin* 1859: 98 (1860). Lectotype (Bruyns 2019): South Africa, Cape of Good Hope, by the Klipplaat River, near Shiloh, Queenstown div., 13 Nov. 1832, *Drège 3562* (K). *Euphorbia striata* var. *cuspidata* Boiss. in DC., *Prodr.* 15 (2): 170 (1862). *Euphorbia cuspidata* Bernh., *Flora* 28: 86 (1845), *nom. illegit. non* Bertol. (1843). Type: South Africa, Natal, summit of Tafelberg, 2000-3000', Sept. 1839, *Krauss 441* (MO, holo.; BM, BOL, K, M, TCD, TUB, iso.).

Bisexual spineless glabrous sparingly branched non-succulent herb 0.15–0.5 m tall branching annually from top of slightly thicker subterranean perennial rootstock (often made up of strings of slender tubers) bearing fibrous roots. *Branches* erect, simple or branched into panicle at synflorescence, usually deciduous, cylindrical, 0.1–0.3 m × 1–2 mm, without tubercles, grey-green or glaucous often becoming red near base; *leaves* usually few and widely separated along branches (absent in lower parts), alternate, (8) 12–50 (75) × 1–6 (8) mm, erect or ascending, deciduous, glaucous, narrowly linear-ovate and tapering towards base, glabrous, flat to slightly channelled above, tapering gradually to sharply acute tip, sessile, estipulate. *Synflorescences* terminal, glabrous, consisting of very shortly pedunculate (peduncle 1–2 mm long, slightly pubescent) bisexual cyathium surrounded by 3–5 simple to rebranched terete very slender rays 25–100 mm long, rays subtended by whorl of 3–5 glaucous semi-circular to rhomboid-ovate bracts 4–10 × 6–14 mm and each tipped with a bisexual cyathium also surrounded by 2–4 sessile semi-circular to rhomboid-ovate bracts with further rays developing in their axils; *cyathia* cupular, glabrous, 3–5 mm broad (± 2 mm long below insertion of glands), with 5 small lobes with finely toothed margins, grey-green; *glands* 4, ± crescent-shaped to nearly rectangular, 1.2–2.5 mm broad, spreading and contiguous, green to reddish above, ± flat and slightly pitted above, outer margins sometimes produced into two small spreading horns at their ends and hollowed out between them (giving slightly crescent-shape), inner margins flat; *stamens* with finely pubescent pedicels, bracteoles few, finely filiform, pubescent; *ovary* globose and

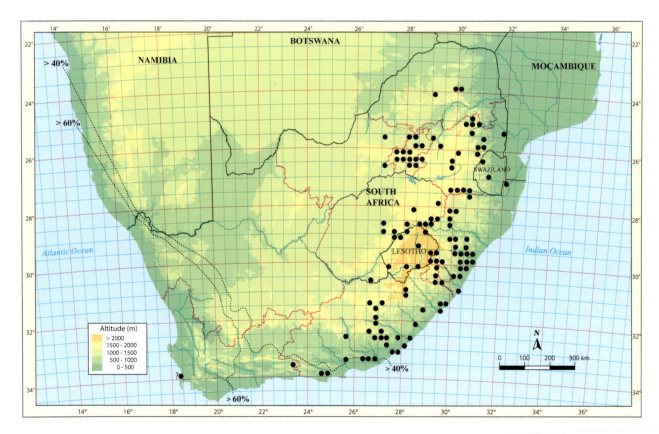

Fig. 4.74. Distribution of *Euphorbia striata* (© PVB).

4.2 Sect. *Esula*

Fig. 4.75. *Euphorbia striata*, plant ± 0.4 m tall consisting of single slender stem with conspicuous bracts around tiny cyathia, *PVB 13072*, Barkly Pass, southern Drakensberg, South Africa, 17 Dec. 2015 (© PVB).

Fig. 4.76. *Euphorbia striata*, plant ± 0.3 m tall with three slender stems barely visible among the grass, in rocky patch in grassland, Garden Castle, near Underberg, South Africa, 22 Dec. 2015 (© PVB).

strongly 3-angled, glabrous, soon exserted and pendulous on decurved and later erect pedicel 2–3 mm long; styles 0.8–1.2 mm long, branched and spreading to near base (tips again bifid). *Capsule* 3.5–4.5 mm diam., obtusely 3-lobed, glabrous, exserted from cyathium on gradually erect pedicel 2–3 mm long.

Distribution & Habitat

Euphorbia striata is widely distributed on the eastern side of Southern Africa. It is recorded from Haenertsburg in the former eastern Transvaal, from near Mbabane in Swaziland through Natal to the Eastern Cape, as far as Humansdorp and the Uniondale district. It is also common in Lesotho (Hargreaves 1992c), though few specimens have been made from there, and records extend into the Free State and to the Witwatersrand, west of Johannesburg.

Euphorbia striata is almost exclusively a plant of grasslands in both flat and sloping areas and sometimes grows in slightly bare patches, with rocks projecting from the ground. It often occurs on firm to hard, relatively deep and fertile loam not far from the banks of rivers, but in exposed spots which become periodically dry and are regularly burnt. Plants are usually common but are often very inconspicuous.

Fig. 4.77. *Euphorbia striata*, slender stems arising from small tubers connected by delicate rhizomes, Kokstad, South Africa, 21 Dec. 2015 (© PVB).

Diagnostic Features & Relationships

The plant in *E. striata* arises from a series of small tubers held together in slender strings beneath the ground to a depth of 30 cm or more and the whole plant retreats to these tubers in the dry season so that it is a geophyte. Each tuberlet appears to be able to grow independently and this makes the

Fig. 4.78. *Euphorbia striata*, top of plant with bracts and cyathia, South Africa, ± 1986 (© S.P. Fourie).

underground parts of the plant able to withstand a lot of disturbance from animals and even ploughing (Hargreaves 1992c). Above the ground, each plant consists of a few, slender, erect, deciduous branches that arise directly from just above the rootstock and regularly reach 30 cm tall without dividing. The distinctly greyish leaves are quite widely spaced along the branches (never densely clustered) and are usually held in an ascending attitude. Mostly without any petiole, they are narrowly linear-ovate, tapering fairly abruptly to their bases and tapering gradually above into a fine tip.

In *E. striata* the bracts under the cyathia are especially broad and short, often almost semi-circular and in pairs forming an almost circular 'plate'. Although they mostly share the grey-green colour of the rest of the plant, their margins may be tinged with red. There is usually some branching from the axils just above these bracts (to form rays bearing further cyathia), where otherwise the main branches are simple below them. The cyathia are inconspicuous and

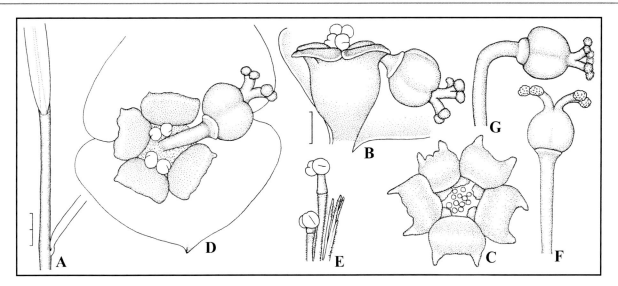

Fig. 4.79. *Euphorbia striata*. **A**, portion of branch with bases of leaves (scale 2 mm). **B**, bisexual cyathium from side (scale 1 mm, as for **C–G**). **C**, central male cyathium from above. **D**, bisexual cyathium from above. **E**, anthers and bracteoles. **F, G**, female floret. Drawn from: **A, B, D, E, G**, *PVB*, near Underberg, South Africa. **C, F**, *K.W. Grieve 2141*, Umtamvuna, Port Edward, South Africa (© PVB).

barely noticeable inside these bracts, with the capsules also inconspicuous.

Euphorbia striata is particularly distinctive for the sparingly branched plant with the leaves widely spaced out along the branches, in the shape of the leaves (tapering both to their bases and tips) and in the broad bracts under the cyathia that are much broader than the leaves.

History

Euphorbia striata was first collected by C.P. Thunberg. No locality was given for this collection either in Thunberg (1800) or on the sheet in his herbarium. It is difficult to reconstruct where this first collection may have been made since *E. striata* is known, for example, in the Long Kloof, where some members of Thunberg's expedition passed late in 1772 and again late in 1773.

Note: Several members of subg. *Esula* are widespread weeds in the northern hemisphere. Some of these have been introduced to southern Africa and have become weeds here too. Three of these weedy species are regularly encountered in southern Africa, namely *E. helioscopia* L. (of sect. *Helioscopia*), *E. peplus* L. (of sect. *Tithymalus*) and *E. terracina* L. (of sect. *Pachycladae*). No other members of these sections are present naturally in southern Africa. These European species are now common, especially in the southwestern parts of South Africa with a Mediterranean climate (wet winters and dry summers). Here they are found mainly in disturbed areas around the cities. *Euphorbia helioscopia* and *E. peplus* are annuals, while *E. terracina* is a small, diffuse shrub that may live for several years.

Fig. 4.80. *Euphorbia helioscopia*, a usually single-stemmed annual 120–200 mm tall with dense head of bracts, Rondebosch, South Africa (© PVB).

Fig. 4.81. *Euphorbia helioscopia*, bracts with denticulate margins and capsules smooth, Rondebosch, South Africa (© PVB).

Fig. 4.82. *Euphorbia peplus*, similar to though usually more delicate than *E. helioscopia*, bracts with entire margins and capsules ridged, Claremont, South Africa (© PVB).

Fig. 4.83. *Euphorbia terracina*, shrub around 0.5 m tall alongside the road, Camps Bay, Cape Town, South Africa (© PVB).

Fig. 4.84. *Euphorbia terracina*, flowering parts, Rondebosch, South Africa (© PVB).

Euphorbia subg. Euphorbia

Euphorbia subg. **Euphorbia**
Arthrothamnus Klotzsch & Garcke (see under sect. *Tirucalli*).
Monadenium Pax (see under sect. *Monadenium*).
Tirucalia Raf. (see under sect. *Tirucalli*).

Bisexual sometimes unisexual, sometimes large spineless shrubs with slender cylindrical branches or often dwarf to large spiny succulent shrubs or trees with slender to stout angled (cylindrical) branches often around a stouter stem or trunk with more angles than branches or becoming cylindrical. *Leaves* alternate to opposite, ovate-lanceolate and fleshy to fleeting rudiments, usually arising on raised tubercles, stipules rarely glandular usually in form of small prickles, rarely absent. *Synflorescences* many near tips of branches in axils of tubercles or leaves on short peduncles, each a solitary (usually male) cyathium surrounded by opposite (whorled) scale-like bracts in whose axils further short peduncles usually arise each terminated by a further (usually bisexual) cyathium (rarely further rays developing around these cyathia – *E. knuthii*), bracts differently shaped to leaves. *Cyathia* bisexual or unisexual (in unisexual plants male cyathia larger than females); glands 1 and cupular with entire margins or 4–8 (12) and elliptic and flat with entire margins, without petaloid appendages; male florets with glabrous pedicels. *Capsule* deeply 3-lobed to obtusely 3- to 6-lobed and ± spherical, smooth (papillate), glabrous to pubescent, 3–30 mm diam., ± sessile to exserted, dehiscing explosively or occasionally drying up and falling apart on ground. *Seeds* 3–6 per capsule, ellipsoidal to spherical, smooth to slightly rugulose, rarely carunculate.

Of the 21 sections recognized in subg. *Euphorbia* by Dorsey et al. (2013), three are represented in southern Africa, namely sect. *Euphorbia* (with most of the species), sect. *Monadenium* (one species) and sect. *Tirucalli* (four species).

Fig. 5.1. Relationships between southern African members of subg. *Euphorbia*. The sections represented in southern Africa are red (those not represented are green). The length of the vertical side of each of the triangles is proportional to the number of species in that section (adapted from Dorsey et al. 2013).

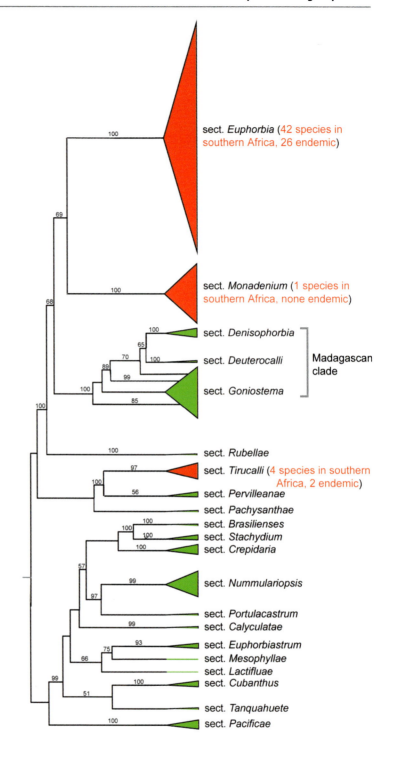

Table 5.1. The sections of *Euphorbia* subg. *Euphorbia* occurring naturally in southern Africa (*Euphorbia tirucalli* is not included for southern Africa).

Section	Namibia (endemic)	South Africa (endemic)	Southern Africa (endemic)	Elsewhere (total)	Annuals	Perennial, non-succulent herbs	Woody shrub to tree	Succulents	Geophytes
Euphorbia	9 (3)	35 (20)	42 (26)	± 300 (± 340)	0	0	0	42	0
Monadenium	0 (0)	1 (0)	1 (0)	± 90 (± 90)	0	0	0	1	0
Tirucalli	4 (0)	2 (0)	4 (2)	22 (24)	0	0	0	4	0

Key to the sections of subg. *Euphorbia* in southern Africa:

1. Branches spineless, terete to slightly ridged beneath each leaf, new growth tomentose towards apices, stipules often present as small, rounded dark glands..sect. **Tirucalli**
1. Branches usually with spines around bases of leaves (though sometimes absent on upper branches of large trees or obsolescent), terete to distinctly winged (angled), new growth glabrous, stipules spine-like and (1) 2 spines usually inserted just below each leaf..2
2. Cyathial gland continuous in rim around cyathium except for gap on one side, leaves alternate to spirally arranged, not surrounded by horny 'spine-shield', seeds with caruncle..sect. **Monadenium**
2. Cyathial glands usually 5 (occasionally more) and always distinct, leaves decussately arranged, surrounded by horny 'spine-shield' on which spines are inserted just below leaves, seeds without caruncle..................................sect. **Euphorbia**

5.1 Sect. Euphorbia

Euphorbia sect. **Euphorbia**

Euphorbia sect. *Aculeatae* Haw., *Philos. Mag. Ann. Chem.* 1: 275 (1827). Type: *Euphorbia caerulescens* Haw.

Bisexual densely branched rarely dwarf succulent shrub, large shrub to tree becoming woody and covered with grey bark on stem or trunk towards base in larger plants. *Branches* grey-green to bright green, glabrous, with tubercles in 4–6 rows or joined into 4–6 angles (sometimes more rows on stem which loses its angles and green bark with age). *Leaves* minute and caducous (sometimes to 80 mm long on stem of young trees), opposite, sessile, surrounded by grey to brown hardened spine-shield often fused into continuous hardened margin along angles with (1) 2 (3) spines below leaf and two small scale-like stipules or stipular prickles at bases of leaf-margins. *Synflorescences* in short axillary cymes near tips of branches on peduncles 1–30 mm long, 1–5 per axil, glabrous, each cyme with (1) 3 (5) cyathia 3–12 mm diam. subtended by minute scale-like bracts, central cyathium usually functionally male, outer cyathia bisexual, glands 5 (–12), entire and ± ellipsoidal, without appendages. *Capsule* obtusely to deeply 3-lobed or ± spherical, 3–24 mm diam., well exserted to sessile, glabrous, usually dehiscing explosively (except in *E. ingens* and *E. virosa*). *Seeds* 3–6 per capsule, smooth to tuberculate, 2–12 mm long, ± spherical to ovoid, grey sometimes streaked with brown, without caruncle.

The 344 species of this section (Dorsey et al. 2013) are found in Africa, the Arabian Peninsula and in south-eastern Asia from Pakistan to Indo-China. The greatest diversity is encountered in the drier, though not desertic, tropical parts of East and north-east Africa. In southern Africa 42 species occur. Of these, 26 are endemic, several are closely related to Angolan species (whose relationships are mainly among southern African species) while others, such as *E. ingens*, are related to similar species from East and north-east Africa (Bruyns et al. 2011).

Boissier called this sect. *Diacanthium*. Pax (1904) split sect. *Diacanthium* informally into five 'groups' *Monacanthae*, *Diacanthae*, *Triacanthae*, *Tetracanthae* and *Intermediae*, depending on how many spines arose on each spine-shield. Leach often referred to 'the very natural' group 'sect. *Tetracanthae* Pax' (e.g. Leach 1976a, c) but this was not used in Carter and Leach (2001). Unpublished results from molecular data show that 1- and 3-spined plants arose several times among 2- and 4-spined plants (with the single dorsal spine known in several cases to be formed by progressive fusion of two dorsal spines, Fig. 1.32 in volume 1). These results also show that there is a natural grouping that corresponds closely to Pax's '*Tetracanthae*'. In southern Africa this contains *E. aeruginosa, E. clivicola, E. kaokoensis, E. louwii, E. lydenburgensis, E. otjipembana, E. pisima, E. schinzii, E. steelpoortensis, E. subsalsa* and *E. venteri*. In all of these the stipular prickles (which provide the two extra spines to make each spine-shield 4-spined) are small relative to the dorsal spines. Here the plants mainly form small shrubs in which the stem is morphologically indistinguishable from the branches, the branches are unsegmented, the spine-shields hardly ever form a continuous margin along the angles (only occasionally doing so in *E. lydenburgensis*), the cyathia are always transversely (horizontally) disposed in the cymes (which are solitary in each axil) and the seeds are verrucose. North of southern Africa (where there are far more species of this type), there are many, such as *E. isacantha* (Tanzania) and *E. scitula* (Angola), where the dorsal spines and stipular prickles are more or less equal in size. These plants more obviously justify the name '*Tetracanthae*' than those in southern Africa.

The remainder of sect. *Euphorbia* is made up of two large, well-supported groups (mostly two-thorned or '*Diacanthae*'; namely clades C1 and part of C2 of Bruyns et al. 2011) and several groups with very few species (e.g. those containing *E. canariensis* and *E. gymnocalycioides*), which may be remnants of lineages that were once richer in species (Bruyns et al. 2011). The species of sect. *Euphorbia* in southern Africa apart from those in '*Tetracanthae*' belong to these two large groups and one of them also contains the species from SE Asia and the type of *Euphorbia*, namely *E. antiquorum*. Most of these are large shrubs or trees, but some are small and there are also a few geophytes among them. Most of them possess a trunk-like stem (or tuber in a few

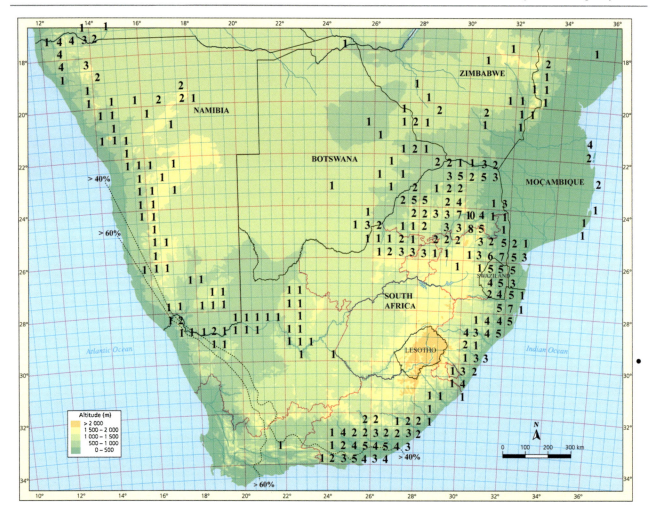

Fig. 5.2. Distribution of *Euphorbia* sect. *Euphorbia* in southern Africa, showing number of species per half-degree square (© PVB).

cases) which is thicker than and has more angles than the branches, the branches are often distinctly segmented, with the spine-shields often fusing to form a continuous, hard margin along the angles, the pair of spines is accompanied by stipular structures that are more rarely spine-like and more usually form small and irregular excrescences, the cyathia are borne transversely or vertically in the cymes (which may be several per axil) and the seeds are smooth. *Euphorbia griseola*, *E. knuthii* (both included by Pax in 'Tetracanthae') and *E. watербergensis*, which are smaller in stature and lack the obviously different central stem, also belong here. Since these members of sect. *Euphorbia* in southern Africa belong to two distinct lineages, they do not form a natural group and so no subsections are recognized here.

Key to the species of sect. *Euphorbia* in southern Africa:

1. Geophyte-like dwarf succulents, shrubs to trees with stem morphologically different from the branches (subterranean in geophyte-like forms as *E. stellata* and in *E. knuthii*, to trunk-like in shrubs and in trees), branches often stout, often segmented and often > 25 mm thick, cyathia in cymes vertically or transversely disposed..................2.
1. Shrubs or dwarf succulents with many branches but without morphologically different (much thicker) stem, branches slender, mostly < 25 mm thick (to 45 mm thick in *E. otjipembana*), cyathia in cymes transversely disposed........................30.
2. Low, geophyte-like dwarf succulents with thick central stem to dwarf or small shrubs with tuber, branches often in rosette or shrub-like clump from apex of large subterranean tuber-like stem (branches not prominently segmented).................3.
2. Shrubs (often large) usually with prominent trunk and stout branches usually much greater than 25 mm thick or trees (with slender to thick often segmented branches)..8.

3. Capsules exserted from cyathium, cymes usually consisting of 3 transversely ('horizontally') disposed cyathia, peduncle of cyathium ± 2 mm thick..4.
3. Capsules sessile and included in cyathium, cymes usually consisting of 3 vertically disposed cyathia (rarely transversely disposed), peduncle of cyathium 3–6 mm thick..5.
4. Plant with many slender erect to trailing branches, often somewhat rhizomatous and spreading from apex of tuber, not forming rosette...**E. knuthii**
4. Plant with few to many branches spreading on surface of ground to ascending, radiating from and forming rosette around apex of subterranean tuber-like stem..**E. stellata**
5. Spine-shield extending for less than 1 mm above paired spines (axillary bud less than 1 mm from adjacent leaf-rudiment)..6.
5. Spine-shield extending for at least 3 mm above paired spines (axillary bud 3–5 mm from adjacent leaf-rudiment)......7.
6. Cyathia in groups of three per cyme, peduncles 5–10 mm long, ovary almost cylindrical and tapering gradually into styles, styles 4.5–6 mm long, cyathium bright yellow...**E. enormis**
6. Cyathia solitary per cyme, peduncles 1.5–4 mm long, ovary distinctly swollen below styles; styles 2–4 mm long, cyathium mostly deep red..**E. vandermerwei**
7. Branches twisted spirally along their length, tubercles deeply indented along angles so that angles not continuous..**E. tortirama**
7. Branches not twisted spirally along their length, tubercles shallowly indented along angles so that angles continuous..**E. clavigera**
8. Plants forming succulent trees (occasional specimens remaining shrubby while most in population are trees) with conspicuous trunk and usually with all branches well above ground...9.
8. Plants forming robust (sometimes very large) succulent shrubs usually with some branches remaining close to ground..19.
9. Stem not branching for first 1.5–2 m or more..10.
9. Stem not branching for first 0.1–0.4 m only..11.
10. Young stem bearing prominent leaves 20–35 mm long at apex, branches persistent and not shed from trunk with age, cymes on peduncles 2–8 mm long, ovary sessile and 'calyx' extended into several filiform lobules around ovary, capsule globose, fleshy and not dehiscing explosively..**E. ingens**
10. Young stem bearing leaf-rudiments to 15 mm long at apex, branches not persistent and shed from trunk with age, cymes on peduncles 10–30 mm long, ovary raised on pedicel 1 mm long and 'calyx' not extended around ovary, capsule obtusely 3-angled and dehiscing explosively..**E. eduardoi**
11. Ultimate branches usually 65 mm or more thick (measured across wings), deeply and prominently segmented with segments broadest below middle, ovary on short pedicel < 0.5 mm long..**E. cooperi**
11. Ultimate branches usually 20–60 mm thick (measured across wings), shallowly segmented with segments broadest around middle or not noticeably segmented, ovary raised on pedicel at least 1 mm long..12.
12. Cymes usually solitary (rarely more) in each axil..13.
12. Cymes usually 3 (rarely fewer) in each axil..17.
13. Ultimate branches reaching maximum of 25 (30) mm thick, cymes with 3 transversely disposed cyathia....................14.
13. Ultimate branches reaching maximum of 50 (60) mm thick, cymes with 5–9 cyathia or with 3 vertically disposed cyathia..16.
14. Spine-shields joined into continuous margin along angles on ultimate branches, leaf-rudiments ± 0.5 × 0.25 mm..**E. sekukuniensis**
14. Spine-shields remaining separate along angles on ultimate branches, leaf-rudiments at least 1 mm long and usually more than 1 mm broad..15.
15. Crown of tree broadly bowl-shaped, ultimate branches mostly 4-angled and square in cross-section, leaf-rudiments deltoid, cyathia 5–8 mm broad, capsules obtusely 3-angled..**E. tetragona**
15. Crown of tree narrowly bowl-shaped to V-shaped or irregular, ultimate branches mostly 2- to 3-angled, leaf-rudiments ovate, cyathia 3.5–5 mm broad, capsules deeply 3-angled with somewhat flattened sides........................**E. grandidens**
16. Branches 6- to 9-angled (rarely 5-angled), cymes of 3 vertically disposed cyathia, ovary much thicker than subtending pedicel, styles united in column for at least short distance above base..**E. zoutpansbergensis**
16. Branches mainly 4-angled (rarely 3- or 5-angled), cymes of 5–9 cyathia, with central cyathium bearing two transversely disposed and two more vertically disposed (additional cyathia sometimes in axils of bracts on these laterals), ovary more or less as thick as the subtending pedicel, styles free right to base..**E. excelsa**

17. Leaf-rudiments narrowly lanceolate (material from around Komatipoort), cyathia usually > 6 mm (6–8 mm) diam., capsules strongly 3-angled with somewhat flattened sides, 8–10 mm diam..**E. confinalis**
17. Leaf-rudiments ± ovate, cyathia usually < 6 mm (4–6 mm) diam., capsules globose and very obtusely 3-angled with inflated sides, 5–8 mm diam..18.
18. Branches 3-angled, bright green, leaf-rudiments usually slightly broader than long, anthers yellow, styles deeply divided for about half of length...**E. triangularis**
18. Branches 4- to 5-angled, grey-green, leaf-rudiments usually longer than broad, anthers red, styles only slightly divided at apex..**E. keithii**
19. Cymes solitary (rarely 2) in each axil..20.
19. Cymes usually in groups of 3 or 4 (rarely fewer or more) in each axil..25.
20. Branches 7- to 8-angled, cyathia with 7–12 glands, ovary longer than broad, capsule spherical and berry-like and not dehiscing explosively, usually with more than 3 locules..**E. virosa**
20. Branches 4- to 6-angled, cyathia with 5 glands, ovary ± as long as broad, capsule deeply 3-angled and dehiscing explosively, 3-locular..21.
21. Ovary raised on short pedicel ± 0.75 mm long with toothed calyx at apex of pedicel, styles with short united portion joined to abruptly swollen ± spherical ovary...**E. rowlandii**
21. Ovary sessile or raised to 0.5 mm on short pedicel, without toothed calyx at base, styles with united portion gradually widening towards base into slightly swollen ovary...22.
22. Spine-shield without small prickles alongside axillary bud, plants from northern Namibia...........................23.
22. Spine-shield with several small prickles 1–4 mm long alongside axillary bud, plants from NE South Africa along Olifants River valley...24.
23. Plant with stout, spirally 7- to 9-angled central stem 0.2–1.0 m tall, spine-shields united into continuous hard margin 4–8 mm broad along angles, styles divided to near base just above ovary..**E. otjingandu**
23. Stem not conspicuous and not more than 0.1 m tall, spine-shields united into continuous horny grey to white margin 2–4 mm broad along angles, styles united to well above middle with cylindrical portion above ovary..........................
....**E. otavibergensis**
24. Plant small and densely branched, to 20 × 30 cm, branches 20–35 mm thick and bright green....................**E. restricta**
24. Plant a diffuse shrub to 0.6 × 1.5 (5) m, branches 30–80 mm thick and often yellow-green.......................**E. barnardii**
25. Ovary and capsule ± sessile, spine-shield with small prickles alongside axillary bud....................................26.
25. Ovary and capsule raised on pedicel 1 mm long or more, spine-shield usually without prickles alongside axillary bud.........27.
26. Branches up to 50 mm thick and shallowly constricted into segments..**E. pseudocactus**
26. Branches up to 150 mm thick and very deeply constricted into segments...**E. grandicornis**
27. Spine-shields united into continuous margin and ± uniformly broad (not contracting noticeably below leaf-rudiment to axil below)...**E. avasmontana**
27. Spine-shields separate or irregularly united into ± continuous margin and then not uniformly broad (contracting noticeably between leaf-rudiment and axil below)..28.
28. Branches usually green with yellow-green mottling, styles branched to near base, capsule deeply 3-angled, later exserted on pedicel 6–7 mm long, extra prickles sometimes present alongside axillary bud........................**E. knobelii**
28. Branches uniformly green or grey-green, styles branched at most to middle, capsule obtusely 3-angled, later exserted on pedicel 2–5 mm long, extra prickles absent alongside axillary bud...29.
29. Plant usually with some rhizomatous branches around its perimeter, branches deeply articulated into almost spherical segments, calyx often extended into some teeth around ovary..**E. radyeri**
29. Plant without rhizomatous branches around its perimeter, branches indistinctly articulated into elongated cylindrical segments, calyx not extended around ovary..**E. caerulescens**
30. Leaf-rudiments 4–15 mm long, seeds smooth...31.
30. Leaf-rudiments 0.5–2 mm long, seeds mostly verrucose (smooth in *E. waterbergensis*)...............................32.
31. Peduncles of cymes and cyathia ≥ 2 mm long, capsules deeply and sharply 3-angled, branches 5–12 mm thick, 4-angled
....**E. knuthii**

31. Peduncles of cymes and cyathia ≤ 1 mm long, capsules obtusely 3-angled, branches 8–18 mm thick, 4- to 6-angled......
....**E. griseola**
32. Ovary raised on pedicel 1–1.5 mm long, spine-shields sometimes joined into continuous hard dark grey margin along angles (c.f. *E. pisima*)..33.
32. Ovary sessile or raised on pedicel up to 0.5 mm long, spine-shields always remaining separate along angles...............36.
33. Branches (4) 5- or 6-angled, grey-green, styles with distinct cylindrical fused zone at base, seeds smooth...**E. waterbergensis**
33. Branches 4-angled, bright green or yellow-green, styles branched ± to base, seeds verrucose......................................34.
34. Branches 12–20 mm thick, spine-shields usually forming continuous brownish grey margin along angles, flowering April and May...**E. lydenburgensis**
34. Branches 5–12 (15) mm thick, spine-shields usually remaining separate along angles, flowering August to October..........35.
35. Branches pea-green, plant not rhizomatous..**E. pisima**
35. Branches grey-green, plant usually rhizomatous...**E. steelpoortensis**
36. Most spine-shields with 5 spines (2 longer spines, 2 stipular prickles and 1 additional spine at base of shield), styles very slender and divided right to base...**E. louwii**
36. Spine-shields (almost always) with 4 spines (2 spines and 2 stipular prickles), styles sometimes with thickened tips and sometimes fused into column near base..37.
37. Branches (12) 16–45 mm thick, 4- to 9-angled, cyathial glands often ascending to erect and often concave above, bright yellow to suffused with pink..38.
37. Branches 5–12 (20) mm thick, 4-angled, cyathial glands spreading and slightly concave to slightly convex inside, bright yellow to occasionally suffused with brown..39.
38. Stipular prickles much shorter than spines behind leaf-rudiment (< half their length), branches ± flat to shallowly grooved between angles, 4- to 6-angled, female floret raised on pedicel ± 0.5 mm long......................................**E. otjipembana**
38. Stipular prickles usually nearly as long as spines behind leaf-rudiment, branches with deeply V-shaped grooves between angles, 5- to 8 (9)-angled, female floret ± sessile ....**E. kaokoensis**
39. Branches cylindrical or sub-cylindrical with tubercles hardly raised from surface and angles not clearly visible (except for arrangement of spine-shields into four rows, usually slightly spiralling), young spines and spine-shields distinctly reddish brown (copper-coloured), spine-shield sometimes with additional small spine at base...................**E. aeruginosa**
39. Branches with distinctly raised tubercles decussately arranged into four rows or fused vertically into four angles, young spines purple, black or grey, spine-shield without additional spine at base..40.
40. Tubercles not joined vertically into angles but arranged into 4 rows, opposite pairs forming broad shoulders above and tapering to narrower waist below, spine-shields pale grey and shortly deltate, scarcely decurrent below spines and usually acute at base, stipular prickles ± 0.25 mm long..**E. clivicola**
40. Tubercles joined vertically into four distinct angles along branches, spine-shields dark, prominently decurrent below spines and usually rounded at base, stipular prickles 0.25–3 mm long..41.
41. Bracts on lateral cyathia ± 1 mm long and mostly not reaching to bases of cyathial glands, cyathial glands slightly concave above, plants from north-western Namibia..**E. subsalsa** ssp. **otzenii**
41. Bracts on lateral cyathia ± 2 mm long and mostly nearly equalling cyathial glands, cyathial glands flat to slightly convex above, plants from north-eastern side of southern Africa..42.
42. Tubercles projecting and joined into distinct angles along branches with shallow grooves between these angles, cyathial glands bright yellow..**E. schinzii**
42. Tubercles standing out slightly from branches and angles very indistinct, cyathial glands dark yellowish green to brownish green sometimes tinged with red...**E. venteri**

Euphorbia aeruginosa Schweick., *Bull. Misc. Inform.* 1935: 205 (1935). Lectotype (Bruyns 2012): South Africa, Transvaal, Soutpan, Soutpansberg, 12 Apr. 1934, *Schweickerdt & Verdoorn 688* (K; PRE, iso.).

Bisexual spiny glabrous dense succulent shrub 0.05–0.3 (0.4) × 0.15–1 m, with many branches from similar stem with small cluster of fibrous roots, densely branched at and slightly below ground level and frequently rebranching.

Branches erect to spreading, 20–300 × 5–9 mm, subcylindrical to obscurely 4-angled (rarely 5-angled), not constricted into segments, smooth, pale bluish green; *tubercles* arranged into 4 (–5) obscure slightly spiralling rows with surface flat or convex to sometimes indistinctly concave between rows, low-conical and truncate, projecting 1–2 mm from angles, with spine-shields 5–7 (10) mm long, 1–2 × 3–4 mm above spines and 2–6 (8) mm long below spines but remaining well separated from next, bearing 2 spreading to slightly deflexed initially reddish brown (later dark brown) spines 6–20 mm long; *leaf-rudiments* on tips of new tubercles towards apices of branches, 0.7–1.5 × 0.7–1.5 mm, erect, fleeting, ovate, sessile, with brown stipular prickles 0.7–4 mm long (sometimes another small prickle ± 1 mm long near base of spine-shield). *Synflorescences* many per branch towards apex, each a solitary cyme in axil of tubercle, on very short peduncle < 1 mm long, each cyme with 3 transversely disposed cyathia, central male, lateral 2 bisexual and developing later each on short peduncle 1–2 × 1–2 mm, with 2 ovate to lanceolate bracts 1–3 × 1–3 mm subtending lateral cyathia; *cyathia* shallowly conical-cupular, glabrous, 3–4 mm broad (2 mm long below insertion of glands), with 5 obovate lobes with deeply incised margins, bright yellow; *glands* 5, transversely rectangular and contiguous, 2–2.5 mm broad, yellow, ascending to spreading, inner margins slightly raised, outer margins entire, surface between two margins dull; *stamens* glabrous, bracteoles enveloping groups of males, with finely divided tips, glabrous; *ovary* obtusely 3-angled, glabrous, slightly reddish green, raised on very short pedicel < 0.3 mm long; styles 2.5–3.5 mm long, branched to well below middle (fused in lower third) and ascending to widely spreading. *Capsule* 3–4 mm diam., obtusely 3-angled, glabrous, sessile.

Distribution & Habitat

So far only recorded in South Africa, *Euphorbia aeruginosa* is widely known in the northern parts of the former Transvaal. It is particularly associated with the Soutpansberg, from near Soutpan in the north-west to the village of Masisi at the eastern tip of the range. It has also been collected further eastwards on low sandstone ridges and is known to continue into the Kruger National Park to near Pafuri. To the west of the Soutpansberg it is recorded on the Blouberg and to the south in the northern parts of the Waterberg.

Typically, *E. aeruginosa* grows tightly wedged into crevices in sloping and north-facing or almost flat sandstone out-

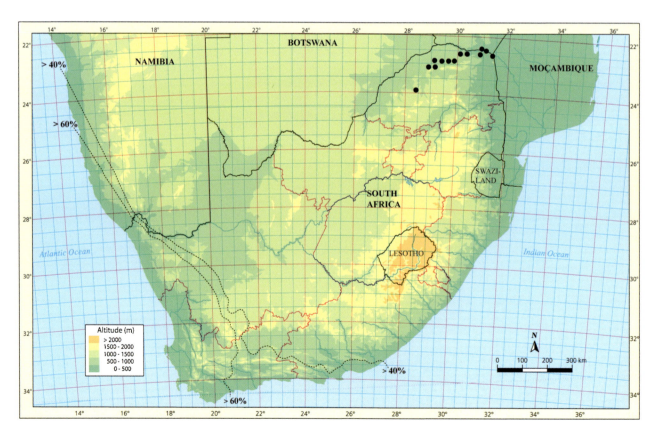

Fig. 5.3. Distribution of *Euphorbia aeruginosa* (© PVB).

5.1 Sect. Euphorbia

Fig. 5.4. *Euphorbia aeruginosa*, in crevices among sandstone rocks, ± 15 cm tall, above Soutpan, Soutpansberg, South Africa, 30 Dec. 1996 (© PVB).

crops at very variable altitudes (from the foot of the mountains at 700–800 m to their upper plateaux at 1300–1500 m). These outcrops of sandstone are often practically bare of other plants, though it was found growing together with a small form of *E. griseola* and with *Ceropegia* (*Huernia*) *nouhuysii* and *C. whitesloaneana* on one such spot. It has also been observed in the quite different habitat of very hot, gravelly, quartzitic, gently sloping areas among *mopane* near the border of the Kruger National Park at an altitude of only 350 m. Here it occurred together with *E. schinzii* subsp. *bechuanica*.

Diagnostic Features & Relationships

Plants of *Euphorbia aeruginosa* may be up to 1 m in diameter and these larger specimens consist of enormous numbers of tightly packed, erect, mostly relatively short branches. In flat areas with *mopane*-bush, more spreading and laxly branched plants may develop. The branches are comparatively slender and rarely exceed 8 mm in thickness. In *E. aeruginosa* the tubercles are hardly raised from the surface of the branch at all and are arranged neatly into four (often slightly spiralling) rows, so that strongly defined angles do not develop on the branches. The surface between these rows is flat to very slightly convex (when plants are turgid) and this also adds to the indistinctness of the angles. Occasionally (such as in plants from the Blouberg) hardly any angles are visible at all and they then have particularly rounded branches. The surface of the branches between the spines always has a distinctive blue-green hue and it is from this that the name is derived. The spines and spine-shields are bright red-brown when young and soon mature to a very

Fig. 5.5. *Euphorbia aeruginosa*, in crevices on sandstone outcrops, ± 30 cm tall, *PVB 7473*, Nwanedi, South Africa, 4 Nov. 2011 (© PVB).

Fig. 5.6. *Euphorbia aeruginosa*, flat area among *mopane* trees (with *E. schinzii* ssp. *bechuanica*), *PVB 12060*, Tshikondeni Mine, South Africa, 4 Nov. 2011 (© PVB).

Fig. 5.7. *Euphorbia aeruginosa*, *PVB 7002*, Ga-Kibi, near Blouberg, South Africa, 5 Oct. 2006 (© PVB).

dark brown. Here the spine-shields are often noticeably short and round, almost elliptic, and are comparatively widely spaced out along the edges of the branches. This feature is quite variable and sometimes they are much longer, though they are never continuous along the angles. Schweickerdt (1935) mentioned the presence of a small fifth spine near the base of the spine-shield, but this is not always there and was absent in the material illustrated here.

Although occasional flowers appear in plants in habitat until early November, the main flowering period in *E. aeruginosa* is between August and October. Then the upper 30–80 mm of many of the branches become covered with bright yellow cyathia. There is usually only one cyme in the axil of each tubercle. All three cyathia in each cyme develop in relatively quick succession and thereby add to the brightness of the display that they provide. As in most of the vegetatively similar species, the three cyathia in each cyme are relatively small and tightly clustered and the female florets have long, slender styles that remain fused near their bases.

The combination of colours of the spines (and spine-shields) and the surface of the branches as well as the very slightly angled, almost cylindrical branches (with especially low tubercles) are the main features distinguishing *E. aeruginosa* from *E. schinzii*. *Euphorbia schinzii* subsp. *bechuanica* also differs very clearly by its much more robust branches, which are often thicker and considerably longer and are much more clearly 4-angled. In subsp. *bechuanica* the spines are generally more robust and longer, the stipular prickles are larger and more spine-like and the leaf-rudiments are also longer and broader. *Euphorbia*

5.1 Sect. Euphorbia

Fig. 5.8. *Euphorbia aeruginosa*, *PVB 7002*, Ga-Kibi, near Blouberg, South Africa, 5 Oct. 2006 (© PVB).

aeruginosa shares with *E. louwii* the colour of the spines and spine-shields and the rather rounded branches as well as the occasional occurrence of a small fifth spine on each shield (which is always present in *E. louwii* but only occasionally present in *E. aeruginosa*).

History

Euphorbia aeruginosa was discovered in November 1932 by I.C. Verdoorn, A.A. Obermeijer and H.G.K. Schweickerdt during an expedition to Soutpan, at the north-western base of the Soutpansberg.

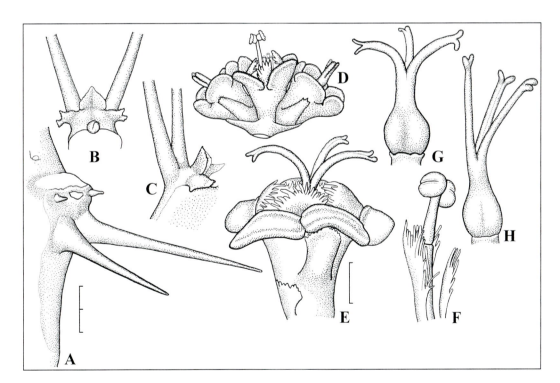

Fig. 5.9. *Euphorbia aeruginosa*. **A**, spine-shield and spines (scale 2 mm, as for **B**–**D**). **B**, young spines and leaf-rudiment from above. **C**, young spines and leaf-rudiment from side. **D**, side view of cyme in first male stage. **E**, female stage of cyathium (scale 1 mm, as for **F**–**H**). **F**, anther with bracteoles. **G**, **H**, female floret. Drawn from: **A**–**G**, *PVB 7473*, Nwanedi, South Africa; **H**, *PVB 7101a*, above Waterpoort, Soutpansberg, South Africa (© PVB).

Euphorbia avasmontana Dinter, *Sukk. Forsch.* 2: 96 (1928). Lectotype (Bruyns 2012): Namibia, near Windhoek, Auas Mtns, *Dinter* (PRE).

Euphorbia volkmanniae Dinter, *Sukk. Forsch.* 2: 124 (1928). Type: Namibia, near Otavi, Auros, Feb. 1925, *Dinter* (B, photo only).

Euphorbia hottentota Marloth, *S. African J. Sci.* 27: 336 (1930). Lectotype (Bruyns 2012): South Africa, Cape, Richtersveld, Kubus Kloof, 300 m, 29 Aug. 1925, *Marloth 12520* (PRE).

Euphorbia kalaharica Marloth, *S. African J. Sci.* 27: 338 (1930). Lectotype (Bruyns 2012): South Africa, Cape, Neusberg, near Kakamas, 700 m, 15 Aug. 1928, *Marloth 14039* (PRE).

Euphorbia sagittaria Marloth, *S. African J. Sci.* 27: 337 (1930). *Euphorbia avasmontana* var. *sagittaria* (Marloth) A.C.White, R.A.Dyer & B.Sloane, *Succ. Euphorb.* 2: 817 (1941). Lectotype (Bruyns 2012): South Africa, Cape, 12 miles south of Upington towards Prieska, Aug. 1929, *Marloth 14035* (PRE).

Euphorbia venenata Marloth, *S. African J. Sci.* 27: 337 (1930). Type: Namibia, Tsarris Mtns, 1500 m, west of Maltahöhe, Oct. 1910, *Marloth 4687* (PRE, holo.; K, iso.).

Bisexual spiny large glabrous succulent shrub 0.5–2 (2.5) × 1–3 m with many branches at and just above ground level from central stem 0.2–0.5 m tall tapering rapidly into a short taproot bearing fibrous roots. *Branches* initially spreading then erect, mostly simple, 0.2–2 m × 20–70 mm, 4- to 6 (8)-angled, often slightly constricted into ellipsoidal segments (becoming evenly narrower towards base and top), often dying off and wearing away with age, smooth, slightly greyish to yellowish green; *tubercles* fused into 4–6 (8) wing-like angles with deep to flattened grooves between angles, laterally flattened and projecting 4–6 mm from angles, with spine-shields united into continuous horny grey margin along angles 2–3 mm broad above spines and 1–3 mm broad below them (rarely separate), bearing 2 spreading and widely diverging grey spines 3–15 mm long; *leaf-rudiments* on tips of new tubercles towards apices of branches,

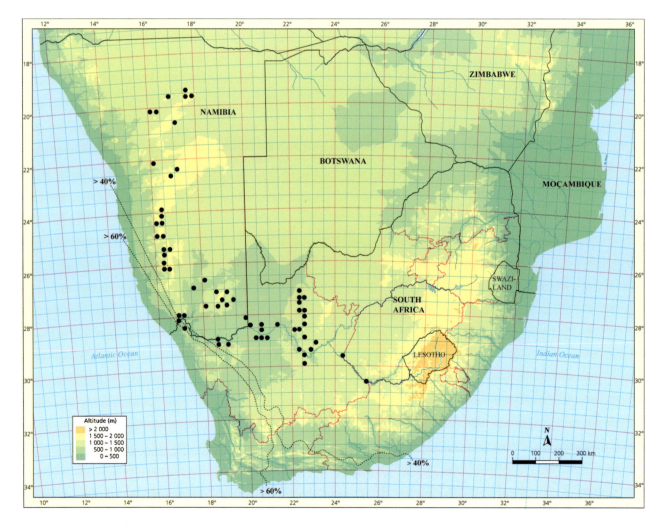

Fig. 5.10. Distribution of *Euphorbia avasmontana* (© PVB).

1–3 × 2–3 mm, erect, fleeting, ovate, acute to obtuse, sessile, with small irregular obtuse brown stipules. *Synflorescences* many per branch towards apex, in groups of (2) 3–5 cymes in axil of tubercle, each cyme on peduncle 1–3 mm long, with 3 vertically disposed cyathia (sometimes more developing on sides), central male to bisexual, lateral 2 bisexual and developing later each on short peduncle 1–2 × 1–2 mm, with 2 ovate yellow bracts 1–3 × 2–3 mm subtending lateral cyathia; *cyathia* shallowly conical-cupular, glabrous, 4–5 (7) mm broad (3 mm long below insertion of glands), with 5 obovate lobes with deeply incised margins, bright yellow to greenish yellow; *glands* 5, transversely elliptic and contiguous, 3 mm broad, yellow to greenish yellow, ascending to spreading, inner margins not raised, outer margins entire and slightly ascending; *stamens* glabrous, bracteoles enveloping groups of males, with finely divided tips or filiform, glabrous; *ovary* obtusely 3-angled, glabrous, green, exserted and projecting upwards or downwards on pedicel 3–4 mm long; styles 2–3 mm long, branched to above middle or in upper third. *Capsule* 5–8 mm diam., acutely 3-angled, glabrous and green to shiny red, exserted on decurved and later erect pedicel 3–7 mm long.

Distribution & Habitat

Euphorbia avasmontana is widely distributed in Namibia from the mountains between Grootfontein and Tsumeb, in the Auas Mountains around Windhoek to some of the hills around Rehoboth and in the Tsarris Mountains west of Maltahöhe. In the southern part of Namibia it is very common indeed in the foothills and even on some of the upper slopes of the Great Karas Mountains in the south-east and it also occurs in the foothills of the Hunsberge of the south-west, mainly in outliers of these mountains near the Orange River and along its tributary the Fish River. In South Africa it is mainly known close to the Orange River, from near Kubus and Sendelingsdrift in the Richtersveld, to around Pella (near Pofadder) and then further east from Riemvasmaak to Upington and along the river to the dry hills around Prieska. There are some more isolated populations in the Langberg and Skurweberg, south and south-west of Olifantshoek.

Over this vast area it occurs mainly on stony, east-, north- and west-facing slopes at altitudes of between 300 and 1600 m. In the Great Karas Mountains there are some enormous, almost forest-like populations in the bases of some of the valleys, as, for example, around Narudas, in the south-east of these mountains. The aridity of these habitats varies greatly. Between Sendelingsdrift and Prieska plants often grow on arid schistose, gneissic or quartzitic slopes, with very little other vegetation apart from *Portulacaria namaquensis*, *Kleinia longiflora* and a few other larger species of *Euphorbia* such as *E. dregeana*, *E. gregaria* and occasionally also *E. virosa*. In contrast to this, in the Langberg they occur among quartzite rocks with a fairly dense cover of trees and shrubs of *Acacia*, *Boscia* and *Croton*. It is also found among fairly open bushland (with few other succulents) on schistose slopes around Windhoek, while in the dolomitic and limestone outcrops in the mountains between Tsumeb and Grootfontein the plants may be quite difficult to see among the fairly dense cover of trees and other shrubs (as in Fig. 5.14).

Diagnostic Features & Relationships

In *E. avasmontana* the plant forms a large and quite spectacular shrub, with the branches arising from a thicker, central stem which remains much shorter than the branches. Here the branches arise near the apex of the stem, initially spread and then become strongly erect. In *E. avasmontana* the older branches do not remain alive but die off, to collapse and decay around the sides of the plant. The major part of these old branches usually wears away, but sand-filled remnants of their bases often persist in somewhat untidy piles around the lower parts of the stem. Young plants consist of a single stem until they are about 7–10 cm tall and this stem is initially faintly mottled with yellowish on green. For the first 2 cm or so the stem is 4-angled, but after that the number of angles rapidly increases. Generally, the branches are very regularly but only faintly constricted into short segments. The spines are of variable length, even on one plant. They are longest towards the apex of the branch (often particularly long and sharp on young plants) and often wear away towards the base so that the bases of the older branches may be spineless. In the arid parts of the distribution the spine-shields are continuous and uniformly broad, while in the moister regions they are narrower below the spines than above them, though they are usually still continuous.

Flowering in *E. avasmontana* usually takes place between September and December. Large numbers of cyathia are produced on the upper segment of the branches, usually with three to five cymes (rarely fewer than three but often variable in number on one plant) developing in each axil, each cyme usually with three vertically arranged cyathia. The cyathia are quite variable in diameter, larger in the south-east and smallest in plants from the hills between Outjo and Tsumeb. The male florets have long pedicels and filaments so that the cyathia appear to have a 'fuzz' of yellow above them. The female florets are borne on fairly long, decurved or ascending (but rarely straight) pedicels which soon place them just outside the cyathium, with those in the upper cyathium (of the three in each cyme) facing upwards and those in the lower cyathium facing downwards (thereby keeping them well away from the pollen of the central cyathium). Each of them has an obscurely 3-angled ovary with a slender style divided only near its apex. The exserted capsules are either

Fig. 5.11. *Euphorbia avasmontana*, in the driest habitat where it occurs, schistose slopes with *Aloe gariepensis*, *PVB 10063*, east of Namusberg, Rosh Pinah, Namibia, 15 Jul. 2005 (© PVB).

Fig. 5.12. *Euphorbia avasmontana*, *PVB 11349*, south of Dassiefontein, Great Karas Mountains, Namibia, 4 Apr. 2009 (© PVB).

Fig. 5.13. *Euphorbia avasmontana*, on schists with grass and scattered trees, near Avis Dam, Windhoek, Namibia, 7 Nov. 2008 (© PVB).

5.1 Sect. Euphorbia

Fig. 5.14. *Euphorbia avasmontana*, among dense bush in the wettest area where it occurs, Otavi Mountains, between Tsumeb and Grootfontein, Namibia, 2 Jan. 2006 (© PVB).

pendulous or inclined upwards and are shiny green to bright red. They are 3-angled but without the margins of each locule being curved inwards (or only very slightly so) and so have a distinctly triangular shape (when viewed from above) with fairly acute apices to the triangle. They dehisce explosively, sending the small seeds some distance from the parent plant.

Euphorbia avasmontana looks superficially similar to *E. virosa*. Vegetatively the two differ in that the branches of *E. virosa* are much thicker and, while the branches in both are equally regularly and clearly constricted into segments, the manner in which the angles are sinuate makes this less clear in *E. virosa*. They are also mostly 7-angled (and up to 8-angled) in *E. virosa* and mostly 5- or 6-angled (and rarely more than 7-angled) in *E. avasmontana*. Another feature regularly found in *E. virosa* and much rarer in *E. avasmontana* is that the branches may rebranch above, especially in older specimens. Where they occur together, the yellowish green colour of the branches of *E. avasmontana* immediately separates the two, although this is not a constant feature of this species and in some areas the branches are also greyish green. The dying off of the older branches that is common in *E. avasmontana* is not found at all in *E. virosa*, where even the oldest branches seem to persist as long as the plant lives. In *E. virosa* the branches bear generally much more robust thorns (often also longer than in *E. avasmontana*) and the angles are broader, paler and distinctly sinuate with a quite deep furrow between them. In *E. avasmontana* the angles are regular and straight, often with a flat area between them. In both, the cyathial glands are contiguous around the edge of the cyathium but the whole structure is much larger in *E.*

Fig. 5.15. *Euphorbia avasmontana*, in first male stage, *PVB 11122*, Naroep, west of Pofadder, South Africa (© PVB).

Fig. 5.16. *Euphorbia avasmontana*, second male stage, *PVB 10063*, east of Namusberg, Rosh Pinah, Namibia (© PVB).

Fig. 5.17. *Euphorbia avasmontana*, second male stage, *PVB 12872*, 10 km west of Outjo, Namibia (© PVB).

virosa, where the cyathia are 8–12 mm broad and usually have between 8 and 10 glands. They differ also in their respective fruit. In *E. virosa* the capsule is spherical, 10–24 mm in diameter, usually 4-locular and it disintegrates (that is, dehisces without exploding) to release the large, pea-sized seeds. On the other hand, in *E. avasmontana*, the capsule is deeply and clearly 3-angled, it is much smaller (reaching at most 8 mm in diameter), it has three locules and it explodes when ripe. Consequently, under plants of *E. avasmontana*, one does not find the parts of old capsules and both sterile and fertile seeds that usually accumulate under a large plant of *E. virosa*.

As Leach (1969) pointed out, the closest relative of *E. avasmontana* is *E. gracilicaulis*, from southern Angola, with which it shares the relatively long cyathia on fairly long peduncles, the much exserted female florets and deeply lobed capsules. The two differ mainly in the more prominent stem that regularly develops into a trunk from which the branches arise well above the base in *E. gracilicaulis*. In *E. gracilicaulis* the branches are also more slender than the most slender forms of *E. avasmontana* (i.e. those from the area between Outjo and the Otavi Mountains).

History

Euphorbia avasmontana was apparently collected in the Auas Mountains of central Namibia by Dinter in 1924 (White et al. 1941: 815), though it is not surprising that Dinter (1928: 96) mentioned that this plant had been known to them already for many years. H.W. Pearson collected specimens of *E. avasmontana* in the Great Karas Mountains in December 1912. In South Africa it was recorded in May 1922 by Hunter in Griqualand and in August 1925 by Marloth in the Richtersveld (the type of *E. hottentota*).

5.1 Sect. Euphorbia

Fig. 5.18. *Euphorbia avasmontana*, brightly coloured triangular capsules, *PVB 12822*, Namtib, Tiras Mountains, Namibia, 17 Dec. 2014 (© PVB).

However, these collections are probably not the first records of *E. avasmontana*. Among the illustrations in Raper and Boucher (1988: fig. 45) and Panhuysen (2011: afb. 10) on the travels of Robert J. Gordon, there is a reproduction of a painting, almost certainly by Johannes Schumacher, who was Gordon's illustrator during that expedition, which shows a *Euphorbia* very like *E. avasmontana* on the rocks near the Augrabies Falls on the Orange River, which they visited between August and September 1779 on the way to near the present-day Prieska. Among the paintings by François le Vaillant is one which has been considered to belong to *E. virosa* (Jordaan 1973) but, because of its cyathia with 5 glands, could be claimed to be of *E. avasmontana*. However, a closer examination of this painting shows that it is a copy of part of the figure of *E. virosa* which appeared in Paterson (1789, 1790) and that the cyathia were modified to have just five glands.

Marloth (1930) described four further 'species' that were vegetatively quite similar to *E. virosa*. These were *E. hottentota*, *E. kalaharica*, *E. sagittaria* and *E. venenata*. When describing these species, he supplied very little of the vital data necessary to elucidate their relationships to *E. virosa* and the few specimens that have ever been made of these 'species' have been placed fairly arbitrarily under these names, though most of them have ended up being sunk under the name *Euphorbia avasmontana*, following White et al. (1941). Marloth's description of *E. venenata* is vague about facts like the size of the cyathia and the number of glands on each cyathium. However, the type specimen at PRE has a few separate female florets (with quite long pedicels), a photograph of the tip of a flowering branch (showing five glands

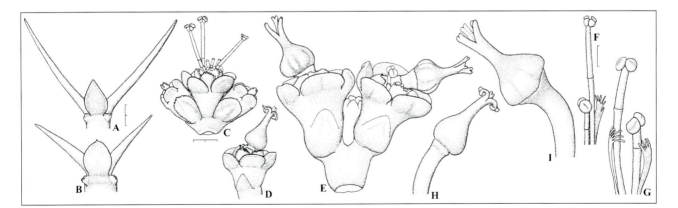

Fig. 5.19. *Euphorbia avasmontana*. **A, B**, young spines and leaf-rudiment from above (scale 2 mm, as for **B**). **C**, side view of cyme in first male stage (scale 1 mm, as for **D, E**). **D, E**, female stage of cyathium and cyme. **F, G**, anthers and bracteoles (scale 1 mm, as for **G–I**). **H, I**, female floret. Drawn from: **A, C, D, F, H**, *PVB 11122*, Naroep, west of Pofadder, South Africa; **B**, *PVB 10063*, east of Namusberg, Rosh Pinah, Namibia. **E, G, I**, near Avis Dam, Windhoek, Namibia (© PVB).

per cyathium) and some dissected cyathia (also with five glands per cyathium). From these, it is clear that *E. venenata* is not the same as *E. virosa* (with which the name has sometimes been associated), but that it is referable rather to *E. avasmontana*.

Another of Marloth's names, *E. hottentota,* was maintained as distinct from *E. avasmontana* in Bruyns et al. (2006). Marloth (1930: 335) separated *E. avasmontana* and *E. hottentota* by the number of angles on the branches (7-angled in *E. avasmontana*; 5- to 6-angled in *E. hottentota*) but White et al. (1941: 824) pointed out that 'some of Marloth's herbarium specimens do not agree entirely with the typical form' so that the identity of this 'species' is less clear than Marloth thought. Over the large area where *E. avasmontana* occurs, branches are frequently 4-angled and may have up to eight angles and no clear separation into 5- to 6-angled plants on the one hand and 7-angled plants on the other hand is possible (as Dyer pointed out also to W.F. Bayer in letter of 20 Dec. 1938, PRE records). No differences in the floral structures have been detected on which two distinct species could be recognised and consequently only one species is now recognised.

Euphorbia barnardii A.C.White, R.A.Dyer & B.Sloane, *Succ. Euphorb.* 2: 965 (1941). Type: South Africa, Transvaal, Sekukuniland, farm Driekop, east of Leoleo (Lulu) Mountain, 3000', 6 Jan. 1936, *Barnard 449* (PRE, holo.; K, MO, iso.).

Bisexual spiny glabrous succulent shrub 0.1–0.6 × 0.3–1.5 (5) m, much branched at and just above ground level from short central stem tapering rapidly into a short taproot bearing fibrous roots. *Branches* ascending or initially spreading then erect, occasionally rebranched, 100–500 × 30–80 mm, 5- to 6-angled, deeply constricted into almost cordate segments 40–80 (100) mm long, smooth, green or yellowish green; *tubercles* fused into 5–6 very prominent wing-like angles with deep grooves between angles, laterally flattened and deltate projecting 4–10 mm from angles, with spine-shields united into ± continuous horny initially brown and later grey margin along angles ± 2–4 mm broad above spines and 1–2 mm broad below them (rarely separate), bearing 2 spreading and widely diverging initially red and later grey spines 4–11 mm long and usually two prickles 1–2 mm long alongside flowering axil (arising dorsally on bracts); *leaf-rudiments* on tips of new tubercles towards apices of branches just above pair of spines, 2–3.5 × 2–3.5 mm, ascending to spreading, fleeting, ovate, sessile, with small irregular brown stipular prickles 0.5–1 mm long. *Synflorescences* many per branch towards apex on last 1–2 segments, each a solitary cyme in axil of tubercle, very shortly peduncled, each cyme with 3 vertically (to transversely) disposed cyathia or with single cyathium, central usually male sessile and often deciduous, lateral 2 bisexual and developing later each on short peduncle 1–4 × 3–4 mm, with 2 ovate red-brown bracts 1–3 mm long and 2–3 mm broad subtending lateral cyathia; *cyathia* cupular, glabrous, 4–9 mm broad (3 mm long below insertion of glands), with 5 obovate lobes with deeply incised margins, yellow-green; *glands* 5, transversely elliptic and contiguous, 2–4 mm broad, yellow or yellow-green, spreading, inner margins raised, outer margins entire and ascending to spreading, surface between two margins covered with copious nectar; *stamens* glabrous, bracteoles enveloping groups of males, with finely divided tips, glabrous; *ovary* only slightly swollen from base of styles, ellipsoidal, glabrous, green, sessile; styles 3.5–5 mm long, branched from middle. *Capsule* 6–9 mm diam., deeply 3-angled, red above and green below, glabrous and shiny, sessile.

Distribution & Habitat
Euphorbia barnardii is found in Sekukuniland, where it occurs at altitudes of 850–1100 m on a few relatively isolated hills that rise from the plains to the east of the larger Leoleo Mountain. These hills are strewn with brown, noritic boulders, with short, scattered shrubs of *Acacia* and other stunted trees and *Aloe castanea* and *A. wickensii* among these rocks. *Euphorbia barnardii* grows on the lower slopes of these hills and also in the flats immediately below them in some cases, where housing has not yet destroyed them.

Diagnostic Features & Relationships
Euphorbia barnardii forms clumps which are usually between 0.5 and 1 m in diameter. Within each clump, the branches are densely clustered and obscure the short stem from which they arise. They usually spread somewhat away from the stem, after which they become erect for anything up to 0.5 m. They have a yellow-green colour which is quite conspicuous if the plant is exposed. The branches are prominently and regularly constricted into short, cordate segments that are usually broader than long and often increase in breadth towards the apex of the branch. They are armed with short pairs of spines that rarely exceed 10 mm long.

Flowering in *E. barnardii* takes place between the beginning of July and mid-August. When in full bloom, the uppermost segment of each branch has a dense arrangement of the bright yellow cyathia along the angles and such plants are very beautiful among the dry grass and dead twigs. One cyme is produced in the axil of each tubercle and from this between one and three cyathia develop. The lateral cyathia are usually produced in a vertical position (relative to the

5.1 Sect. Euphorbia

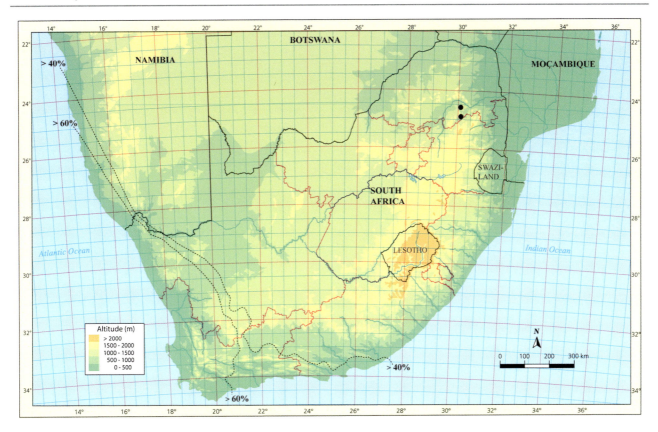

Fig. 5.20. Distribution of *Euphorbia barnardii* (© PVB).

Fig. 5.21. *Euphorbia barnardii*, on gentle slope among rocks and dry grass with *Aloe castanea* and *A. wickensii*, west of Steelpoort, South Africa, 22 Jul. 2011 (© PVB).

Fig. 5.22. *Euphorbia barnardii*, on gentle slope among rocks and dry grass, west of Steelpoort, South Africa, 22 Jul. 2011 (© PVB).

axis of the branch) but may be produced transversely as well, even on the same plant as others which are vertically disposed. The cyathia are cup-shaped and comparatively deep, with sessile ovaries that remain embedded in the cyathium after fertilization and the rest of the cyathium dries up around the expanding capsule.

In many respects *E. barnardii* is close to *E. restricta* and is also similar to *E. grandicornis*, sharing the similarly

Fig. 5.23. *Euphorbia barnardii*, in flower on gentle slope among rocks and dry grass, west of Steelpoort, South Africa, 22 Jul. 2011 (© PVB).

shaped female florets and boldly segmented branches, though the segments are not nearly as large in *E. barnardii*, nor are the spines nearly as long. In *E. barnardii* the cymes are solitary in each axil (as for *E. restricta*), rather than in groups of three, as is typical of *E. grandicornis*. *Euphorbia barnardii*, *E. grandicornis* and *F. restricta* all share the extra pair of spines found adjacent to the axillary bud, which may be particularly prominent in *E. grandicornis* but are occasionally absent in *E. barnardii*.

5.1 Sect. Euphorbia

Fig. 5.24. *Euphorbia barnardii*, *PVB 11896*, west of Steelpoort, South Africa, 22 Jul. 2011 (© PVB).

History

According to White et al. (1941), *Euphorbia barnardii* was discovered by W.G. Barnard in 1935 and the type collection was made in January 1936. There are few records of the species and it is considerably threatened in habitat by ever-expanding and wasteful urban sprawl in Sekukuniland.

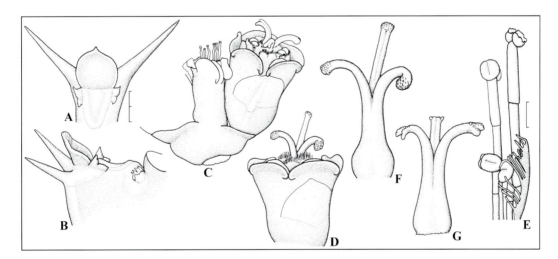

Fig. 5.25. *Euphorbia barnardii*. **A**, young spines and leaf-rudiment from above (scale 2 mm, as for **B–D**). **B**, young spines and leaf-rudiment from side. **C, D**, side view of cyme and cyathium in female stage. **E**, anthers and bracteole (scale 1 mm, as for **F, G**). **F, G**, female floret. Drawn from: *PVB 11896*, west of Steelpoort, South Africa (© PVB).

Euphorbia caerulescens Haw., *Philos. Mag. Ann. Chem.* 1: 276 (1827). *Euphorbia virosa* var. *caerulescens* (Haw.) A.Berger, *Sukkul. Euphorb.*: 81 (1906). Lectotype (designated here): South Africa, Cape of Good Hope, 1823, *Bowie*, (painting number 292/428, 'drawn from the plant from which Haworth described' by G. Bond at K).

Euphorbia ledienii A.Berger, *Sukkul. Euphorb.*: 80 (1906). Type: South Africa, fl. & fr. Aug. 1906, received from collection of F. Ledien (NY, holo.).

Euphorbia ledienii var. *dregei* N.E.Br., *Fl. Cap.* 5 (2): 366 (1915). Lectotype (Bruyns 2012): South Africa, near Port Elizabeth, received 9 Sept 1912, *I.L.Drège* (K).

Bisexual spiny glabrous succulent shrub 1–3 × 1–3 m, branching extensively mainly from base of similar stem with woody and fibrous roots, without spreading rhizomes around edge of plant. *Branches* 20–60 mm thick, slightly constricted into elongated slightly elliptic (but mainly parallel-sided) segments, smooth, green to grey-green; *tubercles* fused into 3–7 wing-like sometimes slightly sinuate angles, laterally flattened, rounded, projecting 3–7 mm from angles, spine-shields around apex and united into continuous horny and later somewhat corky brown to grey or black margin, 2–5 mm broad in upper part tapering to 1.5–3 mm below, usually bearing 2 spreading and widely diverging brown to grey spines (2–) 6–10 mm long; *leaf-rudiments* on tips of new tubercles towards apices of branches and stem, 1–4 × 2–4 mm, spreading, fleeting, broadly ovate, obtuse, sessile, with brown ± pyramidal stipules ± 1 mm long. *Synflorescences* in large numbers per branch towards apex, each a group of 1–8 cymes in axil of tubercle, on peduncle 2–6 × 2–3 mm, each cyme with 3 ± vertically disposed cyathia, central male, outer 2 female only (or bisexual) and developing later, with 2 ovate bracts 1–1.5 × 1.5–2 mm subtending cyathia; *cyathia* cupular-conical, glabrous, 3.5–6 mm broad (2–3 mm long below insertion of glands), with 5 lobes with deeply incised margins, bright yellow; *glands* (3–)5, transversely oblong to kidney-shaped or rectangular, 2–3 mm broad, bright yellow, ascending-spreading, slightly convex to concave above, outer margins entire and slightly raised; stamens glabrous, bracteoles palmate and enveloping groups of stamens, deeply and finely divided, glabrous; *ovary* globose, glabrous, included to slightly exserted on erect pedicel 1.5–2.5 mm long and soon becoming slightly exserted, calyx slightly extended around base; styles 2–4 mm long, branched in upper third. *Capsule* 5–6 mm diam., very obtusely 3-angled, glabrous, slightly shiny pale red above and green towards base, exserted on erect pedicel 2–6 mm long.

Distribution & Habitat

Euphorbia caerulescens is most plentiful in the dry south-eastern part of South Africa, between Humansdorp, Uitenhage and Enon. In this region its distribution extends northwards towards Steytlerville, where isolated patches occur surrounded by *E. radyeri*. *Euphorbia caerulescens* is also plentiful in the low-lying area north of Grahamstown along the Fish River, between Carlisle Bridge and Committees Drift.

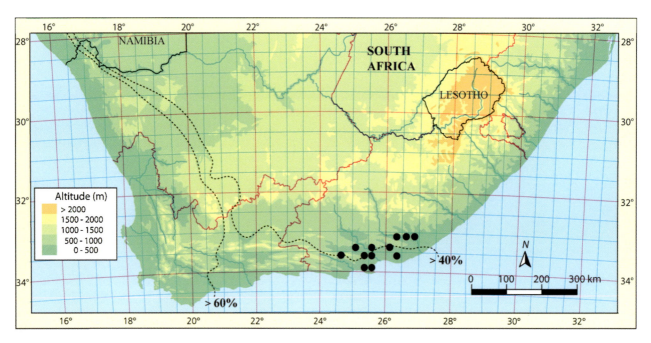

Fig. 5.26. Distribution of *Euphorbia caerulescens* (© PVB).

Euphorbia caerulescens generally grows on stony hills and flat areas on ground derived from shales or alluvium. In the Fish River Valley it occurs with a profusion of other succulents, including many species of *Euphorbia*. Here it may become the dominant element in the vegetation, forming a 'noorsveld' very similar to that of *E. radyeri* in the Jansenville district. In the area between Enon, Addo and Uitenhage it inhabits thickets of 'Addo Bush', forming tall, scattered plants among very spiny, nearly impenetrable scrub, often with a remarkably wide selection of other small succulents, especially in any open patches between the denser bush.

5.1 Sect. Euphorbia

Fig. 5.27. *Euphorbia caerulescens*, in dense scrub with *Aloe africana*, NE of Uitenhage, South Africa, 25 Oct. 2012 (© PVB).

Diagnostic Features & Relationships

Depending on its habitat, *E. caerulescens* may form a tall, spindly and relatively diffuse shrub to 2 m or even 3 m tall (reaching its greatest length when sheltered and growing in dense bush, where some branches sometimes project from the top of other large shrubs), or it may be a much more compact and densely branched shrub 1–1.5 m tall when growing in more open situations. Typical specimens are between one and three metres in diameter.

Fig. 5.28. *Euphorbia caerulescens*, flowering in dense scrub NE of Uitenhage, South Africa, 25 Oct. 2012 (© PVB).

The branches of *E. caerulescens* are mostly 4- to 5-angled, with very slight constrictions at varying intervals after each year's growth. They are grey-green in the drier areas and greyish with green young growth in the wetter parts. All branches are armed with stout greyish spines, which arise on a more or less continuous horny or corky margin along the angles. On new growth the spines are red and the small, rounded leaf-rudiments just above the spines are quite obvious for a short period. The spine-shields are also initially red but soon become pale brown, changing to grey or black with age. As they age they become covered with a corky excrescence which forms a continuous, somewhat irregular ridge along the angles.

Fig. 5.29. *Euphorbia caerulescens*, flowering, NE of Uitenhage, South Africa, 25 Oct. 2012 (© PVB).

Fig. 5.30. *Euphorbia caerulescens*, first male stage, *PVB 10618*, NE of Uitenhage, South Africa, 24 Jul. 2018 (© PVB).

Flowering takes place between October and mid-November. Cymes arise on the parts of the branch that developed during the previous growing season, i.e. usually in the uppermost segment or the upper 50–150 mm of the branch, where a cluster of cymes breaks through the corky spine-shield in the axil of each spine-pair. Each cluster consists of one to five or more cymes and each cyme is in turn made up of three faintly scented, bright yellow but fairly narrow cyathia. The central functionally male cyathium in each cyme is flanked by two lateral ones that may be functionally female only in *E. caerulescens*. Male florets arise on long pedicels and the females are also significantly pedicellate, so that the ovary often partly protrudes from the cyathium.

The two similar and very closely related species *E. caerulescens* and *E. radyeri* are fairly easily separated. Branches around the perimeter of most plants of *E. radyeri* are usually rhizomatous and this phenomenon is unknown in *E. caerulescens*. The branches are thicker and tend to have a more bluish green colour in *E. radyeri* than in *E. caerulescens*, though their colour varies greatly in the latter, with green branches on plants from sheltered spots and much greyer branches on specimens that are more exposed. The branches of *E. radyeri* are more deeply articulated into almost spherical segments, while those of *E. caerulescens* are generally only indistinctly articulated into elongated, cylindrical segments. The tubercles are often much longer and broader and the leaf-rudiments are somewhat larger in *E. radyeri* than in *E. caerulescens*.

Florally *E. caerulescens* and *E. radyeri* are very similar. In *E. caerulescens* the cyathia are often slightly narrower and become more abruptly narrow beneath the glands, while the female florets are borne on a slightly longer pedicel and are usually without the slightly elongated calyx that is generally present in *E. radyeri*. Flowering appears to take place slightly earlier in *E. caerulescens* than in *E. radyeri*, with the latter just beginning to flower when flowering in the former is almost over. The capsules in *E. radyeri* are mostly a little larger than those of *E. caerulescens* and are more conspicuously 3-angled.

White et al. (1941) mentioned that *E. caerulescens* and *E. radyeri* 'come into contact' around 32 km north-east of Uitenhage towards Steytlerville. In line with this, an abrupt change from the one to the other over a distance of around

5.1 Sect. Euphorbia

Fig. 5.31. *Euphorbia caerulescens*, second male stage with females already exserted, Piggott's Bridge, South Africa, 24 Oct. 2018 (© PVB).

Fig. 5.32. *Euphorbia caerulescens*, with pale red capsules, Piggott's Bridge, South Africa, 24 Oct. 2018 (© PVB).

100 m was observed along this road to Steytlerville and also north of Perdepoort towards Waterford. Dyer (1931: 101) also mentioned that some apparent hybrids between the two species are found in this transitional zone between them.

History

Euphorbia caerulescens was described by Haworth from material brought by James Bowie from the Cape to Kew in 1823. It is not known precisely where this material came from, but it must have originated somewhere between Grahamstown, Port Elizabeth and Uitenhage, where Bowie did much collecting between October 1821 and the middle of 1822. However, this is not the first record of this species. A specimen of *E. caerulescens* was collected in the Port Elizabeth district, most probably in December 1773 when Thunberg and Masson were in this area. Thunberg seems to have believed that this collection belonged to *E. canariensis*, which is similar-looking but endemic to the Canary Islands. This led to *Euphorbia canariensis* being listed in his *Prodromus plantarum Capensium* (*Prodr. Fl. Cap.* 2: 86 (1800)). *Euphorbia caerulescens* was also collected along the Swartkops River near Uitenhage by J.F. Drège and his brother Carl in December 1831 (*Drège 8210* at S), but this was considered to be *E. tetragona* by Boissier (1862). There is some evidence that it was also gathered by J.F. Drège in the valley of the Fish River around 1830 (specimen at S). Berger treated it as a variety of *E. virosa*.

The name of this species is usually spelt '*coerulescens*', but this is not how Haworth (1827) originally spelt it. Klotzsch (1860a, b) are among the earliest references where the incorrect spelling '*coerulescens*' appeared and this has been followed by almost all authors since then, including Berger (1906), Prain (1909), Brown (1915), Dyer (1931) and White et al. (1941), while Carter (2002) returned to the original and correct spelling.

In Bruyns (2012) a specimen made in 1876 by N.E. Brown and deposited at Kew was selected as the lectotype. This was from a plant in cultivation at Kew that had been used by Haworth for his description of *E. caerulescens*. However, this specimen was not examined by Haworth and so it is not

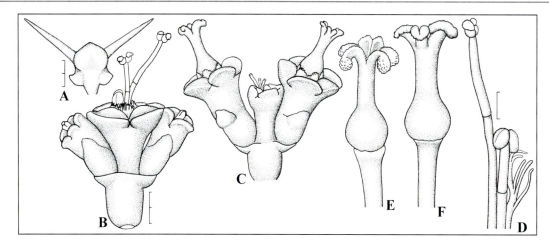

Fig. 5.33. *Euphorbia caerulescens*. **A**, young spines and leaf-rudiment from above (scale 2 mm, as for **C**). **B**, side view of cyme in first male stage (scale 2 mm). **C**, side view of cyme in female stage. **D**, anthers and bracteole (scale 1 mm, as for **E**, **F**). **E**, **F**, female floret. Drawn from: **E**, Fort Brown, NE of Grahamstown, South Africa. **A–D**, **F**, *PVB 10618*, NE of Uitenhage, South Africa (© PVB).

suitable as a lectotype. The illustration by George Bond (left hand part of Fig. 5.34) that is selected here was made from the plant from which Haowrth described *E. caerulescens* and seems to have been made before he described it, so it is more suitable as the lectotype. Another illustration was made by Duncanson (right hand part of Fig. 5.34) but that is of a much greyer-branched plant and is not indicated as being from the material seen by Haworth.

Berger distinguished *E. ledienii* as a separate species from *E. caerulescens* in 1906. He based his rather meagre description on cultivated material (whose precise origin was unknown) sent to him by F. Ledien of Dresden, Germany in August 1906. Berger's description was supplemented by one drawn up by N.E. Brown, based on cultivated plants that had been at Kew since 1868, and published in an account of *E. ledienii* by Prain (1909). Here, Prain mentioned that the branches of *E. ledienii* differed from those of *E. caerulescens* by 'the more numerous and less sinuate angles, its more distant and less marked constrictions and its shorter spines'. Brown (1915: 244) distinguished *E. caerulescens* and *E. ledienii* on the basis of the glaucous or bluish green stems, with spines 6–12 mm long in the former; the green, not glaucous stems, with spines 2–6 mm long in the latter. However, in many populations wide variation will be found in most of these features, largely dependent on the degree of exposure of the plants and, as Dyer (1931) and White et al. (1941) realised, these criteria are not useful in distinguishing two species in this region.

Fig. 5.34. Water-colours of *Euphorbia caerulescens*, made from collections of James Bowie which arrived at Kew in 1823. Left: 'Drawn from the plant from which Haworth described' by George Bond 292/428, the lectotype of *E. caerulescens*. Right: 'Dec. 23 1823, Introduced from the Cape of Good Hope by Mr Bowie', by Thomas Duncanson 802/504 (© RBG Kew).

Dyer (1931) and White et al. (1941) maintained that there were indeed two large, shrubby, paired-spine-bearing species in this region. One is the slender-branched, non-rhizomatous species which is found in the area around Port Elizabeth and Uitenhage and also in the valley of the Fish River between Carlisle Bridge and Committees Drift; the other has thicker branches and a rhizomatous habit and occurs in the drier, north-western sector of the distribution of the two, namely between Jansenville, Steytlerville and Somerset East and extending towards Uitenhage from Jansenville. As Dyer (1931) and Lotsy and Goddijn (1928) explained, the rhizomatous plants of the Jansenville district were called *E. caerulescens*, a tradition that was started by Marloth (1925: 137, fig. 91) and was accepted since no locality was given for the type. The non-rhizomatous plants were referred to as *E. ledienii*.

However, the feature of rhizomatous branches was not mentioned by Haworth. It is also not visible in either the specimen pressed by N.E. Brown from Bowie's plant or in the drawing made by Bond from this plant (the left hand part of Fig. 5.34). In Bond's illustration the branch is square in cross-section and it does not show the deep constrictions that are typical of the rhizomatous plants from the Jansenville district. Furthermore, N.E. Brown knew the plant from which Haworth had described *E. caerulescens*, which was still alive at Kew in November 1876 when he pressed some pieces from it. It did not have a rhizomatous habit. Therefore, the association of rhizomatousness with *E. caerulescens* is erroneous (as emphasized also by Lotsy and Goddijn (1928)) and this name must refer to the non-rhizomatous plants. Consequently, also, *E. ledienii* is a synonym of *E. caerulescens*. This was not recognised in Bruyns et al. (2006), where *E. ledienii* was treated as a separate species from *E. caerulescens*. From Bruyns (2012), the rhizomatous plants were treated as a separate species, *E. radyeri*.

Euphorbia clavigera N.E.Br., *Fl. Cap.* 5 (2): 362 (1915). Type: Swaziland, near Bremersdorp (Manzini), 1800', 5 Jan. 1905, *Burtt-Davy 3010* (K, holo.).

Euphorbia persistens R.A.Dyer, *Fl. Pl. South Africa* 18: t. 713 (1938). Type: Moçambique, East of Ressano Garcia, July 1936, *F.Z.van der Merwe E14 sub PRE 23395* (PRE, holo.; K, PRE, iso.).

Euphorbia umfoloziensis Peckover, *Aloe* 28: 37 (1991). Type: South Africa, Natal, near Dingaanstat, 10 Apr. 1981, *Peckover* (PRE, holo.).

Bisexual spiny glabrous succulent 30–200 mm tall with 1–several short turnip-shaped stems from central thick carrot-shaped mainly subterranean stem up to 300 × 50–150 mm from which fibrous roots arise, each stem with rosette of angled and spiny branches around its apex. *Branches* ascending to erect, 30–200 × 15–30 mm, clavate and not segmented to constricted into 2–6 ± obovate wedge-shaped segments, smooth, dark to pale grey-green (sometimes reddish when young) with fine cream or paler green mottling along groove between angles and extending to near axillary buds; *tubercles* fused into 3–4 (5) prominent and sometimes slightly spiralling wing-like angles along branches, deltate, laterally flattened and projecting 2–10 mm from angles, with spine-shield 1–2 mm broad around apex spreading up along edge of angle to axil of tubercle and down for 3–6 mm below spines, bearing 2 spreading and widely diverging yellow-brown to grey spines 6–10 (15) mm long and 2 minute prickles ± 1 mm long alongside axillary buds; *leaf-rudiments* on tips of new tubercles towards apices of branches, 0.5–1.5 × 1–1.5 mm, erect, fleeting, ovate and slightly channelled down middle, apiculate, sessile, with small irregular stipular prickles 0.5–1 mm long. *Synflorescences* 1–8 per branch towards apex on terminal segment, each a solitary cyme in axil of tubercle, on peduncle 2–5 × 3–5 mm, each cyme with 3 vertically disposed cyathia or only one cyathium, central male or bisexual, lateral 2 bisexual and developing later each on stout peduncle 2–5 mm long and thick, with 2 ovate bracts 1–4 × 3–4 mm subtending cyathia; *cyathia* cupular, glabrous, 5–7 mm broad (2–3 mm long below insertion of glands), with 5 lobes with deeply incised margins, bright yellow to yellow-green or green; *glands* 5, transversely oblong, very short, 2.5–4 mm broad, yellow, spreading, inner margins sometimes raised, outer margins entire rounded and slightly incurved at middle; *stamens* glabrous, bracteoles enveloping groups of males, with finely divided tips, glabrous; *ovary* 3-angled, glabrous, raised on very short pedicel to sessile; styles 3.5–5 mm long, branched in upper two thirds. *Capsule* 7–10 mm diam., green with red at apex and along edges, obtusely and deeply 3-angled, glabrous, sessile.

Distribution & Habitat

Euphorbia clavigera is known mainly around and within Swaziland, from near Ressano Garcia in Moçambique to near the capital, Mbabane. Further south there are more isolated records in Kwazulu-Natal around Nongoma and Ulundi. However, there are much more distant records suggesting a considerably wider but still poorly documented distribution: Fourie (1988) mentioned plants near Punda Milia in the northern part of the Kruger National Park, of which there is no herbarium record and a specimen (*Bruyns 7676* (BOL)) from the mouth of the Save River near Mambone in central Moçambique also falls within this species.

Plants are found in a wide variety of habitats. These range from patches of shallow soil on and filling crevices in flat, rhyolite outcrops, as along the western foot of the Lebombo Mountains slightly north of the border of Swaziland, here occurring with many other succulents (including *Euphorbia*

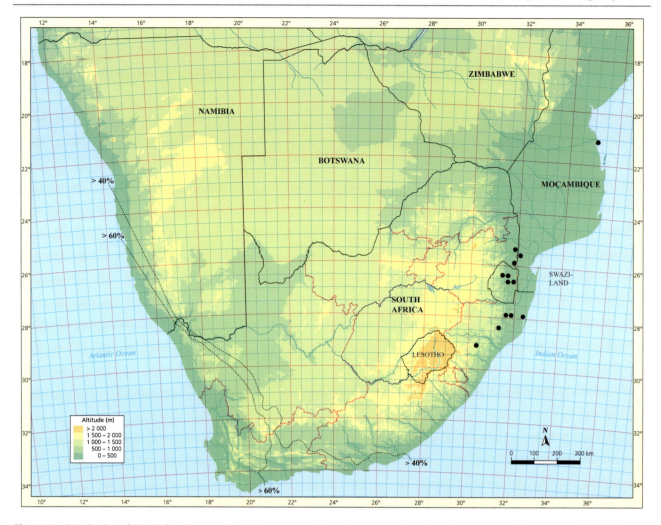

Fig. 5.35. Distribution of *Euphorbia clavigera* (including one record from the mouth of the Save River in central Moçambique) (© PVB).

cooperi, *E. schinzii* and *Pachypodium saundersiae*), to comparatively deep but stony and gravelly loam with much quartz-gravel on the surface among short trees (as near Mbabane, Swaziland) and to very eroded and overgrazed patches among gravel and between scattered, spiny bushes (as near Nongoma, Natal). They are often very common where they occur, but seem to form localised colonies of plants.

Diagnostic Features & Relationships
A typical specimen of *Euphorbia clavigera* consists of a mainly subterranean stem that is usually between 8 and 10 cm in diameter at its maximum, though it may reach 15 cm thick on occasion in older plants. Around its apex the stem bears a dense rosette of branches that hide it completely. When exposed, the stem is usually covered with a corky bark and shows no photosynthetic activity. The length of this stem varies greatly with the depth of the soil and it may even run horizontally beneath the surface when plants grow in shallow soil on slabs of rock. With time the stem may divide and develop additional, similar-looking, subterranean branches near its apex, each of which also bears a rosette of branches around its apex.

Each branch is joined to the stem by means of a very slender base. From this they usually spread outwards, then become erect. Generally, the branches last for 2–4 years, after which the outer ones die off and are replaced by new branches at the apex of the stem. Occasionally all the branches die off and are replaced in the next growing season. The branches are very variable in length, though they are often short and may be only 30–50 mm long. Their length appears to depend on the age of the plant, with longer ones on older specimens and also on well-sheltered plants. Most branches have three angles, though specimens with three, four or even five angles on some or all of their branches may occur. From their slender bases, the branches become broader towards their somewhat truncate apices and they may be segmented along the way, depending on how many years they have lasted. In many plants the branches are attractively marked by curved cream or pale green bars

Fig. 5.36. *Euphorbia clavigera*, near type-locality, *PVB 11864*, near Luve, Swaziland, 7 Jan. 2011 (© PVB).

Fig. 5.37. *Euphorbia clavigera*, *PVB 11902*, Mbangwane, south of Komatipoort, South Africa, 24 Jul. 2011 (© PVB).

Fig. 5.38. *Euphorbia clavigera*, *PVB 11836*, between Nongoma and Mkuzi, South Africa, 31 Dec. 2010 (© PVB).

that run from the axillary bud to the centre of the groove between the angles and then continue down this groove (usually fading towards the base of the branch). Although the much flattened tubercles are fused into conspicuous angles along the branch, they still project from the angles and remain quite prominent, usually with the largest just after the narrowest portion of a segment. Each tubercle bears a pair of spreading grey spines (initially red) which vary greatly in length, but they are mostly 5–10 mm long. The spine-shield is small, usually restricted to the summit of the tubercle, and only occasionally forms a continuous margin along the angles.

Fig. 5.39. *Euphorbia clavigera*, *PVB 11836*, between Nongoma and Mkuzi, South Africa, 31 Dec. 2010 (© PVB).

Flowering in *E. clavigera* generally takes place in early summer (though it may occur later as well), once new growth has occurred on the branches and cyathia then develop in the axils of recently produced tubercles. Each cyme is solitary in the axil of a tubercle and bears one or three cyathia (in the latter case vertically disposed). The cyathia are borne on relatively long but stout peduncles in plants from Swaziland and northwards. In material from Natal (Nongoma and Ulundi) the peduncle is very short. In plants from around Ulundi only a single bisexual cyathium develops on each cyme (and these are what gave rise to the name *E. umfoloziensis*) but around Nongoma the usual three cyathia develop. The cyathia are broad and shallow and vary between bright yellow and green, with a quite large, deeply 3-angled capsule developing partly inside them after fertilization.

Fig. 5.40. *Euphorbia clavigera*, *PVB 11902*, Mbangwane, south of Komatipoort, South Africa, flowering in cultivation, Dec. 2016 (© PVB).

Euphorbia umfoloziensis was distinguished from *E. vandermerwei* (with which it shared the single cyathia in each cyme) by various floral features, including the reddish hue of most parts of the cyathium in *E. vandermerwei* and the longer peduncles. However, its real relationships are with *E. clavigera* and the presence of this species as far south as Nongoma in Natal makes the likelihood of another such similar species slightly further south more doubtful. Apart from

Fig. 5.41. *Euphorbia clavigera*, single cyathium per cyme, *PVB 11323*, Dingaanstat, near Ulundi, South Africa, 31 Dec. 2010 (© PVB).

the fact that the cyathia are solitary in each cyme (with the lateral cyathia remaining rudimentary) and the single cyathium is then bisexual, the styles may be slightly longer and the peduncles slightly thicker and shorter but no other significant differences between *E. clavigera* and *E. umfoloziensis* could be found. Consequently, despite being recognised as distinct in Bruyns 2012, this name has been placed now in synonymy.

Fig. 5.42. *Euphorbia clavigera*, with capsules, *PVB 11836*, between Nongoma and Mkuzi, South Africa, 31 Dec. 2010 (© PVB).

Euphorbia clavigera is closely related to *E. enormis*, *E. tortirama* and *E. vandermerwei* and the differences between them are discussed under *E. enormis* and *E. tortirama*.

History

Euphorbia clavigera was first recorded by Joseph Burtt-Davy near the present-day Manzini (then known as Bremersdorp) in Swaziland in January 1903. Some material was also sent by Gerstner to Marloth from the Nongoma area of Natal in January 1928 and these constitute the first records of this species from Natal, where some collections were later distinguished as '*E. umfoloziensis*. Plants from near Moamba in Moçambique were distinguished as *E. persistens* by R.A. Dyer but Leach (1976c: 17) first suggested that these were conspecific with *E. clavigera*.

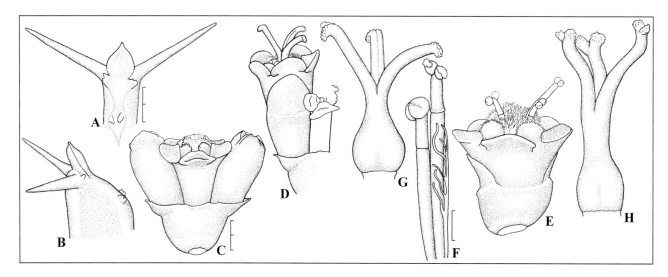

Fig. 5.43. *Euphorbia clavigera*. **A**, young spines and leaf-rudiment from above (scale 2 mm, as for **B**). **B**, young spines and leaf-rudiment from side. **C**, side view of cyme just before first male stage (scale 2 mm, as for **D**, **E**). **D**, side view of cyme in female stage. **E**, side view of solitary cyme. **F**, anthers and bracteole (scale 1 mm, as for **G**, **H**). **G**, **H**, female floret. Drawn from: **A**, **B**, *PVB 11836*, between Nongoma and Mkuzi, South Africa. **C**, **D**, **G**, *PVB 11902*, Mbangwane, south of Komatipoort, South Africa. **E**, **F**, **H**, *PVB 11323*, Dingaanstat, near Ulundi, South Africa (© PVB).

Euphorbia clivicola R.A.Dyer, *Bothalia* 6: 221 (1951). Type: South Africa, Transvaal, Lunsklip, 20 miles north of Potgietersrust, 13 Sept. 1946, *Plowes sub PRE 28386* (PRE, holo.; K, iso.).

Bisexual spiny glabrous succulent 0.03–0.15 (0.3) × 0.15–0.5 m, with many branches congested into dense clump from somewhat thickened and mainly subterranean stem 50–150 × 20–60 mm, with fibrous roots arising from stem, densely branched at and slightly below ground level. *Branches* erect, 20–60 × 5–15 mm, subcylindrical and sometimes thicker towards base, not constricted into segments, smooth, pale bluish green (yellow-green or reddish); *tubercles* in almost decussate pairs arranged into 4 very obscure weakly spiralling rows, conical and sloping gradually below spines, truncate above, projecting 3–6 mm from angles, with spine-shields 2–8 (10) × 1–2 mm and 2–3 mm broad above spines (just encircling leaf-rudiment) and 1–3 (6) mm long and deltate to very slender below spines but remaining well separated from next, bearing 2 spreading initially reddish brown (later grey) spines 3–10 (12) mm long; *leaf-rudiments* on tips of new tubercles towards apices of branches, ± 0.5 × 0.7–1.2 mm, erect, fleeting, ovate to semi-circular, sessile, with tiny stipular prickles ± 0.25 mm long. *Synflorescences* several per branch usually towards apex, each a solitary cyme in axil of tubercle, sessile, each cyme with 3 transversely disposed cyathia, central male, lateral 2 bisexual and developing later each on short peduncle 1–2 mm long and thick, with 2 ovate bracts 1–2 × 1–2 mm subtending lateral cyathia; *cyathia* shallowly cupular, glabrous, 3–4 mm broad (2 mm long below insertion of glands), with 5 pale yellow or cream obovate lobes with deeply incised margins, bright yellow; *glands* 5, transversely rectangular and contiguous, 1–1.5 mm broad, bright yellow to brownish yellow, spreading, inner margins slightly raised to flat, outer margins entire and spreading, surface between two margins dull; *stamens* glabrous, bracteoles enveloping groups of males, with finely divided tips, glabrous; *ovary*

5.1 Sect. Euphorbia

3-angled, glabrous, slightly reddish green near top, sessile; styles 2–2.5 mm long, branched to slightly below middle. *Capsule* 3–4 mm diam., obtusely 3-angled, glabrous, sessile.

Two subspecies are recognized, separated as follows:

1. Branches not rhizomatous, uniformly coloured, spines 4–10 mm long, cyathial glands bright yellow...subsp. **clivicola**
1. Branches often somewhat rhizomatous, with paler stripe along groove between rows of tubercles, spines to 12 mm long, cyathial glands brownish yellow...subsp. **calcritica**

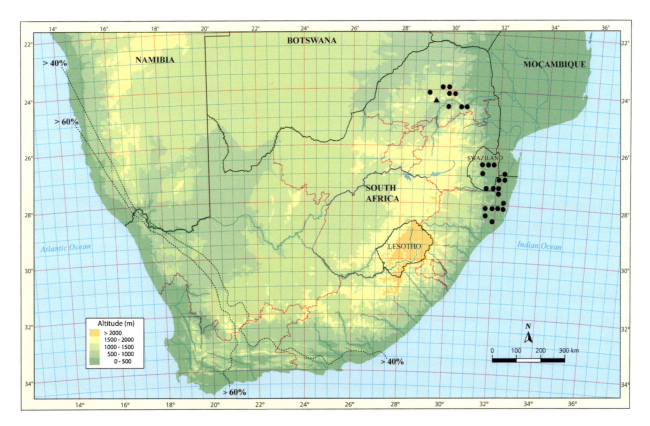

Fig. 5.44. Distribution of *Euphorbia clivicola* (● = ssp. *clivicola*; ▲ = ssp. *calcritica*) (© PVB).

Euphorbia clivicola subsp. clivicola

Bisexual succulent 0.05–0.15 (0.3) × 0.15–0.5 m, without rhizomatous branches, stem often somewhat swollen underground. *Branches* 20–150 × 5–15 mm, smooth, dull pale bluish green to dark green; *tubercles* in decussate pairs (to alternating) arranged into 4 rows, with spine-shields 3–5 mm long, ± 1 × 2 mm above spines and 3–4 mm long and deltate below spines (remaining well separated from next), spines 3–10 (12) mm long; *leaf-rudiments* ± 1–1.5 × 1 mm, with stipular prickles ± 0.25 mm long. Styles branched to around or slightly below middle.

Fig. 5.45. *Euphorbia clivicola* ssp. *clivicola*, dense clump with very short branches Percy Fife Nature Reserve, near Potgietersrus, South Africa, c. 1986 (© S.P. Fourie).

Fig. 5.46. *Euphorbia clivicola* ssp. *clivicola*, branches around 8 cm long, *PVB 12115*, near Dikgale, NE of Boyne, South Africa, 22 Feb. 2012 (© PVB).

Distribution & Habitat

Subsp. *clivicola* is known between Potgietersrust and some 30 km north-east of Polokwane (Pietersburg) and from there into the mountains along the Olifants River and some of the low-lying areas closer to this river. Further populations are known in Swaziland and Kwazulu-Natal. In Kwazulu-Natal they are widespread in the north from the low-lying coastal area to the dry foothills below the escarpment.

Subsp. *clivicola* occurs in stony to rocky habitats, on gently sloping hillsides derived from schists and granites, often with large amounts of quartz pebbles strewn on the surface. In some areas it grows between stones. In others, plants are wedged tightly into crevices between larger rocks, while in some parts of coastal Kwazulu-Natal it occurs in gravelly sand with no obvious rocks. Plants are usually found among short bushes or grasses, often in relatively barren spots where the grass-cover is minimal.

Fig. 5.47. *Euphorbia clivicola* ssp. *clivicola*, among tufts of grass in gravelly sand, *PVB 11851*, Makhatini Flats, Kwazulu-Natal, South Africa, 2 Jan. 2011 (© PVB).

5.1 Sect. Euphorbia

Fig. 5.48. *Euphorbia clivicola* ssp. *clivicola*, plant covered with cyathia, *PVB 13743*, just north of Ohrigstad, South Africa, 31 Jul. 2019 (© PVB).

Diagnostic Features & Relationships

Plants of subsp. *clivicola* are usually mound-like, with many short, undivided branches arising from the ground in a compact group mostly up to about 30 cm in diameter. Excavation of this structure reveals a thick, mostly underground, eventually somewhat swollen stem at the apex of which there are many short branches which branch again close to the surface. It is further ultimate branches from these lower ones that appear above the ground. The branches that appear above the ground generally remain short and

Fig. 5.49. *Euphorbia clivicola* ssp. *clivicola*, bright yellow cyathia, *PVB 13743*, just north of Ohrigstad, South Africa, 31 Jul. 2019 (© PVB).

Fig. 5.50. *Euphorbia clivicola* ssp. *clivicola*, *PVB 11851*, Makhatini Flats, Kwazulu-Natal, South Africa, flowering 9 Dec. 2018 (© PVB).

stout (though their thickness is very variable) and have relatively few, quite widely spaced tubercles that are weakly arranged into four rows. Some of these branches may be deciduous.

Flowering times are variable across the range of this subspecies, with this happening in July-August in the former Transvaal and as late as December-January in the low-lying areas of Kwazulu-Natal.

Previously the concept of *Euphorbia clivicola* was largely confined to the type. However, apart from their more slender branches and their habit of somctimes flowering in midsummer, no other significant differences have been found between the type of *E. clivicola* and plants found over a wide area. In some areas (such as Makhatini Flats, *Bruyns 11851* (BOL), Burgersfort, *Bruyns 13555* (BOL)) the branches are dull green and not the usual pale bluish green, but no other differences were noted between these and plants from around Polokwane (Pietersburg). Consequently, these are now all included in subsp. *clivicola*.

History

Euphorbia clivicola was described from plants brought to R.A. Dyer by D. Plowes in September 1946. Dyer (1951) mentioned another collection by Kirsten from 'near Pietersburg', which may have been made earlier. Plants appear to have become very rare at the type locality (which now lies within the Percy Fife Nature Reserve) and those around Pietersburg are greatly threatened by the sprawling urban expansion that has taken place there.

Fig. 5.51. *Euphorbia clivicola* ssp. *clivicola*, *PVB 12115*, near Dikgale, NE of Boyne, South Africa, flowering 20 Jan. 2018 (© PVB).

5.1 Sect. Euphorbia

Fig. 5.52. *Euphorbia clivicola* ssp. *clivicola*, with capsules, *PVB 11851*, Makhatini Flats, Kwazulu-Natal, South Africa, 2 Jan. 2011 (© PVB).

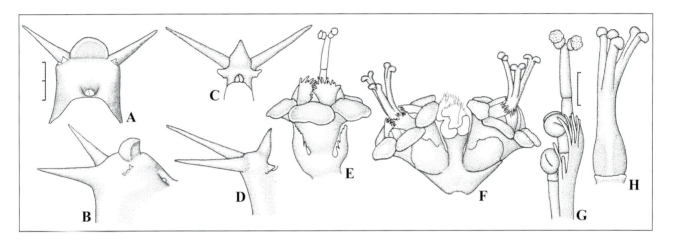

Fig. 5.53. *Euphorbia clivicola* ssp. *clivicola*. **A**, **C**, young spines and leaf-rudiment from above (scale 2 mm, as for **B–F**). **B**, **D**, young spines and leaf-rudiment from side. **E**, side view of cyme at early first male stage. **F**, side view of cyme in female stage. **G**, anthers and bracteoles (scale 1 mm, as for **H**). **H**, female floret. Drawn from: **A**, **B**, *PVB*, outskirts of Pietersburg, South Africa. **C–H**, *PVB 11329*, Ulundi, Kwazulu-Natal, South Africa (© PVB).

Euphorbia clivicola subsp. **calcritica** Bruyns, *Phytotaxa* 436: 207 (2020). Type: South Africa, between Zebediela and Roedtan, 1030 m, 31 Oct. 2011, *Bruyns 12045* (BOL, holo.).

Slightly rhizomatous succulent forming small clusters of branches to 150 mm diam. *Branches* 40–60 (150) × 6–10 mm thick, subcylindrical, grey-green to brownish green with pale stripe between rows of tubercles; *tubercles* with spine-shields 5–10 × 1–3 mm and ± 3 mm broad above spines (just encircling leaf-rudiment) and 4–8 mm long and very slender below spines, bearing 2 spreading initially reddish brown (later grey) spines 5–12 mm long; *leaf-rudiments* ± 0.5 × 0.7–1.2 mm, erect, fleeting, ovate, with tiny stipular prickles < 0.25 mm long. *Cyathia* and *glands* slightly brownish yellow with reddish bracts.

Fig. 5.54. *Euphorbia clivicola* ssp. *calcritica*, branches slightly rhizomatous, *PVB 12045*, near Zebediela, South Africa, 22 Feb. 2012 (© PVB).

Distribution & Habitat

This subspecies is common around Zebediela and southwards to near Roedtan, which is the only area from where there are records (and it may well occur more widely). At the type locality it occurs in flat areas with much calcritic rubble on and beneath the surface among short tufts of grass. There is also a scanty cover of other small bushes (many not exceeding 15 cm tall) among larger Acacias, with *Cynanchum viminale* and *Kleinia longiflora* in stony calcareous soil with many calcrete chips on and beneath the surface.

Diagnostic Features & Relationships

Subsp. *calcritica* shares with ssp. *clivicola* the manner in which the prominent tubercles are not joined into angles on the branches. Its branches may be slightly rhizomatous, while those of ssp. *clivicola* are usually not rhizomatous at all. It differs from subsp. *clivicola* by the longer spines (though some plants included here under subsp. *clivicola* may have spines to 12 mm long and these may be very 'mean-looking'), the pale stripe along the middle of the branches and the slightly brownish yellow cyathia.

Subsp. *calcritica* also shares some features with *E. venteri*. These are the slightly rhizomatous and spreading habit of the branches, their similar marking with fine pale lines between the tubercles and the dull slightly brownish yellow of the cyathial glands. However, its geographical distance from *E.*

Fig. 5.55. *Euphorbia clivicola* ssp. *calcritica*, *PVB 12045*, near Zebediela, South Africa, flowering 12 Oct. 2012 (© PVB).

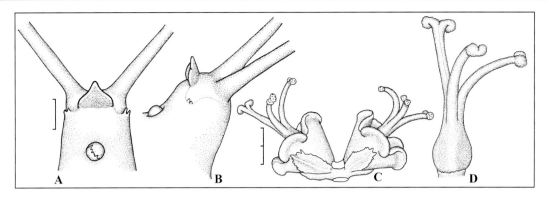

Fig. 5.56. *Euphorbia clivicola* ssp. *calcritica*. **A**, young spines and leaf-rudiment from above (scale 1 mm, as for **B**, **D**). **B**, young spines and leaf-rudiment from side. **C**, side view of cyme at female stage (scale 2 mm). **D**, female floret. Drawn from: *PVB 12045*, near Zebediela, South Africa (© PVB).

venteri and the arrangement of the tubercles suggests that it is closer to *E. clivicola*.

In cultivation, subsp. *calcritica* flowers mainly in October.

History

This unusual form of *E. clivicola* was first observed by Frank Bayer in pebbly ground at a place called Yella Mara (alt. ± 2500') in the Potgietersrust district on 12 Nov. 1947 (b/w photo in PRE). At PRE there is also a colour photograph of a collection by F. Venter from Roedtan but no specimens were made from either of these records.

Euphorbia confinalis R.A.Dyer, *Bothalia* 6: 222 (1951). Type: South Africa, Transvaal, Kruger Nat. Park, 2 miles east of 'The Gorge Camp', 900', 20 May 1949, *Codd & De Winter 5580* (PRE, holo.; K, NH, iso.).

Bisexual spiny glabrous succulent tree 3–10 m tall, with gradually deciduous branches forming rounded crown and arising in upper part of cylindrical erect solitary or sparingly forked trunk-like stem 3–8 m tall, trunk naked except for spines below crown, with many fibrous roots spreading from base. *Branches* spreading then ascending and rarely rebranching, older branches drying up and falling off stem after a few years, 0.3–1.5 m × 35–60 (100) mm, (3-) 4- (5-) angled, shallowly constricted into segments 50–200 mm long, smooth, pale green; *tubercles* very evenly fused into (3) 4 (5) prominent wing-like angles with surface deeply concave between angles, low-conical, laterally flattened and projecting 2–5 mm from angles, spine-shields continuous forming hard grey to corky grey to black margin 2–3 mm broad along angles to projecting downwards 3–8 mm below spines and not joined up, bearing 2 spreading and widely diverging brown to grey spines 3–12 mm long to spineless; *leaf-rudiments* on tips of new tubercles towards apices of branches, (1) 3–7 × 1.5–2.5 mm, spreading, deciduous, ovate (E. of Espungabera) to narrowly lanceolate (young plant from Komatipoort), sessile, with small irregular brown stipules ± 1 mm long. *Synflorescences* many per branch towards apex (mostly on last segment of branch), with 1–3 cymes in axil of each tubercle, shortly peduncled, each cyme with 3 vertically disposed cyathia, central male, outer two bisexual and developing later each on short peduncle 2–3 × 2–3 mm thick, with 2–4 small broadly ovate bracts 1–2 × 2 mm subtending lateral cyathia; *cyathia* cupular, glabrous, 5–8 mm broad (2 mm long below insertion of glands), with 5 obovate lobes with deeply toothed margins, yellow to yellow-green; *glands* (4) 5, broadly cuneate, contiguous, 1.5–2.5 mm broad, yellow, slightly spreading, inner margins slightly raised, outer margins entire and slightly raised, surface between two margins dull; *stamens* with glabrous pedicels, anthers yellow suffused with red (especially around pore), bracteoles enveloping groups of males, with finely divided tips, glabrous; *ovary* 3-angled with slightly flattened sides, glabrous, green, raised but included on erect pedicel ± 1 mm long later increasing to 2–3 mm and pushing capsule out of cyathium; styles ± 2 mm long, spreading, branched to near middle. *Capsule* 8–10 mm diam., clearly 3-angled, glabrous, pale green suffused with pink to red, exserted on short erect pedicel 5–8 mm long.

Leach (1966) divided *E. confinalis* into two subspecies, subsp. *confinalis* and subsp. *rhodesica*. These are distinguished on the number of angles in the trunk of a young plant, the number of angles in the branches and differences in the bracts below the cyathia. Of these, only subsp. *confinalis* occurs within our area and the description above and what follows below apply to this subspecies only (though the known distributions of both are shown on Fig. 5.57).

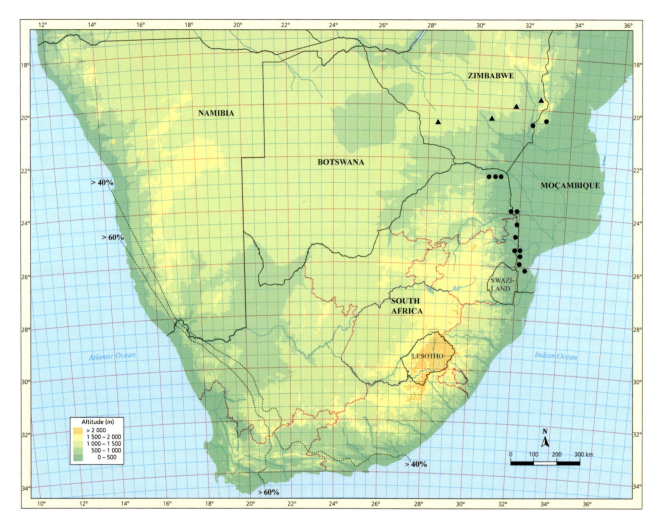

Fig. 5.57. Distribution of *Euphorbia confinalis* (● = ssp. *confinalis*, ▲ = ssp. *rhodesica*) (© PVB).

Distribution & Habitat
Euphorbia confinalis is mainly confined to the slopes of the Lebombo Mountains in Moçambique, South Africa and Swaziland. It is also found further north, beyond this range, in Moçambique and in eastern Zimbabwe. In South Africa it also occurs in the eastern end of the Soutpansberg. Here it is common, usually along the tops of ridges, from the villages of Muswodi and Thengwe eastwards towards the Kruger National Park.

5.1 Sect. Euphorbia

Euphorbia confinalis usually occurs in small to substantial colonies on slopes among low, rocky outcrops, where the cover of grass is scantier and where fires then pose less of a threat to younger plants. Although the dominant cover of these slopes consists of scattered trees and grasses, other succulents like *Aloe*, *Cissus*, *Sansevieria* and *Sarcostemma* are usually present. In the eastern Soutpansberg, it occurs among trees and shrubs on stony, sandstone outcrops, with occasional plants of *E. ingens* and often near or among stands of *Androstachys johnsonii*.

Diagnostic Features & Relationships

Trees of *E. confinalis* are imposing and often very tall, with large numbers of ascending branches clustering at the top of the undivided or sometimes forked trunks to form almost spherical heads. In very young plants the stem is 4-angled and may be attractively mottled with cream on green, but this is lost as it grows older. The branches in young plants are often nearly erect and may have distinctly wavy angles but, with age, they have a more spreading habit and straighter angles. They mostly have a pale green colour. When the plant is young the branches and the stem are protected by conspicuous spines, but the stem soon loses its spines and, once they are sufficiently far from the ground, the branches are also usually weakly spined or spineless. When spines are

Fig. 5.58. *Euphorbia confinalis* ssp. *confinalis*, *PVB 12061*, eastern Soutpansberg, South Africa, 4 Nov. 2011 (© PVB).

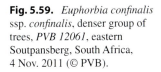

Fig. 5.59. *Euphorbia confinalis* ssp. *confinalis*, denser group of trees, *PVB 12061*, eastern Soutpansberg, South Africa, 4 Nov. 2011 (© PVB).

present the spine-shield is grey (often with a fine black margin) and continuous but once the spines disappear the spine-shield becomes a blackish and corky ridge which is still continuous but is especially broad around each node. Considerable variation has been observed in the tiny leaf-rudiments, which were comparatively long and narrowly lanceolate in young plants from around Komatipoort but shorter and ovate in plants from east of Espungabera in Moçambique. The stipules also varied and were more slender in the plants from Komatipoort than in those from east of Espungabera.

Fig. 5.60. *Euphorbia confinalis* ssp. *confinalis*, young tree, north of Moamba, Moçambique, 11 Jan. 2004 (© PVB).

Fig. 5.61. *Euphorbia confinalis* ssp. *confinalis*, mature tree with *Aloe marlothii* alongside dense bush, north of Moamba, Moçambique, 11 Jan. 2004 (© PVB).

Flowering in *E. confinalis* occurs from June to early August. In many plants the cyathia arise in huge numbers on the last segment of most of the branches, with as many as three cymes in the axil of each tubercle. The cyathia emit a faint, sourish-sweet odour and their yellow glands lend a yellowish colour to the tips of the branches when a tree is in full bloom. When the styles are mature, the ovary is included within the cyathium, but the pedicel lengthens gradually to push it out and ultimately the pedicel reaches a length of around 6 mm. The ovary is somewhat triangular in the early stages and, while it remains triangular, the angles become more rounded with age. Once they are mature, the comparatively large capsules are usually red or pink above, becoming paler towards their bases. Male flo-

5.1 Sect. Euphorbia

rets have distinctly reddish anthers, with the blush of red darkest around the pore.

Euphorbia confinalis and *E. triangularis* are not easily separated and the differences between them are discussed under the latter. Even though they are similar, they are not closely related and the closest relative of *E. confinalis* is *E. candelabrum* from Angola. The shape of the trees is very similar in both, the capsules are similar in shape and in *E. confinalis* the anthers are suffused with red. This is unusual among the trees in southern Africa but similar to the anthers of *E. candelabrum*, which are purplish red.

Fig. 5.62. *Euphorbia confinalis* ssp. *confinalis*, flowering branch, *PVB 11764*, Komatipoort, South Africa, 24 Jul. 2011 (© PVB).

Fig. 5.63. *Euphorbia confinalis* ssp. *confinalis*, fruiting branch, *PVB 11764*, Komatipoort, South Africa, 12 Sep. 2010 (© PVB).

History

Euphorbia confinalis was collected first by F.Z. van der Merwe in 1936 near Ressano Garcia in Moçambique. White et al. (1941: 895–8, fig. 1023) provided a brief description and a photograph of *E. confinalis* under their account of *E. triangularis* and expressed the view that it was unlikely to be 'more than varietally distinct' from *E. triangularis*. However, with the accumulation of further collections, mostly from exploration of the Kruger National Park by L.E. Codd and others, it became clear that it was a distinct species and it was finally described by R.A. Dyer in 1951.

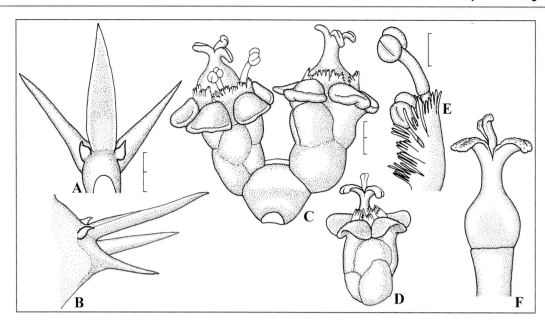

Fig. 5.64. *Euphorbia confinalis* ssp. *confinalis*. **A**, young spines and leaf-rudiment from above (scale 2 mm, as for **B**). **B**, young spines and leaf-rudiment from side. **C**, side view of cyme at female stage (scale 2 mm, as for **D**). **D**, side view of cyathium in female stage. **E**, anthers and bracteoles (scale 1 mm, as for **F**). **F**, female floret. Drawn from: *PVB 11764*, Komatipoort, South Africa (© PVB).

Euphorbia cooperi N.E.Br. ex A.Berger, *Sukkul. Euphorb.*: 83 (1906). Lectotype (Bruyns 2012): South Africa, Natal, Umgeni Valley?, 1862, *Cooper*, cultivated plant at Kew Gardens, pressed Sept. 1899 by N.E. Brown (K).

Bisexual spiny glabrous succulent tree 3–7 m tall with grey-brown cylindrical trunk (at first 4–6-angled, green and often brightly banded with green and cream) 0.5–4 m tall from extensive system of woody and fibrous roots, branching and rebranching extensively to form roughly hemispherical crown. *Branches* spreading then becoming erect (so curving upwards only gradually from base), older branches usually drying up and falling off stem after a few years (especially in taller plants), (35) 60–90 mm thick, deeply constricted into conical-ovate segments, smooth, green; *tubercles* fused into (3) 4–6 prominent wing-like angles with deep triangular channels between them, laterally flattened and rounded and projecting 1–2 mm from angles, with spine-shields united into continuous sometimes slightly sinuate horny grey margin 4–6 mm broad along angles bearing 2 spreading and widely diverging grey spines 3–10 mm long; *leaf-rudiments* on tips of new tubercles towards apices of branches and stem, 1–2 × 1–2 mm, spreading, fleeting, broadly ovate and slightly concave above, sessile, usually with small irregularly shaped stipular prickles < 0.5 mm long. *Synflorescences* in large numbers per branch towards apex (mostly on last segment of branch), with 1–3 cymes in axil of each tubercle, on very short peduncle 1–2 × 2–3 mm, each cyme with 3 vertically disposed cyathia, central male, outer 2 bisexual and developing later, with several ovate bracts 1–4 × 2–4 mm subtending cyathia; *cyathia* deeply conical, glabrous, 5–8 mm broad (3 mm long below insertion of glands), with 5 lobes with deeply incised margins, yellow-green; *glands* 5, transversely rectangular or elliptic, 3–4 mm broad, yellow to green, slightly convex above, outer margins entire and spreading; stamens with glabrous pedicels, bracteoles finely divided nearly to base and enveloping groups of males; *ovary* ellipsoidal, glabrous, on very short pedicel < 0.5 mm long, often suffused with red towards top; styles 2–5.5 mm long, free nearly to base to branched in upper third. *Capsule* 9–14 mm diam., triangular to deeply 3-lobed, glabrous, usually shiny dark red, exserted on short slightly curved later erect pedicel to 5 mm long.

Leach (1970a) divided *E. cooperi* into three varieties (var. *cooperi*, var. *calidicola* from Malawi, Moçambique, Zambia and Zimbabwe and var. *ussanguensis* from Malawi, Tanzania and Zambia). These are distinguished by the degree of exsertion of the capsule from the cyathium, the number of angles on the branches and the thickness of these angles. Only var. *cooperi* occurs in southern Africa and so the description above and what follows apply to this variety only.

Distribution & Habitat
Euphorbia cooperi is widely distributed from the north-eastern parts of South Africa (the former Natal and northern parts of the Transvaal), northern Moçambique, Swaziland

5.1 Sect. Euphorbia

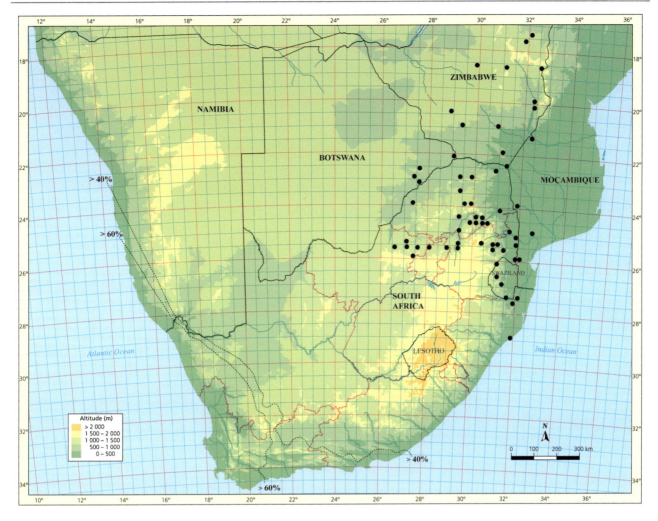

Fig. 5.65. Distribution of *Euphorbia cooperi* var. *cooperi* in Africa south of 17° S (© PVB).

and Botswana, through Zimbabwe to northern Tanzania, where it is found as far as the southern shores of Lake Victoria.

Plants generally form island-like, often fairly dense colonies which may be a considerable distance from the next such group. They are especially associated with domes of granite and other rocky outcrops, where they usually grow among large rocks on the upper parts of hills or in shallower soils around the edges of exposed sheets of granite. They will sometimes even be found in shallow soils on the domes themselves (with *Xerophyta* and succulents such as *Aloe*, *Plectranthus*, *Sansevieria*), though the soil must be deep enough to support the tree and its rootstock. *Euphorbia cooperi* is known at altitudes of between 200 and 1500 m.

Diagnostic Features & Relationships

Euphorbia cooperi is one of the most characteristic of the thick-stemmed and thick-branched, 'chandelier-shaped', tree-forming species of *Euphorbia* in the warmer parts of southern Africa. In a given habitat, plants usually occur in a wide range of sizes, from the large shrubs of younger specimens to trees in older plants. Large individuals generally have a very stout, almost cylindrical trunk which is usually bare of branches for a length of 1–2 m above the ground where the branches have fallen off. Towards the top of this trunk, the very conspicuous, comparatively thick branches with their very broadly curving, ascending-erect habit arise to form an almost hemispherical crown. In most specimens the branches have four or five angles. Another conspicuous feature is that the branches are always deeply segmented and each segment is always much broader towards its base than towards its apex (where the spines are also closer together). The wings along the branches are comparatively thick (unlike those of, say, *E. halipedicola*, which are thin by comparison), with a continuous, grey-brown, horny margin along them on which the relatively short spines arise. The stems of young plants may be somewhat mottled with paler on darker green, but they soon lose this distinctive colouring and many do not show it at all. The leaf-rudiments are always small, even in young specimens.

In *E. cooperi*, flowering takes place between April and August, when the last segment on most of the upper branches becomes covered along the angles with a dense beard of cyathia which may reach 15 mm broad. Mostly three cymes develop in the axil of each tubercle. The cyathia are typically yellow-green, lending a yellowish hue to the tips of the branches and they emit a faint sweetish scent. The capsules remain embedded in the cyathia after fertilisation, with only slight lengthening of their pedicels as they mature. Towards the end of July, the cyathia start to dry up and the brightly shiny, dark red capsules begin to appear, often remaining like this until September before dehiscing.

Though it more frequently occurs in dense colonies than *E. ingens* does, *Euphorbia cooperi* never attains quite the stature and bulk that *E. ingens* may achieve. The two species can usually easily be separated by the much more spreading, hemispherical crown with far fewer, very distinctly segmented branches of *E. cooperi*. Young plants of *E. cooperi* are less conspicuously mottled than those of *E. ingens* and they lose this mottling sooner. They also lack the much longer juvenile leaves of *E. ingens*. In *E. cooperi*, the female florets lack the somewhat extended calyx that is usually found in *E. ingens*. Leach (1970a) believed that *E. cooperi* is closely related to *E. grandicornis*, differing by the shorter spines and taller stature. Evidence from unpublished molecular data is that it is close to the Angolan *E. faucicola* and to *E. enormis* and *E. tortirama*.

Fig. 5.66. *Euphorbia cooperi* var. *cooperi*, foothills of Soutpansberg, near Soutpan, South Africa, 30 Dec. 1996, with John Foord (© PVB).

Fig. 5.67. *Euphorbia cooperi* var. *cooperi*, with *Aloe marlothii*, *PVB 11892*, Zion City, near Boyne, South Africa, 21 Jul. 2011 (© PVB).

Fig. 5.68. *Euphorbia cooperi* var. *cooperi*, with *Aloe globuligemma*, near Boboneng, Botswana, 3 Jan. 2013 (© PVB).

Fig. 5.69. *Euphorbia cooperi* var. *cooperi*, in shallow soil on steep granitic dome 40 km NE of Molocue, Moçambique, 22 Dec. 1998, with Cornelia Klak (© PVB).

Fig. 5.70. *Euphorbia cooperi* var. *cooperi*, flowering almost over, Penge, South Africa, 23 Jul. 2011 (© PVB).

Fig. 5.71. *Euphorbia cooperi* var. *cooperi*, second male stage, *PVB 11904*, Gutschwa Kop, north of Nelspruit, South Africa, 27 Jul. 2011 (© PVB).

Fig. 5.72. *Euphorbia cooperi* var. *cooperi*, capsules deep red, just north of Burgersfort, South Africa, 10 Sep. 2010 (© PVB).

History

At the suggestion of N.E. Brown, *Euphorbia cooperi* was named after his father-in-law. Thomas Cooper, who collected it in the 'Umgeni Valley', in the hinterland of Durban in Natal in June or July of 1862. Cooper brought back plants to Britain, some of which were grown at Kew and some at La Mortola, near Ventimiglia on the Italian Riviera. One of these was still at Kew in September 1899, when Brown made two herbarium specimens from it. Brown mentioned *E. cooperi* for the first time in a 'Hand-list of Tender Dicotyledons' published in 1900 by Kew Gardens. It was formally described in 1906 by Alwin Berger, who mentioned that by this stage the largest plant at Kew was about 3 m tall. Under *E. cooperi*, Berger (1906: fig. 21) included a photograph of a tree in habitat, but the plant depicted was actually *E. ingens*.

Cooper mentioned to N.E. Brown that he had seen 'only a few plants' along the valley of the Umgeni River. Today, *E. cooperi* is not known in that part of Kwazulu-Natal, so one must assume that Cooper found an isolated, small group of plants a little beyond its normal range.

5.1 Sect. Euphorbia

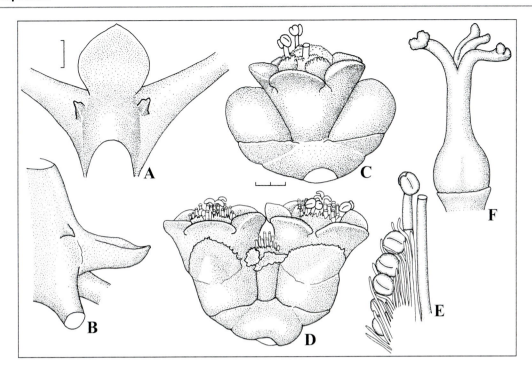

Fig. 5.73. *Euphorbia cooperi* var. *cooperi*. **A**, young spines and leaf-rudiment from above (scale 2 mm, as for **B**, **E**, **F**). **B**, young spines and leaf-rudiment from side. **C**, side view of cyme at first male stage (scale 2 mm, as for **D**). **D**, side view of cyathium in second male stage. **E**, anthers and bracteoles. **F**, female floret. Drawn from: *PVB 11904*, Gutschwa Kop, north of Nelspruit, South Africa (© PVB).

Euphorbia eduardoi L.C.Leach, *Bol. Soc. Brot.*, sér. 2, 42: 161 (1968). Type: Angola, Namibe distr., Posto experimental de Caraculo (Dois Irmãos), 550 m, 5 May 1960, *Mendes 3959* (LISC, holo.; BM, LUAI, PRE, iso.).

Bisexual spiny glabrous succulent tree 3–10 (12) m tall with shiny yellow-brown cylindrical trunk (at first 3-, 4- or 5-angled and brightly banded with dark green and cream) 5- to 7-angled and 2–8 m tall from extensive system of

Fig. 5.74. *Euphorbia eduardoi*, large colony near top of steep slopes with *Commiphora* and with masses of grey-brown shrubs of *Myrothamnus* covering the ground, Otjihipa, Namibia, 23 Dec. 1999 (© PVB).

woody and fibrous roots, with older branches drying up and falling off stem after a few years so that crown mostly remains small and ± hemispherical near apex of stem. *Branches* usually simple, ascending-spreading, (0.5) 1–3 m × 50–100 (120) mm, slightly and unevenly constricted into ± elliptic segments of variable length, smooth, pale to dark green; *tubercles* fused into 4–5 prominent slightly wavy stoutly wing-like angles with concave to V-shaped channels between them, laterally flattened and rounded and hardly projecting from angles (usually not projecting more than 2 mm), spine-shields red-brown to yellow-brown and becoming grey, 8–12 × 6–8 mm, obovate, joined into continuous ± evenly broad horny margin on stem of young plants, later (on branches) usually broadest around spines and sometimes narrowing to 2 mm broad below them rarely separate (usually becoming somewhat corky with age), around apex bearing 2 spreading and widely diverging stout grey spines 6–15 mm long (initially red or black); *leaf-rudiments* on tips of new tubercles towards apices of branches and stem, 2–3 (–15) × 2–4 (–6) mm, spreading, fleeting, deltate (linear-lanceolate in young plants), sessile, with irregular dark brown or blackish stipules to 1.5 mm long. *Synflorescences* many per branch near apex, with 1–3 cymes in axil of tubercle, on green to dark red peduncle 10–30 × 4–9 mm (thicker towards apex), each cyme with 3–5 vertically to transversely disposed cyathia, central male, outer cyathia bisexual and

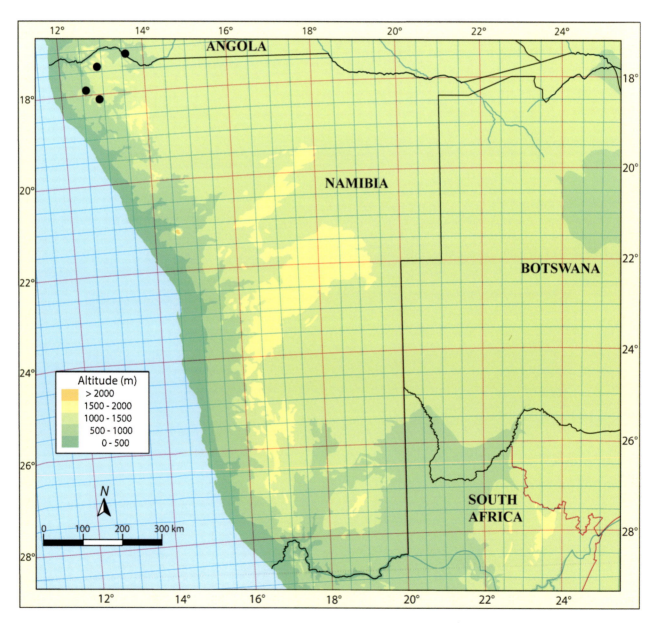

Fig. 5.75. Distribution of *Euphorbia eduardoi* in southern Africa (© PVB).

developing later on slightly flattened dark red to green peduncles 14–18 mm long, with 2–5 ovate bracts 3–5 × 3–6 mm subtending cyathia (sometimes further cyathia developing in axils of bracts around outer cyathia); *cyathia* shallowly cupular, glabrous, 10–16 mm broad (± 2 mm long below insertion of glands), with 5 broad and short glabrous lobes with finely incised margins, green suffused with red towards base; *glands* 5, transversely oblong-elliptic and contiguous, 5–8 mm broad, yellow-green, slightly concave above, outer margins entire; stamens glabrous with pale filaments and red anthers, bracteoles enveloping groups of males, with finely divided tips, glabrous; *ovary* very obtusely 3-angled, glabrous, on pedicel 1 mm long and soon increasing to 5 mm and just pushing capsule out of cyathium; styles (2) 2.5–3 mm long, branched to around middle. *Capsule* 16–20 mm diam., red above becoming pale green below, obtusely 3-angled, glabrous, on short pedicel ± 6 mm long and only just protruding from remains of cyathium.

Distribution & Habitat

Euphorbia eduardoi is widely distributed in southern Angola from north-east of Lobito (where it is found within a few kilometres of the coast) and east of Benguela to Iona in the mountains of the southern-most part. Across the border in Namibia it is known only in the Kaokoveld. Here it occurs on the slopes of the Otjihipa, Baynes Mountains and Zebra Mountains and southwards to slopes north and east of Orupembe.

Generally growing on steep slopes among gneissic, granitic or schistose rocks, *E. eduardoi* is often quite scattered in occurrence, though some denser, almost forest-like colonies may develop (as in Fig. 5.74). Plants grow wedged among rocks in the company of scattered, deciduous trees (especially species of *Commiphora*) and very little undergrowth, in scanty soils. Occasionally in Angola they are found in the flats between hills, possibly where some were planted, but generally their typical habitat is among boulders on slopes.

Diagnostic Features & Relationships

The inordinately tall, telephone-pole-like trees of *E. eduardoi* are some of the most unusual and striking among the succulent species of *Euphorbia*. Young specimens consist of a single, erect stem. Plants less than 100 mm tall are slender and 3- or 4-angled with slender leaves and they soon change to become 5- to 7-angled. At a height of 0.5 m tall they are between 100 and 200 mm thick and form a pole-like plant that could easily be mistaken for a cactus. Branching begins once the stem is around 2 m tall. In all small plants the surface of the stem is brightly banded with cream and dark green, but by the time the branches appear this mottled colour has changed into uniform dark green and this gradually changes into an often yellowish, shiny bark on a more or less cylindrical trunk. The initial branches are steeply ascending and it is only once the stem is much longer that the branches have a more spreading attitude and the crown may become more hemispherical. The stem continues growing and may reach over 10 m in length, though in some cases it remains much shorter (usually then with longer branches). The branches are relatively short-lived and dry up and fall off so that there is mostly only a small cluster of them in a crown around the apex of the tree. Most of the branches are 4- or 5-angled and dull green in colour. Spines on the young stem are stout and close together, with a continuous, broad, hard, horny margin on the angles formed by the more or less uniformly thick spine-shields. They are initially bright red and soon change to brown and later grey. On the branches the spines continue to be produced but are much shorter and the spine-shield becomes much more slender below the spines, though it is usually still continuous along the angles. The leaves are also longer in the young plant than they are on the branches.

Flowering of *E. eduardoi* takes place between December and May. Although the branches may be more than 10 m from the ground, when flowering they have an obvious halo of cymes around their tips. Large trees in full flower have clusters of cyathia on the tips of every branch, but the cyathia are only present near the tip of the last segment of each branch. Plants become mature at around 3 m tall and on these the cyathia can sometimes be observed at closer quarters. They are borne on quite long, but stout, red or dark green peduncles, of which there may be up to three at each node. Each peduncle bears a large, central male cyathium and three or more usually slightly smaller bisexual cyathia arranged variably around it. These lateral cyathia also arise on stout, but considerably shorter peduncles and may themselves bear further cyathia from the axils of their subtending bracts. All cyathia are comparatively large, shallowly plate-like and yellow-green, with large glands touching each other around the edges. Although the ovary is initially included in the cyathium, it is later pushed out just beyond the lobes, where the capsules are held until they are ripe and dehisce explosively.

Where it occurs in southern Africa, *E. eduardoi* cannot be confused with any other species. In Angola it occurs as far north as the southernmost populations of *E. candelabrum* Welw. It has much thicker branches than *E. candelabrum* and the young plants (when they consist of a stem only) are also much thicker in *E. eduardoi*. Such young plants are 3-angled in both but in *E. eduardoi*, by the time it is 20 cm tall it is usually already 5- to 7-angled. In *E. candelabrum* the 3- or 4-angled stage usually persists until it is over 0.5 m tall and

Fig. 5.76. *Euphorbia eduardoi*, trees nearly 10 m tall on steep slopes with grey-brown shrubs of *Myrothamnus* on dolomite, Otjihipa, Namibia, 23 Dec. 1999 (© PVB).

Fig. 5.77. *Euphorbia eduardoi*, *PVB 5588*, Okombambi, west of Baynes Mountains, with Geoff Tribe, Namibia, 20 Feb. 1993 (© PVB).

Fig. 5.78. *Euphorbia eduardoi*, very tall trees on basaltic rock-fall, NE of Oncocua, Angola, 15 Dec. 2016 (© PVB).

Fig. 5.79. *Euphorbia eduardoi*, trees with all branches broken off by baboons, 17 km south of Sao Nicolau, Angola, 18 Mar. 2017 (© PVB).

Fig. 5.80. *Euphorbia eduardoi*, flowering tip of branch, *PVB 12847*, east of Orupembe, Namibia, 25 Dec. 2014 (© PVB).

Fig. 5.81. *Euphorbia eduardoi*, male cyathia with pink anthers and green peduncles, *PVB 12847*, east of Orupembe, Namibia, 25 Dec. 2014 (© PVB).

branching begins when it is around 1 m tall. Such young plants have fairly prominent ovate leaves 10–20 × 5–7 mm, while in *E. eduardoi* the leaves on the young plant are linear-lanceolate and rarely exceed 10 × 4 mm.

Fig. 5.82. *Euphorbia eduardoi*, male cyathia with yellow anthers and reddish peduncles, *PVB 12847*, east of Orupembe, Namibia, 25 Dec. 2014 (© PVB).

DNA-data was used (Bruyns et al. 2011) to show that *E. eduardoi* and *E. vallaris* are closely allied. Neither of them is closely related to *E. ingens*, as suggested by Leach (1968b), but are related rather to *E. zoutpansbergensis*. A rather surprising result is that *E. eduardoi* and *E. candelabrum* are not as closely related as one might expect, in view of their close geographical proximity and the unusually long peduncles in both. It has been found that *E. candelabrum* is very closely allied to *E. confinalis*.

History

Major C.H. Hahn, at the time working in Ondangua, sent photographs of *E. eduardoi* to Dyer at Pretoria. These were taken near Otjipemba in the Kaokoveld of Namibia ('Okonjombo, Osato Road' and were not photos by Max Otzen as Leach 1968b, thought) and also 'on the western border of Ovamboland, 25 miles south of the Oru-Wa-Hakana Falls of the Cunene'. Numbering on the back of some of them ('No 41/1, 41/2 and 41/3') suggests that they

Fig. 5.83. *Euphorbia eduardoi*, mature capsules raised only very slightly above the remains of the cyathia, *PVB 13410*, south of Catengue, Angola, 13 Mar. 2017 (© PVB).

were taken in 1941. Otzen (in a letter at PRE) referred to it as 'the Tree Euphorbia for which Mr White has reserved the name Otzenii'. The first herbarium record was made by Merxmüller & Giess in the Kaokoveld at the end of January 1958. This specimen was listed tentatively by P.G. Meyer (1967) under the name *E. conspicua* N.E.Br. (= *E. candelabrum* Welw.). Nevertheless, Meyer noted several important and obvious differences between the material that he had before him and N.E. Brown's species and he was clearly not convinced that they were the same. *Euphorbia eduardoi* was first collected in Angola by Eduardo José Mendes in 1960 and Leach named it after him.

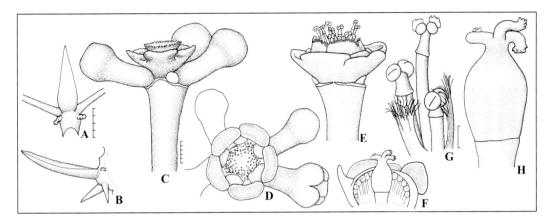

Fig. 5.84. *Euphorbia eduardoi*. **A**, young spines and leaf-rudiment from above (scale 4 mm, as for **B**, **E**, **F**). **B**, young spines and leaf-rudiment from side. **C**, side view of cyme at first male stage (scale 5 mm, as for **D**). **D**, face view of cyathium at first male stage. **E**, side view of male cyathium. **F**, side view of dissected bisexual cyathium. **G**, anthers and bracteoles (scale 1 mm, as for **H**). **H**, female floret. Drawn from: *PVB 12847*, east of Orupembe, Namibia (© PVB).

It is curious that, despite their wide travels in Angola, neither Welwitsch nor Gossweiler made any specimens of *E. eduardoi*. The only reasonable explanation is that both believed this plant to belong to *E. candelabrum*, of which they had already made enough specimens.

Euphorbia enormis N.E.Br., *Fl. Cap.* 5 (2): 362 (1915). Type: South Africa, Pietersburg, Sept. 1905, *Marloth 5144* (PRE, holo.; K, iso.).

Bisexual spiny glabrous succulent 50–300 mm tall with 1–several short cylindrical stems from central carrot-shaped mainly subterranean stem up to 300 × 150 mm from which fibrous roots arise, each stem with rosette of angled and spiny branches around its apex. *Branches* ascending to erect, 50–300 (500) × 15–50 mm, clavate to constricted into 2–6 ± obovate diamond- or wedge-shaped segments, smooth, pale to dark green with fine cream mottling along groove between angles and extending to near axillary buds; *tubercles* fused into 3–4 prominent and sometimes slightly spiralling somewhat wing-like angles, laterally flattened and projecting 2–10 mm from angles, with spine-shield 2–3 mm broad around apex spreading up along edge of angle to axil of tubercle and down for 3–8 mm below spines (near apex of branch often continuous along edges of angles), bearing 2 spreading and widely diverging brown to grey spines 4–10 mm long and 2 minute prickles ± 1 mm long alongside axillary buds; *leaf-rudiments* on tips of new tubercles towards apices of branches, ± 1 × 2–3 mm, erect, fleeting, broadly ovate and channelled down middle, sessile, with small irregular stipular prickles 0.5–1 mm long. *Synflorescences* 1–12 per branch towards apex on terminal segment, each a solitary cyme in axil of tubercle, on peduncle 5–10 × 3–6 mm, each cyme with 3 vertically disposed cyathia, central male, lateral 2 bisexual and developing later each on stout peduncle 3–6 × ± 4 mm, with 2 ovate bracts 2–4 mm long and 3–4 mm broad subtending cyathia; *cyathia* somewhat cylindrical-cupular, glabrous, 5–8 (12) mm broad (3–5 mm long below insertion of glands), with 5 lobes with deeply incised margins, bright yellow; *glands* 5 (6), transversely oblong, short, 2.5–5 mm broad, yellow, spreading, inner margins sometimes raised, outer margins entire rounded and spreading; *stamens* glabrous, bracteoles enveloping groups of males, finely divided, glabrous; *ovary* almost cylindrical and tapering gradually into styles, glabrous, almost sessile; styles 4.5–6 mm long, widely spreading and branched in upper half. *Capsule* 8–10 mm diam., green with red at apex and along edges, obtusely 3-angled, glabrous, sessile.

Distribution & Habitat

Euphorbia enormis is known in the semi-arid, eastern parts of the former Transvaal in the area often referred to as Sekukuniland, between Malipsdrift and Steelpoort, with a few records further to the north-east near Mica and others not far from Zebediela.

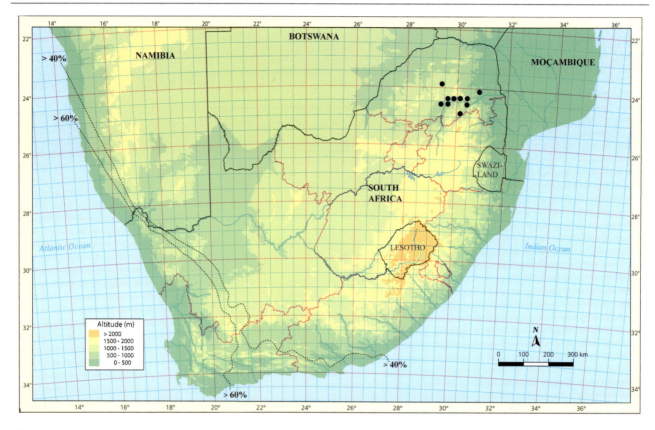

Fig. 5.85. Distribution of *Euphorbia enormis* (© PVB).

Plants are usually found in dry, flat, often gravelly, somewhat overgrazed areas among short bushes, often with a wide variety of other small succulents such as *Euphorbia clivicola*, *E. maleolens*, plenty of Aloes, *Ceropegia* (*Duvalia*) *polita*, *Ceropegia* (*Huernia*) *stapelioides* and *Ceropegia* (*Stapelia*) *gigantea*, with occasional trees of *Euphorbia ingens* and *E. excelsa* and various Acacias scattered over the landscape.

Fig. 5.86. *Euphorbia enormis*, plant in stony ground with low bushes, *PVB 12098*, north of Ohrigstad, South Africa, 1 Jan. 2012 (© PVB).

5.1 Sect. Euphorbia

Fig. 5.87. *Euphorbia enormis*, plant nearly 0.5 m diam., with *E. steelpoortensis* and *Aloe* in an overgrazed patch in a village, south of Penge, South Africa, 13 Jan. 1996 (© PVB).

Diagnostic Features & Relationships

In *E. enormis* the plant arises from a stout, largely subterranean stem that may sometimes be divided near its apex into several similar, stem-like branches. Only the tip of the stem protrudes a short distance from the ground, or it may be entirely subterranean. Around the apex of the stem (or subdivisions of it), an ultimately dense rosette of ascending branches develops, so that the whole plant forms a cluster close to the surface of the soil. The branches are mostly fairly slender and appear to last on the plant for 2–3 years, after which they dry up and fall off. They are 3- or 4-angled and strongly and regularly segmented every 20–50 mm of their length (where each segment usually corresponds to a year's growth). Pairs of small but stout and hard, spreading spines are borne along the angles on a hard, sometimes continuous, pale whitish grey, horny margin formed by the spine-shields. The surface between the spine-shields and between the angles is usually bright green, with a striking mottling of irregular, fine cream lines that run long the groove between the angles and continue out towards each node. Particularly tiny leaf-rudiments are produced on new growth, accompanied by very small stipular rudiments, with similar rudiments present around the adjacent axillary bud.

In *Euphorbia enormis*, flowering takes place mostly between November and January. The cyathia are produced on new growth around the tips of the branches. They are borne on comparatively long and stout green peduncles and are themselves relatively large and bright yellow, with spreading yellow glands. The male florets are borne on long pedicels and the bracteoles are finely divided more or less to

Fig. 5.88. *Euphorbia enormis*, first male stage, *PVB 6619*, Penge, South Africa, Dec. 2011 (© PVB).

Fig. 5.89. *Euphorbia enormis*, female stage, *PVB 6619*, Penge, South Africa, Dec. 2011 (© PVB).

Fig. 5.90. *Euphorbia enormis*, cyathia and capsules, south of Penge, South Africa, 13 Jan. 1996 (© PVB).

their bases. In the female floret the ovary is not clearly distinct from the styles (nor is it noticeably 3-angled either) and it tapers gradually into them, the styles are deeply divided with the divided parts widely spreading above the glands.

The habit combined with the comparatively large capsules and cyathia, the thick peduncles beneath the cyathia and the almost sessile ovary all suggest that *E. enormis* is related to *E. clavigera*, *E. tortirama* and *E. vandermerwei*, as well as to *E. barnardii* and more distantly to *E. grandicornis* (to which the very indistinct ovary suggests a relationship). *Euphorbia enormis* generally lacks the conspicuously twisted branches and the prominent tubercles and spines of *E. tortirama*. It has also much longer peduncles and longer, more cylindrical cyathia than *E. tortirama* as well as longer male and female florets. With *E. clavigera* it shares the straight, mostly untwisted branches (though they are very differently coloured in these two species). The most obvious differences between them lie in the longer spines (with considerably elongated spine-shield above the spines reaching to the quite distant axillary bud) and the more deeply constricted, more prominently and

5.1 Sect. Euphorbia

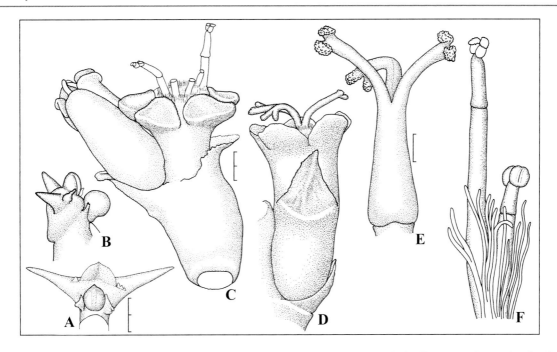

Fig. 5.91. *Euphorbia enormis*. **A**, young spines and leaf-rudiment from above (scale 4 mm, as for **B**). **B**, young spines and leaf-rudiment from side. **C**, side view of cyme at first male stage (scale 2 mm, as for **D**). **D**, side view of cyathium at female stage. **E**, female floret (scale 1 mm, as for **F**). **F**, anthers and bracteoles. Drawn from: *PVB 6619*, Penge, South Africa (© PVB).

sharply angled branches in *E. clavigera*. *Euphorbia vandermerwei* also has the spines and the neighbouring axillary bud very close together and this species differs by the considerably shorter peduncles, the smaller male florets and the smaller females which are more abruptly swollen at their bases and are without the gradually tapering ovary of *E. enormis*.

History
Euphorbia enormis was described from a plant collected by Rudolf Marloth, ostensibly near Pietersburg, though the species does not occur around this town but in the lower country considerably further to the south-east. The specimen at PRE is dated September 1905, so that he found it a considerable time before it was described.

Euphorbia excelsa A.C.White, R.A.Dyer & B.Sloane, *Succ. Euphorb.* 2: 966 (1941). Type: South Africa, Transvaal, Lydenburg distr., hills near Olifants River, Apr. 1938, *Van der Merwe 1677a* (PRE, holo.).

Bisexual spiny glabrous succulent tree 3–15 m tall, with gradually deciduous branches forming rounded crown and arising in upper part of cylindrical erect solitary (very rarely forked) trunk-like stem 3–12 m tall, trunk naked except for spines below crown, with many fibrous roots spreading from base. *Branches* spreading then ascending (often strictly ascending in younger plants) and often rebranching, older branches drying up and falling off stem after a few years, 0.3–1.5 m × 15–40 mm, (3) 4 (5)-angled and usually square in cross-section, slightly constricted into segments 50–200 mm long, smooth, grey-green; *tubercles* very evenly fused into 4 (rarely 3 or 5) not at all wing-like angles with surface flat to very slightly concave between angles, low-conical, laterally flattened and projecting 1–2 mm from angles, spine-shields continuous forming hard grey to corky black margin 2–3 mm broad along angles, bearing 2 spreading and widely diverging brown to grey spines 3–8 mm long; *leaf-rudiments* on tips of new tubercles towards apex of branch, 1–2 × ± 0.5 mm, spreading, deciduous, lanceolate, sessile, with minute irregular brown stipules up to 0.1 mm long. *Synflorescences* many per branch towards apex (usually only on last segment), with usually only 1 cyme in axil of each tubercle, with very short peduncle ± 1 mm long, each cyme usually with 5 cyathia (occasionally up to 9 cyathia), central male (rarely bisexual), outer four (2 transversely and 2 vertically disposed) usually bisexual and developing later each on short peduncles 1–2 × 2 mm, with 4 small broadly ovate bracts 1–2 × 2–3 mm subtending lateral cyathia (2 bracts on each of lateral peduncles in

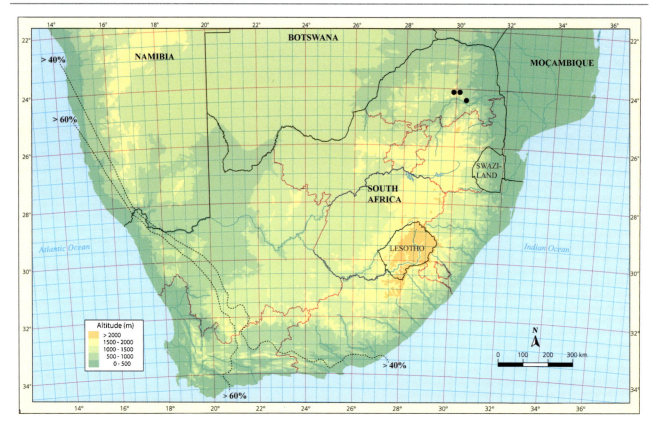

Fig. 5.92. Distribution of *Euphorbia excelsa* (© PVB).

whose axils occasionally further bisexual cyathia may develop); *cyathia* conical-cupular, glabrous, 5–7 mm broad (2–3 mm long below insertion of glands), with 5 obovate lobes with finely toothed margins, yellow to yellow-green; *glands* 5, transversely elliptic, contiguous to slightly separated, 2–2.5 mm broad, yellow, spreading, inner margins not raised, outer margins entire and spreading, surface between two margins dull; *stamens* with glabrous pedicels, anthers yellow, bracteoles enveloping groups of males, with finely divided tips, glabrous; *ovary* 3-angled, glabrous, green, raised and partly exserted on erect pedicel ± 1.5 mm long (soon pushed out of cyathium); styles 1.5–2 mm long, widely spreading, branched to near base. *Capsule* 8–9 mm diam., 3-angled (occasionally 4-angled), glabrous, green suffused with red above, exserted on erect pedicel 5–8 mm long.

Distribution & Habitat

Euphorbia excelsa is common but very local in Sekukuniland, in the dry hills and slopes leading down to the Olifants River from about 30 km east of Chunies Poort to near Burgersfort.

Euphorbia excelsa occurs with various species of *Acacia*, *Dichrostachys* and other, mostly deciduous trees in semi-arid scrub, often with *E. cooperi* nearby or even together with it. In these areas, *E. excelsa* forms large, dense colonies and may become dominant (then lending a greyish hue to the slopes or other areas where it occurs), sometimes in heavy loam in the flats or among rocks on the lower slopes of hills. They also occur as more scattered, often taller trees on steep slopes among large rocks and other scattered, deciduous trees.

Diagnostic Features & Relationships

Euphorbia excelsa forms an impressive tree that may reach a height of 10 m or more, though most specimens are between three and five metres tall. There is almost always only a single stem which branches well away from the ground. The tallest specimens have very widely spreading branches in an almost hemispherical crown. Shorter trees have more strictly ascending branches that give them an elliptic outline from the side. Most of the trees inherit a distinctive, grey-green colour from their branches, which are almost always 4-angled and nearly square in cross-section. On young plants the spines are very well-developed, but they gradually become shorter as the plants become larger and they disappear entirely in many of the mature trees, whose branches are then spineless.

5.1 Sect. Euphorbia

Fig. 5.93. *Euphorbia excelsa*, grove of trees (with *Aloe marlothii*), many of them young, near Penge, South Africa, 1 Jan. 2012 (© PVB).

Fig. 5.94. *Euphorbia excelsa*, in dry conditions, *PVB 11771*, west of Burgersfort, South Africa, 9 Sep. 2010 (© PVB).

Fig. 5.95. *Euphorbia excelsa*, tree in flower, Malipsdrift, west of Burgersfort, South Africa, 21 Feb. 2012 (© PVB).

Flowering in *E. excelsa* takes place between January and March, during the hottest months of summer. When in full flower, the uppermost segment of most of the branches is encrusted with cyathia and the tree assumes a slightly yellowish hue. In *E. excelsa* usually only one cyme arises in the axil of each tubercle. As is usual, the central cyathium in each cyme is generally male. An unusual feature of *E. excelsa* is that usually four cyathia develop from the axils of the bracts on the central cyathium. This corresponds to the presence of four bracts at the apex of the peduncle of each cyme. After the central cyathium has matured the four lateral cyathia develop and these are usually bisexual (initially female and then male). Two of them are placed transversely and two vertically relative to the axis of the branch and all appear to develop concurrently. Further cyathia may develop in the axils of the bracts on the peduncles of these lateral cyathia so that up to eight or even nine may arise on the cyme around its central cyathium. The cyathia are relatively tall and narrow, with small, spreading yellow-green glands. The female floret matures on a relatively long pedicel but is still partly included within the cyathium. Soon the pedicel lengthens to push it out so that the capsule matures outside the cyathium. The capsules remain on the trees for the rest of summer, for autumn and winter and ripen only in late September to early October, at the beginning of the next summer.

According to Bruyns et al. (2011), the closest known relative of *E. excelsa* is not one of the other tree-forming species but the shrubby *E. waterbergensis*, while the two of them are related to the tree-forming *E. grandidens* and *E. sekukuniensis*. *Euphorbia excelsa* and *E. waterbergensis* are the only species that regularly produce more than three cyathia in each cyme and may have more than two bracts at the tip of the peduncle of the cyme. Vegetatively, they are very

Fig. 5.96. *Euphorbia excelsa*, flowering branches, Malipsdrift, west of Burgersfort, South Africa, 21 Feb. 2012 (© PVB).

Fig. 5.97. *Euphorbia excelsa*, tip of flowering branch, many cymes with more than three cyathia, *PVB 11771*, west of Burgersfort, South Africa, 9 Sep. 2010 (© PVB).

different-looking, but they also share the small leaf-rudiments with particularly tiny stipular prickles close to the leaf-rudiment on the swollen part of the spine-shield. The respective cyathia are similar, with deeply divided, rather thick styles and partly exserted ovaries, though the ovary is more rounded in *E. waterbergensis* than in *E. excelsa* and *E. waterbergensis* has reddish stamens.

Euphorbia excelsa is readily separated from the other stouter-branched, tree-forming species (*E. confinalis*, *E. keithii*, *E. triangularis* and *E. zoutpansbergensis*) by the larger clusters of cyathia that make up the cymes. The ovary is also smaller than in any of the others and the leaf-rudiments are narrower than in all except *E. confinalis*.

History

Euphorbia excelsa is a very striking tree and the first record of it consists of photographs taken by A.O.D. Mogg in Sekhukuniland 'between Maliepsdrift and Winters', in April 1928 (PRE records). Some material was also seen by Marloth in April 1931. It was first collected by W.G. Barnard on 3 June 1935 (K, PRE). Further collections were made by F.Z. van der Merwe in July 1936, July 1937 and again in 1938, all in the dry area near the Olifants River in Sekukuniland. His collection of 1938 provided the type material.

Fig. 5.98. *Euphorbia excelsa*, tip of fruiting branch, *PVB 11771*, west of Burgersfort, South Africa, 9 Sep. 2010 (© PVB).

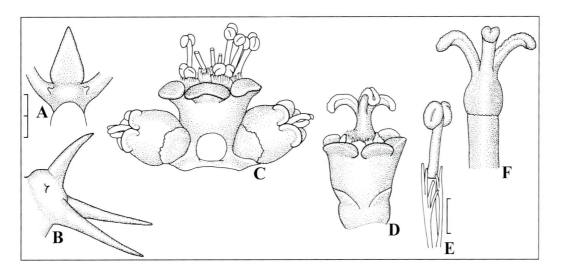

Fig. 5.99. *Euphorbia excelsa*. **A**, young spines and leaf-rudiment from above (scale 2 mm, as for **B–D**). **B**, young spines and leaf-rudiment from side. **C**, side view of cyme at first male stage. **D**, side view of cyathium at female stage. **E**, anther and bracteoles (scale 1 mm, as for **F**). **F**, female floret. Drawn from: *PVB 11771*, west of Burgersfort, South Africa (© PVB).

Euphorbia grandicornis A.Blanc, *Catalogue and Hints on Cacti*, ed. 2: 68 (1888). Lectotype (Bruyns 2012): Illustration on left hand side of figure on page 68 of A.Blanc, *Catalogue & Hints on Cacti*, ed. 2 (1888).

Euphorbia grandicornis K.I.Goebel, *Pflanzenbiol. Schilderungen* 1: 42, fig. 15 (1889), nom. illegit., non A.Blanc (1888). Lectotype (designated here): *Pflanzenbiol. Schilderungen* 1: 42, fig. 15 (1889).

Euphorbia grandicornis J.E.Weiss, *Neubert's Deutsch. Gart.-Mag.* 46: 291 (1893), nom. illegit., non A.Blanc (1888). Lectotype (designated here): Illustration on lower right hand side of *Neubert's Deutsch. Gart.-Mag.* 46: 291 (1893).

Euphorbia grandialata R.A.Dyer, *Fl. Pl. South Africa* 17: t. 641 (1937). Type: South Africa, Transvaal, Penge mine, *Van der Merwe 1002 sub PRE 21372* (PRE, holo.; K, W, iso.).

Bisexual spiny large glabrous succulent shrub 0.5–2 × 1–4 (5) m, much branched at and just above ground level from short central stem bearing fibrous roots. *Branches* initially spreading then erect, occasionally rebranched, 0.2–2 m × 50–150 mm, 3- to 4 (5)-angled, deeply constricted into almost hemispherical segments 50–150 (300) mm long, smooth, greyish green often with horizontal paler green bands; *tubercles* fused into 3–4 very prominent wing-like and slightly undulating angles with deep grooves between angles, laterally flattened and projecting 5–15 mm from angles, with spine-shields united into continuous horny initially red and later grey margin along angles ± 3–5 mm broad above spines and 2–4 mm broad below them (rarely separate), bearing 2 spreading and widely diverging initially red and later grey spines (5) 15–70 mm long and 2 smaller prickles (0.5) 3–7 mm long alongside axillary buds with slight brown ridge at base projecting over these buds; *leaf-rudiments* on tips of new tubercles towards apices of branches, 2–6 × 2–6 mm, spreading and slightly recurved towards tip, fleeting, ovate, sometimes acute, sessile, with brown stipular prickles 1–2 mm long. *Synflorescences* many per branch on last 1–3 segments, each of dense cluster of (1–) 3 cymes transversely arranged in axil of tubercle, very shortly peduncled, each cyme with 3 vertically disposed cyathia (rarely more developing on sides), central usually male sessile and often deciduous, lateral 2 bisexual and developing later each on short peduncle 1–2 (4) × 3–4 mm, with 2 ovate red-brown bracts 1–3 × 2–3 mm subtending lateral cyathia; *cyathia* shallowly conical-cupular, glabrous, 6–8 (11) mm broad (3 mm long below insertion of glands), with 5 obovate lobes with deeply incised margins, bright yellow; *glands* 5, transversely elliptic and contiguous, 4–5 mm broad, yellow, spreading, inner margins very slightly raised, outer margins entire and spreading, surface between two margins covered with copious nectar; *stamens* glabrous, bracteoles enveloping groups of males, with finely divided

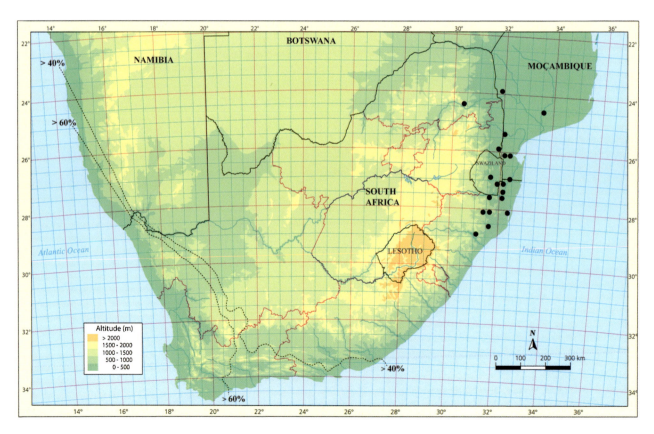

Fig. 5.100. Distribution of *Euphorbia grandicornis* ssp. *grandicornis* (© PVB).

tips or filiform, glabrous; *ovary* only slightly swollen from base of styles, ellipsoidal, glabrous, green, sessile (to peduncle 0.5 mm long); styles 5–6 mm long, branched in upper third. *Capsule* 10–15 mm diam., deeply 3-angled, pink to red above and along angles and green between them, glabrous and shiny, sessile.

Leach (1970a) divided *E. grandicornis* into two subspecies, subsp. *grandicornis* and subsp. *sejuncta* (from Moçambique). These were separated by differences in their size and the numbers of angles of the branches. Of these only subsp. *grandicornis* occurs within our area and the description above and what follows below apply to this subspecies only.

Fig. 5.101. *Euphorbia grandicornis* ssp. *grandicornis*, enormous clump ± 5 m diam. with long spines, Okhukhu, near Ulundi, South Africa, 16 Jan. 1991 (© PVB).

Distribution & Habitat

Euphorbia grandicornis is mainly found in the low-lying areas of what was formerly called Zululand in northern Natal, where it is known from around the Ndumu Reserve to Hluhluwe and into the hills along the Tugela River as far west as Middledrift. It is also known in Moçambique as far north as Moamba and Massingir in the Maputo district, with slight intrusions from here into the former Transvaal in valleys in the Lebombo Mountains. It is also found in some of the low-lying areas of south-eastern Swaziland close to the border with Natal and Moçambique.

Fig. 5.102. *Euphorbia grandicornis* ssp. *grandicornis*, relatively short-spined, Okhukhu, near Ulundi, South Africa, 16 Jan. 1991 (© PVB).

Euphorbia grandicornis usually grows in low-lying, loamy areas where the rainfall is lower than in the surrounding, higher areas, but is sufficient for seasonal inundation to occur regularly. Here a dense cover of succulents may develop, in which it often forms the dominant component.

This community of succulents (of which *Aloe marlothii* is often also a member) may develop with scattered *Acacia* trees into spiny thickets that are nearly impenetrable. Open patches in these thickets harbour smaller succulents such as *E. knuthii* and various stapeliads.

Fig. 5.103. *Euphorbia grandicornis* ssp. *grandicornis*, long-spined, *PVB 11824*, near Ulundi, South Africa, 29 Dec. 2010 (© PVB).

Diagnostic Features & Relationships

A very distinctive species, *E. grandicornis* is usually readily separated from other spiny species in southern Africa by its shrubby habit (i.e. not forming a tree and not developing a tall and substantial main trunk), the roughly hemispherical and comparatively short but very broad segments of the branches that become abruptly narrow below before merging into the next segment, particularly deeply winged, broadly undulating angles and the especially long spines, which are usually between 30 and 50 mm long but may reach 70 mm on occasion. In a wider context, *E. grandicornis* is one of several similar species with a broadly shrubby habit and broad branches with particularly prominent wings along them: others are *E. breviarticulata* (north-east and east Africa, where the spines may even be longer than in *E. grandicornis*), *E. cactus* (north-east Africa and Arabia) and *E. bougheyi* (Malawi and Moçambique).

In *E. grandicornis* the spine-shield is initially red and soon becomes grey, remaining like this for the remainder of its life. It is broadest around the leaf-rudiment, becoming much more slender between the largest spines and the next leaf-axil below to which it is joined. While they are developing, the spines are also soft and red, though they rapidly harden off to their typical grey colour. On each spine-shield there are six spines (as seen in Fig. 5.108A). The two longest ones lie just behind and below the leaf-rudiment and spread outwards to form a formidable protection along the stem, covering much of the area between the angles. On either side at the base of the leaf-rudiment there are two very small stipular prickles (usually around 1 mm long) that arise just above the slightly swollen base of the longest spines. Some distance above this there are two more spines, usually at least twice as long as the prickles subtending the leaf-rudiments. These develop on the edge of the spine-shield alongside the axil of the tubercle (in which three tiny buds are usually visible – the primordia of the cymes that mature later) and around their bases on the inside, they have a small, brown, toothy fringe which rises somewhat over these buds to afford them some protection. In the early stages of the spine-complex, the leaf-rudiments are quite conspicuous, usually with a grey-green blade with a reddish margin (that contrasts strongly with the red of the spines and spine-shield). Very soon they shrivel up and fall off, so that they are only present at the tips of the branches.

Flowering usually takes place between April and August. Large numbers of cyathia develop on the upper segments of many of the branches. In the axil of each tubercle there are mostly three cymes, arranged transversely relative to the axis of the branch and each of these cymes usually gives rise to three vertically disposed cyathia. The cyathia are bright yel-

Fig. 5.104. *Euphorbia grandicornis* ssp. *grandicornis* (formerly *E. grandialata*), with *Aloe castanea*, growing in crevices between outcrops of dolomite, Penge, South Africa, 13 Jan. 1996 (© PVB).

low and comparatively large, with broad, contiguous, bright yellow glands and they make quite a show among the spines. When young the ovary is small and inconspicuous at the base of the large styles but it develops into a particularly impressive, deeply 3-angled capsule that is usually shiny red above and green towards its base. It swells up within the cyathium, which eventually splits up around it and falls off once it has dried out.

In habit and the shape of the branches, *E. grandicornis* bears a resemblance to *E. bougheyi*. This differs by the shorter spines, the more prominently undulating angles, the taller stem and the denser and longer cymes of cyathia in the

Fig. 5.105. *Euphorbia grandicornis* ssp. *grandicornis*, first male stage on short-spined plant, Tugela River, near Greytown, South Africa, 7 May 2014 (© PVB).

leaf-axils. It shares the long spines with *E. breviarticulata* (which often also forms large shrubs, as do also the somewhat similar and weakly allied *E. ballyi* and *E. cactus* from NE Africa) and the two differ mainly in that in *E. breviarticulata* up to six cymes may develop in each axil, though this is not always reliable. *Euphorbia grandicornis* is also closely related to *E. pseudocactus,* under which this is discussed in more detail.

Fig. 5.106. *Euphorbia grandicornis* ssp. *grandicornis*, female stage, Tugela River, near Greytown, South Africa, 10 Jun. 2012 (© PVB).

Here *E. grandialata* is reduced to synonymy under *E. grandicornis*. White et al. (1941) gave the differences between *E. grandicornis* and *E. grandialata* as the shorter spines, more frequently developed secondary spines and the involucres larger by one third than those of *E. grandicornis*. However, the cyathia have been found to be 7–11 mm broad in *E. grandialata* (against 6–8 mm in *E. grandicornis*) and so this does not reliably separate them. In *E. grandialata* the spines are 8–25 mm long (15–70 mm long in *E. grandicornis*). So again, there is considerable overlap and there are other areas (e.g. in the valley of the Tugela River) where the spines of *E. grandicornis* are regularly only 15–30 mm long. Consequently *E. grandialata* cannot readily be distinguished and so it is reduced to synonymy.

Fig. 5.107. *Euphorbia grandicornis* ssp. *grandicornis*, capsules, Tugela River, near Greytown, South Africa, 5 Jun. 2016 (© PVB).

History

Euphorbia grandicornis was introduced to Kew around 1876 by J. Benjamin Stone M.P., who brought plants to England from Zululand, in what is today part of Kwazulu-Natal. These plants flourished to form large shrubs, though they seem to have struggled to flower and only the male cyathia ever developed fully (Brown 1915). In January 1896 N.E. Brown made two sheets of dried specimens from them (actually three specimens on the two sheets). Brown also published two figures in *Hooker's Icones Plantarum* (Brown 1897), one of which shows a well-developed shrub with the typical habit of this species. Here he also gave the first detailed description of *E. grandicornis*, which he based on the plants in cultivation at Kew.

However, this was not the first appearance of this species in cultivation in the West. Cuttings, some possibly originat-

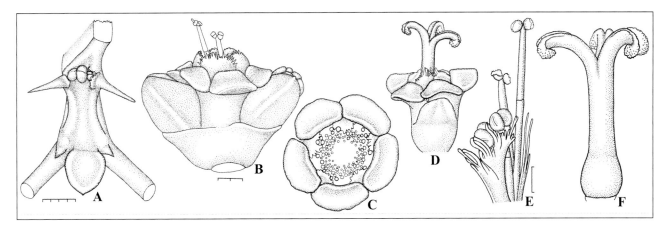

Fig. 5.108. *Euphorbia grandicornis* ssp. *grandicornis*. **A**, young spine-complex and leaf-rudiment from above (scale 4 mm). **B**, side view of cyme at first male stage (scale 2 mm, as for **C**, **D**). **C**, male cyathium from above. **D**, female cyathium from side. **E**, anthers and bracteoles (scale 1 mm, as for **F**). **F**, female floret. Drawn from: *PVB 4467*, near Ndumu, South Africa (© PVB).

ing from these plants at Kew and others clearly introduced even earlier into cultivation, seem to have been widely distributed both in Europe and in the USA. The name *Euphorbia grandicornis* was mentioned as early as 1865 (Oudemans 1865: 100) in *Neerland's Plantentuin* and a figure also appeared, among a selection of plants grown by a Mr Peacock in *The Gardeners' Chronicle and Agricultural Gazette* in June 1873, thus both of these well before the plants grown at Kew were brought from Natal. Another figure of it ostensibly appeared in 1887 in the *Catalogue* of A. Blanc & Co., Philadelphia, Pennsylvania, USA (according to White et al. 1941), but I have been unable to trace any copies of this. It appeared again in Blanc's *Catalogue and Hints on Cacti* of 1888, where Blanc was keen to interest potential buyers in 'some 50 varieties' of *Euphorbia* that he had available, some of which had been 'procured from South Africa at great expense'. In this Catalogue of 1888, the name was first validly published by the brief mention of a few distinctive features of the species. Several sketches of cuttings were published by Karl I.E. Goebel (1889, here fig. 15 was labelled '*Euphorbia grandidens*' but discussed in the text as *E. grandicornis* and the same figure was used again as fig. 29 and there correctly called '*Euphorbia grandicornis*') and the brief mention of certain distinctive features of the species again makes the name validly published here. A further figure of a planted cutting was published by Johann E. Weiss (1893) together with a brief description of the species. These 'accounts' of Blanc, Goebel and Weiss all appeared well before Brown's description, all of them used the name *E. grandicornis* and each of them validly published the name *Euphorbia grandicornis* (even though, for example, Weiss' brief description appeared under the title of '*Empfehlenswerte Cacteen*'). Thus, the correct name of this species is *Euphorbia grandicornis* A.Blanc, rather than *E. grandicornis* Goebel ex N.E.Br., as it is often given.

Brown (1915) suggested that *E. grandicornis* was possibly not distinct from the East African *E. breviarticulata*. This opinion was accepted by White et al. (1941). Leach (1970a: 41–2) contested this and gave the main differences between them as the manner in which the 1–3 cymes of cyathia are transversely arranged in the axil of the tubercle in *E. grandicornis*, while the up to 6 cymes are 'haphazardly' arranged in *E. breviarticulata*. In each cyme the central male cyathium was said to be deciduous in *E. breviarticulata* and 'usually, if not always, to be persistent' in *E. grandicornis*. Gilbert (1995), without giving details or evidence for it, expressed the view that these criteria do not separate the two and certainly the second criterion is not reliable in South African plants of *E. grandicornis*, where the central, male cyathium may be deciduous.

Euphorbia grandidens Haw., *Philos. Mag. J.* 66: 33 (1825). Lectotype (Bruyns 2012): South Africa, Cape of Good Hope, received 1822, *Bowie* (no specimen preserved: painting number 807/323 by T. Duncanson at K).

Euphorbia evansii Pax, *Bot. Jahrb. Syst.* 43: 86 (1909). Type: South Africa, Transvaal, Lowveld, near Barberton, *Pole Evans* (missing).

Bisexual spiny glabrous succulent tree 3–18 m tall, with many cylindrical ascending branches each bearing numerous secondary branches each ending in cluster of slender branchlets at apex, arising on cylindrical or faintly 6- to 8-angled erect solitary or sparingly forked trunk-like stem 1–8 (12) m tall, trunk naked except for scattered spines, with many fibrous roots spreading from base. *Secondary branches* becoming like trunk; *branchlets* ascending, older branchlets drying up and falling off stem after a few years, 0.3–1.5 m × 10–20 mm, (2-) 3- (4-) angled, scarcely constricted into segments, smooth, green; *tubercles* very evenly fused into (2) 3 (4) prominent wing-like angles with surface slightly concave between angles, conical, laterally flattened and projecting 3–5 mm from angles, spine-shields grey to pale brown, 1–2 mm broad, separate and projecting downwards 3–5 mm below spines to slender point, usually bearing 2 spreading and widely diverging brown to grey spines 0.5–10 mm long; *leaf-rudiments* on tips of new tubercles towards apex of branch, 2–2.5 × 2–3 mm, spreading, deciduous, broadly ovate, sessile, mostly with small irregular brown stipular ridges or prickles up to 1 mm long. *Synflorescences* many per branch towards apex, with 1 cyme in axil of each tubercle, with short peduncle 1–2 × 2–3 mm, each cyme with 3 transversely disposed cyathia, central male, outer two bisexual and developing later each on short peduncle 1–2 × ± 2 mm, with 2 small broadly ovate bracts 1–2 × 2 mm subtending lateral cyathia; *cyathia* cupular to slightly urceolate, glabrous, 3.5–5 (6) mm broad (2 mm long below insertion of glands), with 5 obovate lobes with finely toothed margins, green; *glands* 5, transversely elliptic, contiguous to well separated, 1.5–2.5 mm broad, green to faintly suffused towards inside with red, spreading, inner margins not raised, outer margins entire and spreading, surface between two margins dull to finely pitted and convex; *stamens* with glabrous pedicels, anthers reddish yellow (darker red around pores), bracteoles enveloping groups of males, with finely divided tips, glabrous; *ovary* globose, glabrous, green, raised and partly exserted on erect pedicel ± 1.5 mm long; styles 1.5–2 mm long, spreading, branched to near base. *Capsule* 7–10 mm diam., somewhat obtusely 3-angled, glabrous, mainly red above and pale green below but often wholly pale green, exserted on decurved and later erect pedicel 2–6 mm long.

Distribution & Habitat

Euphorbia grandidens is a particularly widely distributed species within southern Africa. The northernmost record is from the Abel Erasmus Pass in the Olifants River Valley of the former eastern Transvaal. From here it has been sporadically recorded in some of the sheltered, steep-sided valleys near Nelspruit and Barberton, along the eastern side of Swaziland and into the neighbouring parts of Moçambique (near

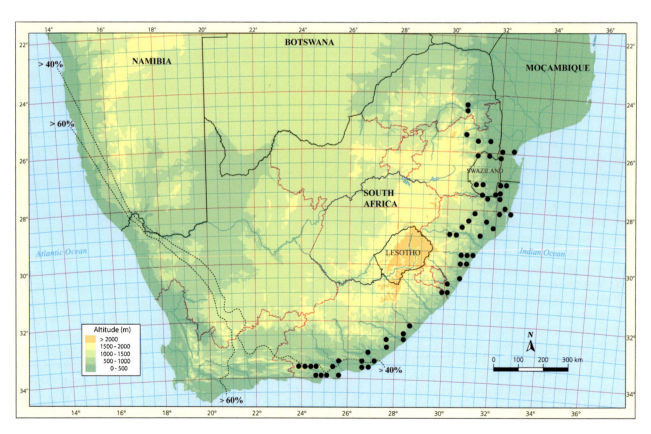

Fig. 5.109. Distribution of *Euphorbia grandidens* (© PVB).

Namaacha, Carter and Leach 2001: but also an odd record from near Marracuene on the coast of Moçambique). Southwards it continues (again sporadically) though Natal and into the Eastern Cape. In the Eastern Cape it occurs as far south as the valleys of the Kabeljouws and the Gamtoos Rivers, near Humansdorp. Along the Gamtoos River it occurs with *E. triangularis* near Hankey and Patensie but extends alone for a considerable distance up the Baviaanskloof (at least as far as the settlement of Studtis) and in the deep valley of the Groot River (a tributary of the Gamtoos) towards Cockscomb Peak and in the direction of Steytlerville.

As with *E. tetragona* and *E. triangularis*, *E. grandidens* is mainly found on steep, rocky slopes (often north- or west-facing) among a dense, thicket-like or even forest-like mass of entangled vegetation, often with various other succulents and especially with *Portulacaria afra*. These dense, subtropical thickets develop especially on the warm slopes of deep valleys of rivers, on shallow, humus-rich soils derived from both sandstones and shales. Occasionally *E. grandidens* occurs in coastal forest on sand. It is often found growing near to or together with *E. triangularis* and, more rarely, together with *E. tetragona*, but will sometimes be found in denser and moister situations than either of these two other species will tolerate. As with both of these species, trees of *E. grandidens* may form dense, often quite isolated colonies.

Fig. 5.110. *Euphorbia grandidens*, tall trees with lichen-covered stems in dense bush, Gamtoos River Valley, east of Hankey, South Africa, 4 Dec. 2006 (© PVB).

Fig. 5.111. *Euphorbia grandidens*, old, exposed trees, above Fish River Valley, east of Grahamstown, South Africa, 19 Oct. 2008 (© PVB).

Diagnostic Features & Relationships

Plants of *E. grandidens* form impressive trees with a stout, usually completely cylindrical and spineless trunk that may reach 30 cm thick or more. This starts off in young specimens as a 3- or 4-angled, quite slender stem, somewhat mottled with cream on green between the angles (though this soon changes to a greyish, corky bark) and armed with pairs of spines on the distinct tubercles along the angles. The number of angles on the stem rapidly increases to near 8 and these are soon only visible at the apex of the stem, which is cylindrical lower down. The first branchlets appear on the stem after it has reached 5–20 cm tall (often forming a drooping, spiny mass that must afford the stem some protection) but these mostly fall off with time and it is only later when perennial branches develop, mostly after the stem is at least 1 m tall. In mature specimens there are often several major branches to the trunk. These branches have a strongly ascending habit and also become cylindrical with age (as in Fig. 5.119 and 5.111). Each of these major branches and the main trunk has a cluster of branchlets at the apex. The cluster of branchlets is rather mop-like, with very large numbers of untidily disposed branchlets in young plants, it forms a bowl-shaped crown in plants of medium height, while in old specimens there are fewer branchlets in a much neater, mainly ascending and considerably sparser cluster. Generally, the branchlets (which may branch again towards their apices) are quite slender and many of them are only 10 mm thick, quite pliable, with the quite prominent tubercles widely spaced along them. They vary from flat (i.e. 2-angled) to 3- or 4-angled and are bright green with a variable armature of quite stout to entirely obsolete spines (in which case the spine-shield is usually somewhat corky). Most of the branchlets fall off after a year or two but a few develop further into stouter branches and themselves later bear a crown of branchlets at their apex. The tubercles are widely spaced along these branches and, while very prominent in young plants, become less so on older specimens. Each tubercle is initially tipped by a very small ovate leaf-rudiment and two spines (often missing on the upper branches on big trees, though not on branches that occasionally develop on a big tree close to the ground) all of which are seated on a small, elliptic spine-shield.

Fig. 5.112. *Euphorbia grandidens,* young tree in dense bush, Gamtoos River Valley, east of Hankey, South Africa, 4 Dec. 2006 (© PVB).

Fig. 5.113. *Euphorbia grandidens,* cyathia in female stage, *PVB 6870a,* Springs, NE of Uitenhage, South Africa, 24 Apr. 2016 (© PVB).

In *E. grandidens* the cyathia appear in (April to) August to November, usually well after those of *E. triangularis*, with populations in the northern areas flowering first and those in the south flowering sometimes as late as October and November. In years of good rains the uppermost branchlets become covered with cyathia, which release copious nectar during the last male stage and then also seem to give off a sweetish odour (this not being detectable in earlier stages). Each cyme consists of a single cyathium in the axil of the leaf-rudiment, from which two further transversely disposed, lateral cyathia develop relatively quickly. Since the cyathia are comparatively small, widely separated along the branch and green in colour, with green to faintly reddish glands, they are mostly fairly inconspicuous against the green of the branchlets. The anthers are suffused with red (usually darkest around the pores) and the female floret in each lateral cyathium is raised on a pedicel almost out of the cyathium. After the female floret has matured, its pedicel elongates slightly further so that eventually the capsule is well exserted from the cyathium on a spreading to erect pedicel. Seeds are released in late November to early December.

Analysis of DNA-data (Bruyns et al. 2011) showed that *E. grandidens* is not closely allied to the other large, tree-forming, spiny species *E. tetragona* and *E. triangularis*, with which it frequently occurs (and which are themselves closely related), but to another group of species that includes the enormous *E. ingens* and also the small geophytes *E. decidua* and *E. stellata*. At present the closest known relative of *E. grandidens* is *E. sekukuniensis*. Both share the very variable number of angles on the fairly slender branchlets and the greenish, rather inconspicuous and almost cupular cyathia as well as the 3-angled capsules that are eventually well exserted from the cyathium. Differences between them are given under *E. sekukuniensis*.

Euphorbia grandidens frequently occurs together with *E. triangularis* but the two do not flower at the same time, with flowering commencing in *E. grandidens* after it is over in *E. triangularis*. *Euphorbia grandidens* and *E. tetragona* flower at similar times but are rarely seen together and no hybrids

Fig. 5.114. *Euphorbia grandidens*, cyathia in second male stage, Gamtoos River Valley, south of Hankey, South Africa, 31 Oct. 2012 (© PVB).

Fig. 5.115. *Euphorbia grandidens*, capsules, *PVB 6870a*, Springs, NE of Uitenhage, South Africa, 16 Jul. 2018 (© PVB).

have been recorded. They are not so easily separated morphologically and the differences between them are discussed under *E. tetragona*.

History

The first recorded plants of *Euphorbia grandidens* were living specimens sent in 1822 by James Bowie from Cape Town to the Botanic Gardens at Kew (Fig. 5.117), where they were cultivated and where Haworth saw them. He based his description of the species on these plants. Bowie probably collected them around Port Elizabeth, late in 1821 or early in 1822. According to N.E. Brown (1915), some of the descendents of these plants were still thriving at Kew in 1915, almost 100 years later.

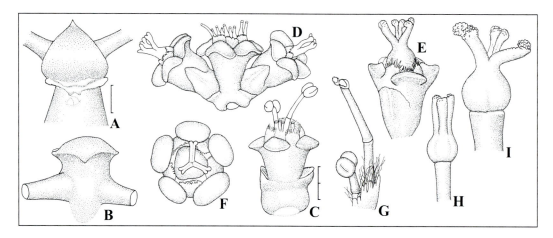

Fig. 5.116. *Euphorbia grandidens*. **A**, young spines and leaf-rudiment from above (scale 1 mm, as for **B**, **G–I**). **B**, young spines and leaf-rudiment from behind. **C**, cyathium at first male stage from side (scale 2 mm, as for **D–F**). **D**, side view of cyme at female stage. **E**, side view of female cyathium. **F**, female cyathium from above. **G**, anthers and bracteole. **H**, **I**, female floret. Drawn from: **A**, **B**, **C**, **E**, **G**, **H**, **I**, *PVB 6870a*, Springs, NE of Uitenhage, South Africa. **D**, *PVB 11765*, Spago Dam, near Malelane, South Africa. **F**, *PVB*, south of Hankey, South Africa (© PVB).

Fig. 5.117. *Euphorbia grandidens*, South Africa, Cape of Good Hope, received 1822, *Bowie*. Watercolour 807/323 by T. Duncanson (© RBG Kew).

Although various photographs have appeared that ostensibly represent *E. grandidens* but actually represent other species, the identity of *E. grandidens* has generally been clear, except for whether or not the name *E. evansii* represents a distinct species. When Pax (1909) described *Euphorbia evansii*, he stated that it had a very isolated position in the *Tetracanthae* on account of its treelike habit. He believed it to be related to *E. tenuispinosa* (*E. taitensis*) from East Africa. Clearly, he had no idea that he might have described something very close to and barely distinguishable from Haworth's *E. grandidens*. White et al. (1941) found that *E. evansii* differed from *E. grandidens* in being a shorter tree (reaching a maximum of 10 m as opposed to 16 m), with 3- to 4-angled secondary branches with gently sinuate margins (as opposed to 3-angled or rarely 2- to 4-angled in *E. grandidens* with more prominently toothed margins) and with the spines lacking the pairs of prickles at their bases, which are often present in *E. grandidens*. None of these differences is clear-cut and I have found it impossible to separate the known collections into two distinct species. Consequently, the name *E. evansii* was placed in synonymy in Bruyns (2012), although it was kept separate from *E. grandidens* in Bruyns et al. (2006).

5.1 Sect. Euphorbia

Euphorbia griseola Pax, *Bot. Jahrb. Syst.* 34: 375 (1904). Type: Botswana, Lobatsi, *Marloth 3413* (missing). Neotype (Leach 1967): Botswana, 2 miles north of Lobatsi, 16 Jan 1960, *Leach & Noel 121* (SRGH, BR, G, K, LISC, PRE, iso.).

Bisexual spiny glabrous shrub-forming succulent 0.15–2.00 m tall (taller to the north of our area) with many branches above ground level and sometimes below ground as rhizomes from short stem from which fibrous roots arise. *Branches* ascending to erect, 50–750 × 8–18 mm, 4- to 6-angled, not or only very slightly constricted into segments but narrowing towards slender base, smooth, green with paler cream markings between angles, often rebranching; *tubercles* fused into 4–6 angles with surface distinctly concave between angles, conical and truncate, laterally flattened and projecting 2–6 mm from angles so that angles have distinctly wave-like profile, deltate, with grey spine-shields 2–3 mm broad and forming continuous horny margin along angles (rarely separate), bearing 2 grey spines 4–8 mm long; *leaf-rudiments* on tips of new tubercles towards apices of branches, 4–8 × 1.5–2 mm, erect, fleeting, lanceolate, sessile, with spreading brown stipular prickles 1–2 mm long. *Synflorescences* 1–15 per branch towards apex, each a solitary cyme in axil of tubercle, subsessile, each cyme with 1–3 transversely disposed cyathia, central male or bisexual, lateral 2 bisexual and developing later each on short peduncle 1–2 × 1–2 mm, with 2 almost rectangular apically toothed bracts 1.5–2 × 2 mm subtending cyathia; *cyathia* conical-cupular, glabrous, 3–4 mm broad (1–1.5 mm long below insertion of glands), with 5 obovate lobes with deeply incised margins, green, sometimes red on sides; *glands* 5, transversely elliptic and contiguous, 1.5–2 mm broad, yellow to green, spreading, flat above, outer margins entire and spreading, surface between two margins dull; *stamens* glabrous, bracteoles enveloping groups of males, with finely divided tips, glabrous; *ovary* 3-angled, glabrous, green, on erect pedicel 2–3 mm long and later exserted from cyathium on decurved pedicel to 5 mm long; styles 1–2 mm long, branched at least to below middle. *Capsule* 4–5 mm diam., deeply 3-angled, glabrous, grey-green, exserted from cyathium on decurved and later erect pedicel 5–7 mm long.

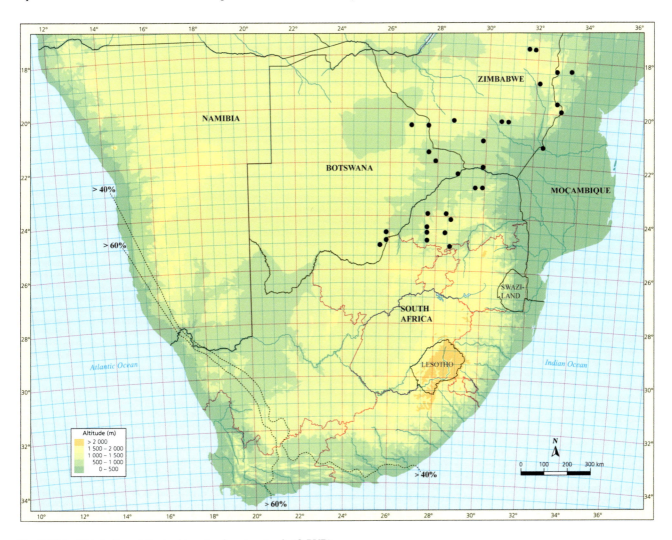

Fig. 5.118. Distribution of *Euphorbia griseola* ssp. *griseola* (© PVB).

Leach (1967) divided *E. griseola* into three subspecies, subsp. *griseola*, subsp. *mashonica* (from Malawi, Moçambique, Zambia and Zimbabwe) and subsp. *zambiensis* from Zambia only. These were separated on the presence of a distinct stem, presence of constrictions of the branches and the extent of the continuity of the spine-shields along the angles. Of these only subsp. *griseola* occurs within our area and the description above and what follows below apply to this subspecies only.

Fig. 5.119. *Euphorbia griseola* ssp. *griseola*, large shrubs to 1.5 m tall among granitic boulders, *PVB 12399*, Mmadinare, eastern Botswana, 31 Dec. 2012 (© PVB).

Distribution & Habitat

Euphorbia griseola subsp. *griseola* is of widespread and rather scattered occurrence in Botswana, Malawi, Moçambique and Zimbabwe. In Botswana it is mainly found in the south-east from Lobatse to Gaberones and Francistown in the east. In South Africa it is only found in the former Transvaal, especially on the western side of the Waterberg (being much rarer on the eastern flank). It is also known in the Soutpansberg and as far east as near Rust de Winter, near Pretoria. In Zimbabwe it is widespread in the southern half, from the Matopos to near Marandellas. It is very poorly known in Moçambique, where it has been recorded near Errego in Zambézia Province and is also known from a few hills in Manhica Province, west of Beira.

Fig. 5.120. *Euphorbia griseola* ssp. *griseola*, small shrub (flowering sparsely) to 0.3 m tall with *Aloe globuligemma* among sandstone boulders, near Strydom Dam, Waterberg, South Africa, 1 Jan. 1996 (© PVB).

In Botswana and South Africa subsp. *griseola* is found in shallow soil in crevices in flat expanses of exposed sandstone or granitic domes or in shallow soils among large rocks on somewhat exposed sandstone or granitic outcrops (sometimes sheltered under trees). These habitats are often relatively bare and form a locally much harsher environment than the surrounding countryside. In such spots in the Waterberg, subsp. *griseola* may be found with other succulents such as *E. schinzii* and *E. waterbergensis*, *Cotyledon barbeyi*, several species of *Crassula* and *Kalanchoe* and several stapeliads. In Zimbabwe and in Moçambique it is known to occur in shallow soil on granitic domes, again with many other succulents.

Fig. 5.121. *Euphorbia griseola* ssp. *griseola*, small shrub to 0.15 m tall with *Euphorbia aeruginosa* and dry grasses in crevices in sandstone boulders, *PVB 7101*, above Waterpoort, Soutpansberg, South Africa, 31 Dec. 1996 (© PVB).

Diagnostic Features & Relationships
In southern Africa, subsp *griseola* generally forms dense, very spiny shrubs which may reach 2 m in height and at least 1 m in diameter. In some populations, however, the plants are only 150 mm tall, spreading by means of short underground rhizomes. Such small plants have been seen in the Soutpansberg (*PVB 7101*) and were recorded by Leach (1967) from Zimbabwe. However, even these very small plants show the characteristic colour of the branches (green with paler markings between the angles) and the tough, sharp spination that is typical of subsp. *griseola*. The branches are usually only slightly or not at all constricted into segments and are most frequently 5-angled, though both 4- and 6-angled individuals can be found. They are typically around 15 mm thick, erect to slightly spreading and sparingly re-branched above the base.

The leaf-rudiments of subsp. *griseola* are quite distinctive, usually erect and considerably longer than broad, with finely papillate margins towards their apex. The stipular prickles at their bases are mostly small and have a slightly raised horny collar around their bases which continues into a few papillae on the surface of the spine-shield between them. The spines are notably hard and sharp, usually 8–10 mm long, broadly spreading and greyish brown.

Flowering in subsp. *griseola* usually takes place between October and December. The cyathia are produced in large numbers near the tips of the branches, one group of three in the axil of each tubercle. They are typically very small (roughly half the diameter of those in *E. schinzii*, for example) and usually yellowish green, though sometimes suffused with red on the sides. Male florets are borne on long pedicels and the anthers may be faintly reddish. The female florets are subtended by a relatively long pedicel which, already from early stages, pushes the ovary out of the cyathium so that the very short styles are adequately exposed. Once pollination has taken place, the pedicel elongates even

Fig. 5.122. *Euphorbia griseola* ssp. *griseola*, *PVB 12387*, east of Tonata, Botswana, 11 Dec. 2018 (© PVB).

Fig. 5.123. *Euphorbia griseola* ssp. *griseola*, central cyathium bisexual, *PVB 7101*, above Waterpoort, Soutpansberg, South Africa, 26 Sep. 2007 (© PVB).

further to push the deeply triangular capsule well beyond the cyathium.

Analysis of DNA-data (Bruyns et al. 2006) showed that subsp. *griseola* was closely related to *E. knuthii* and *E. mlanjeana*. All of these are species with small cyathia, strongly exserted ovaries, short styles and deeply 3-angled capsules with smooth, subglobose seeds. Other close relatives in tropical Africa are *E. jubata* and *E. persistentifolia*. This group of species is deeply nested among larger trees and is not closely related to the other shrubby, spiny species of the *E. schinzii*-complex.

History

Euphorbia griseola subsp. *griseola* was first recorded by Marloth on 2 November 1903 in the 'Lobatsi Plain'. His specimens disappeared and they are presumed to have been destroyed in World War II at the Berlin Herbarium or at Breslau. The next recorded collection was made by Maria Wilman near Lobatse in January 1925. Although Dyer made a few collections of it, it was really the investigations of L.C. Leach that showed how widely distributed and how variable it was. He was also able to show that some of the material referred in White et al. (1941) to *E. heterochroma* (e.g. their fig. 873) was subsp. *griseola*.

5.1 Sect. Euphorbia

Fig. 5.124. *Euphorbia griseola* ssp. *griseola*, more reddish cyathia, central male only, *PVB 7101*, above Waterpoort, Soutpansberg, South Africa, 1 Nov. 2008 (© PVB).

Fig. 5.125. *Euphorbia griseola* ssp. *griseola*, capsules fully grown in some cyathia and immature in others, *PVB 12387*, east of Tonata, Botswana, 11 Dec. 2018 (© PVB).

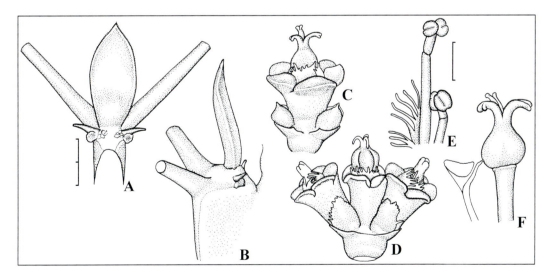

Fig. 5.126. *Euphorbia griseola* ssp. *griseola*. **A**, young spines and leaf-rudiment from above (scale 2 mm, as for **B–D**). **B**, young spines and leaf-rudiment from side. **C**, young central bisexual cyathium from side (lateral cyathia not yet developed). **D**, side view of cyme where central and lateral cyathia are bisexual. **E**, anthers and bracteoles (scale 1 mm, as for **F**). **F**, female floret (in part of dissected cyathium). Drawn from: *PVB 7101*, above Waterpoort, Soutpansberg, South Africa (© PVB).

Euphorbia ingens E.Mey. ex Boiss. in DC., *Prodr.* 15 (2): 87 (1862). Lectotype (Bruyns 2019): South Africa, Natal, in woods near Port Natal (Durban), 100', Apr. 1832, *Drège 4614* (S; K, P, iso.).

Euphorbia similis A.Berger, *Sukkul. Euphorb.*: 69 (1906). Type: South Africa, Natal ? (missing).

Bisexual spiny glabrous succulent tree 3–10 m tall with grey-brown cylindrical trunk (at first 3- then 4- to 5-angled, brightly banded with shiny dark green and cream) 1.5–2 m tall from extensive system of woody and fibrous roots, branching and rebranching extensively to form spherical to obconical crown. *Branches* ascending directly from base (so not curved upwards), persistent and not drying up and falling off with age, 50–110 mm thick, constricted into ± elliptic segments, smooth, dark dull green; *tubercles* fused into 4–5 prominent wing-like angles with deeply triangular channels to flat area between them, laterally flattened and rounded and projecting 3–10 mm from angles, with poorly developed spine-shield (usually becoming somewhat corky with age) around apex bearing 2 spreading and widely diverging grey spines 1–6 mm long (always present on young trees but gradually disappearing as tree becomes taller); *leaf-rudiments* on tips of new tubercles towards apices of branches and stem, 2–3 × 2–4 mm, spreading, fleeting, ovate (± 20–80 × 10–30 mm, linear- to oblanceolate in young plants), sessile, pale green, with tooth-like dark brown stipules ± 1 mm long. *Synflorescences* in large numbers per branch towards apex, with 2–6 cymes in axil of each tubercle, on peduncle 2–21 × 2–7 mm, each cyme with 3 transversely to vertically disposed cyathia, central male, outer 2 bisexual and developing later on peduncles 1–6 × 2–4 mm, with several ovate bracts 3–5 × 4–6 mm subtending cyathia; *cyathia* cupular, glabrous, 8–10 mm broad (2–3 mm long below insertion of glands), with 5 lobes with deeply incised almost filiform margins, pale green; *glands* 5, transversely oblong to rectangular, 3–4 mm broad, yellow to pale green, slightly convex above, outer margins entire; stamens glabrous, bracteoles enveloping groups of males, with finely divided tips, glabrous; *ovary* ellipsoidal, glabrous, almost sessile and with 'calyx' extended around ovary into several filiform lobules; styles 3 mm long, branched almost to base. *Capsule* (8)

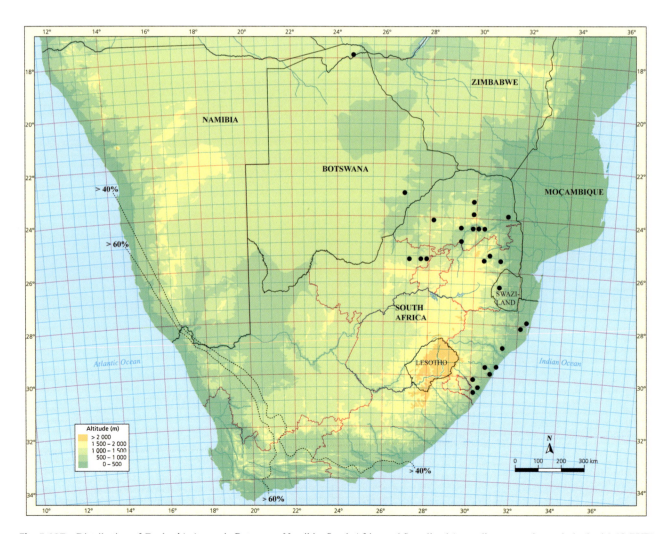

Fig. 5.127. Distribution of *Euphorbia ingens* in Botswana, Namibia, South Africa and Swaziland (according to specimens in herbaria) (© PVB).

5.1 Sect. Euphorbia

12–18 mm diam., initially spherical or very obtusely 3-angled and fleshy, later drying out to 7–12 mm diam. and deeply 3-angled then breaking up, glabrous, shiny red above and pale green below, exserted on short erect pedicel 2–5 mm long.

Distribution & Habitat

Essentially a tropical species, *E. ingens* is recorded from Durban in Natal northwards into the tropical parts of the former Transvaal, south-eastern Angola, Kenya, Malawi, Moçambique, Swaziland, Tanzania, Zambia and Zimbabwe and as far north as southern Ethiopia and Somalia (Bruyns and Berry 2019). Although it is plentiful in South Africa, it is less common in Botswana and in Namibia it is only known from the eastern end of the Caprivi Strip.

Plants are found in flat areas and on gently sloping hills, much more rarely in steep places and even more rarely in precipitous spots. They often grow among *Acacia* and other trees and often (though not always) in places where there are few other succulents. Many specimens are widely scattered among the other trees, but they sometimes form dense stands that may even dominate the vegetation over short stretches. *Euphorbia ingens* frequently occurs together with *E. cooperi*, though it does not grow in shallow soils on outcrops of rock as *E. cooperi* often does.

Fig. 5.128. *Euphorbia ingens*, young tree next to *Acacia*, near Zion City, near Boyne, South Africa, 30 Jul. 2019 (© PVB).

Fig. 5.129. *Euphorbia ingens*, large tree in fruit, near Zion City, near Boyne, South Africa, 30 Jul. 2019 (© PVB).

Diagnostic Features & Relationships

Euphorbia ingens is one of the best-known and most characteristic species of *Euphorbia*, popularly known as the '*naboom*'. Each plant starts off as an erect, single-stemmed, initially 3-angled and later usually 4- or 5-angled, robustly spiny succulent, in which the stem is brightly banded with cream on a shiny dark green background. Branching begins after a height of 1.5–2 m is reached and after this the plant changes in appearance greatly as it grows older. Branches do not die off and are not shed, persisting for the entire life of the tree. As branching increases, the trunk gradually swells to be more or less cylindrical, with its bark changing as well

Fig. 5.130. *Euphorbia ingens*, two large trees close together, near Zion City, near Boyne, South Africa, 30 Jul. 2019 (© PVB).

Fig. 5.131. *Euphorbia ingens*, seedling ± 1 m tall with attractively mottled stem, near Zion City, near Boyne, South Africa, 21 Jul. 2011 (© PVB).

to a grey, fissured cork-like texture. The trunk becomes topped by a crown that is soon usually much taller than the length of the trunk. In younger, but mature trees the crown is roughly V-shaped, but it may become almost spherical in older trees. The crown is formed by repeated branching of the branches off the trunk and a large tree has a particularly large number of end-branches. These ultimate branches are a dull dark green and are segmented, as in most other species, though the segments are elliptic in outline and are often relatively inconspicuous. Young plants of *E. ingens* are well-armed with spines and hardened spine-shields along the angles. As the plant grows larger, the spines become smaller and they usually disappear entirely on the ultimate branches of large trees. Here the spine-shields become corky and cover an irregularly shaped patch around the leaf-rudiment. They do not extend from one leaf down to the next so that patches of green are left along the edges of the angles. They often become yellow-brown in colour and have an irregular, almost warty surface. The leaves also change considerably with the age of the plant. On a young plant leaves around the apex are conspicuous and up to 35 mm long but they decrease in size as the plant grows older and, on the ultimate branches of large trees, they are scale-like and very inconspicuous indeed.

In *E. ingens* large numbers of cyathia are produced (usually early in winter between April and June) along the upper 0.5 m of the ultimate branches. Up to six cymes break through the corky spine-shield in a leaf-axil and then one may observe rows of these quite dense clusters of cyathia along the angles. The peduncles of the cymes are variable in length: those in South Africa are mostly short (only 2 or 3 mm long, though 5–15 mm long in some collections from the Kruger National Park, e.g. *Codd & De Winter 5587*) but on the northern edge of our region (as in the Caprivi Strip,

Namibia), they are up to 20 mm long, as they are also further north. Their pale green colour contrasts somewhat against the darker green of the branches. As usual, a functionally male cyathium develops first and later two bisexual cyathia arise in the axils of the slightly husk-like bracts on its side. These may be aligned anywhere between vertically and transversely relative to the axis of the branch. The anthers are exserted on quite long pedicels and are usually bright pinkish red. The cyathium is particularly densely packed with bracteoles and stamens (almost like those of *E. gregaria*) and it is quite difficult to expose the ovary. When the ovary is eventually revealed, it is almost sessile, somewhat longer than broad and one finds that the normally rudimentary 'calyx' is extended into slender processes, some of which are almost as long as the ovary itself. The comparatively large, often brightly shiny, red berry-like capsules are very conspicuous on the trees. As they ripen and dry out they decrease slightly in diameter, after which they break up and the seeds fall to the ground.

Fig. 5.132. *Euphorbia ingens*, tip of branch covered with cyathia, cultivated plant, Cape Town, South Africa, 17 May 2014 (© PVB).

Fig. 5.133. *Euphorbia ingens*, cyathia with pinkish anthers, cultivated plant, Cape Town, South Africa, 28 Apr. 2014 (© PVB).

A very similar species, *E. abyssinica*, is found in northeast Africa and yet another superficially rather similar species, *E. ammak*, occurs in Arabia. However, whereas *E. abyssinica* and *E. ingens* are closely related (and related closely also to *E. ampliphylla*), *E. ammak* is not a close relative, despite the similar appearance of the plants. *Euphorbia abyssinica* differs in various features from *E. ingens*, but most obviously in the greater number of angles in the stems of the young plants. Over much of tropical Africa, *E. ingens* has recently become known as *E. candelabrum* Kotschy (e.g.

Carter 1988) but this has been shown to be erroneous (Bruyns and Berry 2019) and all of these are referable to the very widely distributed *E. ingens*.

History

The type of *Euphorbia ingens* was collected by J.F. Drège and his brother Carl in forests and woody places near Durban (known then as Port Natal) on 11 April 1832, during an expedition led by Andrew Smith to Natal. It is curious that there are no earlier records of this very conspicuous plant.

The name *E. similis* A.Berger was placed as a synonym of *E. ingens* by White et al. (1941). Berger described this 'species' from some cultivated plants usually referred to in horticultural circles as '*E. natalensis*', which was an illegitimate name (there being an earlier *E. natalensis* Bernh. from 1846). N.E. Brown made two specimens from plants cultivated at Kew under the name *Euphorbia similis*. On one of these specimens he mentioned that, to be certain of its identity, he had sent a branch to Berger who had confirmed that this was what he named *E. similis*. However, many of the pressed branches on the two specimens at Kew bear leaves 15–80 mm long and so they cannot represent either *E. ingens* or *E. similis* but are more likely to belong to *E. ampliphylla*, as P.R.O. Bally suggested on one of these specimens.

Fig. 5.134. *Euphorbia ingens*, tip of branch with dense cluster of pale capsules, 50 km west of Polokwane (Pietersburg), South Africa, 11 Jun. 2019 (© PVB).

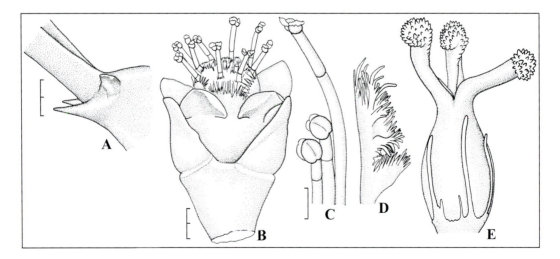

Fig. 5.135. *Euphorbia ingens*. **A**, young spines and leaf-rudiment from side on young plant 0.5 m tall (scale 2 mm). **B**, central male cyathium from side (lateral cyathia not yet developed, scale 2 mm). **C**, anthers (scale 1 mm, as for **D**, **E**). **D**, bracteole. **E**, female floret. Drawn from: **A**, *PVB 11331*, Ulundi, South Africa. **B–E**, cultivated plant, Kirstenbosch, Cape Town, South Africa (© PVB).

5.1 Sect. Euphorbia

Euphorbia kaokoensis (A.C.White, R.A.Dyer & B.Sloane) L.C.Leach, *Dinteria* 12: 33 (1976). *Euphorbia subsalsa* var. *kaokoensis* A.C.White, R.A.Dyer & B.Sloane, *Succ. Euphorb.* 2: 965 (1941). Type: Namibia, Kaokoveld, Kauas Okawe, 28 Nov. 1939, *C.H. Hahn sub Otzen 3* (PRE, holo.).

Bisexual spiny glabrous succulent 0.2–1 m tall with many branches from similar stem with small cluster of fibrous roots, much branched just above ground level. *Branches* erect, 50–500 × 15–25 mm, 5- to 8 (9)-angled, not constricted into segments, smooth, grey-green; *tubercles* arranged into 5–8 (9) low and slightly wing-like angles with surface slightly concave between angles, conical and truncate, laterally flattened and projecting 2–4 mm from angles, with spine-shield 6–10 mm long and 1.5–3 mm broad around apex and spreading down but just (by 1–2 mm) remaining separate from next, bearing 4 spreading to slightly deflexed initially red-brown (later dark brown) spines, lower longer 2 spines 7–24 mm long; *leaf-rudiments* on tips of new tubercles towards apices of branches, 1–2 × 2 mm, erect, fleeting, ovate, sessile, with red-brown (later dark brown) stipular prickles 3–14 mm long. *Synflorescences* 1–20 per branch towards apex, each a solitary cyme in axil of tubercle, on short peduncle 1.5–2 mm long, each cyme with 3 transversely disposed cyathia, central male, lateral 2 bisexual and developing later each on short peduncle 2 mm long and thick, with 2 ovate to rectangular bracts 2 × 1–2 mm subtending lateral cyathia; *cyathia* shallowly conical-cupular, glabrous, 3.5–4.5 mm broad (2.5 mm long below insertion of glands), with 5 obovate slightly reddish lobes with deeply incised margins, pale orange-yellow; *glands* 5, transversely rectangular and often contiguous, 1.5–2 mm broad, pale pinkish yellow, erect and slightly concave inside, inner margins not raised, outer margins entire and erect, surface between two margins shiny with secretion; *stamens* glabrous, with reddish to cream filaments, bracteoles enveloping groups of males, with finely divided tips, glabrous; *ovary* obtusely 3-angled, glabrous, pale green, nearly sessile; styles 3–3.5 mm long, branched for about two thirds of length. *Capsule* 3–5 mm diam., obtusely 3-angled, glabrous, pale green somewhat suffused with red below, sessile.

Distribution & Habitat

Euphorbia kaokoensis is only known in the Kaokoveld of north-western Namibia. It is most common on the high

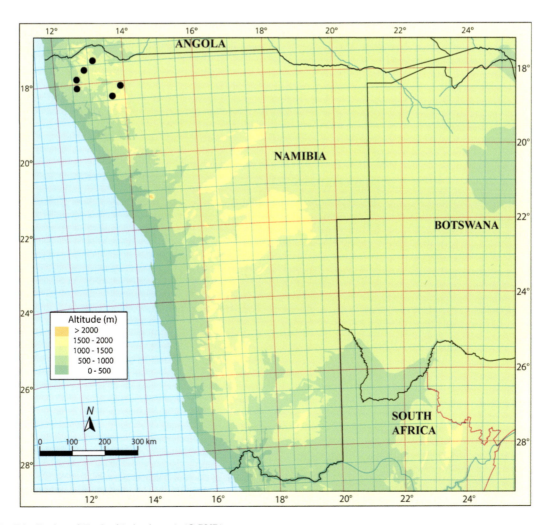

Fig. 5.136. Distribution of *Euphorbia kaokoensis* (© PVB).

ground west of Okonguati and it is also known further to the south-west at a few places in the mountains north and east of Orupembe.

In the higher areas of the Kaokoveld, *E. kaokoensis* is found in gently sloping places among granitic hills, in gravelly, slightly calcareous granitic ground among short trees of *mopane* and *Terminalia prunioides*, while nearer the coast it occurs in crevices in limestone, marble and granite, with small shrubs of *mopane*, various species of *Commiphora* and *Myrothamnus* (here with many other *Euphorbia* including *E. eduardoi* and *E. virosa*).

Fig. 5.137. *Euphorbia kaokoensis*, large plant ± 2 m diam. among *mopane* trees, *PVB 8033*, near Otjipemba, Namibia, 21 Dec. 1999 (© PVB).

Fig. 5.138. *Euphorbia kaokoensis*, low shrub ± 0.5 m diam., among gneissic rocks, *PVB 8057*, east of Orupembe, Namibia, 29 Dec. 2014 (© PVB).

Diagnostic Features & Relationships

Euphorbia kaokoensis forms dense and spiky shrubs that are usually considerably broader than they are tall. The branches are mostly 5- to 8-angled, with the angles only slightly raised out of the surface so that older branches have a tessellate appearance. Here the stipular prickles are substantial and are usually around half the length of the spines. The four spines on the spine-shield then combine to form a formidable armament and the plants are difficult to handle.

5.1 Sect. Euphorbia

Fig. 5.139. *Euphorbia kaokoensis*, plant ± 0.3 m diam., in crevices among gneissic boulders, *PVB 8057*, east of Orupembe, Namibia, 29 Dec. 2014 (© PVB).

Flowering in *E. kaokoensis* usually takes place between October and December, when the upper parts of the branches become covered with cyathia. In their early stages the cymes are pale red but as they mature the cyathia become yellow, though their bases and the peduncles and bracts remain suffused with red. Always produced in transverse groups of three in each cyme, the cyathia bear pinkish yellow glands which are erect and concave above, forming a cup-like structure around the edge of the cyathium.

In the northern Kaokoveld, *Euphorbia kaokoensis* occurs within 20 km of the closely related *E. otjipembana*. In *E. otjipembana* the branches are thicker and 4- to 6-angled and are less strongly spined (the branches are more slender, more sharply spined and with longer stipular prickles in *E. kaokoensis*). *Euphorbia otjipembana* also flowers earlier than *E. kaokoensis*. Florally, with its quite thin, erect and somewhat pinkish cyathial glands *E. kaokoensis* is most similar to subsp. *fluvialis* (and less so to subsp. *otjipembana*). In subsp. *fluvialis* the plant is taller, with the branches more erect from their bases and they are always 4-angled. Leach (1976a: 6–7) separated *E. kaokoensis* from subsp. *fluvialis* by the branches < 15 mm thick in the latter, but in subsp. *fluvialis* they may also reach 25 mm thick.

History

Euphorbia kaokoensis was first introduced into cultivation by Max Otzen from a collection of Major C.H. Hahn, which was made in November 1939 in the Kaokoveld of north-west Namibia. The locality 'Kauas Okawe' that Otzen gave, has not been traced on modern maps. At this locality it grew 'to huge bunches in rock clefts' (PRE records). Another speci-

Fig. 5.140. *Euphorbia kaokoensis*, cyathia in second male stage, *PVB 8057*, east of Orupembe, Namibia, 3 Nov. 2003 (© PVB).

Fig. 5.141. *Euphorbia kaokoensis*, capsules, *PVB 8057*, east of Orupembe, Namibia, 29 Dec. 2014 (© PVB).

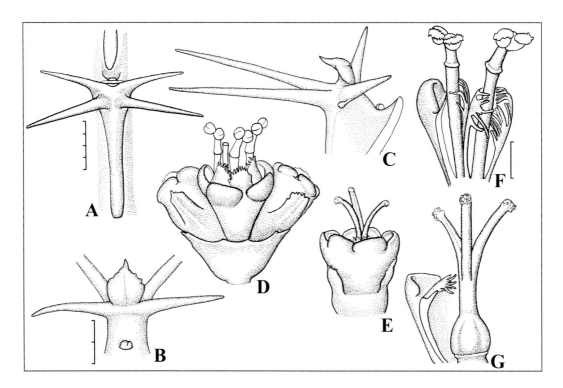

Fig. 5.142. *Euphorbia kaokoensis*. **A**, fully developed spine-complex (scale 4 mm). **B**, young spines and leaf-rudiment from above (scale 2 mm, as for **C–E**). **C**, young spines and leaf-rudiment from side. **D**, cyme from side in first male stage. **E**, side view of female cyathium. **F**, anthers and bracteoles in part of dissected cyathium (scale 1 mm, as for **G**). **G**, female floret (in part of dissected cyathium). Drawn from: **A–D**, **F**, *PVB 8057*, east of Orupembe, Namibia. **E**, **G**, *PVB 8033*, near Otjipemba, Namibia (© PVB).

men was collected in January 1955 by an 'extension officer J.E.V. Joubert near Ohopoho on limestone ridge' (PRE records). White et al. (1941) treated it as a variety of *E. sub-* *salsa*, which is widespread in southern Angola, but Leach (1976a) decided that it ought to be recognised as a distinct species.

5.1 Sect. Euphorbia

Euphorbia keithii R.A.Dyer, *Bothalia* 6: 223 (1951). Type: Swaziland, western edge of Lebombo Mtns, near Stegi, fl. 1949, *Keith sub PRE 28423* (PRE, holo.; GRA, K, NH, S, SRGH, iso.).

Bisexual spiny glabrous succulent shrub or tree, 1–6 m tall, with gradually deciduous branches forming rather untidy almost hemispherical crown near top of cylindrical erect solitary (rarely forked) trunk-like stem 0.5–6 m tall, trunk naked except for spines below crown, with many fibrous roots spreading from base. *Branches* spreading then ascending and often rebranching, most older branches drying up and falling off stem after a few years, 0.3–2 m × 30–40 mm, (3-) 4- to 5- (6-) angled, constricted into segments 50–250 mm long, smooth, grey-green; *tubercles* evenly fused into (3) 4–5 (6) wing-like angles with surface deeply concave between angles, low-conical, laterally flattened and projecting 2–3 mm from angles, spine-shields continuous forming hard grey to corky black margin 2–3 mm broad along angles, bearing 2 spreading and widely diverging brown to grey spines 3–8 mm long; *leaf-rudiments* on tips of new tubercles towards apices of branches, 3–6 × 2–3.5 mm, spreading, deciduous, ovate-cordate, sessile, with minute irregular brown tooth-like stipules < 0.5 mm long. *Synflorescences* many per branch towards apex (usually on last segment only), with (1–) 3 cymes in axil of each tubercle, shortly peduncled, each cyme with 3 vertically disposed cyathia, central male, outer two bisexual and developing later each on short peduncle 1–3 × 2 mm, with 2 small broadly ovate bracts 1.5–2.5 × 1–1.5 mm subtending lateral cyathia; *cyathia* conical-cupular, glabrous, 4–5 mm broad (2 mm long below insertion of glands), with 5 obovate lobes with finely toothed margins, slightly yellowish green; *glands* 5, transversely elliptic, well separated, 1.5–2 mm broad, yellowish green, spreading, inner margins not raised, outer margins entire and spreading, surface between two margins dull and smooth; *stamens* with glabrous pedicels, anthers dark red (filaments reddish green, pedicels green), bracteoles enveloping groups of males, with finely divided tips, glabrous; *ovary* globose and very obtusely 3-angled, glabrous, green, exserted on erect pedicel 2–5 mm long; styles 2–3.5 mm long, spreading, branched in upper third. *Capsule* 5–7 mm diam., very rounded 3-angled, glabrous, green suffused with red above, exserted on decurved and later erect pedicel ± 6 mm long.

Distribution & Habitat

Euphorbia keithii is known in Swaziland along the western flank of the Lebombo Mountains, where it has been recorded from near the borderpost (with Moçambique) of Namaacha to south of Siteki, usually at altitudes of 300–500 m. There is

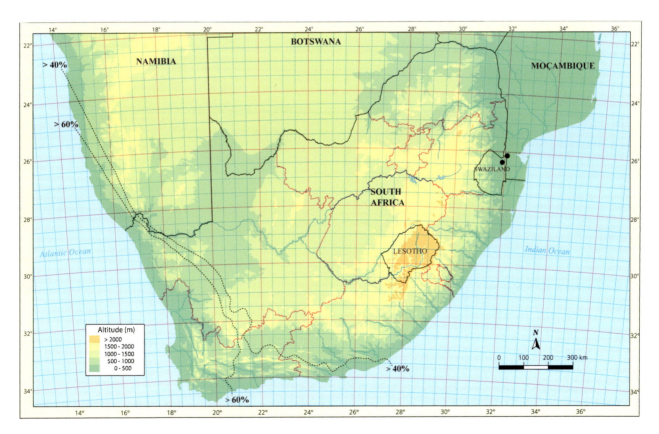

Fig. 5.143. Distribution of *Euphorbia keithii* (© PVB).

also a single known collection from the eastern flank of the Lebombo Mountains at Goba in Moçambique at an altitude of 150 m.

Plants form small, rather scattered colonies near the summit of the mountains on reddish outcrops of rhyolite between patches of denser bush and forest. On these outcrops they occur with *Xerophyta* and a diverse assortment of succulents such as *Aloe sessiliflora*, *Ceropegia cimiciodora*, *C. linearis*, *C. stapeliiformis* and several stapeliads, *Cotyledon barbeyi*, *Crassula swaziensis*, *Cynanchum gerrardii*, *C. viminale*, *Eulophia petersii*, *Pachypodium saudersiae*, several species of *Plectranthus* and *Tetradenia*.

Fig. 5.144. *Euphorbia keithii*, large tree ± 4 m tall in shallow soil on steep rocky SW-facing slopes with *Aloe sessiliflora*, *PVB 11868*, Lebombo Mountains, north of Siteki, Swaziland, 7 Jan. 2011 (© PVB).

Fig. 5.145. *Euphorbia keithii*, ± 3 m tall, on steep rocky slopes, *PVB 11868*, Lebombo Mountains, north of Siteki, Swaziland, 7 Jan. 2011 (© PVB).

Diagnostic Features & Relationships

Euphorbia keithii is a shrub- to tree-forming species which, in venerable specimens, may reach 6 m tall. However, most plants remain between 2 and 5 m tall, with some not exceeding 1 m at maturity. Many of them have a somewhat shrubby appearance with the branches persisting in a rather untidy arrangement towards the top of the tree. Only in very tall specimens is the trunk free of branches towards its base, with a clearly defined crown of branches near the top. The branches are mostly 4- to 5-angled, with distinct constrictions into segments and they have a noticeably grey-green colour, often with some blackening by fine lichens in older specimens. In *E. keithii* the leaf-rudiments are comparatively large and fairly conspicuous on new growth. Peculiar about them is their often wavy margins.

Flowering in *E. keithii* takes place between late October and late December, with the capsules persisting into

5.1 Sect. Euphorbia

Fig. 5.146. *Euphorbia keithii*, shrubby plant ± 1 m tall with *Xerophyta*, *PVB 11868*, Lebombo Mountains, north of Siteki, Swaziland, 7 Jan. 2011 (© PVB).

Fig. 5.147. *Euphorbia keithii*, flowering branch, *PVB 11868*, Lebombo Mountains, north of Siteki, Swaziland, 7 Jan. 2011 (© PVB).

Fig. 5.148. *Euphorbia keithii*, capsules, *PVB 11868*, Lebombo Mountains, north of Siteki, Swaziland, 7 Jan. 2011 (© PVB).

January. The cyathia are relatively small and well spaced out along the angles of the branches so that they do not alter its appearance greatly. They are green with a faint tinge of yellow and are sweetly scented. The male florets contrast quite strongly with the green of the cyathia, with their reddish filaments and deep red anthers, while the female florets are inconspicuous (matching the green of the cyathium closely), with the ovary slightly exserted from the cyathium when the styles mature and becoming further exserted on fertilization.

Analysis of DNA-data (Bruyns et al. 2011) showed that *E. keithii* is most closely allied to *E. caerulescens*, *E. tetragona* and *E. triangularis*. Of these it is most similar to *E. triangularis* (which also occurs within 20 km of where *E. keithii* is known), differing from it principally by forming smaller trees and shrubs with persistent, mostly 4- or 5-angled branches along the trunk and the grey colour of the branches. Other differences are the longer leaf-rudiments with smaller stipules, the generally smaller cyathia, the red anthers with reddish filaments and the broader, less stalked divisions of the style. *Euphorbia keithii* also flowers some months after *E. triangularis*.

History

Euphorbia keithii was discovered by Donald R. Keith. He first brought it to the attention of R.A. Dyer in November and December 1942 (when Dyer wrote down a description) but some was also sent to B.L. Sloane, who took photographs of these in September 1942. Dyer believed it 'was probably related to *E. pseudocactus*'. From a plant growing in his garden, Keith provided Dyer with flowering material in Nov.-Dec. 1949 and fruiting material in January 1950 and it was from these that Dyer described it. Keith also appears to have taken Dyer to plants in habitat 7–8 miles south-east of Siteki (c. 7 miles NNE of Keith's farm Ravelston) early in June 1947, for Dyer (1951) included photographs of specimens in their natural habitat. The species has never been well-known and the first published photographs in colour of it appear to be those in the Swaziland Tree Atlas (Loffler and Loffler 2005).

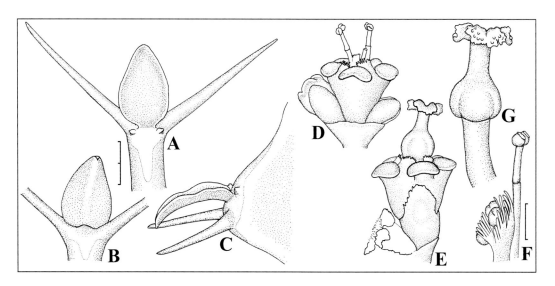

Fig. 5.149. *Euphorbia keithii*. **A**, young spines and leaf-rudiment from above (scale 2 mm, as for **B–E**). **B**, young spines and leaf-rudiment from rear. **C**, young spines and leaf-rudiment from side. **D**, cyme from side in first male stage. **E**, side view of female cyathium. **F**, anthers and bracteole (scale 1 mm, as for **G**). **G**, female floret. Drawn from: *PVB 11868*, Lebombo Mountains, north of Siteki, Swaziland (© PVB).

Euphorbia knobelii Letty, *Fl. Pl. South Africa* 14: t. 521 (1934). Type: South Africa, Transvaal, Enselsberg near Zeerust, Sept. 1933, *Knobel sub PRE 15854* (K, holo.).

Euphorbia perangusta R.A.Dyer, *Fl. Pl. South Africa* 18: t. 716 (1938). Type: South Africa, Transvaal, Koedoesrant, north of Zeerust, Jan. 1936, *Louw 99* (sub *PRE 23399*) (PRE, holo.; BOL, GRA, K, MO, P, SRGH, iso.).

Bisexual spiny glabrous succulent shrub 0.15–1 × 0.3–1.5 m, much branched at and just above ground level from short central stem tapering rapidly into a short taproot bearing fibrous roots. *Branches* ascending or initially spreading then erect, occasionally rebranched, 100–500 mm long, 20–60 mm thick, (3-, 4-) 5- to 7-angled, weakly constricted into roughly spherical to cylindrical or slightly cordate segments 15–50 (100) mm long, smooth, green with variably prominent light green to yellow bands radiating from centre; *tubercles* fused into 5–7 prominent wing-like sometimes slightly spiralling and undulating angles with deep grooves between angles, laterally flattened and deltate, projecting

4–10 mm from angles, with spine-shields 4–7 (10) mm long, 1–2 × 1–3 mm above spines and 2–6 (8) mm long and more slender below spines, from well separated from next below to joined into continuous red-brown and later grey horny margin along angles, bearing 2 spreading to slightly deflexed initially reddish brown (later grey) spines 6–13 mm long (usually much shorter in constrictions of branches) occasionally with 2 further prickles alongside synflorescences; *leaf-rudiments* on tips of new tubercles towards apex of branch, 1.5–2 × ± 1.5 mm, erect, fleeting, ovate, sessile, with small brown stipular prickles 0.7–1 mm long. *Synflorescences* many per branch usually on last two segments, (1–) 3 cymes in axil of each tubercle, on peduncle 2–3 × 1.5–2 mm, each cyme with 3 vertically disposed cyathia, central male, lateral 2 bisexual and developing later each on short peduncle 1–2 × 1–2 mm, with 2 ovate bracts 1–1.5 × 1.5–2 mm subtending lateral cyathia; *cyathia* conical, glabrous, 4–5 mm broad (2 mm long below insertion of glands), with 5 pale yellow or cream obovate lobes with deeply incised margins, bright yellow-green faintly suffused with red; *glands* 5, transversely oblong and contiguous, 1.5–2 mm broad, bright yellow-green, ascending, inner margins slightly raised, outer margins entire and ascending, surface between two margins dull; *stamens* glabrous, finely divided bracteoles enveloping groups of males, with finely divided tips, glabrous; *ovary* 3-angled, glabrous, green, partly exserted from cyathium on pedicel 1–2 mm long later becoming decurved and 5–7 mm long; styles 1.5–2 mm long, branched to near base. *Capsule* 5 mm diam., deeply 3-angled, glabrous, exserted on decurved and later erect pedicel 6–7 mm long.

Distribution & Habitat

Euphorbia knobelii is found in the western part of the former Transvaal, close to the border with Botswana. Here it is known on the northern slopes of Enselsberg, a series of low mountains

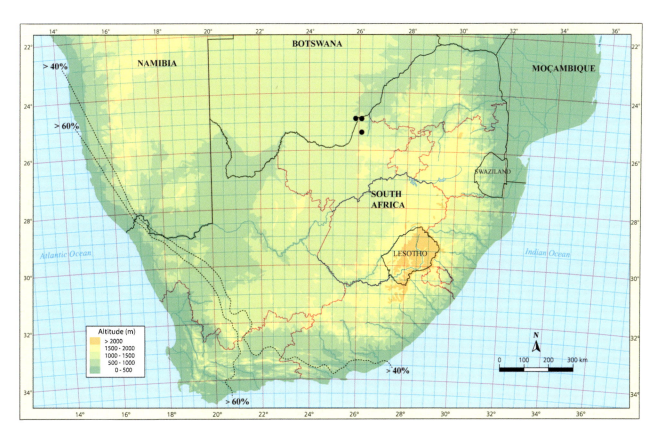

Fig. 5.150. Distribution of *Euphorbia knobelii* (© PVB).

lying to the north of Zeerust and on another even lower series of ridges, the Koedoesrand, some 80 km to the north.

In both these areas plants are found in shallow pockets of soil in narrow crevices on often nearly flat or gently sloping, relatively bare, sandstone outcrops. In the Enselsberg, they mainly occur on the north-facing and drier slopes among scattered grasses, stunted trees and clumps of *Myrothamnus*. In the Koedoesrand, they occur on very similar-looking

Fig. 5.151. *Euphorbia knobelii*, shrub ± 1.5 m diam. in shallow soil on bare sandstone slabs with grasses and brown clumps of *Myrothamnus*, *PVB 12077*, Enselsberg, north of Groot Marico, South Africa, 23 Dec. 2011 (© PVB).

Fig. 5.152. *Euphorbia knobelii*, in shallow soil on sandstone slabs, *PVB 12077*, Enselsberg, north of Groot Marico, South Africa, 23 Dec. 2011 (© PVB).

slopes, among scattered trees and tufts of grass and *Myrothamnus*, though here they also occur on the southern aspect of the hills.

Diagnostic Features & Relationships

Plants of *Euphorbia knobelii* form impressive shrubs in which the branches spread and then ascend around a small central stem, which is hidden and hard to observe but rarely seems to exceed 10 cm tall. In young plants this central stem is 3- to 4-angled while most of the branches (even in young specimens) are 5-angled. Most branches are around 3–4 cm thick, with somewhat irregular, almost spherical but not deeply incised segments that, in habitat, are often shorter than broad. In cultivation, this appears to change and longer, often considerably thicker segments are produced, perhaps reflecting the harshness of the habitat and its nutritionally poor soils (with richer soils and more evenly distributed watering provided in cultivation).

5.1 Sect. Euphorbia

Fig. 5.153. *Euphorbia knobelii*, in shallow soil on sandstone slabs, *PVB 12078*, ± 80 km north of Enselsberg, north of Groot Marico, South Africa, 25 Dec. 2011 (© PVB).

In *E. knobelii* the green surface of the branches between and on the sides of the angles is often attractively marked with pale green to yellow bars which run from the axil of each tubercle to the centre of the branch. Along the branches the angles are straight to gently spiralling, often somewhat undulating and are armed with pairs of spines of exceedingly variable length, commonly varying between 12 mm (on the most exposed spots in the middle of the segments) and 3–4 mm long (in the indentations between successive segments). They often seem to be longer, too, towards the bases of the branches than towards their tips. In the uppermost segment they are reddish brown and this changes to grey with age. The spine-shields are also very variable in length and even on individual branches vary from continuous to discrete.

Flowering takes place between August and October, when the upper one or two segments on each branch become yellowish from the many cyathia that they bear, usually in groups of three in each axil. The cyathia are borne on relatively long peduncles (though still only around 3 mm long) and are comparatively small, with somewhat longer-than-usual lobes and long male florets. In the female florets, when the styles mature, the ovary is usually just contained within the cyathium on a pedicel around 1.5 mm long but, on fertilization, this suddenly elongates to 6–7 mm long and bends over to hold the maturing capsule well outside the cyathium. The capsules are usually bright red, shiny and deeply 3-angled and become erect just before exploding.

Fig. 5.154. *Euphorbia knobelii*, flowering, *PVB 12077*, Enselsberg, north of Groot Marico, South Africa, 1 Nov. 2012 (© PVB).

Fig. 5.155. *Euphorbia knobelii*, flowering with capsules developing, *PVB 12078*, ± 80 km north of Enselsberg, north of Groot Marico, South Africa, 9 Nov. 2012 (© PVB).

History

Euphorbia knobelii was discovered by J.C.J. Knobel and the type specimen has the date Sept. 1933 on it, possibly indicating when it was first collected, but possibly also indicating when the specimen was pressed at the National Herbarium (PRE). The plants described by Dyer as *E. perangusta* were discovered by W.J. Louw in January 1936.

These two names were maintained as distinct species in Bruyns (2012) but they are sometimes difficult to distinguish (Fourie 1987: 64) and consequently they are more likely to be local forms of a single species. Traditionally they were distinguished by the relative lack of secondary branches, the more frequently 6-angled branches with less obvious mottling, the more regularly continuous spine-shields and the narrower angles in *E. perangusta*. Their respective habitats are notably similar and plants in both areas prove to be somewhat variable. When branches from both places are compared, it is more or less impossible to distinguish them. Since no differences were found in the flowers, they are treated as belonging to one species here.

White et al. (1941) showed pictures taken by Louw of very large plants of '*E. perangusta*' growing with grasses and trees, apparently on sandstone outcrops. It would appear that grazing pressures have greatly reduced them, though Fourie (1987) still showed some fairly large specimens and here the bareness of the habitat is very clear.

Fig. 5.156. *Euphorbia knobelii*, flowering with capsules developing, *PVB 12077*, Enselsberg, north of Groot Marico, South Africa, 9 Nov. 2012 (© PVB).

5.1 Sect. Euphorbia

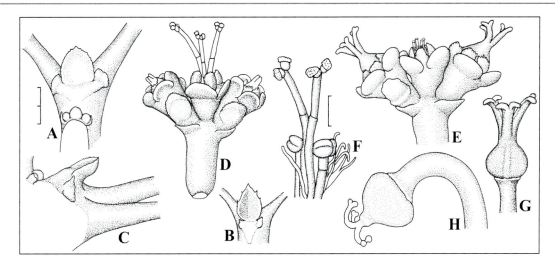

Fig. 5.157. *Euphorbia knobelii.* **A**, **B**, young spines and leaf-rudiment from above (scale 2 mm, as for **B–E**). **C**, young spines and leaf-rudiment from side. **D**, **E**, cyme from side in different stages. **F**, anthers and bracteoles (scale 1 mm, as for **G**, **H**). **G**, **H**, female floret at different stages. Drawn from: **A**, **B**, **D**, **F–H**, *PVB 12077*, Enselsberg, north of Groot Marico, South Africa. **C**, **E**, *PVB 12078*, ± 80 km north of Enselsberg, north of Groot Marico (© PVB).

Euphorbia knuthii Pax, *Bot. Jahrb. Syst.* 34: 83 (1904). Lectotype (Bruyns 2012): Moçambique, Ressano Garcia, 1000', 27 Dec. 1897, *Schlechter 11949* (K; BOL, BM, BR, G, GRA, HBG, PRE, WAG, iso.).

Bisexual spiny glabrous small succulent 50–200 mm tall with many branches from neck of turnip-like swollen subterranean stem from which cluster of swollen tuberous roots and fibrous roots arise as well as spreading underground

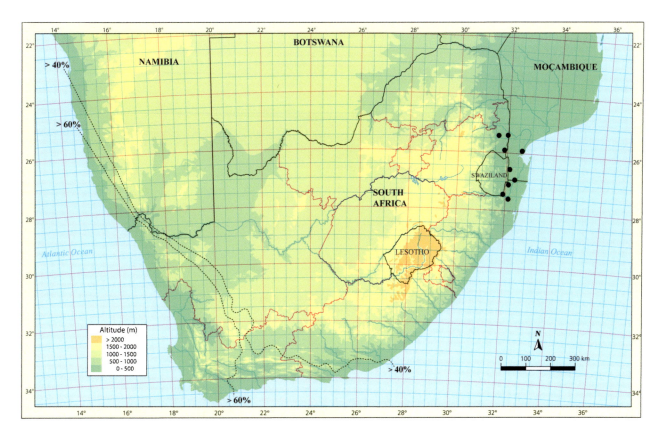

Fig. 5.158. Distribution of *Euphorbia knuthii* ssp. *knuthii* (© PVB).

Fig. 5.159. *Euphorbia knuthii* ssp. *knuthii*, plant in open ground between trees, with branches protruding ± 15 cm from ground and connected underground by rhizomes, north of Moamba, Moçambique, 11 Jan. 2004 (© PVB).

Fig. 5.160. *Euphorbia knuthii* ssp. *knuthii*, growing tip of branch with leaf-rudiments, *PVB 4466*, near Ndumu, South Africa (© PVB).

Fig. 5.161. *Euphorbia knuthii* ssp. *knuthii*, flowering branch, *PVB 4466*, near Ndumu, South Africa, 18 Jan. 2012 (© PVB).

cylindrical rhizomes giving rise to solitary branches or small clusters of branches. *Branches* erect, 50–350 × 5–12 mm, (3) 4 (5)-angled, not constricted into segments but narrowing towards slender base, branching sparingly at base and rarely above, smooth, pale to dark green with grey or cream markings between angles; *tubercles* fused into (3–) 4 narrow angles with surface distinctly concave between angles, conical and truncate, laterally flattened and prominent and projecting 2–6 mm from angles so that angles have distinctly wave-like profile, deltate, with spine-shield 2–8 (10) mm long and 1–2 mm broad around apex and spreading down often for only a short distance and remaining separate from next, bearing 2 spreading initially brown (later grey) spines 4–9 mm long; *leaf-rudiments* on tips of new tubercles towards apices of branches, (3) 4–15 × 1–3.5 mm, erect, fleeting, lanceolate, acuminate, sessile, usually with spreading brown stipular prickles 1–2 mm long. *Synflorescences* 1–6 per branch towards apex, each a solitary cyme in axil of tubercle, on short peduncle ± 2 × 2 mm, each cyme with solitary cyathium of 3 transversely disposed cyathia, central male or bisexual, lateral 2 bisexual and developing later each on short peduncle 2–4 × ± 2 mm (sometimes with further 1–4 cyathia developing from axils of bracts), with 2 ovate to almost rectangular apically toothed bracts 1.2–2 × 1 mm subtending cyathia; *cyathia* shallowly conical-cupular, glabrous, 3–5 mm broad (2 mm long below insertion of glands), with 5 obovate lobes with deeply incised margins, green; *glands* 5, transversely oblong or rectangular and contiguous, 1.5–2.3 mm broad, green, spreading, flat above, outer margins entire and spreading, surface between two margins dull; *stamens* glabrous, bracteoles enveloping groups of males, with finely divided tips, glabrous; *ovary* 3-angled, glabrous, green, at first partly exserted on erect pedicel nearly 2 mm long later fully exserted on decurved pedicel; styles 2–2.5 mm long, branched to just below middle. *Capsule* 4–7 mm diam., deeply 3-angled, glabrous, exserted from cyathium on decurved and later erect pedicel 6–10 mm long.

Leach (1973b) divided *E. knuthii* into two subspecies, subsp. *knuthii* and subsp. *johnsonii* (only known in coastal Moçambique between the Buzi and Save Rivers), see also Leach (1963). These were separated on the rhizomatous habit with subsidiary tubers of subsp. *knuthii* with its mainly 4-angled branches (versus mostly non-rhizomatous plant with branches from a single tuber and branches mostly 3-angled in subsp. *johnsonii*). What follows applies to subsp. *knuthii*.

Distribution & Habitat
Euphorbia knuthii is found in South Africa and Moçambique in low-lying areas between the coast around Maputo and the Lebombo Mountains and southwards to northern Natal between Ndumu and Mkuzi. The distribution extends slightly west of the Lebombo Mountains in the low-lying areas west of Komatipoort, but it has not been recorded in similar places in Swaziland (Leach 1973b).

Generally *E. knuthii* grows in flat, seasonally quite moist areas in black turf soils, usually among *Acacia* trees or with large clumps of *Euphorbia grandicornis*. In the south of Moçambique and around Ndumu, it grows together with the very rhizomatous stapeliad *Ceropegia paradoxa* among scattered trees and large clumps of *E. grandicornis*.

Diagnostic Features & Relationships
Euphorbia knuthii is distinctive for the rhizomatous branches spreading away from a central main tuber, with smaller, turnip-like tubers often developing at intervals along these rhizomes. Each rhizome is cylindrical beneath the ground and usually ends in a short, erect branch or a small cluster of branches above the ground. These above-ground portions are mostly 4-angled and thicker than the subterranean part, with prominent, distinct tubercles that are not completely merged into the angles so that, when viewed from the side, the angles have a noticeably undulating profile. The area between the angles is concave and usually mottled with cream on dark green, whereas the angles are dark green. Quite prominent leaf-rudiments are produced on young growth, each with a tiny stipular prickle on either side just in front.

Flowers appear in *E. knuthii* over much of the summer and have been observed (in cultivation) between October and March. They appear towards the tips of the branches and are relatively inconspicuous and small, with green glands. Each cyme may consist only of a solitary cyathium, others bear the usual three cyathia and yet others continue to branch from the axils of the bracts to produce a succession of up to four cyathia on a knobbly 'peduncle', though such long 'peduncles' as shown by Fourie (1989) are not recorded in habitat. One of the notable features of the cyathia is the manner in which the pedicel of the female floret lengthens after fertilization and curves sideways to push the ovary well out beyond the circumference of the cyathium.

The branches of *E. knuthii* are superficially suggestive of the complex around *E. schinzii*, while the tubers are sugges-

Fig. 5.162. *Euphorbia knuthii* ssp. *knuthii*, flowering branch with some bisexual central cyathia, *PVB 4466*, near Ndumu, South Africa, 12 Mar. 2013 (© PVB).

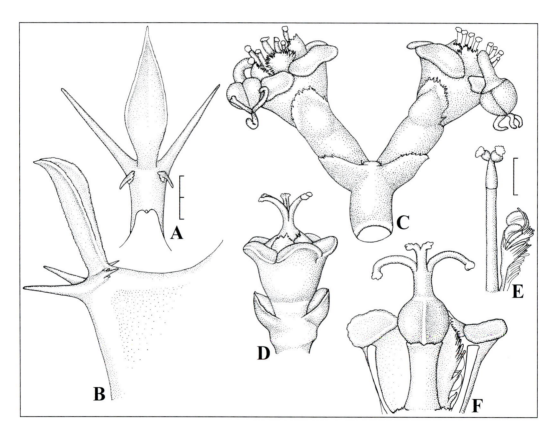

Fig. 5.163. *Euphorbia knuthii* ssp. *knuthii*. **A**, young spines and leaf-rudiment from above (scale 2 mm, as for **B–D**). **B**, young spines and leaf-rudiment from side. **C**, cyme from side in late stage. **D**, female cyathium from side. **E**, anthers and bracteole (scale 1 mm, as for **F**). **F**, female floret in dissected cyathium. Drawn from: **A**, **B**, **D–F**, *PVB 9385*, near Komatipoort, South Africa. **C**, *PVB 4466*, near Ndumu, South Africa (© PVB).

tive of *E. stellata*. Consequently White et al. (1941) placed *E. knuthii* between these two, following the arrangement of N.E. Brown (1915). Its rhizomatous habit is unique among all the southern African members of sect. *Euphorbia* and features such as the occasional production of more than the standard three cyathia per cyme, the long-pedicelled, deeply 3-angled capsules and the smooth seeds suggest that its affinities lie elsewhere. Leach (1973b, 1992) rejected any affinity of *E. knuthii* with *E. schinzii*, on account of the ± sessile capsules and verrucose, ellipsoidal seeds of the latter. Instead, he suggested it was related to *E. stellata*, with which it shares many of these features (Leach 1973b). In Leach (1992) he suggested affinities with *E. griseola*, *E. jubata* and *E. richardsiae* (Fig. 5.162). Analysis of DNA-data (Bruyns et al. 2006, 2011) showed that *E. knuthii* was indeed closely allied to *E. griseola* and *E. mlanjeana* (and later to *E. jubata*, while *E. richardsiae* has not been analysed in this way yet). *Euphorbia mlanjeana* also shares the characteristics of exserted, deeply 3-angled capsules and smooth, subglobose seeds with *E. griseola* and *E. knuthii*, so that Leach's suggestions relating to characters of the capsules and seeds are corroborated exactly by the new results from DNA-data.

History
Euphorbia knuthii was discovered around the end of December 1897 by Rudolf Schlechter at Ressano Garcia, near the border between Moçambique and South Africa. This was during his last expedition in southern Africa, which covered parts of Moçambique to Beira and Umtali in Zimbabwe. Schlechter seems to have realized that he had found something new, annotating his sheets with a proposed name, but F. Pax decided to name it for the German botanist Paul Knuth. Many important and diagnostic details of it (such as the rhizomatous habit, the prominent tubercles and leaves, the exserted capsules) were shown for the first time by Phillips (1929b), though unfortunately under the name '*Euphorbia kunthii*'.

Euphorbia louwii L.C.Leach, *J. S. African Bot.* 46: 207 (1980). Type: South Africa, Transvaal, c. 14 km east of Marken, 900 m, 1 Nov. 1975, *Leach et al. 15555* (PRE, holo.; K, SRGH, iso.)

Bisexual spiny glabrous succulent shrub 0.15–1 × 0.15–0.5 (1) m, with many branches from similar stem with small cluster of fibrous roots, densely branched at and slightly above ground level and only occasionally rebranching. *Branches* erect, 0.15–0.5 (1) m × 8–14 mm, 5- to 7-angled, occasionally slightly constricted into segments, smooth, pale bluish grey-green and paler to creamy between angles; *tubercles* fused into 5–7 slightly wing-like angles with surface concave between angles, low-conical and truncate, projecting 2–3 mm from angles, with spine-shields 5–10 mm long, 1–2 × 2–3 mm above spines and 4–7 mm long below spines but remaining well separated from next, bearing 2 spreading initially reddish brown (later dark brown) fine spines 5–10 mm long (with 2 stipular prickles 0.7–4 mm long and often a further prickle 1–3 mm long near base, i.e. 5 spines per shield); *leaf-rudiments* on tips of new tubercles towards apices of branches, 0.7–2.2 × ± 0.6 mm, erect, fleeting, subulate, sessile, with brown stipular prickles 0.7–4 mm long. *Synflorescences* many per branch mainly towards apex, each a solitary cyme in axil of tubercle, on short peduncle 0.5–1.5 mm long, each cyme with 3 transversely disposed cyathia, central male, lateral 2 bisexual and developing later each on short peduncle 1–2 mm long and thick, with 2 slender bracts 1–2 × 1 mm subtending lateral cyathia (and sometimes giving rise to further bisexual cyathia in their axils); *cyathia* shallowly conical-cupular, glabrous, 3.5–4 mm broad (1.5 mm long below insertion of glands), with 5 obovate lobes with deeply incised margins, bright yellow to yellow-green often suffused with red outside towards base; *glands* 5, transversely rounded-rectangular and nearly contiguous, 1.5–2 mm broad, yellow to yellow-green, ascending to spreading, inner margins not raised, outer margins entire, surface between two margins dull; *stamens* glabrous, bracteoles enveloping groups of males, with finely divided tips, glabrous; *ovary* obtusely 3-angled, glabrous, green, raised on short pedicel to 0.5 (0.75) mm long; styles 2.5–3 mm long, branched almost to base. *Capsule* 3–4 mm diam., obtusely 3-angled, glabrous, sessile.

Distribution & Habitat
Euphorbia louwii is known from a small area around the hamlet of Marken, north-east of the Waterberg in the former Transvaal. Here it is fairly common, especially among conglomerates on the hotter, northern aspects of low hills, where it grows among large rocks, various deciduous trees (especially *Combretum*, *Croton* and *Ochna*), with *Selaginella* and *Xerophyta* and several succulents such as *Aloe*, *Kalanchoe luciae*, *Sansevieria* and some stapeliads.

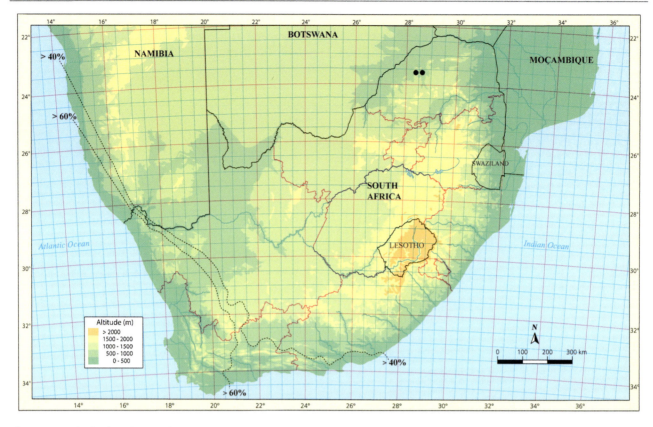

Fig. 5.164. Distribution of *Euphorbia louwii* (© PVB).

Fig. 5.165. *Euphorbia louwii*, plants to ± 40 cm tall among conglomerate boulders with *Sansevieria aethiopica*, *PVB 12051*, south of Marken, South Africa, 2 Nov. 2011 (© PVB).

Diagnostic Features & Relationships

Plants of *E. louwii* form shrubs mostly around 0.5–0.7 m tall with lots of branches arising near the ground, close to the base of the indistinguishable stem. The branches re-branch sparingly above and are very slightly constricted into segments. They are distinctively 5- to 7-angled, usually around 10 mm thick and have a pale grey-green colour with a paler colour in the grooves between the angles. Along the angles,

5.1 Sect. Euphorbia

Fig. 5.166. *Euphorbia louwii, PVB 12051*, south of Marken, South Africa, 2 Nov. 2011 (© PVB).

the branches are armed with many pairs of quite long but fairly weak, brownish spines that arise on slender spine-shields. Above the spines at the bases of the leaves (i.e. in a stipular position), each spine-shield bears two further small prickes that are usually much shorter than the main spines but may reach half their length. Each spine-shield often also bears another solitary prickle near its base that is of very variable length, though usually around the same length as the stipular prickles. Altogether then, each spine-shield usually bears five spines and prickles.

Flowering in *E. louwii* usually takes place in August and September, when the uppermost parts of many of the branches become covered with the small, brightly coloured cyathia. The axil of each tubercle bears a solitary cyme of

Fig. 5.167. *Euphorbia louwii*, with mats of *Selaginella*, in shallow gritty soils, *PVB 12051*, south of Marken, South Africa, 2 Nov. 2011 (© PVB).

Fig. 5.168. *Euphorbia louwii,* just beginning first male stage, *PVB 12051*, south of Marken, South Africa, 11 Aug. 2019 (© PVB).

three cyathia. Usually the cyathia are yellow or greenish but they may be suffused with red, especially towards their bases. The long and slender styles are divided more or less right to their bases above the ovary.

Euphorbia louwii is closely related to *E. aeruginosa* and *E. schinzii*. It is most easily distinguished from them by the 5- to 7-angled branches and the considerably longer and more slender leaf-rudiments. The presence of five spines and prickles per spine-shield that is common in *E. louwii* is also found occasionally in *E. aeruginosa*. The two species also have a very similar colour of the branches and spine-shields, though the pale streak between the angles in *E. louwii* is missing in *E. aeruginosa*. In all these species the cyathia are very similar and mainly differ by the more deeply divided styles of *E. louwii*, which it shares with *E. schinzii* and *E. venteri*.

History

Euphorbia louwii was first collected in October 1938 by F.Z. van der Merwe and was recorded again in May 1948 by W.F. Bayer. Sometime after 1969, W.J. Louw sent plants of it to L.C. Leach and in November 1975 Leach visited the type locality with Louw and Rossouw. It was described 5 years later from material collected on this occasion.

Fig. 5.169. *Euphorbia louwii,* first male stage, *PVB 12051*, south of Marken, South Africa, 11 Sep. 2012 (© PVB).

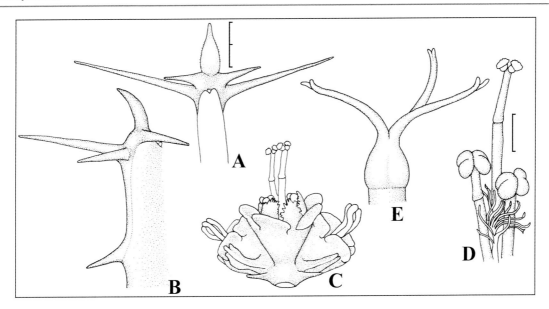

Fig. 5.170. *Euphorbia louwii*. **A**, young spines and leaf-rudiment from above (scale 2 mm, as for **B**, **C**). **B**, young spines and leaf-rudiment from side. **C**, cyme from side in female stage. **D**, anthers and bracteole (scale 1 mm, as for **E**). **E**, female floret. Drawn from: *PVB 12051*, south of Marken, South Africa (© PVB).

Euphorbia lydenburgensis Schweick. & Letty, *Fl. Pl. South Africa* 13: t. 486 (1933). Lectotype (Bruyns 2012): South Africa, Transvaal, Steelpoort Valley, 30 miles north of Lydenburg, 7 July 1932, *Van Balen & De Wyn sub PRE 14398* (PRE; K, iso.).

Bisexual spiny to spineless glabrous succulent shrub 0.15–1.5 (2.5) × 0.1–1 m, with relatively few branches in lower half of similar erect unbranched stem 0.3–1.5 m long becoming slightly swollen and nearly cylindrical towards base and covered there with grey bark, or densely branched around short not clearly visible stem (to 0.2 m long), with many fibrous roots on stem only. *Branches* erect and rarely rebranching, 0.3–1.5 m × 12–20 mm or spreading then erect in small clump and 0.1–0.3 m × 12–20 mm, 4 (5)-angled, not constricted into segments, smooth, green or yellow-green becoming brown and slightly cylindrical towards base with age; *tubercles* very evenly fused into 4 (–5) low wing-like angles with surface slightly concave between angles, conical and truncate, laterally flattened and projecting 1–3 mm from angles, spine-shields forming ± continuous hard grey to brown margin 1–2 mm broad along angles, bearing 2 spreading and widely diverging initially red (soon becoming grey) spines 3–9 mm long (spines sometimes absent) and sometimes further small spine-like protrusion near base; *leaf-rudiments* on tips of new tubercles towards apices of branches, ± 0.5 × 1 mm, erect, fleeting, broadly ovate and ridge-like, sessile, mostly with small stipular prickles up to 1 mm long. *Synflorescences* many per branch towards apex, each a solitary cyme in axil of tubercle, on peduncle 0.5–1 mm long, each cyme with 3 transversely disposed cyathia (occasionally more developing), central male, outer two bisexual and developing later each on short peduncle 1–1.5 × 2 mm, with 2 small broadly ovate bracts ± 2 × 2 mm subtending lateral cyathia; *cyathia* conical-cupular, glabrous, 3–4.5 mm broad (2 mm long below insertion of glands), with 5 narrowly obovate lobes with finely incised margins, yellow; *glands* 5, transversely rectangular and contiguous, 1.5–2.5 mm broad, yellow, slightly spreading, inner margins not raised, outer margins entire and spreading, surface between two margins dull; *stamens* with glabrous pedicels, anthers yellow, bracteoles enveloping groups of males, with finely divided tips, glabrous; *ovary* obtusely 3-angled, glabrous, green, raised on pedicel 1–1.5 mm long; styles 2 mm long, widely spreading, branched to near base. *Capsule* 3–4 mm diam., obtusely 3-angled, glabrous, dull dark green to red, raised on ascending pedicel to 2 mm long.

Euphorbia lydenburgensis is restricted to the drier northern parts of the Lydenburg district in Sekukuniland, where it occurs especially along the valley of the Olifants River (and side-valleys joining it) from west of Penge to the Abel Erasmus Pass and eastwards to near Branddraai.

Euphorbia lydenburgensis is related to members of the *E. schinzii*-complex (including *E. clivicola*, *E. schinzii* and *E. venteri*) and to *E. malevola* (Bruyns et al. 2006, 2011) with which it shares the included ovaries (sessile or raised on pedicel up to 1 mm long) and the verrucose seeds. The two varieties recognized are very closely related to each other and also to *E. pisima*. Most members of the *E. schinzii*-complex have more slender branches and all of them lack the distinct stem that is usually present at the base of the plant in var. *lydenburgensis*. Their leaf-rudiments are longer and usually more subulate than those of *E. lydenburgensis*, which are particularly small and ridge-like. In the *E. schinzii*-complex the cyathia are very similar but the ovary has a much shorter pedicel (around 1 mm long in *E. lydenburgensis*). *Euphorbia lydenburgensis* and *E. pisima* share the bright green colour of the branches. This colour changes to an equally bright yellow during the dry season when the plant is under stress and returns to the usual green when the plant receives water and swells up again.

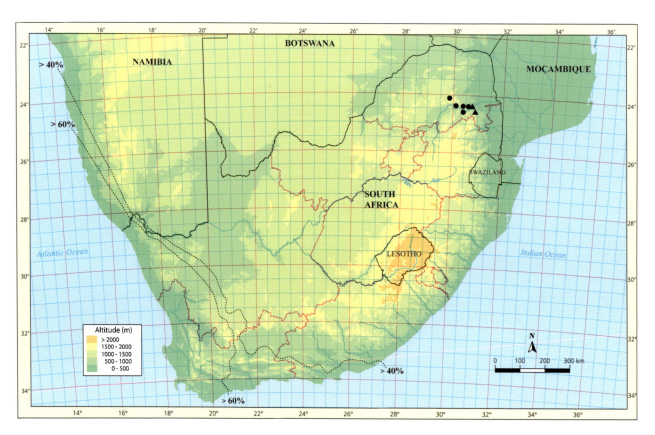

Fig. 5.171. Distribution of *Euphorbia lydenburgensis* (● = var. *lydenburgensis*; ▲ = var. *minor*) (© PVB).

5.1 Sect. Euphorbia

Two varieties are recognised here and may be separated as follows:

1. Stem developing into distinct, short trunk at base with branches ascending from stem into shrub 0.3–1.5 × 0.3–1 m, branches erect and rarely rebranching, 0.3–1.5 m long..var. **lydenburgensis**
1. Plant without visible trunk at base, forming shrublet 0.15–0.3 × 0.15–0.5 m, branches spreading close to ground then erect, 0.1–0.3 m long...var. **minor**

Fig. 5.172. *Euphorbia lydenburgensis* var. *lydenburgensis,* shrub ± 0.8 m tall, Penge, west of Burgersfort, South Africa, 22 May 1980 (© PVB).

Euphorbia lydenburgensis var. lydenburgensis

Spiny to spineless glabrous succulent shrub 0.3–1.5 × 0.3–1 m diam., with relatively few branches in lower half of similar erect unbranched stem 0.3–1.5 m tall, becoming slightly swollen and cylindrical towards base. *Branches* erect and rarely rebranching, 0.3–1.5 m × 15–20 mm.

Distribution & Habitat

Var. *lydenburgensis* is mainly found from Penge to the Abel Erasmus Pass and becomes less common eastwards towards Branddraai.

Plants are usually found on rocky slopes among bushes or deciduous trees on soils derived from dolomites or sandstones, occasionally on gentle, shaly slopes among a spiny vegetation with *Acacia tortilis* and *Ziziphus mucronatus*.

Diagnostic Features & Relationships

Var. *lydenburgensis* forms neat, erect shrubs usually 1–1.5 m tall that are mostly sparingly branched somewhat above the base of a slightly stouter stem. In particularly bushy situations on gentle slopes the plants may be sparsely branched but, when growing in shallow soils on rocky outcrops, they are considerably more densely branched. When it is still short (under 15 cm long), the stem is mottled with transverse dark green or reddish bars on a pale green background, but this is soon lost and the stem then assumes the distinctive, somewhat shiny, uniformly green or yellow-green colour that the branches later assume too. With time, the distinctive colour of the branches and of the stem is lost lower down, as these parts become covered with a

Fig. 5.173. *Euphorbia lydenburgensis* var. *lydenburgensis*, shrub ± 0.8 m tall in yellowish dry state, *PVB 11900*, NW of Penge, west of Burgersfort, South Africa, 22 May 1980 (© PVB).

smooth, corky grey or brown, sometimes slightly shiny bark.

In var. *lydenburgensis* the stem and branches are very regularly 4-angled (though the stem may become somewhat rounded with age). The tubercles are especially evenly joined into the angles so that these are continuous and hardly interrupted along the branches by constrictions, with a very shallow, slightly concave area between them in their upper parts. In the lower half of the branches (and the stem) this concave area becomes flat and even may be slightly convex, in which case the branches are almost cylindrical (and a little swollen) towards their bases. In young plants the spines are prominent, initially red and soon turning grey. On older plants they become variable in size, but are usually weak and fairly soft and plants are periodically encountered in which they are entirely absent. In these the angles are lined only with a slender horny margin provided by the spine-shields, which is sometimes discontinuous along the branches. The leaf-rudiments in var. *lydenburgensis* are exceptionally small, only rising slightly above the level of the spine-shield and usually remaining for a while as a minute scale. They are mostly flanked by two small stipular prickles from which a slight ridge sometimes extends further downwards along the edge of the spine-shield. There is sometimes a further spine-like enation near the base of the spine-shield, though this is not usually sharp-tipped.

Fig. 5.174. *Euphorbia lydenburgensis* var. *lydenburgensis*, spineless shrub ± 1 m tall, near Mahlashi, north of Burgersfort, South Africa, 1 Jan. 2012 (© PVB).

5.1 Sect. Euphorbia

Fig. 5.175. *Euphorbia lydenburgensis* var. *lydenburgensis,* very diffuse shrubs to ± 2 m tall, near Mahlashi, north of Burgersfort, South Africa, 1 Jan. 2012 (© PVB).

Fig. 5.176. *Euphorbia lydenburgensis* var. *lydenburgensis,* first male stage on plant with spines, flowering when in the dry yellowish state, *PVB 6615*, Kromellenboog, NW of Burgersfort, South Africa, 1 Jun. 2019 (© PVB).

Fig. 5.177. *Euphorbia lydenburgensis* var. *lydenburgensis,* female stage on spineless plant, *PVB 6615*, Kromellenboog, NW of Burgersfort, South Africa, 1 Jun. 2019 (© PVB).

Flowering in var. *lydenburgensis* takes place in April and May, when the upper parts of the branches become covered with large numbers of cyathia. These are small and yellow, with only one cyme of three transversely disposed cyathia developing in the axil of each leaf-rudiment. After flowering is over, the capsules persist until late October or November, after which the seeds are released. On some plants the capsules are dark green, while they are red on others.

History

Var. *lydenburgensis* was described from material gathered in 1932 at an undisclosed locality in the Lydenburg district by Jan C. van Balen & G.J. de Wijn. This was the first recorded collection of the species.

Fig. 5.178. *Euphorbia lydenburgensis* var. *lydenburgensis*, mature capsules on spineless plant, *PVB 6615*, Kromellenboog, NW of Burgersfort, South Africa, 2 Oct. 2012 (© PVB).

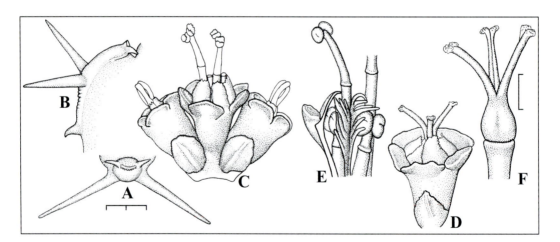

Fig. 5.179. *Euphorbia lydenburgensis* var. *lydenburgensis*. **A**, young spines and leaf-rudiment from above (scale 2 mm, as for **B–D**). **B**, young spines and leaf-rudiment from side. **C**, cyme from side at beginning of female stage. **D**, cyathium at female stage. **E**, anthers and bracteoles (scale 1 mm, as for **F**). **F**, female floret. Drawn from: *PVB 6615*, Kromellenboog, NW of Burgersfort, South Africa (© PVB).

5.1 Sect. Euphorbia

Euphorbia lydenburgensis var. **minor** Bruyns, *Haseltonia* 25: 55 (2018). Type: South Africa, Lydenburg distr., near Abel Erasmus Pass, California 228, 1300 m, 30 Oct. 1980, *N.H.G. Jacobsen 3530* (PRE, holo.).

Spiny glabrous succulent shrublet 0.15–0.3 × 0.15–0.5 m, with many short branches close to ground around and hiding stem up to 0.2 m tall. *Branches* spreading then erect and rarely rebranching, 0.1–0.3 m × 15–20 mm.

Distribution & Habitat

Euphorbia lydenburgensis var. *minor* is known on several hillsides between Abel Erasmus Pass and Branddraai, northeast of Ohrigstad. Here they occur in shallow pockets of soil on exposed dolomitic outcrops at 1000–1400 m with other much stunted vegetation, often with the similarly green-branched *E. pisima* nearby and various other succulents.

Diagnostic Features & Relationships

Var. *minor* forms dense clusters of branches, usually only 0.1–0.2 m tall but occasionally reaching up to 30 cm tall, with the plant reaching maturity and flowering at a height of 0.1 m or less. These low shrublets may be up to 30 cm or more in diameter. There is no short, central stem visible in these plants at all and the branches are often distinctly spreading, then becoming erect. At 12–20 mm thick, they are similar in thickness to var. *lydenburgensis* and also, when overly exposed, turn bright yellow-green as var. *lydenbur-*

Fig. 5.180. *Euphorbia lydenburgensis* var. *minor*, shrublet ± 30 cm broad in crevices between rocks with *Kleinia stapeliiformis*, plant in the green state of the rainy season, *PVB 7102*, west of Abel Erasmus Pass, north of Ohrigstad, South Africa, 1 Jan. 1997 (© PVB).

Fig. 5.181. *Euphorbia lydenburgensis* var. *minor*, branches < 15 cm tall, *PVB 12099*, Leboeng, east of Abel Erasmus Pass, north of Ohrigstad, South Africa, 1 Jan. 2012 (© PVB).

Fig. 5.182. *Euphorbia lydenburgensis* var. *minor*, shrublet ± 20 cm diam., *PVB 12099*, Leboeng, east of Abel Erasmus Pass, north of Ohrigstad, South Africa, 1 Jan. 2012 (© PVB).

gensis does too. Some of them have a small prickle at the base of the spine-shield (these are not always continuous along the angles) as one may find in var. *lydenburgensis* and they share the very short, relatively broad leaf-rudiments typical of var. *lydenburgensis* (and unlike the more slender ones of *E. schinzii*).

Florally the two varieties are not distinguishable. Both share the features of relatively few male florets, fairly tall cyathial lobes bent around these male florets and ovaries raised on pedicels just over 1 mm long. Var. *minor* usually flowers between April and July.

History

Euphorbia lydenburgensis var. *minor* was first observed by F.Z. van der Merwe in July 1938 near Branddraai, according to a specimen in PRE. Van der Merwe collected both varieties near Branddraai. White et al. (1941: 783) mentioned the particularly large plants that he had found at Branddraai but did not refer to this short-branched variety at all. Although White et al. intended publishing the short-branched variety (it was to have appeared on page 888, according to the 'Final Proof' at PRE), Dyer seems to have decided ultimately to delete it. Their concept of var. *minor* was based on plants grown in the garden at the Division of Botany from the collection *F.Z. van der Merwe 1662* (cult. as 3728/3/38, floweried 18 Sept. 1940, from Elandsfontein, 20 miles north of Pilgrim's Rest). In his notes, Dyer also mentioned another collection (De Bad, near Waterval, between Lydenburg and Burgersfort, July 1936, *F.Z. v. d. Merwe 12*) but no specimen was preserved of this. It is possible that he was unsure what to do with some other, yet more slender-branched forms that van der Merwe had also collected at Branddraai: (namely *v.d. Merwe sub 3854/6/38*

Fig. 5.183. *Euphorbia lydenburgensis* var. *minor*, *PVB 12099*, Leboeng, east of Abel Erasmus Pass, north of Ohrigstad, South Africa, 26 Aug. 2015, the yellow colour of the plant is a sign of stress during the dry season (© PVB).

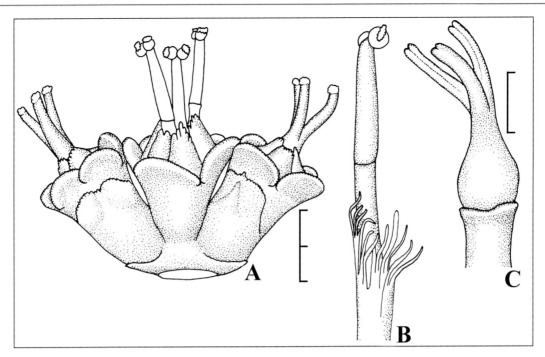

Fig. 5.184. *Euphorbia lydenburgensis* var. *minor*. **A**, cyme from side at beginning of female stage (scale 2 mm). **B**, anther and bracteole (scale 1 mm, as for **C**). **C**, female floret. Drawn from: *PVB 12099*, Leboeng, east of Abel Erasmus Pass, north of Ohrigstad, South Africa (© PVB).

and *v.d. Merwe 1694* of July 1938) which are now included in *E. pisima*.

During a herpetological survey of the northern Drakensberg in October 1980, it was re-collected by Niels H.G. Jacobsen, not far from where van der Merwe had discovered it.

Euphorbia otavibergensis Bruyns, *Haseltonia* 25: 50 (2018). Type: Namibia, Tsumeb distr., Otjikoto Lake, *De Winter 3678a* (PRE, holo.; K, WIND, iso.)
Euphorbia otavimontana Swanepoel, *Namibian J. Environm.* 3: 11 (2019). Type: Namibia, Otavi Mountains, farm Auros 595, 1880 m, 1 Jul. 2008, *Swanepoel 349* (WIND, holo.; PRE, iso.).

Bisexual spiny large glabrous succulent shrub 0.5–2 × 0.5–2 m with many branches at and just above ground level from central stem 0.2–0.5 m tall tapering rapidly into a short taproot bearing fibrous roots. *Branches* initially spreading then erect, mostly simple, 0.2–2 m × 20–70 mm, (4-) 5- to 6-angled, strongly constricted into short segments broadest near their bases and narrower towards their apex, smooth, greyish green; *tubercles* fused into (4) 5–6 wing-like angles with deep grooves between angles, laterally flattened and hardly projecting from angles, with spine-shields united into continuous horny grey to white margin along angles 2–4 mm broad above spines and 2–4 mm broad below them, bearing 2 spreading and widely diverging grey to whitish spines 3–10 mm long; *leaf-rudiments* on tips of new tubercles towards apices of branches, 1–3 × 2–3 mm, erect, fleeting, ovate, acute to obtuse, sessile, with small irregular obtuse brown stipules ± 0.5 mm long. *Synflorescences* many per branch towards apex, each a solitary cyme (rarely 2–3) in axil of tubercle, on peduncle 1–3 mm long, each cyme with 3 usually vertically disposed cyathia, central usually male to bisexual, lateral 2 bisexual and developing later each on short peduncle 1–2 × 1–2 mm, with 2 ovate yellow bracts 1–3 × 2–3 mm subtending lateral cyathia; *cyathia* shallowly conical-cupular, glabrous, 5–8 mm broad (1 mm long below insertion of glands), with 5 obovate lobes with deeply incised margins, bright greenish yellow; *glands* 5, transversely elliptic and contiguous, 3 mm broad, greenish yellow, spreading, inner margins not raised, outer margins entire and slightly ascending; *stamens* glabrous, bracteoles enveloping groups of males, with finely divided tips or filiform, glabrous; *ovary* obtusely 3-angled, glabrous, green, ± sessile; styles 2–3 mm long, branched in upper third above stout column. *Capsule* 5–8 mm diam., obtusely 3-angled, glabrous and green to shiny red, sessile.

Distribution & Habitat
Euphorbia otavibergensis is found in the mountains between Otavi, Tsumeb and Grootfontein and is probably best-known to the north-west of them around Lake Otjikoto. It is also recorded westwards of this in the Etosha Park and as far as 100 km west of Otavi.

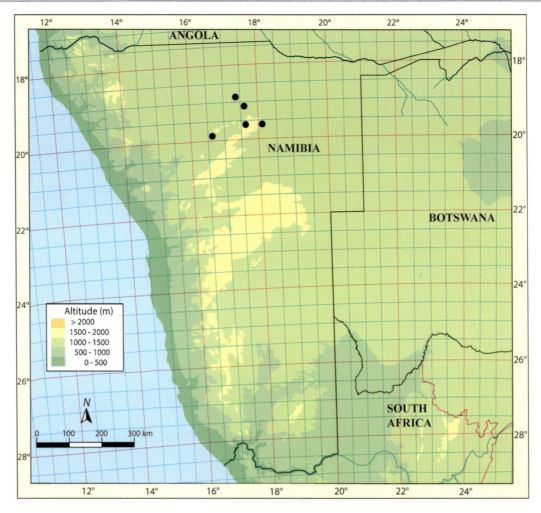

Fig. 5.185. Distribution of *Euphorbia otavibergensis* (© PVB).

Fig. 5.186. *Euphorbia otavibergensis*, at its most impressive, ± 2 m tall with *Aloe litoralis* on outcrops of limestone, *PVB 12873*, Lake Otjikoto, Namibia, 5 Jan. 2015 (© PVB).

Generally, *E. otavibergensis* inhabits shallow soil among calcareous rocks on the lower slopes or at the bases of hills or in very shallow soils overlaying calcareous outcrops in flat spots. It does not grow together with *E. avasmontana*. The two have been seen on the same slopes, with *E. avasmontana* among trees between large rocks around the tops of the hills and *E. otavibergensis* at the foot of the slope, also among rocks but in open patches with fewer trees.

Fig. 5.187. *Euphorbia otavibergensis*, ± 1.5 m tall, *PVB 13582*, 90 km west of Otavi, Namibia, 22 Dec. 2018 (© PVB).

Diagnostic Features & Relationships

Euphorbia otavibergensis is another species where the plants may form large shrubs. In some spots the plants remain small, as near Grootfontein (*PVB 4114*), where they do not exceed 0.5 m tall, but they may develop into impressive bushes nearly 2 m tall and the same in diameter, as is well-known around Lake Otjikoto. They always have a greyish colour, even on young growth. The branches persist on the stem and are very distinctly constricted into segments which (as in *E. cooperi*) are usually much broader towards their bases, though occasionally they are circular in outline.

Fig. 5.188. *Euphorbia otavibergensis*, ± 0.5 m tall, *PVB 4114*, Mariabron, Grootfontein, Namibia, 2 Jan. 2006 (© PVB).

Flowering usually takes place in September and October. Many plants only produce a single cyme in the axil of each tubercle and this is visible already some months before it matures as a hemispherical bud. Some plants of *E. otavibergensis* may produce more cymes per axil but the outer ones are usually slightly behind the middle one in development.

Among Namibian species, florally *E. otavibergensis* is distinctive for its quite broad cyathia which are shallow (not distinctly cylindrical for about 2 mm below the insertion of the glands as in *E. avasmontana*) and the sessile female floret. The pedicel of the female floret does not elongate even after pollination so that the capsule remains sessile ontop of the remains of the cyathium. The fused part of the styles is somewhat stout and the styles are divided only in their upper third (the whole structure is altogether more slender in *E. avasmontana*, where the ovary is also noticeably exserted from the cyathium already at an early stage). The capsule is rather more obtusely 3-angled than in *E. avasmontana*. All these differences suggest that it is not closely allied to *E. avasmontana* and unpublished results from molecular data place it together with *E. virosa*, *E. otjingandu* and other large shrubs with more or less sessile capsules such as *E. ingenticapsa*, *E. strangulata* and *E. dispersa* from Angola. Leach (1971) compared *E. ingenticapsa* with what he called *E. venenata* from Tsumeb, with little doubt the same as

Fig. 5.189. *Euphorbia otavibergensis*, first male stage, *PVB 4114*, Mariabron, Grootfontein, Namibia, 20 Sep. 2006 (© PVB).

Fig. 5.190. *Euphorbia otavibergensis*, capsules, *PVB 4114*, Mariabron, Grootfontein, Namibia, 5 Nov. 2006 (© PVB).

E. otavibergensis. Vegetatively *E. otavibergensis* resembles *E. faucicola*, which also occurs closest to it in Angola (in the gorge of the Cuchi River, a tributary of the Okavango River), but this species is less closely related and has more exserted capsules (Leach 1977).

History

Plants belonging to *Euphorbia otavibergensis* were featured by White et al. (1941) under the name *E. venenata*. This is not correct, since *E. venenata* conforms to typical *E. avasmontana* (in the longer, more slender cyathia and exserted capsules, among other details), but this usage appeared also in Leach (1971). Dinter (1928: 124) observed what he thought were two distinct species on the farm Auros in the Otavi Mountains, where he spent some time between February and March 1925 at the invitation of the owner, Mrs Margaret Volkmann. One (with distinctly constricted, blue-green branches) grew at the foot of the hills and reminded him most of '*E. dinteri*' (= *E. virosa*) while another (with slender dark green much taller branches) grew along the summits of the hills. The latter he named *E. volkmanniae*. The former clearly corresponds to what is now *E. otavibergensis* (and was shown in Fig. 910 and 925 of White et al. 1941) and the latter to *E. avasmontana* (as in Fig. 923, 924 and 926 of White et al. 1941). The two species still occur close together in this area on several farms, such as Auros and Gauss. Elsewhere only one of them has been observed: *E. otavibergensis* grows on its own to the north-west into Etosha and eastwards towards Grootfontein and *E. avasmontana* is alone to the west around Outjo and further south.

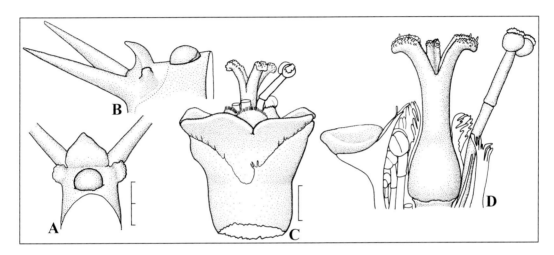

Fig. 5.191. *Euphorbia otavibergensis*. **A**, face view of young spines and leaf-rudiment (scale 2 mm, as for **B**, **C**). **B**, side view of young spines and leaf-rudiment. **C**, side view of cyathium at second male stage. **D**, side view of dissected cyathium with anthers, bracteoles and female floret (scale 1 mm). Drawn from: *PVB 4114*, Mariabron, Grootfontein, Namibia (© PVB).

Euphorbia otavibergensis was published on 28 December 2018 and *E. otavimontana* on 10 January 2019 so that within 1 month this species had been given two names.

Euphorbia otjingandu Swanepoel, *S. African J. Bot.* 75: 497 (2009). Type: Namibia, Kunene Region, along Van Zyl's Pass 1 km west of Otjihende, 1305 m, 1 May 2007, *Swanepoel 268* (WIND, holo.; PRU, iso.).

Bisexual spiny large glabrous succulent shrub 0.5–3 (4) × 1–4 m, with many branches developing from 200 to 900 mm above ground level from short often strongly spirally 7- to 9-angled central stem 0.2–1.0 m tall tapering rapidly into a short taproot bearing woody and fibrous roots. *Branches* ascending then erect, mostly simple but occasionally branching again near tips, 0.2–3 m × 45–190 mm, 4- to 6- (8-) angled, frequently but irregularly constricted into ovate segments, smooth, uniformly greyish green; *tubercles* fused into 4–6 (8) slightly sinuate wing-like angles with initially V-shaped and soon rounded grooves 10–30 mm deep between angles, laterally flattened and projecting 4–10 mm from angles, with spine-shields united into continuous horny sinuate grey-brown margin along angles 4–8 mm broad, bearing 2 spreading and widely diverging initially glossy dark red and later grey spines 5–12 (22) mm long; *leaf-rudiments* on tips of new tubercles towards apices of branches, 2.5–4.5 × 3.5–5.5 mm, erect, fleeting, ovate-deltate, sessile, with small obtuse reddish to brown irregular stipules ± 1 mm long. *Synflorescences* many per branch towards apex, 1–3 (5) in axil of tubercle, on stout green peduncle 2.5–7 × 5–8 mm, each cyme with 3 usually vertically disposed cyathia, central male, lateral 2 bisexual and developing later each on short peduncle 3–6 × 4–6 mm, with 2 broadly ovate yellow bracts 2–4 × 5–6 mm subtending lateral cyathia; *cyathia* broadly conical-cupular, glabrous, 6–10 mm broad (3–4 mm long below insertion of glands), with 5 obovate lobes with deeply incised margins, pale yellow-green; *glands* (4) 5 (6), transversely elliptic, 3–5 mm broad, green to yellow-green, spreading, contiguous, inner margins sometimes raised, outer margins entire, surface between two margins with some nectar; *stamens* glabrous, bracteoles enveloping groups of males, with finely divided tips, glabrous; *ovary* obscurely trigonous, glabrous, green, raised slightly on pedicel ± 0.5 mm long and not exserted from cyathium, with calyx slightly extended around ovary into short obtuse lobules; styles 3–3.5 mm long, divided to below middle. *Capsule* 11–15 mm diam., obtusely 3-angled, glabrous, shiny pale green to bright red, slightly exserted on stout pedicel ± 5 mm long.

Distribution & Habitat
Euphorbia otjingandu is known only from the Kaokoveld of north-western Namibia, close to the edge of the escarpment, from Kaoko Otavi to west of Opuwa and to the

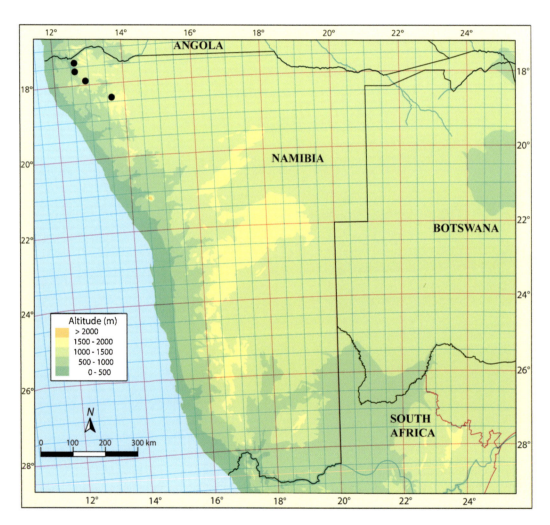

Fig. 5.192. Distribution of *Euphorbia otjingandu* (© PVB).

Otjihipa Mountains, just south of the Cunene River. Swanepoel (2009b) also recorded it from west of Ruacana. However, this specimen lacks both fruit and flowers and is more likely to belong to *E. virosa*, which occurs sporadically in this area.

Euphorbia otjingandu occurs on rocky slopes, often facing west and often on schists with much quartz-gravel, among a fairly sparse cover of trees and shrubs. This consists of species typical of the area such as *mopane*, *Catophractes*, various *Commiphora*, *Euphorbia guerichiana*, *Cissus curro-*

5.1 Sect. Euphorbia

Fig. 5.193. *Euphorbia otjingandu*, tree ± 3 m tall among *mopane* of similar height, angles on trunk spiralling to right, *PVB 12866*, west of Etanga, Namibia, 31 Dec. 2014 (© PVB).

Fig. 5.194. *Euphorbia otjingandu*, shrub in front ± 2 m tall, trunk without obvious angles, *PVB 12866*, west of Etanga, Namibia, 31 Dec. 2014 (© PVB).

rii, *Pachypodium lealii* and *Sesamothamnus* 'leistneri', with little undergrowth.

Diagnostic Features & Relationships

Plants consist of a short, 7- to 9-angled stem that is usually just under 1 m tall and is exposed and clearly visible below the first branches, often with the angles very conspicuously twisted up it. Above this stem there is a large, dense head of strongly ascending, thick branches, usually slightly taller than broad, sometimes spreading out and sometimes rising very steeply to form a more narrow but very substantial shrub or small tree. The branches are armed with short, stout spines and most have between four and six angles.

Flowering has been recorded from June to August but may begin as early as May. Clusters of greenish cyathia develop near the tips of the branches, each cyathium usually with five glands.

In appearance *Euphorbia otjingandu* is very similar to *E. virosa*, which occurs nearby (though usually at lower altitudes) and is easily be mistaken for it. Young plants in both species are very heavily spined and lack the attractive mottling of the surface found in such species as *E. eduardoi*.

Fig. 5.195. *Euphorbia otjingandu*, unusual specimen ± 5 m tall, shedding lower branches, *PVB 12866*, west of Etanga, Namibia, 31 Dec. 2014 (© PVB).

Fig. 5.196. *Euphorbia otjingandu*, cyathia before first male stage, *PVB 12866*, west of Etanga, Namibia, 6 Jun. 2019 (© PVB).

Vegetatively large plants of the two species are separated by the obvious, short trunk that is clearly visible in *E. otjingandu* but is generally hidden from view by the slightly descending lower branches in *E. virosa*. The spines in *E. otjingandu* are shorter than those of *E. virosa* and are not as protectively distributed along the branches, so that the surface between the angles is more readily accessible than it is in *E. virosa*. Florally the two are easily distinguished by the far fewer glands and the trigonous capsules that dehisce explosively in *E. otjingandu*. The male florets are also not as densely packed in the cyathium, while the female florets are differently shaped and are not exserted from the cyathium as they are in *E. virosa*.

Unpublished analyses of DNA-data show that *E. otjingandu* is related to robust shrubs in the region such as *E. otavibergensis* and *E. virosa* from Namibia and to others from Angola such as *E. dispersa*, *E. ingenticapsa* and *E. strangulata*. All of these have deeply segmented branches, broad and fairly shallow cyathia mostly in a single cyme per axil and more or less sessile, obtusely 3-angled capsules.

History

Euphorbia otjingandu occurs in very remote parts of Namibia and was observed during surveys conducted on trees of Namibia during the period 2000–2005. Before this, plants were photographed in February 1993 by the present author and Geoff Tribe at Kaoko Otavi (Fig. 5.199). It is possible that a photo of a small plant by Max Otzen, from Kaoko Otavi, taken in 1940 (of his 40/17, PRE), is also of this species.

5.1 Sect. Euphorbia

Fig. 5.197. *Euphorbia otjingandu*, cyathia in first male stage, *PVB 12866*, west of Etanga, Namibia, 14 Jun. 2019 (© PVB).

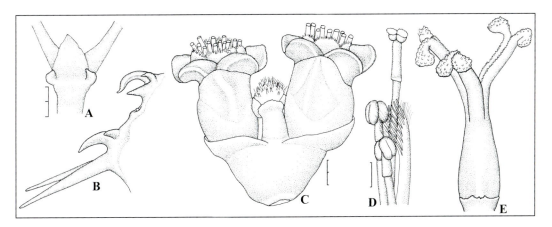

Fig. 5.198. *Euphorbia otjingandu*. **A**, face view of young spines and leaf-rudiment (scale 2 mm, as for **B**). **B**, side view of young spines and leaf-rudiments. **C**, side view of cyathium after second male stage (scale 2 mm). **D**, anthers and bracteole (scale 1 mm, as for **E**). **E**, female floret. Drawn from: *PVB 12866*, west of Etanga, Namibia (© PVB).

Fig. 5.199. *Euphorbia otjingandu*, Kaoko Otavi, Namibia, 19 Feb. 1993, with author (© Geoff Tribe).

Euphorbia otjipembana L.C.Leach, *Dinteria* 12: 29 (1976). Type: Namibia, Kaokoveld, north of Otjipemba, 21 July 1973, *Leach & Cannell 15044* (PRE, holo.; BM, K, LISC, M, MO, SRGH, WIND, iso.).

Bisexual spiny glabrous succulent shrub 0.2–0.8 × 0.2–1 m, with few to many branches from shorter central stem bearing fibrous roots, branching above ground. *Branches* ascending to erect, 0.15–0.5 m × 15–45 mm, 4- to 6-angled, not or only slightly constricted into segments 20–50 mm long, smooth, bright green to grey-green above becoming pale brown below; *tubercles* fused into 4–6 prominent sometimes slightly sinuate angles with flat to shallowly grooved area between angles, laterally flattened and projecting 2–4 mm from angles, with spine-shields 6–10 mm long, spreading down to just above next axillary bud but initially just remaining separate from next (later often fusing into continuous horny margin) 2–3 mm broad above spines and 1–2 mm broad below them, bearing 2 spreading and widely diverging grey spines 4–14 mm long; *leaf-rudiments* on tips of new tubercles towards apices of branches, 1–2 × 1–2 mm, erect to recurved, fleeting, ovate-acute, sessile, with brown stipular prickles 0.7–6 mm long.

Synflorescences many per branch towards apex, each a solitary cyme in axil of tubercle, on short peduncle 1–2 mm long, each cyme with 3 transversely disposed cyathia, central male sessile and often deciduous, lateral 2 bisexual and developing later each on short peduncle 1–3 × 2–3 mm, with 2 ovate green bracts 1–3 × 1–2 mm subtending lateral cyathia; *cyathia* shallowly conical-cupular, glabrous, 3–5 mm broad (1.5–2.5 mm long below insertion of glands), with 5 obovate lobes with deeply incised margins, pale orange to bright yellow; *glands* 5, transversely elliptic, contiguous to slightly separated, 1.5–2.5 mm broad, yellow to pinkish yellow, erect to spreading, often slightly concave above, outer margins entire and spreading; *stamens* glabrous, bracteoles enveloping groups of males, with finely divided tips, glabrous; *ovary* obtusely and deeply 3-angled, glabrous, green to suffused with red, raised on short thick pedicel slightly more than 0.5 mm long; styles 2–2.5 mm long, branched to below middle. *Capsule* 3–4 mm diam., deeply 3-angled, glabrous and shiny, green, sessile.

Two subspecies are recognized. Subsp. *otjipembana* is endemic to Namibia, while subsp. *fluvialis* is found in both Namibia and in the neighbouring part of southern Angola.

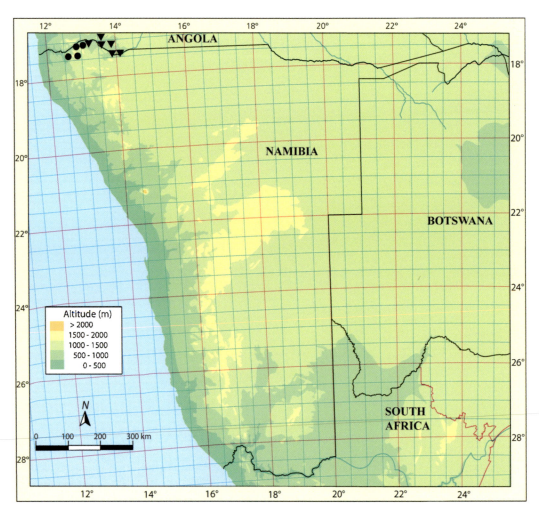

Fig. 5.200. Distribution of *Euphorbia otjipembana* (● = ssp. *otjipembana*; ▼ = ssp. *fluvialis*) (© PVB).

5.1 Sect. Euphorbia

The two subspecies may be distinguished as follows:

1. Branches (4) 5- to 6-angled, spines 4–9 mm long, cyathial glands yellow and slightly concave above, spreading to ± erect..subsp. **otjipembana**
1. Branches 4-angled, spines 9–14 mm long, cyathial glands suffused with pink (rarely yellow), concave above, ± erect (rarely somewhat spreading)..subsp. **fluvialis**

Euphorbia otjipembana subsp. otjipembana

Euphorbia otjipembana subsp. *okakoraensis* Swanepoel, *Phytotaxa* 117: 53 (2013). Type: Namibia, Kaokoveld, north-eastern slopes of Okakora near Okombambi, 930 m, 24 March 2007, *Swanepoel 271* (WIND, holo.; PRE, iso.).

Shrub 0.2–0.5 × 0.5–1 m. *Branches* ascending to erect, 0.15–0.5 m × 15–45 mm, (4) 5- to 6-angled, bright green to greyish green; spines 4–9 mm long; *leaf-rudiments* 1.5–2 × 1.5–2 mm, stipular prickles 0.7–3 mm long. *Cyathia* shallowly conical-cupular, glabrous, 4.5–5 mm broad, bright yellow (faintly red); *glands* transversely elliptic and contigu-

Fig. 5.201. *Euphorbia otjipembana* ssp. *otjipembana*, ± 30 cm diam., above edge of cliff, *PVB 8001*, eastern flank of Baynes Mountains, Namibia, 15 Dec. 1999 (© PVB).

ous to slightly separated, 2.5 mm broad, yellow, erect to spreading, slightly concave above, outer margins spreading; *ovary* green to suffused with red; styles branched to below middle. *Capsule* 3–4 mm diam., glabrous and shiny, green.

Distribution & Habitat

Subsp. *otjipembana* only occurs in the Kaokoveld, the north-western corner of Namibia. In this area it is known in the deep valley north-west of Etengua, between the Baynes Mountains and the Otjihipa. This lies on the western side of the Baynes Mountains, between the places known as Otjipemba and Otjomborombongo and is where the type plant originated. It has also been found on the eastern slopes of the Baynes Mountains, to the east of the type locality. Similar plants were collected in the Okakora and Otjihipa as well, between five and 20 km to the west of the type locality. Despite searching in suitable areas north of the Baynes Mountains in Angola, I have not managed to find it in Angola at all.

In the area north-west of Etengua, subsp. *otjipembana* grows in the foot of the valley, among rocks and around the bases of low *mopane* trees, with *Cyphostemma uter*. Occasional colonies of plants were also encountered in 1993 among rocks and scattered trees on the tops of the mountains to the west, known as Okakora. Plants further west in the Otjihipa were found in December 1999 among short grasses

Fig. 5.202. *Euphorbia otjipembana* ssp. *otjipembana*, plant ± 1 m diam., with the characteristic green branches, among *mopane* shrubs, *PVB 8019*, near Okombambi, western side of Baynes Mountains, Namibia, 19 Dec. 1999 (© PVB).

Fig. 5.203. *Euphorbia otjipembana* ssp. *otjipembana*, among vertically oriented strata in outcrops of schist, *PVB 8019*, near Okombambi, western side of Baynes Mountains, Namibia, 19 Dec. 1999 (© PVB).

and clumps of *Xerophyta* on rocky slopes. On the eastern side of the Baynes Mountains it grows among rocks and short grasses, in treeless patches of shallow ground above cliffs, with clumps of *Xerophyta* nearby.

Diagnostic Features & Relationships

Subsp. *otjipembana* usually has a sparsely to quite densely branched habit and comparatively thick (to 45 mm), spreading to erect branches. Larger specimens form handsome shrubs that are usually broader than tall and are dominated by the bright green colour of the younger parts of the branches. This bright green colour is, however, not present in plants from the Okakora and Otjihipa, where the young growth is slightly greyish green, lending the whole plant a more greyish green colour. Even those from the eastern side of the Baynes Mountains are slightly greyish and not quite as bright green as those from north-west of Etangua. Most of the branches are 5- or 6-angled, but

Fig. 5.204. *Euphorbia otjipembana* ssp. *otjipembana*, somewhat greyer branches, *PVB 8046*, Otjihipa, west of Baynes Mountains, Namibia, 23 Dec. 1999 (© PVB).

occasional 4-angled branches are produced and plants from the eastern side of the Baynes Mountains are quite regularly 4-angled. Between the angles, the branches have broad and shallow grooves and they are only slightly constricted where growth stops each year. The spine-shields extend in a narrow parallel-sided band below the spines to just above the next axillary bud and later often fuse to form a more or less continuous margin along the angles. In comparison to the thickness of the branches, the spines are quite short and they are fairly widely spaced along the angles.

Subsp. *otjipembana* flowers between June and August, a little before the others in the same area come into bloom. The cyathia are produced on the uppermost segment of each branch and their bright yellow colour makes a striking contrast against the green of the stems. All parts of the cyathium are yellow except for the anthers which are reddish and the top of the ovary, which is also suffused with red. The ovary is raised above the base of the cyathium on a stout pedicel about as thick as the ovary itself.

In plants from around the type locality, the bright green colour of their branches usually makes them easily distinguishable from *E. kaokoensis* or subsp. *fluvialis*, which both occur within 20 km of it. However, away from this locality, the branches are not always such a bright green and they may even be grey-green. Then it is their relative stoutness (to 45 mm thick) that is distinctive. Of the four spines on each spine-shield, the two below the leaf-rudiments are shorter in subsp. *otjipembana* than in either *E. kaokoensis* or subsp. *fluvialis*. The stipular prickles of subsp. *otjipembana* are also shorter than in these others.

Fig. 5.205. *Euphorbia otjipembana* ssp. *otjipembana*, female stage, *PVB 8019*, near Okombambi, western side of Baynes Mountains, Namibia, 23 Jun. 2019 (© PVB).

Fig. 5.207. *Euphorbia otjipembana* ssp. *otjipembana*, second male stage, plant with much greyer branches, *PVB 8041*, west of Otjihipa, west of Baynes Mountains, Namibia, 16 Jul. 2019.

Fig. 5.206. *Euphorbia otjipembana* ssp. *otjipembana*, second male stage, plant with very green branches, *PVB 8019*, near Okombambi, western side of Baynes Mountains, Namibia, 24 Jun. 2003 (© PVB).

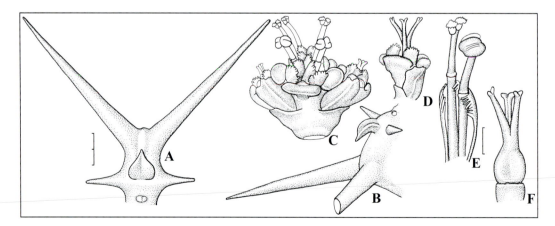

Fig. 5.208. *Euphorbia otjipembana* ssp. *otjipembana*. **A**, young spines and leaf-rudiment from above (scale 2 mm, as for **B–D**). **B**, side view of young spines and leaf-rudiment. **C**, side view of cyathium in first male stage. **D**, side view of cyathium in female stage. **E**, anthers and bracteole (scale 1 mm, as for **F**). **F**, female floret. Drawn from: *PVB 8019*, near Okombambi, western side of Baynes Mountains, Namibia (© PVB).

The cyathia of subsp. *otjipembana* differ from those of *E. kaokoensis* and subsp. *fluvialis* in usually being bright yellow, rather than somewhat pinkish. Their glands are mostly more spreading, rather than nearly erect as in the others. Nevertheless, in the lateral cyathia in some cymes of subsp. *otjipembana*, the glands may be more erect than in the central cyathium and, in plants from the western side of its distribution (e.g. Okakora, *Swanepoel 271* and Otjihipa, *PVB 8041*), the cyathia are often suffused with pink and their glands are more erect. Bright yellow cyathia have occasionally been found in subsp. *fluvialis* too (e.g. Angola, mountains above Iona, *PVB 10387*), but they are unknown in *E. kaokoensis*.

History

Subsp. *otjipembana* was discovered by P.G. Meyer of Munich, Germany, during a plant-collecting trip to the Kaokoveld in June and July of 1969. Meyer saw only a few, scattered plants but was struck by the mostly 5-angled branches and their 'lively green' colour. Leach and Cannell recollected it in July 1973 in the same area where Meyer had discovered it and from these plants Leach described the species in 1976. For some reason Leach (1976a: 30) gave the locality as 'eastern slopes of the Baynes Mountains, north of Otjipemba', although Otjipemba lies at the western foot of the Baynes Mountains. A further subspecies was named from more greyish plants from around 10 km west of the type locality, where it was collected in March 2007 (Swanepoel 2013), but these are treated as part of subsp. *otjipembana* here.

Euphorbia otjipembana subsp. **fluvialis** (L.C.Leach) Bruyns, *Phytotaxa* 436: 210 (2020).
Euphorbia subsalsa subsp. *fluvialis* L.C.Leach, *Dinteria* 12: 29 (1976). Type: Angola, Ruacana Falls, *Leach & Cannell 14509* (LISC, holo.; BM, K, LUAI, M, MO, PRE, SRGH, iso.).

Shrub 0.2–0.7 × 0.2–1 m, much branched just above ground level. *Branches* erect, 50–500 × 15–30 mm, 4-angled, grey-green; spines 9–14 mm long; *leaf-rudiments* ± 1 × 1 mm, erect, stipular prickles 1.5–6 mm long. *Cyathia* shallowly conical-cupular, glabrous, 3–4.5 mm broad, pale orange; *glands* transversely rectangular and often contiguous, 1.5–2 mm broad, pale pinkish yellow, erect and slightly concave inside, outer margins entire and erect; *ovary* pale green; styles branched to near base. *Capsule* 3–4 mm diam.

Fig. 5.209. *Euphorbia otjipembana* ssp. *fluvialis*, shrubs to 0.8 m tall, east of Epupa Falls towards Ruacana, Namibia, 2 Jan. 2015 (© PVB).

Distribution & Habitat

Subsp. *fluvialis* is common along the Cunene River in Namibia from around 20 km west of Epupa Falls eastwards to near Ruacana Falls. It is also common in Angola on the northern bank of the river in the mountains and hills between Iona and Ruacana Falls.

Generally, subsp. *fluvialis* is found in crevices in large outcrops of rock or between large boulders, on calcrete or on darker igneous outcrops. It sometimes occurs on rocky slopes among sparse tufts of grass with *Xerophyta* and a scanty cover of scattered trees, especially *mopane* and various species of *Acacia*.

Fig. 5.210. *Euphorbia otjipembana* ssp. *fluvialis*, shrublet around 0.25 m tall, flowering sporadically, among quartz, *PVB 10387*, Iona, Angola, 4 Dec. 2011 (© PVB).

Fig. 5.211. *Euphorbia otjipembana* ssp. *fluvialis*, shrubs to 0.6 m tall, in shallow soil on low granitic outcrop, *PVB 10786*, south of Otjinjau, Angola, 20 Jan. 2007 (© PVB).

Fig. 5.212. *Euphorbia otjipembana* ssp. *fluvialis*, cyathia strongly suffused with pink, road to Oncocua, Angola, 11 Oct. 2014 (© PVB).

Fig. 5.213. *Euphorbia otjipembana* ssp. *fluvialis*, cyathia almost orange, *PVB 10695*, Serra Techemalinda, Iona, Angola, 16 Nov. 2018 (© PVB).

Diagnostic Features & Relationships

Subsp. *fluvialis* may form very robust clumps up to 1 m in diameter and usually rather less in height. In some areas the branches are characteristically held closely together and parallel to each other in quite neat clumps, but rather more untidy specimens also occur. The branches are stout (usually around 20 mm thick) and 4-angled, regularly parallel-sided, with the surfaces between the angles relatively flat and uniformly pale grey-green. The plants are well-armed with long and sharp spines and stipular prickles, though the prickles are much shorter than the main spines.

Flowering takes place mainly in September and October, when the upper parts of the branches become covered with cyathia. These have a distinctly pink-orange colour, with pinkish glands and reddish stamens, though in the mountains above Iona (in southern Angola) plants were found with yellow glands. At maturity the glands remain somewhat erect and are concave inside, emphasizing the cup-like shape of the cyathia.

With *E. kaokoensis*, subsp. *fluvialis* shares the (mostly) distinctly pinkish cyathia with erect, concave glands. The two differ in that *E. kaokoensis* has 5- to 8-angled branches

Fig. 5.214. *Euphorbia otjipembana* ssp. *fluvialis*, cyathia yellow and with more spreading glands, *PVB 10387*, Iona, Angola, 4 Dec. 2011 (© PVB).

on which the stipular prickles are usually longer (and sometimes nearly as long as the spines). Subsp. *otjipembana* mostly has 5-angled green to greyish green branches and bright yellow cyathia with much more spreading glands. Analysis of DNA-data (Bruyns et al. 2011), showed that subsp. *fluvialis* is close to subsp. *otjipembana* and *E. kaokoensis* (but with only weak support), suggesting that its relationship with *E. subsalsa* is tenuous. Morphologically this is backed up by overlaps in most of the features separating them and also by the longer pedicels beneath the female florets in both subspecies of *E. otjipembana* (see Fig. 5.208F and Fig. 5.216F).

History

Subsp. *fluvialis* was first recorded by Keppel H. Barnard in March 1929 (specimen in SAM). The locality that he gave, namely 'Great Falls to Little Falls' is not very clear, but presumably he meant near Epupa Falls, where it is common even today. Subsp. *fluvialis* was also collected by Max Otzen (*Otzen 4*, PRE) in November 1939. Neither of these early collections was mentioned in White et al. (1941). It was also observed by H. Hall at Epupa Falls in August 1951 and again in January 1956 by E.J. Mendes in Angola on the northern bank of the Cunene River. Leach described it in 1976 as a subspecies of *E. subsalsa*, but it is now placed under *E. otjipembana*.

Fig. 5.215. *Euphorbia otjipembana* ssp. *fluvialis*, capsules, *PVB 10695*, Serra Techemalinda, Iona, Angola, 9 Jan. 2007 (© PVB).

5.1 Sect. Euphorbia

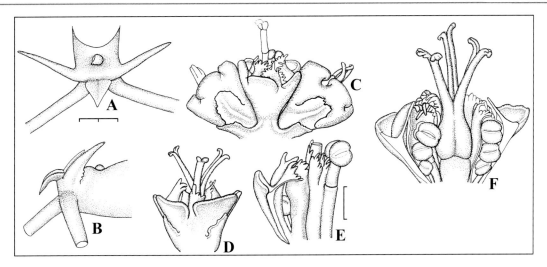

Fig. 5.216. *Euphorbia otjipembana* ssp. *fluvialis*. **A**, young spines and leaf-rudiment from above (scale 1 mm, as for **B–D**). **B**, young spines and leaf-rudiment from side. **C**, side view of cyme ending first male stage. **D**, cyathium in female stage. **E**, anthers and bracteoles. **F**, female floret in dissected cyathium. Drawn from: *PVB*, Ruacana Falls, Namibia (© PVB).

Euphorbia pisima Bruyns, *Phytotaxa* 436: 210 (2020). Type: South Africa, 34 km north-east of Ohrigstad, 1000–1100 m, 11 Jan. 1996, *Bruyns 6607* (BOL, holo.; MO, iso.).

Bisexual spiny glabrous succulent 0.05–0.3 × 0.15–0.4 m with many branches from similar stem, stem with fibrous roots arising from it, densely branched at ground level and not at all rhizomatous. *Branches* erect, 20–170 × 7–10 mm, 4-angled, not constricted into segments, smooth, bright pea-green to yellow; *tubercles* in decussate pairs fused vertically into 4 low angles along branches with surface flat or slightly concave between angles, conical and truncate, projecting 2–4 mm from angles, with spine-shields 4–7 mm long, 1–2 × 2–3 mm above spines and 2–6 (8) mm long and more slender below spines but remaining well separated from next, bearing 2 spreading to slightly deflexed initially reddish brown (later dark brown) spines 6–10 mm long; *leaf-rudiments* on tips of new tubercles towards apices of branches, 0.8–1 × ± 0.5 mm, erect, fleeting, deltoid, sessile, with small stipular prickles 1–2 mm long (rarely another small prickle ± 1 mm long near base of spine-shield). *Synflorescences* many per branch usually towards apex, each a solitary cyme in axil of tubercle, on short peduncle ± 1 mm long, each cyme with 3 transversely disposed cyathia, central male, lateral 2 bisexual and developing slightly later each on short peduncle 1–2 mm long and thick, with 2 ovate bracts ± 1 × 2 mm subtending lateral cyathia; *cyathia* shallowly cupular, glabrous, 4–5 mm broad (1.5 mm long below insertion of glands), with 5 pale yellow obovate lobes with deeply incised margins, bright yellow; *glands* 5, transversely rectangular to nearly square and contiguous, ± 2 mm broad, bright yellow, ascending-spreading, inner margins flat, outer margins entire and slightly ascending, surface between two margins dull; *stamens* glabrous, bracteoles enveloping groups of males, with finely divided tips, glabrous; *ovary* obtusely 3-angled, glabrous, slightly reddish green near top, raised on pedicel 1–1.5 mm long; styles ± 2 mm long, branched to just above base. *Capsule* 3–4 mm diam., obtusely 3-angled, glabrous, slightly raised ± 2 mm inside remains of cyathium.

Distribution & Habitat

At present, *Euphorbia pisima* is known from north of the small town of Ohrigstad, between Abel Erasmus Pass and the village of Moremala. To the north-west, it is found in Bewaarskloof along the Mphogodima River.

In the vicinity of Abel Erasnmus Pass it occurs at around 1300–1700 m on stony dolomitic slopes sometimes among scattered trees, wedged into crevices between rocks, with many other small succulents, small tufts of grass and *Xerophyta*. In Bewaarskloof it grows in crevices on steep outcrops of rock.

Diagnostic Features & Relationships

Euphorbia pisima forms dense ± hemispherical clumps of branches which are rarely more than 10 cm tall, with the tubercles joined into four low angles along each branch. During the growing season, the branches are a bright pea-green and the name is derived from this feature. During the dry season, when the plants are under stress, the branches may become bright yellow. The colour of the branches and their continuous angles separate the plants from the form of *E. clivicola* occurring in the same area, where the branches are greyer and the tubercles are not fused into angles. In the colour if its branches, *E. pisima* is more similar to *E. lydenburgensis* which, however, always has stouter branches (12–20 mm thick, usually round 15 mm) in which the spine-shields are often fused into a continuous margin along the branches (they always remain separate in *E. pisima*). The leaf-rudiments are particularly tiny in *E. pisima* (Fig. 5.224A, B), but they are slightly longer and narrower than those in *E. lydenburgensis*.

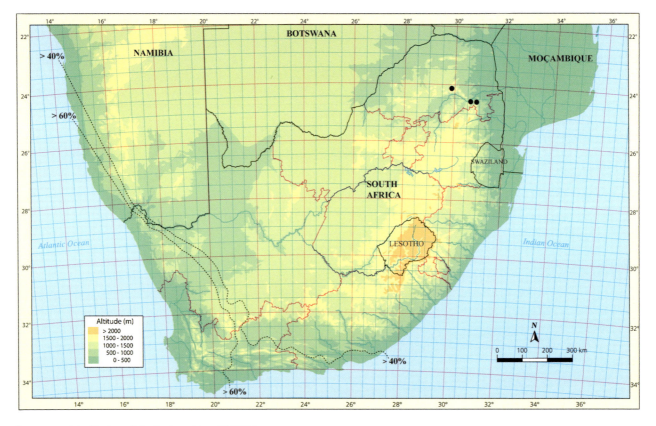

Fig. 5.217. Distribution of *Euphorbia pisima* (© PVB).

Fig. 5.218. *Euphorbia pisima*, plant ± 30 cm broad, among outcrops of dolomite on steep slope, *PVB 6607*, east of Branddraai, NE of Ohrigstad, South Africa, 25 Nov. 2018 (© PVB).

Fig. 5.219. *Euphorbia pisima*, clump ± 30 cm long, with *Aloe cryptopoda* and tufts of grass on gentle slope, *PVB 13559*, Abel Erasmus Pass, north of Ohrigstad, South Africa, 31 Jul. 2019 (© PVB).

Fig. 5.220. *Euphorbia pisima*, clump ± 40 cm broad, among outcrops of dolomite and tufts of grass on gentle slope, *PVB 13559*, Abel Erasmus Pass, north of Ohrigstad, South Africa, 31 Jul. 2019 (© PVB).

Fig. 5.221. *Euphorbia pisima*, second male stage, *PVB 6607*, east of Branddraai, NE of Ohrigstad, South Africa, 6 Sep. 2010 (© PVB).

Fig. 5.222. *Euphorbia pisima*, female stage, *PVB 6607*, east of Branddraai, NE of Ohrigstad, South Africa, 6 Sep. 2018 (© PVB).

Flowering in *E. pisima* usually takes place between July and September, though often as late as October in cultivation. At this time conditions are especially dry and the whole plant is often a bright yellow colour and easily visible from a distance. The bright yellow cyathia superficially resemble those of *E. schinzii* but they differ in the longer peduncles of the lateral cyathia in each cyme, in the short, fused portion of the styles and in the longer pedicel of the female florets. Again, these are all features that are shared with *E. lydenburgensis*.

Although it superficially resembles *E. schinzii* and it sometimes grows together with *E. clivicola* (as at Abel Erasmus Pass), unpublished molecular data shows that *E. pisima* is more closely related to *E. lydenburgensis*. It may occur near to *E. lydenburgensis* var. *minor* and shares with it the low stature and the unusually bright colour of the branches.

History

This brightly green-branched plant is similar in outward appearance to *E. schinzii* and, perhaps as a consequence, it has not been recorded often. The type plants were collected in 1996 near the edge of the Blyde River Canyon and it was also gathered in October 2001 by P.J.D. Winter in Bewaarskloof. Much earlier records were made by F.Z. van der Merwe in June and July 1938 from Branddraai (two specimens at PRE) and Dyer mentioned among his notes that van der Merwe first found this plant in July 1936 (at De Bad, near Waterval, between Lydenburg and Burgersfort). Dyer was uncertain what to do with these collections and considered one of them to belong to *E. lydenburgensis* var. *minor* and the other to be a dwarf variety of *E. lydenburgensis*.

Fig. 5.223. *Euphorbia pisima*, with capsules, *PVB 6607*, east of Branddraai, NE of Ohrigstad, South Africa, 25 Nov. 2018 (© PVB).

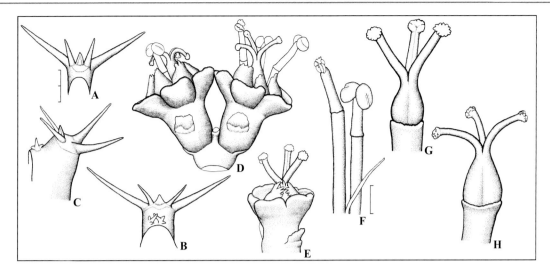

Fig. 5.224. *Euphorbia pisima*. **A, B**, young spines and leaf-rudiment from above (scale 2 mm, as for **B–E**). **C**, side view of young spines and leaf-rudiment. **D**, side view of cyme in second male stage. **E**, side view of cyathium in female stage. **F**, anthers and bracteole (scale 1 mm, as for **G, H**). **G, H**, female floret. Drawn from: *PVB 6607*, east of Branddraai, NE of Ohrigstad, South Africa (© PVB).

Euphorbia pseudocactus A.Berger, *Sukkul. Euphorb.*: 78 (1906). Lectotype (Bruyns 2012): South Africa (originally given as 'country unknown, but probably India', branch from the type plant, received from A. Berger, Oct. 1910 (K).

Bisexual spiny glabrous succulent spreading shrub 0.5–1 (2.5) × 1–2 m, with many branches from much shorter central stem bearing fibrous roots, branching at and above ground level. *Branches* spreading to ascending, 0.2–1 m × 15–50 mm, (3) 4- to 5-angled, irregularly constricted into segments 20–150 mm long and thickest below middle, smooth, dark to light green and often with V-shaped darker green markings radiating from centre against yellow-green background; *tubercles* fused into (3) 4–5 prominent slightly sinuate angles with flat to concave or shallowly grooved area between angles, laterally flattened and projecting 2–8 mm from angles, with spine-shields mostly united into ± continuous horny grey margin along angles 2–3 mm broad above spines and 1–2 mm broad below them, bearing 2 spreading and widely diverging brown to grey spines 4–12 mm long with tiny prickle about 1 mm long above each spine alongside axillary buds and with slight brown ridge at base projecting over these buds; *leaf-rudiments* on tips of new tubercles towards apices of branches, 1.5–2 × 1.5–2 mm, recurved, fleeting, ovate to nearly circular, sessile, with irregular brown stipular prickles 0.5–1.5 mm long. *Synflorescences* many per branch towards apex on last 1–2 segments, with 1 (–3) cymes in axil of each tubercle, very shortly peduncled, each cyme with 3 vertically disposed cyathia, central male on short thick peduncle 1.5–2 × 3–4 mm and often deciduous, lateral 2 bisexual and developing later each on short peduncle 1–2 × 2–3 mm, with 2 broadly ovate brownish edged bracts 1–2 × 2–3 mm subtending lateral cyathia; *cyathia* shallowly conical-cupular, glabrous, 6–8 mm broad (2–3 mm long below insertion of glands), with 5 obovate lobes with deeply incised margins, yellow to yellow-green; *glands* 5, transversely elliptic and contiguous, 3–4 mm broad, yellow to yellow-green, spreading, inner margins raised, outer margins entire and raised, surface between two margins concave; *stamens* glabrous, bracteoles enveloping groups of males, with finely divided tips, glabrous; *ovary* obtusely 3-angled, glabrous, green to suffused with red, very slightly raised on short thick pedicel less than 0.25 mm long; styles 3.5–4 mm long, branched in upper third. *Capsule* 10–15 mm diam., deeply and obtusely 3-angled, yellowish green suffused with red to red above and pale green below, glabrous and shiny, sessile.

Distribution & Habitat

Euphorbia pseudocactus is restricted to the valley of the Tugela River from near Tugela Ferry to Middledrift. It has also been recorded in the valleys of some of the tributaries of the Tugela, such as the Mooi River around Muden, the Bloukrans River around Weenen and the Nadi River between the towns of Nqubevu and Nadi.

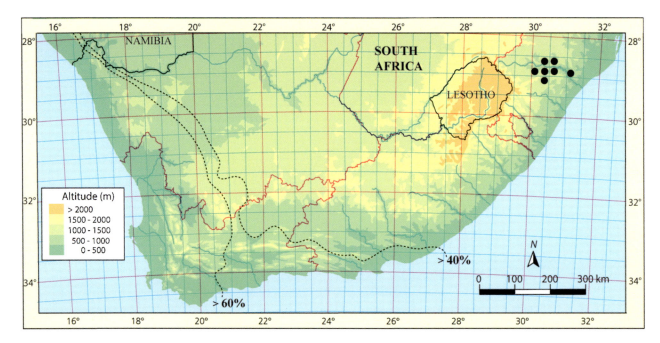

Fig. 5.225. Distribution of *Euphorbia pseudocactus* (© PVB).

Fig. 5.226. *Euphorbia pseudocactus*, many plants in dry scrub with *Acacia tortilis*, *Aloe marlothii* and carpet of *Barleria*, *PVB 9406*, near Muden, South Africa, 13 Jan. 2009 (© PVB).

Generally, *E. pseudocactus* is encountered on the relatively dry, lower slopes of stony hills and in flat areas at the bases of these hills. Here it forms large colonies among scattered trees, on heavy loams derived from shales and dolerites, with a few other succulents and many small, annual clumps of *Blepharis* and other Acanthaceae. Overgrazing appears to enhance the vigour of its populations greatly, presumably by reducing competition from grasses and other small bushes.

Fig. 5.227. *Euphorbia pseudocactus*, shrub ± 0.6 m tall, *PVB 9406*, near Muden, South Africa, 13 Jan. 2009 (© PVB).

Diagnostic Features & Relationships

Typically, *Euphorbia pseudocactus* forms substantial but not particularly dense shrubs that may exceed 1 m in diameter but rarely grow to more than 1 m in height, being more usually around 0.5 m tall. The branches are centred on and joined to an inconspicuous, central stem, from which they generally ascend after spreading for a short distance near their bases. At around 20–40 mm thick, the branches are fairly slender (relative to such species as *E. grandicornis*) and are made up of distinct segments that are broadest near their bases, from where they taperi towards the tip and towards the base of the next segment. Usually the branches have a pale yellow-green background colour on which dark green, V-shaped bands radiate upwards to the axils of the tubercles from a dark green patch running up the groove between the angles. Occasional specimens are encountered in which the branches are uniformly green. The branches are armed with grey spines, generally longest towards the bases of the segments, located on stout grey spine-shields that are mostly fused into a grey margin along the angles.

Flowering in *E. pseudocactus* usually takes place from August to October. Each cyme consists of three vertically disposed cyathia and they are produced in large numbers near the tips of the branches, mostly on the last segment and usually three cymes per axil. They emit a slight, sweet, beeswax-like odour. The cyathia are fairly large, with broad, closely contiguous glands forming an almost continuous ridge around their edge. Both the outer and the inner edges of the glands are raised and the concave patch between them is usually covered with copious nectar. The female floret is raised on a very short pedicel and consists of a slender ovary which gradually tapers via a fairly long neck into the styles. After fertilization, the ovary swells up within the cyathium and the pedicel does not elongate significantly to push it out. The quite large capsules are deeply 3-angled and quite shiny until they ripen.

Fig. 5.228. *Euphorbia pseudocactus*, *PVB 9406*, near Muden, South Africa, 6 Dec. 2002 (© PVB).

Euphorbia pseudocactus is probably most closely allied to *E. grandicornis*, which appears to replace it further north-eastwards and at lower altitudes along the valley of the Tugela River. The two differ in the much broader, 3-angled stems with much larger spines in *E. grandicornis*, though in some collections from the valley of the Nadi River (a tributary of the Tugela) it is not very clear to which of the two species the plants belong. Their respective cyathia are similar in size, though the central, male cyathium of *E. grandicornis* is usually larger. The female floret in *E. pseudocactus* is slightly shorter than that of *E. grandicornis*, but with a slightly broader ovary and a very slightly longer pedicel.

History

Alwin Berger described *Euphorbia pseudocactus* from material in cultivation of unknown origin, though he assumed these plants to be from India. It was long suspected to be a hybrid (Frick 1938). A species fitting Berger's description seems to have been noted first in the valley of the Mooi River near Muden by I.B. Pole Evans in 1916 and again by McClean in 1927 (White et al. 1941). These plants were only positively identified as *E. pseudocactus* in 1938 by R.A. Dyer and G.A. Frick, after material and photographs were again supplied by W.E. Cronwright, who lived nearby in Pietermaritzburg and knew the plants of the Muden Valley especially well.

Fig. 5.229. *Euphorbia pseudocactus*, first male stage, *PVB 9406*, near Muden, South Africa, 15 Sep. 2012 (© PVB).

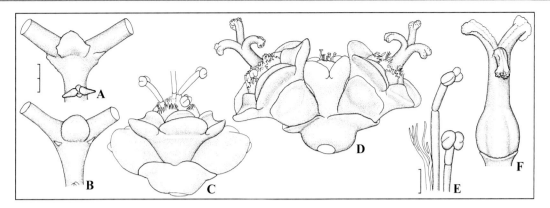

Fig. 5.230. *Euphorbia pseudocactus*. **A**, **B**, young spines and leaf-rudiment from above (scale 2 mm, as for **B–D**). **C**, side view of cyathium in first male stage. **D**, side view of cyathium in female stage. **E**, anthers and bracteole (scale 1 mm, as for **F**). **F**, female floret. Drawn from: *PVB 9406*, near Muden, South Africa (© PVB).

Euphorbia radyeri Bruyns, *Bothalia* 42: 228 (2012). Type: South Africa, Cape, 20 miles from Kendrew towards Jansenville, Jan. 1930, *Dyer 2357* (GRA, holo.; PRE, iso.).

Bisexual spiny glabrous succulent shrub 1–2 × 1–3 m, branching extensively mainly from base of similar stem with woody and fibrous roots, with many peripheral branches spreading underground from plant for up to 0.5 m by rhizomes and then rising erect from soil. *Branches* 30–70 mm thick, strongly constricted into many ± spherical segments, smooth, grey-green; *tubercles* fused into 3–7 wing-like often sinuate angles, laterally flattened and rounded and projecting 3–10 mm from angles, spine-shields around apex and united into continuous horny and later somewhat corky brown to grey or black margin, 4–6 mm broad in upper part tapering to 2–3 mm below, bearing 2 spreading and widely diverging brown to grey spines (2–) 6–15 mm long; *leaf-rudiments* on tips of new tubercles towards apices of branches and stem, 1–4 × 2–4 mm, spreading, fleeting, broadly ovate, obtuse, sessile, with green-brown obtuse ± pyramidal stipules ± 1 mm long. *Synflorescences* in large numbers per branch towards apex, each a group of 1–8 cymes in axil of tubercle, on peduncle 2–4 (6) × 2–3 mm, each cyme with 3 mainly vertically disposed cyathia, central male, outer 2 female only (or bisexual) and developing later, with 2 ovate bracts 1–1.5 × 1.5–2 mm subtending cyathia; *cyathia* cupular-conical, glabrous, 3.5–6 mm broad (2–3 mm long below insertion of glands), with 5 lobes with deeply incised margins, bright yellow; *glands* (3–)5, transversely oblong to kidney-shaped or rectangular, 2–3 mm broad, bright yellow, ascending-spreading, slightly convex to concave above, outer margins entire and slightly raised; stamens entirely glabrous, bracteoles palmate and enveloping groups of stamens, deeply and finely divided, glabrous; *ovary* globose, glabrous, included to slightly exserted on erect pedicel 1.5–2 mm long and soon becoming slightly exserted, calyx slightly extended around base; styles 2–4 mm long, branched in upper third. *Capsule* 6–9 mm diam., obtusely 3-angled, glabrous, somewhat shiny red above and pale green below, erect and exserted on short pedicel 2–5 mm long.

Distribution & Habitat

Euphorbia radyeri is found in a particularly dry part of south-eastern South Africa, between Waterford, Jansenville, Kirkwood and Steytlerville and is especially common and dominant within a radius of about 50 km around Jansenville. Beyond this area there are a few somewhat more scattered and isolated populations towards Somerset East and Pearston. A very isolated, but extensive colony exists on the Klein Karoo near Calitzdorp, between Matjiesvlei and Calitzdorp Dam, which lies some 250 km to the west of the main body of the distribution of *E. radyeri*.

Euphorbia radyeri generally grows on the lower slopes of mountains, on stony hillsides and in flat areas on ground derived from shales. It only rarely ventures into the sandstones of the Witteberg Series in the larger mountains of the area. Noorspoort provides the best example of this, where it is rather unexpectedly encountered on the steep, lower slopes of the poort. *Euphorbia radyeri* frequently forms dense stands and often provides the dominant component of the vegetation. As a consequence, this vegetation derives its name from the *Euphorbia* and is known locally as *noorsveld*, from the popular name '*noors*' for these plants. The huge shrubs that it may develop into, provide shelter for many other small succulents and other shrublets, as well as for many small animals, though it is generally the case that far more small succulents will be found under neighbouring shrubs of *Rhigozum obovatum* than under shrubs of *E. radyeri*. The curious parasite *Hydnora africana* may also be found on it.

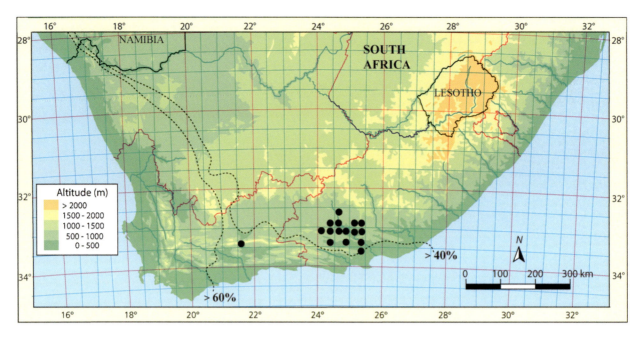

Fig. 5.231. Distribution of *Euphorbia radyeri* (© PVB).

Fig. 5.232. *Euphorbia radyeri*, shrubs ± 1 m tall, with *Aloe striata*, east of Jansenville, South Africa, 3 Sep. 2015 (© PVB).

Fig. 5.233. *Euphorbia radyeri*, ± 1 m tall and obviously rhizomatous around the base, *PVB 10539*, 3 km towards Matjiesvlei, west of Calitzdorp, South Africa, 29 Sep. 2006 (© PVB).

Diagnostic Features & Relationships

Generally, *E. radyeri* forms a densely branched shrub 1–2 m tall, with typical specimens between one and two metres in diameter. *Euphorbia radyeri* exhibits a strong rhizomatous tendency, where some branches around the perimeter of a mature plant spread for up to half a metre as a slender, cylindrical rhizome under the ground before emerging above the surface to become angled, tuberculate and spiny. Branches that develop from rhizomes mostly remain shorter than those on the shrubby part of the plant and many of them do not grow beyond 0.2 m tall, though they often bear flowers, despite being so short.

The branches of *E. radyeri* are mostly 5-angled and are constricted into roughly spherical segments, where each spurt of growth (stimulated by rainfall) gives rise to one such segment. They are grey-green to faintly bluish grey-green and are armed with stout greyish spines often around 10 mm long, which arise on a more or less continuous hardened and shiny or corky margin along the angles formed by the vertically fused spine-shields. On new growth the spines are red and the small, rounded leaf-rudiments just above the spines are conspicuous for a short period before they are shed. The spine-shields are also initially red but soon become pale brown and somewhat irregular and corky, changing to grey or black with age.

Fig. 5.234. *Euphorbia radyeri*, forming shrubs ± 2 m tall here with *Aloe ferox*, *PVB 11521*, south of Jansenville, South Africa, 22 Oct. 2009 (© PVB).

Fig. 5.235. *Euphorbia radyeri* producing a spectacular show of colour in the otherwise drab grey veld, Welgelegen, north of Jansenville, South Africa, 23 Oct. 2009 (© PVB).

Flowering in *E. radyeri* can provide quite a spectacle, transforming the generally drab, grey colour of the *veld* for 2–3 weeks in late October to early December by a flush of bright yellow around the top of each branch in the shrub. However, this spectacle is of comparatively rare occurrence and in most years only a few plants flower and even these may do so only sporadically. Very sporadic flowering has also been observed in late May, well outside the main flowering period. Flowering takes place on the parts of the branch that developed during recent growing seasons, i.e. usually in the upper 50–150 mm of the branch or its last three or four segments. A dense cluster of between one and six (occasionally even more) cyathia breaks through the corky spine-shield in the axil of each spine-pair. Each cyme bears three sweetly scented, bright yellow cyathia (the scent seems to be stronger in the cyathia producing males than in those with female florets), most of which are not much more than 4 mm broad. The lateral cyathia arise in a roughly vertical position (relative to the axis of the branch) but in many of them (especially in larger clusters) this is not very clear and they are quite haphazardly arranged, probably from pressure from the other cymes in the cluster. As is usual, the central cyathium in each cyme is functionally male but the two lateral ones are usually functionally female only. Male florets have long pedicels and the females are also pedicellate, with the ovary usually just contained within the cyathium. After fertilization the pedicel elongates slightly and pushes the maturing capsule outside the cyathium. An unusual feature of *E. radyeri* is that the ovary has a somewhat toothed calyx around its base, whose three longest parts rise up nearly to the middle of the ovary (see Fig. 5.239K, L). Once fertilization has taken place, the pedicel under the ovary elongates slightly so that the capsules mature well above the remains of the cyathium. They are usually fairly shiny red above changing to pale green towards their bases.

Fig. 5.236. *Euphorbia radyeri,* female stage, Welgelegen, north of Jansenville, South Africa, 23 Oct. 2009 (© PVB).

Fig. 5.237. *Euphorbia radyeri,* female stage, *PVB 10539,* 3 km towards Matjiesvlei, west of Calitzdorp, South Africa, 23 Oct. 2008 (© PVB).

5.1 Sect. Euphorbia

Fig. 5.238. *Euphorbia radyeri*, unusually green capsules, Welgelegen, north of Jansenville, South Africa, 23 Oct. 2009 (© PVB).

The differences between *E. caerulescens* and *E. radyeri* are discussed under the former.

History

Euphorbia radyeri, the typical '*Noors*' of the Jansenville district, does not appear to have featured in any accounts of the genus until a figure of it was published by Marloth (1925: fig. 91), who believed that this represented *E. caerulescens*. As explained above, this association of these plants with the name *E. caerulescens* is incorrect and a new name was required for them. R.A. Dyer was, apart from Marloth himself, one of the most dedicated researchers into the genus *Euphorbia*, especially in his detailed and long-term investigations of the species of the Eastern Cape and it seemed appropriate to name this widespread and characteristic species after him.

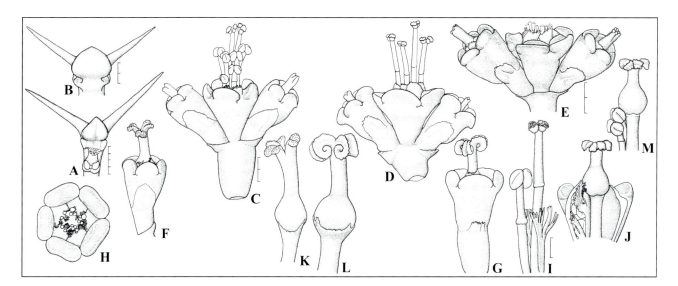

Fig. 5.239. *Euphorbia radyeri*. **A**, **B**, young spines and leaf-rudiment from above (scale 4 mm for **A**; 2 mm for **B**). **C**, **D**, side view of cyme in first male stage (scale 2 mm, as for **D**, **F**, **G**). **E**, side view of cyme at end of first male stage (scale 2 mm, as for **H**). **F**, **G**, female cyathium from side. **H**, male cyathium from above. **I**, anthers and bracteole (scale 1 mm, as for **J–M**). **J**, dissected cyathium. **K–M**, female floret. Drawn from: **A**, **E**, **H**, **J**, **M**, *PVB 10546*, 16 km east of Klipplaat, South Africa. **B**, **C**, **F**, **K**, *PVB 10539*, 3 km towards Matjiesvlei, west of Calitzdorp, South Africa. **D**, **G**, **I**, **L**, Welgelegen, north of Jansenville, South Africa (© PVB).

The colony of *E. radyeri* near Calitzdorp was discovered around 1940 by Max Otzen and was mentioned in White et al. (1941), with photographs by Herbert Krüger.

Euphorbia restricta R.A.Dyer, *Bothalia* 6: 224 (1951). Type: South Africa, Transvaal, The Downs, 4500', 14 Oct. 1947, *Codd & De Winter 3092* (PRE, holo.; GRA, K, NH, PRE, SRGH, iso.).

Dwarf bisexual spiny glabrous succulent 0.06–0.2 × 0.1–0.3 m, much branched at and just above ground level from short partly subterranean central stem 40–80 mm thick in older plants often divided into a few stem-like branches near apex, stem tapering rapidly below into short taproot bearing fibrous roots. *Branches* ascending or spreading, gradually deciduous, occasionally rebranched near tips, 50–160 × 20–35 mm, 4- to 6-angled, constricted into sometimes almost cordate segments 10–30 mm long, smooth, bright green or yellowish green; *tubercles* fused into 4–6 prominent wing-like angles with deep grooves between angles, laterally flattened and deltate projecting 4–8 mm from angles, with spine-shields united into often ± continuous horny initially brown and later grey margin along angles ± 2–3 mm broad above spines narrowing to 1 mm broad below them (sometimes separate), bearing 2 spreading and widely diverging initially red and later grey spines 6–11 mm long and two prickles 1–4 mm long alongside flowering axil (arising dorsally on bracts); *leaf-rudiments* on tips of new tubercles towards apices of branches just above pair of spines, 1–1.5 × ± 0.6 mm, ascending to spreading, fleeting, ovate-lanceolate, sessile, with small irregular brown stipules ± 0.5 mm long. *Synflorescences* many per branch towards apex on last segment, each a solitary cyme in axil of tubercle, very shortly peduncled, each cyme with (2) 3 vertically disposed cyathia, central usually male sessile and often deciduous, lateral (1) 2 bisexual and developing later each on short peduncle 1–2 × 3–4 mm, with 2 ovate red-brown bracts 1–2 × 2 mm subtending lateral cyathia; *cyathia* cupular, glabrous, 5–6 mm broad (2 mm long below insertion of glands), with 5 obovate lobes with deeply incised margins, yellow-green; *glands* 5 (6), transversely elliptic and contiguous, 2 mm broad, yellow, spreading, inner margins raised, outer margins entire and spreading; *stamens* glabrous, bracteoles enveloping groups of males, with finely divided tips, glabrous; *ovary* only very slightly swollen at base of styles, ellipsoidal, glabrous, green, sessile; styles 5–7 mm long, branched from middle and cylindrical below. *Capsule* 6–7 mm diam., 3-angled, red above and green below, glabrous, sessile.

Distribution & Habitat

Euphorbia restricta is found in the mountainous area of the Letaba district, south-west of Tzaneen, mainly on the area

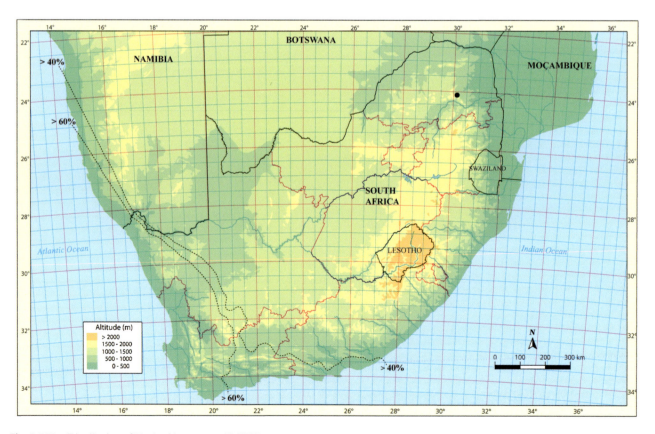

Fig. 5.240. Distribution of *Euphorbia restrtcta* (© PVB).

previously making up the farm 'The Downs'. Here it occurs at an altitude of around 1500 m on low outcrops of dolomite in shallow pockets of soil and in small amounts of compacted earth in crevices between rocks. The shallow soils keep grasses to a minimum (preventing overcrowding and also reducing damage from fire) and succulents are locally abundant, despite the relatively high rainfall. In this habitat there are also small plants of *Aloe castanea*, scattered colonies of *Aloe dolomitica*, specimens of *Euphorbia pulvinata* and *E. clivicola* as well as several species of *Crassula*.

Diagnostic Features & Relationships

Euphorbia restricta forms small, compact plants, often tightly wedged into crevices in rocky outcrops, where its growth is very confined. A relatively thick stem develops from the apex of which the short branches radiate (often subsequently heavily eaten off in habitat). Mostly the branches remain under 20 cm long and around 30 mm thick (though deeply segmented) and they can become very densely packed. They usually have a yellowish green colour with a quite shiny surface between the spines.

Fig. 5.241. *Euphorbia restrtcta,* plant ± 30 cm diam., in crevices in dolomitic outcrop, The Downs, SW of Tzaneen, South Africa, c. 1986 (© S.P. Fourie).

Fig. 5.242. *Euphorbia restrtcta,* plant ± 10 cm diam. and heavily grazed, in crevices in dolomitic outcrop, *PVB 12102*, The Downs, SW of Tzaneen, South Africa, 2 Jan. 2012 (© PVB).

Flowering takes place between April and August on the final segment of some of the branches. The cyathia are bright yellow, contrasting strongly with the green of the branches. The female florets have relatively long styles that are fused into a cylinder above the tiny ovary and spread widely towards their tips.

Euphorbia restricta bears much resemblance to *E. barnardii* and differs from it mainly in being very much smaller in stature, with a more developed subterranean stem and with smaller and narrower leaf-rudiments. The two prickles alongside the base of the cyme (which appear to be dorsal 'spines' on the first pair of bracts that cover the cyme while it develops) are often longer in *E. restricta* than in *E. barnardii*. Florally they differ principally in that the female florets have longer styles with a slightly longer fused cylindrical part in *E. restricta*. In both, the female floret has a particularly short and inconspicuous ovary, suggestive of that in *E. cooperi* too. All these species have very deeply segmented branches and seem to be closely related.

Fig. 5.243. *Euphorbia restricta*, brilliant yellow cyathia in first male stage, *PVB 12102*, The Downs, SW of Tzaneen, South Africa, 2 Jul. 2015 (© PVB).

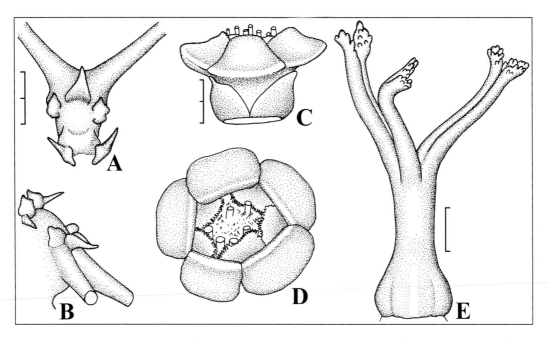

Fig. 5.244. *Euphorbia restricta*. **A**, young spines and leaf-rudiment from above (scale 2 mm, as for **B**). **B**, young spines and leaf-rudiment from side. **C**, side view of cyathium in first male stage (scale 2 mm, as for **D**). **D**, cyathium in first male stage from above. **E**, female floret (scale 1 mm). Drawn from: *PVB 12102*, The Downs, SW of Tzaneen, South Africa (© PVB).

5.1 Sect. Euphorbia

History

Euphorbia restricta was first brought to the attention of R.A. Dyer by the well-known mountaineer and explorer A.H. Crundall in April 1945. It was described from a collection made in the same area by L.E. Codd and B. de Winter in October 1947, when flowering material was gathered for the first time. Few other collections have been made, but the species was at one time heavily exploited by commercial collecting, though some protection is now provided by the fact that most of the known populations lie in the Lekgalamatse Nature Reserve.

Euphorbia rowlandii R.A.Dyer, *Bothalia* 7: 28 (1958). Type: South Africa, Transvaal, Kruger Nat. Park, 8 miles north of Punda Milia, 1600', 25 July 1951, *Rowland Jones 48* (PRE, holo.; K, PRE, SRGH, iso.).

Bisexual spiny glabrous succulent shrub 1–2 × 1–3 m, branching mainly from base of similar but shorter stem with woody and fibrous roots, with many ascending branches. *Branches* 35–50 mm thick, constricted into segments broadest near base and narrowing to 20–30 mm just below next segment, smooth, grey-green; *tubercles* fused

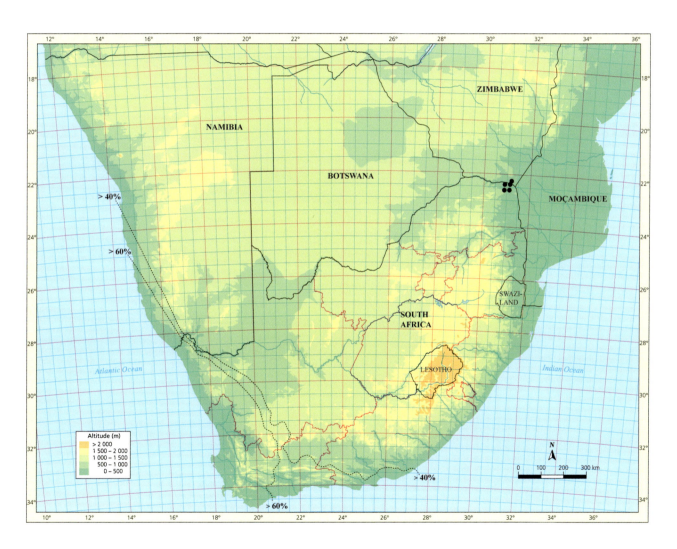

Fig. 5.245. Distribution of *Euphorbia rowlandii* (© PVB).

Fig. 5.246. *Euphorbia rowlandii*, plant ± 1.5 m tall, among reddish sandstone blocks in shallow soil, *PVB 6589a*, Tshikondeni, SE of Messina, South Africa, 4 Nov. 2011 (© PVB).

Fig. 5.247. *Euphorbia rowlandii*, near Tshikondeni, SE of Messina, South Africa, c. 1986 (© S.P. Fourie).

into 5–7 wing-like slightly sinuate angles, laterally flattened and rounded and projecting 3–7 mm from angles, spine-shields around apex and united into horny initially brown and later grey or white margin, 2–3 mm broad, bearing 2 ascending-spreading and widely diverging initially dark brown to black later grey spines 5–12 mm long; *leaf-rudiments* on tips of new tubercles towards apices of branches and stem, 1–2 × 1–2 mm, spreading, fleeting, broadly deltate, obtuse, sessile, with irregular brown stipules 0.5–1 mm long. *Synflorescences* in large numbers per

5.1 Sect. Euphorbia

Fig. 5.248. *Euphorbia rowlandii*, PVB 6589a, Tshikondeni, SE of Messina, South Africa, 4 Nov. 2011 (© PVB).

branch towards apex, each a solitary cyme in axil of tubercle, on peduncle 1–1.5 × 2–3 mm, each cyme with 3 vertically disposed cyathia, central male, outer 2 bisexual and developing later, with 2 ovate bracts 1–2 × 1.5–2 mm subtending cyathia; *cyathia* conical, glabrous, 4–6 mm broad (2–3 mm long below insertion of glands), with 5 lobes with finely incised margins, yellow-green; *glands* 5, transversely elliptic and contiguous, 2–3 mm broad, bright yellow, spreading, slightly convex above, outer margins entire and spreading; stamens entirely glabrous, bracteoles palmate and enveloping groups of stamens, deeply and finely divided, glabrous; *ovary* globose, glabrous, included on stout erect pedicel ± 0.75 mm long, calyx slightly extended around base; styles 1.5–2 mm long, branched to below middle. *Capsule* 8–10 mm diam., deeply and obtusely 3-angled, glabrous, shiny pale green with red on edges, exserted on short stout erect pedicel 2–5 mm long.

Distribution & Habitat

Euphorbia rowlandii is only known in the north-eastern corner of South Africa and in the neighbouring south-eastern corner of Zimbabwe. In South Africa is has been recorded from north-west of Punda Milia in the Kruger National Park westwards to Tshikondeni, at the eastern end of the Soutpansberg.

Fig. 5.249. *Euphorbia rowlandii*, female stage, *PVB 6589a*, Tshikondeni, SE of Messina, South Africa, 4 Nov. 2011 (© PVB).

Fig. 5.250. *Euphorbia rowlandii*, second male stage, *PVB 6589a*, Tshikondeni, SE of Messina, South Africa, 4 Nov. 2011 (© PVB).

This very local species usually grows on the northern aspects of low, sandstone ridges among scattered trees and shrubs, often with open 'forests' of the Lebombo ironwood, *Androstachys johnsonii* filling up much of the space between the larger rocks. Plants grow in shallow pockets of soil between large rocks or in crevices in these rocks.

Diagnostic Features & Relationships
Plants of *E. rowlandii* form imposing, laxly branched shrubs to 2 m tall and up to 3 m in diameter. Larger specimens consist of many branches spreading and then ascending from the generally much shorter stem which is not visible from the side of the plant. Usually around 40 mm thick, the branches are constricted at variable intervals into segments that are roughly triangular in outline, broadest at the base and narrowing towards the apex. Most of the branches are 5- angled and the edges of these angles are armed with strong pairs of somewhat ascending spines that are borne on a more or less continuous horny margin along the angles. Initially the spines are dark brown, almost black and the spine-shields pale brown and with time this changes to a uniform, pale grey in both. Tiny leaf-rudiments appear on new growth but are quickly lost.

Flowering in *E. rowlandii* takes place in July and early August, with the broad and conical cyathia densely clustered towards the tips of the branches. They are bright yellow above from the colour of the glands which are quite closely contiguous around the top of the cyathia and are slightly con-

Fig. 5.251. *Euphorbia rowlandii*, capsules, *PVB 6589a*, Tshikondeni, SE of Messina, South Africa, 4 Nov. 2011 (© PVB).

5.1 Sect. Euphorbia

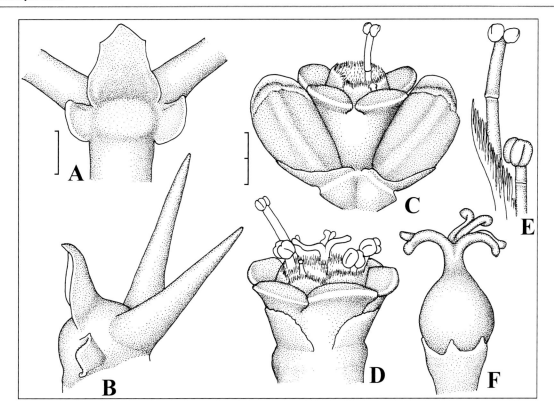

Fig. 5.252. *Euphorbia rowlandii*. **A**, young spines and leaf-rudiment from above (scale 1 mm, as for **B**, **E**, **F**). **B**, young spines and leaf-rudiment from side. **C**, side view of cyme in first male stage (scale 2 mm, as for **D**). **D**, cyathium in female stage. **E**, anthers and bracteole. **F**, female floret. Drawn from: *PVB 6589a*, Tshikondeni, SE of Messina, South Africa (© PVB).

vex above. The ovary remains included in the cyathium on a stout but short pedicel and has a slightly extended calyx around its base (Fig. 5.252F). It is unusually broad and short, with deeply divided styles spreading widely just above it. After fertilization, the capsule is exserted slightly from the remains of the cyathium. It is deeply 3-angled and somewhat shiny red and pale green.

While somewhat similar in appearance to species such as *E. pseudocactus*, with which it shares the deeply lobed capsules borne on short pedicels, *E. rowlandii* is more similar-looking to *E. avasmontana* and *E. otavibergensis*. From the first it differs by the solitary cymes in each axil and the differently shaped capsules which are not exserted to the same degree. From the second it differs by the differently shaped ovary which more abruptly narrows into the styles, is borne on a longer pedicel and has an extended calyx around its base.

History
Euphorbia rowlandii was described from material gathered in July 1951 by M. Rowland Jones, who was then in charge of the northern section of the Kruger National Park. It had initially been observed in this area by L.E. Codd in March 1949 (*Codd 5370* at PRE). Rowland Jones was the first to gather flowering material of it and he forwarded this to R.A. Dyer for the description, which appeared 7 years later.

Euphorbia schinzii Pax, *Bull. Herb. Boiss.* 6: 739 (1898). Lectotype (Bruyns 2012): South Africa, Transvaal, Berea Ridge, Barberton, 3100', 13 Feb. 1891, *Galpin 1297* (BOL; K, iso.).

Bisexual spiny glabrous succulent shrub 0.15–0.5 × 0.05–1 (2) m, profusely branching from near base of similar erect stem (sometimes slightly swollen underground) to 300 mm tall, sometimes rhizomatous, with many fibrous roots. *Branches* erect and often rebranching, 20–300 (500) × 4–12 (20) mm, usually 4-angled (subcylindrical), not constricted into segments, smooth, grey-green to pale bluish green (red, sometimes creamy-green in grooves between angles) becoming pale brown with age;

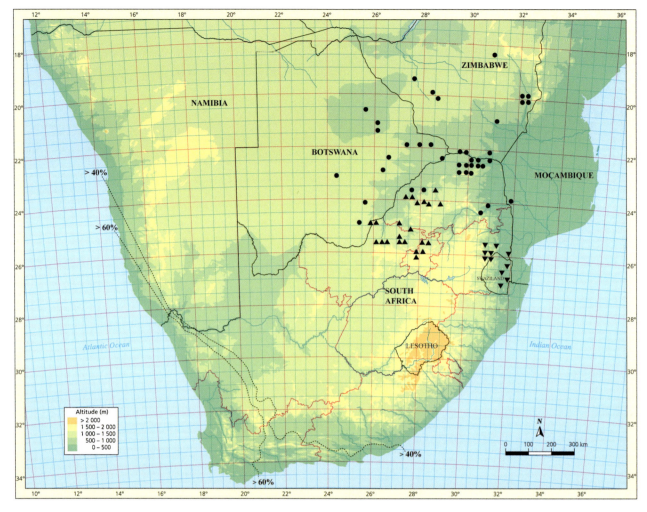

Fig. 5.253. Distribution of *Euphorbia schinzii* (● = ssp. *bechuanica*, ▲ = ssp. *schinzioides*, ▼ = ssp. *schinzii*) (© PVB).

tubercles fused into 4 low wing-like (sometimes obscure) angles with surface flat to slightly concave between angles, conical and truncate, laterally flattened and projecting 1–6 mm from angles, spine-shields 3–6 (18) mm long, well separated (rarely almost continuous), 1–2 mm broad along angles and 2–6 (16) × 1–2 mm long below (tapering to deltate), bearing 2 spreading red to pale brown (later grey) spines 3–12 (18) mm long; *leaf-rudiments* on tips of new tubercles towards apices of branches, 0.5–2.5 × 0.7–1.5 mm, erect ro recurved, fleeting, ovate-lanceolate to elliptic, sessile, with small stipular prickles 0.5–3 mm long. *Synflorescences* many per branch towards apex, each a solitary cyme in axil of tubercle, on very short peduncle to sessile, each cyme with 3 transversely disposed cyathia (occasionally more developing from axils of bracts), central male ± sessile and sometimes deciduous, outer two bisexual and developing later each on short peduncle 1–2 × 2–3 mm, with 2–3 small broadly ovate and truncate and minutely lacerate bracts 1–2 × 1–3 mm subtending lateral cyathia; *cyathia* shallowly conical-cupular, glabrous, 3–5 (6) mm broad (2 mm long below insertion of glands), with 5 pale yellow or cream obovate lobes with finely and deeply incised margins, yellow or pale yellow becoming pale green towards base; *glands* 5, transversely oblong and contiguous, free (sometimes appearing to be laterally fused into cup-like structure), 1.5–3 mm broad, pale to bright yellow, spreading, inner margins flat to only slightly raised, outer margins entire and spreading, surface between two margins dull; *stamens* few, with glabrous pedicels, anthers pale yellow, bracteoles enveloping groups of males, with finely divided tips, glabrous; *ovary* obtusely 3-angled, ± sessile or raised on pedicel < 0.5 mm long, glabrous, green to reddish green; styles 2–5 mm long, spreading, branched for upper two thirds (to around 1 mm from base). *Capsule* 3–4 (5) mm diam., obtusely 3-angled, dull reddish green, glabrous, sessile.

White et al. (1941) had a very broad concept of *E. schinzii*. Under this name, White et al. (1941: fig. 845) depicted a plant from Bulawayo in Zimbabwe and Leach (1991: 136) showed various plants from Zimbabwe. However, the taxonomic status of these was unresolved in Carter and Leach (2001: 427) and their position remains unknown today. There is nevertheless no doubt that *E. schinzii* has close relatives north of southern Africa. *Euphorbia acervata* from the Great Dyke in Zimbabwe is likely to be one such relative, though Carter (2000) only compared it to the little-known *E. tortistyla*. In his informal account of relatives of *E. schinzii*, Hargreaves (1994) included many species from Malawi southwards, while Leach (1964) stated that *E. malevola* is closely related to *E. complexa* (= *E. schinzii*), though it does not occur in southern Africa as defined here.

Three subspecies are recognized and these may be separated as follows:

1. Plant forming a robust shrub 0.15–0.5 × 0.15–2 m, branches 8–12 (20) mm thick, not rhizomatous, spines 6–18 mm long..subsp. **bechuanica**
1. Plant forming a small clump 0.05–0.2 (0.4) × 0.2–0.3 (1) m, branches 6–12 (20) mm thick, often rhizomatous, spines 3–12 (17) mm long...2.
2. Stipular prickles 0.5–1 mm long, spine-shields broad and usually blackish..subsp. **schinzioides**
2. Stipular prickles 1.5–3 mm long, spine-shields slender and usually grey...subsp. **schinzii**

Euphorbia schinzii subsp. schinzii

Euphorbia complexa R.A.Dyer, *Fl. Pl. South Africa* 17: t. 643 (1937). Type: South Africa, Transvaal, road from Louw's Creek to Kaapmuiden, June 1936, *Van der Merwe 100 sub PRE 21373* (PRE, holo.; K, W, iso.)

Small shrub 0.15–0.4 × 0.2–1 m, profusely branching from near base and occasionally rhizomatous. *Branches* 50–300 × 6–12 mm, pale grey-green; *tubercles* fused into 4 low wing-like angles with surface slightly concave between angles, projecting 1–5 mm from angles, with spine-shields separate, grey, 4–8 mm long, 1–2 × 2 mm above spines and 3–6 × ± 1 mm and tapering below spines, spines 3–5 mm long; *leaf-rudiments* 1–1.5 × 1 mm, stipular prickles 1.5–3 mm long. *Cyathia* 3.5–5 mm broad; *glands* pale to bright yellow; styles 2–2.5 mm long, spreading, branched for upper two thirds (to around 1 mm from base).

Distribution & Habitat

Subsp. *schinzii* is found in the mountainous area between Barberton and Nelspruit in the west to near Komatipoort and further east into Swaziland (around Sipofanini) and close to the Moçambican border.

Plants are usually found around the bases of rocky, often granitic or schistose slopes. Here they grow in pockets of soil on slabs of rock or in shallow ground among stones and small clumps of grass, occasionally sheltered by some trees.

Diagnostic Features & Relationships

Plants of subsp. *schinzii* form small shrubs of slender pale grey-green branches. Their angles are mostly quite slender and obvious and their spines are mostly also slender, fine and relatively weak. A particular feature of them is the quite long stipular prickles relative to these spines. However, some collections from along the Lebombo Mountains have rather

Fig. 5.254. *Euphorbia schinzii* ssp. *schinzii*, plant about 30 cm diam., flowering in very dry conditions and partly scorched by fire on right hand side, *PVB*, Barberton, South Africa, 2 Aug. 2019 (© PVB).

Fig. 5.255. *Euphorbia schinzii* ssp. *schinzii*, first male stage, *PVB 11903*, near Swazi border, south-east of Nelspruit, South Africa, 7 Jun. 2013 (© PVB).

Fig. 5.256. *Euphorbia schinzii* ssp. *schinzii*, end of first male stage, *PVB 9390*, Kaalrug, east of Nelspruit, South Africa, 15 Feb. 2014 (© PVB).

stouter spines, though again the noticeable stipular prickles are present.

In habitat subsp. *schinzii* flowers from July onwards and the type was collected in flower in the middle of February of 1891. In cultivation they mostly flower in January and February. The cyathia are the same bright yellow, typical of most forms of *E. schinzii*, but they have fairly short styles compared to those of subsp. *bechuanica*.

History

When he described *E. schinzii*, Pax cited two specimens, *Rehmann 4347* from Pretoria, South Africa (which is missing and was not seen by N.E. Brown either) and *Galpin 1297* from Berea Ridge, which lies immediately behind Barberton. He named the plant after Hans Schinz who, although he had also collected in southern Africa, had nothing to do with any collections of *E. schinzii* in South Africa. Since the specimen *Rehmann 4347* is missing, *Galpin 1297* was selected as the lectotype for *E. schinzii* by Bruyns (2012). However, *Galpin 1297* belongs to the same species as the type of *E. complexa*. Other recent collections from Barberton (e.g. *Kluge 2281* (NBG)), that have been placed in herbaria under *E. schinzii*, are also identical to *E. complexa*. *Euphorbia complexa* is therefore now a synonym of subsp. *schinzii*.

Traditionally, small, clump-forming members of this complex from around Pretoria were taken as typical of *E. schinzii* (White et al. 1941), to the extent that White et al. (1941: 745), Fourie (1988: 91, 92) and Hargreaves (1994: 147) all gave the 'Type locality' of *E. schinzii* as 'Pretoria Distr.; hills near Pretoria'. However, this is not correct any more and the 'type locality' is Barberton. Plants from around Pretoria (and further west and north) are now placed under subsp. *schinzioides*.

Euphorbia complexa was described from material collected by F.Z. van der Merwe in June 1936 near Barberton. Dyer considered that its nearest relative was *E. lydenburgensis* and he compared it also with *E. knuthii*.

Fig. 5.257. *Euphorbia schinzii* ssp. *schinzii*, second male stage, *PVB 11903*, near Swazi border, south-east of Nelspruit, South Africa, 19 Feb. 2019 (© PVB).

Fig. 5.258. *Euphorbia schinzii* ssp. *schinzii*, with capsules, *PVB*, Barberton, South Africa, 2 Aug. 2019 (© PVB).

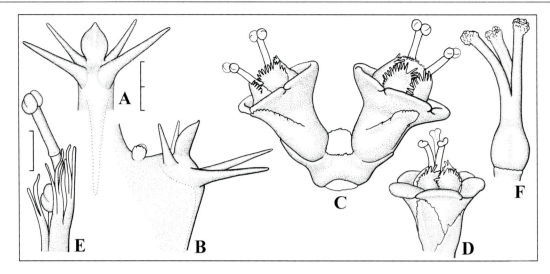

Fig. 5.259. *Euphorbia schinzii* ssp. *schinzii*. **A**, young spines and leaf-rudiment from behind (scale 2 mm, as for **B**–**D**). **B**, young spines and leaf-rudiment from side. **C**, side view of cyme at second male stage. **D**, side view of female cyathium. **E**, anthers and bracteoles (scale 1 mm, as for **F**). **F**, female floret. Drawn from: *PVB 9390*, Kaalrug, east of Nelspruit, South Africa (© PVB).

Euphorbia schinzii subsp. **bechuanica** (L.C.Leach) Bruyns, *Phytotaxa* 436: 214 (2020).
Euphorbia malevola subsp. *bechuanica* L.C.Leach, *J. S. Afr. Bot.* 30: 6 (1964). Type: Botswana, halfway between Palapye and Francistown, Jul. 1937, *Obermeyer* (PRE, holo.; K, PRE, iso.).
Euphorbia limpopoana L.C.Leach ex S.Carter, *Kew Bull.* 54: 960 (2000). Type: Zimbabwe. Fulton's Drift, 25.5 km NNW of Beitbridge, 14 Jan. 1963, *Leach 11582a* (SRGH, holo.).

Ronust shrub 0.15–0.5 × 0.15–2 m, branching strongly from base, not rhizomatous. *Branches* 0.1–0.5 m × 8–12 (20) mm, pale grey-green to tinged with red, sometimes creamy-green in grooves between angles; *tubercles* fused into 4 straight angles with flat to slightly concave area between angles, projecting 1–2 mm from angles, with spine-shields 6–18 mm long, 2–3 × 2 mm above spines and 4–16 × 1–2 mm below spines and usually separate (rarely almost continuous), spines red-brown becoming grey and 6–18 mm long; *leaf-rudiments* 1–2.5 × 0.8–1.2 mm (sometimes with denticulate margins on either side at base), stipular prickles 0.5–3 mm long. *Cyathia* 4–6 mm broad; *glands* yellow; styles 4–5 mm long, branched to base.

Fig. 5.260. *Euphorbia schinzii* ssp. *bechuanica,* plant ± 0.8 m diam., *PVB 12058,* east of Nwanedi, east of Messina, South Africa, 4 Nov. 2011 (© PVB).

Fig. 5.261. *Euphorbia schinzii* ssp. *bechuanica*, plant ± 1 m diam., calcareous grit, *PVB 12336*, southern end of Mgadigadi Pans, Mosu, Botswana, 26 Dec. 2012 (© PVB).

Distribution & Habitat

The most widely distributed of the three subspecies, subsp. *bechuanica* is known in Botswana, Moçambique, South Africa and Zimbabwe. In Botswana it is widely but scantily recorded on the eastern side of the country from the south-eastern edge of the Mkarikari Pans, to between Lobatse and Gaberones and towards Francistown. In Moçambique, South Africa and Zimbabwe it is mainly recorded in the valley of the Limpopo River, though in Zimbabwe it is also known in the drier part of the valley of the Sabi River. In South Africa, collections exist from north-west of the Waterberg to north of the Soutpansberg and records have also been made further eastwards and southwards to the *lowveld* south of Tzaneen.

Subsp. *bechuanica* is often found on the dry, stony lower slopes of hills or in rather bare and heavily overgrazed stony or gravelly to loamy flat areas (sometimes among calcrete outcrops) among scattered trees of *Acacia tortilis* and *Dichrostachys cinerea* or *mopane*. In some such areas it can become the dominant succulent along with scattered plants of *Aloe*, *Cissus* and *Sansevieria*. When it occurs in sufficiently large numbers, it may provide shelter for a wealth of smaller succulents such as various species of *Ceropegia*. In some areas it occurs with *E. aeruginosa*. The two are usually (though not always) at least slightly separated altitudinally, with subsp. *bechuanica* lower down and the other higher up in more quartzitic ground.

Fig. 5.262. *Euphorbia schinzii* ssp. *bechuanica*, spiky mass of branches under *mopane*-trees, *PVB 12058*, east of Nwanedi, east of Messina, South Africa, 4 Nov. 2011 (© PVB).

Fig. 5.263. *Euphorbia schinzii* ssp. *bechuanica*, first male stage almost over, *PVB 7474*, turnoff to Sibasa east of Nwanedi, east of Messina, South Africa, 19 Jul. 2011 (© PVB).

Fig. 5.264. *Euphorbia schinzii* ssp. *bechuanica*, first male stage almost over, *PVB 7776*, Nzhelele Dam, east of Messina, South Africa, 9 Aug. 2019 (© PVB).

Fig. 5.265. *Euphorbia schinzii* ssp. *bechuanica*, female stage showing the fairly long, whitish styles, *PVB 7776*, Nzhelele Dam, east of Messina, South Africa, 19 Jul. 2011 (© PVB).

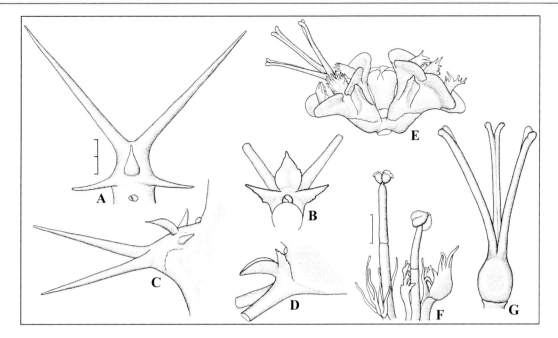

Fig. 5.266. *Euphorbia schinzii* ssp. *bechuanica*. **A**, **B**, young spines and leaf-rudiment from above (scale 2 mm, as for **B**–**E**). **C**, **D**, young spines and leaf-rudiment from side. **E**, cyme from side in female stage. **F**, anthers and bracteoles (scale 1 mm, as for **G**). **G**, female floret. Drawn from: **A**, **C**, *PVB 7776*, Nzhelele Dam, east of Messina, South Africa. **B**, **D–G**, *PVB 7474*, turnoff to Sibasa, east of Nwanedi and east of Messina, South Africa (© PVB).

Diagnostic Features & Relationships

In large colonies, the shrubs of subsp. *bechuanica* are often of very variable size. While many are comparatively small (and sometimes they may all be quite small – not exceeding 15 cm tall and up to 30 cm in diameter), some reach 0.5 m in height and 2 m in diameter. Such huge specimens often form an impenetrable mass around the bases of trees of *Acacia tortilis*, exactly as *E. malevola* does further north. In larger specimens, the branches are often noticeably spreading above an erect base, rather than strictly erect. They are usually around 10 mm thick with grey-green angles and may have a paler, sometimes mottled grey-green colour between the angles. In most cases the branches are difficult to handle because of the long, sharp and rigid spines, which spread straight out from the spine-shields for 15 mm or more and provide a vigorous protection for the branches. This is particularly true of material from overgrazed patches north of the Soutpansberg and from the edge of the Mkarikari Pans in Botswana. Initially often reddish brown (and sometimes at this stage rather similarly coloured to the spines of *E. aeruginosa*), the spines change to dark grey with age.

Flowering takes place mainly in July and August (though it has been recorded between May and early November) and usually the tips of the branches become covered with the bright yellow, sweetly scented cyathia. The female florets have especially long, almost white styles, which give the upper parts of the branch a finely fuzzy appearance.

Subsp. *bechuanica* differs from subsp. *schinzioides* mainly in the lack of any rhizomatous branches and in the stouter, often longer branches with more robust and longer spines. Furthermore, the angles are usually more clearly continuous in subsp. *bechuanica*. However, near Gaberones in Botswana, plants of *E. schinzii* may have the robust branches and spines of subsp. *bechuanica* but the rhizomatous habit of subsp. *schinzioides* (Bruyns et al. 2020). It is also difficult to distinguish subsp. *bechuanica* from more vigorous plants of *E. venteri* (see under *E. venteri*). The differences from *E. aeruginosa* are discussed under that species.

History

Subsp. *bechuanica* was described as a subspecies of *Euphorbia malevola* by L.C. Leach and based on material collected by A.A.Obermeyer in July 1937 in eastern Botswana near Serule, between Palapye and Francistown. However, this was not by any means the first record of it. It was collected by J. Thomas Baines in October 1872, probably north of the Soutpansberg, during one of the two expeditions that he led to Matebeleland (western Zimbabwe) for the South African Gold Fields Exploration Company. It was also gathered north of the Soutpansberg by Jan Gillett in August 1930, during his trip with J. Hutchinson and again in September of 1934 by E.E. Galpin. Brown (1911–12) included Baines' specimen under *E. schinzii*.

White et al. (1941) included several photographs of subsp. *bechuanica* (e.g. Fig. 837, 839, 840, 841). They referred to it as a 'variable form of *E. schinzii*'. Fourie (1988) referred to it as the 'Limpopo valley form of *E. schinzii*'. Hargreaves (1994) discussed the various forms of *E. schinzii* that he had encountered in Botswana. Here he mentioned that he had found members of the *E. schinzii*-complex around Serule, the type locality of *E. malevola* subsp. *bechuanica*, and that these were identical to plants seen in the eastern part of Botswana, that were later described as *E. limpopoana*.

Euphorbia schinzii subsp. **schinzioides** Bruyns, *Phytotaxa* 436: 215 (2020). Type: South Africa, Pretoria distr., Horn's Nek near Hartebeespoort Dam, 27 Aug. 1949, *Prosser 1011* (NBG, holo.; K, PRE, iso.).

Small shrub 0.05–0.15 (0.3) × 0.15–0.3 m, densely branched at and slightly below ground level, often with rhizomatous branches, stem often somewhat swollen underground. *Branches* 50–150 × 7–15 (20) mm, bluish green; *tubercles* fused into 4 angles with surface flat to slightly concave between angles, with spine-shields 5–8 (10) mm long, 1–2 × 2–3 mm above spines and 4–6 (8) mm long and slender below spines (remaining well separated from next), spines 6–10 (17) mm long; *leaf-rudiments* 1–1.5 × 0.7–1.5 mm, with small brown stipular prickles 0.5–1 mm long. *Cyathia* 3–5 mm broad; glands bright yellow; styles 2.5–3 mm long, branched to slightly below middle.

Distribution & Habitat

Subsp. *schinzioides* is found in the mountainous and hilly country of the former Transvaal, South Africa from the Magaliesberg around Pretoria westwards to near Gaberone in Botswana. To the north of Pretoria, it is known in the Waterberg and in the flat areas along its northern base.

Fig. 5.267. *Euphorbia schinzii* ssp. *schinzioides*, robust-branched plant ± 20 cm tall without rhizomes, stony ground alongside young *Aloe marlothii*, *PVB 12291*, Ramotswa, Botswana, 23 Dec. 2012 (© PVB).

Fig. 5.268. *Euphorbia schinzii* ssp. *schinzioides*, rhizomatous plant with branches ± 8 cm tall, *PVB 12291*, Ramotswa, Botswana, 23 Dec. 2012 (© PVB).

Subsp. *schinzioides* is usually found as a small, clump-forming succulent growing in crevices in slabs of rock. Occasionally plants are found in shallow soil in gravelly patches between rocky outcrops.

Diagnostic Features & Relationships
Plants of subsp. *schinzioides* are rarely taller than 150 mm and form densely packed clumps of short branches. The branches mostly have a pale grey-green colour and are armed with small, hard spines arising on small dark spine-shields. Each leaf-rudiment is accompanied by a pair of tiny stipular prickles.

Flowering takes place in subsp. *schinzioides* in August and September, when the tips of the branches become covered with the bright yellow cyathia. These transform the plant from a nondescript grey-green, spiny clump into a bright yellow cluster and give off a wonderfully sweet scent.

Fig. 5.269. *Euphorbia schinzii* ssp. *schinzioides*, densely branched plant with branches ± 10 cm tall, crevices between submerged sandstone rocks, *PVB 12116*, Magaliesberg, Pretoria (© PVB), South Africa, 23 Feb. 2012 (© PVB).

Fig. 5.270. *Euphorbia schinzii* ssp. *schinzioides*, diffusely branched plant with many rhizomatous branches to ± 8 cm tall, in gravelly ground under very dry conditions but covered with flowers, *PVB*, east of Marken, South Africa, 28 Jul. 2019 (© PVB).

Fig. 5.271. *Euphorbia schinzii* ssp. *schinzioides,* branches ± prostrate from desication, in second male stage, *PVB*, east of Marken, South Africa, 28 Jul. 2019 (© PVB).

Perhaps the easiest way of separating subsp. *schinzioides* from the other two subspecies is the tiny stipular denticles combined with the somewhat rhizomatous habit of some branches. In areas where the branches are longer and thicker and the spines longer, it is usually the presence of some rhizomatous branches that separate these from subsp. *bechuanica*.

History

The first collection of subsp. *schinzioides* was made by Joseph Burke, at the 'Orange River' (K), probably late in 1841. This specimen was filed in the herbarium at Kew under the name *E. micracantha*, though it was not cited by N.E. Brown under either *E. micracantha* or *E. schinzii*. Another early record was made by Anton Rehmann (sometimes referred to as Antoni Rehman), a Polish botanist who travelled twice to South Africa and made some of the earliest known collections in the former Transvaal. He collected subsp. *schinzioides* near the present-day Pretoria during his second visit of 1879–1880. Many of Rehmann's collections were sold to Hanz Schinz of Zürich, Switzerland and some remained at the herbarium (Z), though his specimen of subsp. *schinzioides* could not be located there or anywhere else.

Fig. 5.272. *Euphorbia schinzii* ssp. *schinzioides,* female stage, *PVB 12079*, Supingstad, South Africa, 26 Sep. 2019 (© PVB).

Fig. 5.273. *Euphorbia schinzii* ssp. *schinzioides,* female stage, *PVB 12426*, Marulafontein, Waterberg, South Africa, 30 Jul. 2013 (© PVB).

Euphorbia sekukuniensis R.A.Dyer, *Fl. Pl. South Africa* 20: t. 775 (1940). Type: South Africa, Transvaal, Steelpoort River, north of Roossenekal, Aug. 1938, *Van der Merwe 1765* (*sub PRE 25475*) (PRE, holo.; GRA, PRE, SRGH, iso.).

Bisexual spiny glabrous succulent tree 3–8 m tall, with gradually deciduous branches forming almost spherical crown near top of cylindrical erect solitary (rarely forked) trunk-like stem 3–8 m tall, trunk naked except for spines below crown, with many fibrous roots spreading from base.

Branches spreading then ascending and often rebranching, older branches drying up and falling off stem after a few years, 0.3–1 m × 10–20 mm, (3) 4- to 5-angled, slightly constricted into segments 50–200 mm long, smooth, green; *tubercles* very evenly fused into (3) 4–5 slightly wing-like angles with surface flat to concave between angles, conical, laterally flattened and projecting 2–3 mm from angles, spine-shields continuous forming hard grey to corky black margin 2–3 mm broad along angles, bearing 2 spreading and widely diverging brown to grey spines 3–10 mm long; *leaf-rudiments* on tips of new tubercles towards apices of branches,

Fig. 5.274. *Euphorbia sekukuniensis,* grove of trees among outcrops of norite and other trees on low hill, *PVB 13745*, Gethlane Lodge, SE of Burgersfort, South Africa, 1 Aug. 2019 (© PVB).

± 0.5 × 0.25 mm, spreading, deciduous, lanceolate-deltate, sessile, with minute irregular brown ridge-like stipules up to 0.1 mm long. *Synflorescences* many per branch towards apex, with 1 cyme in axil of each tubercle, very shortly peduncled, each cyme with 3 transversely disposed cyathia, central male, outer two bisexual and developing later each on short peduncle 2–3 × 2 mm, with 2 small broadly ovate bracts 1–2.5 × 2 mm subtending lateral cyathia; *cyathia* conical to cupular, glabrous, 4.5–6.5 mm broad (2 mm long below insertion of glands), with 5 obovate lobes with finely toothed margins, yellow to yellow-green; *glands* 5 (6), transversely elliptic, contiguous to well separated, 1.5–2 mm broad, yellow-green, spreading, inner margins not raised, outer margins entire and spreading, surface between two margins dull and flat; *stamens* with glabrous pedicels, anthers yellow, bracteoles enveloping groups of males, with finely divided tips, glabrous; *ovary* triangular-globose, glabrous, green, raised and partly exserted on erect pedicel ± 1.5 mm long; styles 1.5–2 mm long, widely spreading, branched to near base. *Capsule* 8–9 mm diam., deeply 3-angled, glabrous, green suffused with red above, exserted on pendulous to erect pedicel 4–9 mm long.

Distribution & Habitat

Euphorbia sekukuniensis is only found in the part of the former Transvaal known as Sekukuniland, where it has been

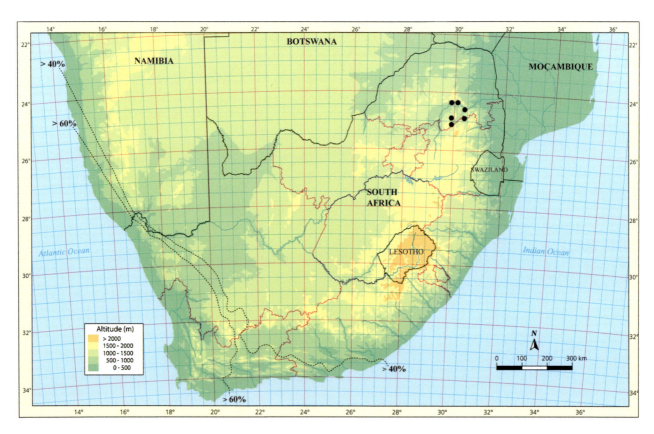

Fig. 5.275. Distribution of *Euphorbia sekukuniensis* (© PVB).

recorded on hills from between Zeekoeigat, Burgersfort, Jane Furse Hospital and a point in the valley of the Steelpoort River about 35 km south-west of Steelpoort. It occurs at altitudes of between 850 and 1300 m.

Plants of *E. sekukuniensis* are occasionally found among trees on stony slopes but most frequently form groups of up to 100 individuals in fairly bare, exposed outcrops of norite around the tops of hills. Here they occur in crevices between boulders in shallow soils usually with much leaf-litter, together with a scanty, deciduous scrub, often with *Myrothamnus* and *Selaginella dregei* and sometimes together with tall trees of *E. cooperi*, and smaller plants of *Commiphora*, *Croton*, *Sterculia* and *Tetradenia*.

Fig. 5.276. *Euphorbia sekukuniensis*, *PVB 11773*, 8 km NE of Burgersfort towards Ohrigstad, South Africa, 11 Sep. 2010 (© PVB).

Fig. 5.277. *Euphorbia sekukuniensis,* among outcrops of norite near summit of hill, *PVB 11897*, SE of Chuniespoort towards Burgersfort, South Africa, 22 Jul. 2011 (© PVB).

Diagnostic Features & Relationships

Euphorbia sekukuniensis forms impressively tall and slender trees, with a stout, smooth-barked trunk. The branches are quite densely grouped at the apex of the tree into an often relatively small, almost spherical, mop-like tuft, but this apical tuft of branches can also be much broader and may even be similar in shape to small-topped plants of *E. cooperi*. Beneath this cluster of living branches there is often a dense, nest-like patch of dead branches that have not yet fallen off. Young specimens begin to develop branches very soon after germination, sometimes when only 30 mm tall, so that they are soon covered with a dense growth of short, spiky, slender branches that partly protects the whole plant from predation (see Fig. 1.74). These branches remain alive until the plant is 1–1.5 m tall, before the lower ones start to fall off and the trunk begins to become bare towards its base. The branches are comparatively slender, often around 15 mm thick and are densely spiny in some trees, becoming almost completely bare of spines in others. This dense spininess often lends the tree a brownish hue when it is seen from a distance. Young tubercles are unusually tall and conical and are tipped by the smallest leaf-rudiments known on any of the southern African tree-forming species.

Flowering in *E. sekukuniensis* takes place between late July and the end of August. Plants appear to become mature after reaching a height of about 1 m but only the larger specimens flower profusely. Here a single cyme of three transversely placed cyathia develops in the axil of each tubercle, usually only on the last segment of each branch. The yellow-green colour of the cyathial glands is similar to the surface of the branches so that flowering does not alter the colour of the tree. Nevertheless, the almost brush-like

Fig. 5.278. *Euphorbia sekukuniensis*, tree among open bush on slopes, *PVB 11768*, SW of Steelpoort, South Africa, 9 Sep. 2010, with Douglas McMurtry (© PVB).

Fig. 5.279. *Euphorbia sekukuniensis,* just before flowering has begin, *PVB 11897*, SE of Chuniespoort towards Burgersfort, South Africa, 22 Jul. 2011 (© PVB).

effect produced by the large numbers of long, slender male florets makes flowering specimens easily spotted. In *E. sekukuniensis* the capsules are deeply 3-lobed and are held well away from the old cyathium on a pedicel that is usually around 6 mm long.

Euphorbia sekukuniensis is closely allied to *E. grandidens* and the two are separated by the unbranched nature of the stem and the denser cluster of branches at the top of the stem in *E. sekukuniensis* as well as the fact that the spine-shields form a continuous horny margin along the angles (even when the spines are absent), while they are much separated in *E. grandidens*. *Euphorbia sekukuniensis* also has especially minute leaf-rudiments and stipular enations.

Florally they differ in the slightly larger, shallower cyathia of *E. sekukuniensis* (especially the central male one) and the considerably more triangular ovary in its early stages (when the styles are receptive).

History

Euphorbia sekukuniensis appears to have been recorded first by W.G. Barnard in May and September 1935 and in August 1938 it was again collected by F.Z. van der Merwe. A specimen gathered by P. Ross Frames in November 1930 may also be of this species (according to Marloth this had an erect main stem like *E. tetragona*, 2 feet high with 5-angled branches at right angles to it).

5.1 Sect. Euphorbia

Fig. 5.280. *Euphorbia sekukuniensis,* first male stage, *PVB 11768,* SW of Steelpoort, South Africa, 1 Aug. 2011 (© PVB).

Fig. 5.281. *Euphorbia sekukuniensis,* capsules, *PVB 11768,* SW of Steelpoort, South Africa, 12 Sep. 2010 (© PVB).

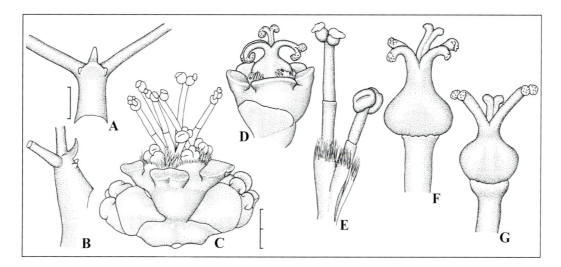

Fig. 5.282. *Euphorbia sekukuniensis.* **A**, young spines and leaf-rudiment from above (scale 1 mm, as for **B**, **E–G**). **B**, young spines and leaf-rudiment from side. **C**, side view of cyme in first male stage (scale 2 mm, as for **D**). **D**, cyathium in female stage from side. **E**, anthers and bracteole. **F**, **G**, female floret. Drawn from: *PVB 11768,* SW of Steelpoort, South Africa (© PVB).

Euphorbia steelpoortensis Bruyns, *Phytotaxa* 436: 216 (2020). Type: South Africa, Steelpoort distr., turnoff to Penge near Burgersfort, flow. 20 Aug. 1973, *Nel 337* (NBG, holo.).

Bisexual succulent 0.1–0.3 (0.4) × 0.15–0.5 m, with outer branches often rhizomatous. *Branches* 50–300 × 5–12 (15) mm, smooth, bluish to purplish green; *tubercles* in decussate pairs fused into 4 occasionally somewhat rounded angles with surface slightly concave to slightly convex between angles, with spine-shields 4–8 (9) mm long, 1–2 × 2–3 mm above spines and 1–4 (5) mm long and often ± equally broad below spines as above, remaining well separated, spines 3–8 (12) mm long; *leaf-rudiments* ± 1 × 1–1.5 mm, with minute brown stipular prickles + 0.5 mm long (rarely another small prickle ± 1 mm long near base of spine-shield). *Synflorescences* many per branch usually towards apex, each a solitary cyme in axil of tubercle, on short peduncle ± 1 mm long, each cyme with 3 transversely disposed cyathia, central male, lateral 2 bisexual and developing slightly later each on peduncle ± 2 mm long and thick, with 2 ovate bracts ± 1 × 2 mm subtending lateral cyathia;.*cyathia* shallowly cupular, glabrous, 3.5–5 mm broad (1.5 mm long below insertion of glands), with 5 pale yellow obovate lobes with deeply incised margins, bright yellow; *glands* 5, transversely rectangular to nearly square

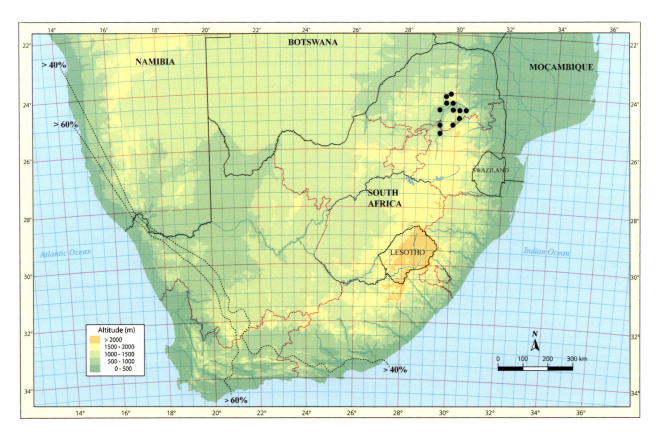

Fig. 5.283. Distribution of *Euphorbia steelpoortensis* (© PVB).

and contiguous, ± 2 mm broad, bright yellow, ascending-spreading, inner margins flat, outer margins entire and slightly ascending, surface between two margins dull; *stamens* glabrous, bracteoles enveloping groups of males, with finely divided tips, glabrous; *ovary* obtusely 3-angled, glabrous, slightly reddish green near top, raised on pedicel 1–1.5 mm long; styles 1–3 mm long, branched to just above base. *Capsule* 3–4 mm diam., obtusely 3-angled, glabrous, slightly raised ± 2 mm inside remains of cyathium and peduncle elongating to 2–5 mm.

Distribution & Habitat

Euphorbia steelpoortensis is mainly found in the valley of the Olifants River. Here it occurs from around Steelpoort and Burgersfort to near Chuniespoort, with an outlying population along the ridge of the Strydpoort Mountains south-west of Haenertzburg.

Mostly, plants are found among rocks and bushes in stony ground in the valleys, but they may also occur on steep slopes among tufts of grass and *Xerophyta*. In the Strydpoort Mountains they grow in shallow pockets of soil in outcrops

5.1 Sect. Euphorbia

Fig. 5.284. *Euphorbia steelpoortensis*, ± 0.3 m diam., in stony ground with bright green *E. enormis*, *Aloe* and a small member of the Commelinaceae, *PVB*, south of Penge, South Africa, 13 Jan. 1996 (© PVB).

Fig. 5.285. *Euphorbia steelpoortensis* looking like *E. aeruginosa*, ± 0.5 m diam., on steep slopes, *PVB 12062*, Jaglist, 82 km SE of Pietersburg towards Burgersfort, South Africa, 5 Nov. 2011 (© PVB).

of dolomite, also with *Xerophyta* and with a wide range of other succulents.

Diagnostic Features & Relationships

Plants in the Strydpoort Mountains are small (rarely more than 50 mm tall) and densely clustered, with little obvious rhizomatous growth and the same is true of many of those occurring on steep slopes in the Olifants River Valley (though here the branches are longer). In flatter areas, usually *E. steelpoortensis* has a rather lax growth, with relatively few branches to 0.2 m long in each clump and many of them spreading away from the centre by underground rhizomes.

In all of them the branches are grey-green, which makes them easy to distinguish from *E. lydenburgensis* and from *E. pisima*. In some populations of *E. steelpoortensis* the branches are relatively rounded, with low tubercles (e.g. *Bruyns 12062* (BOL)) and these may look considerably like *E. aeruginosa*, but they differ from *E. aeruginosa* in the longer pedicel under the female florets. In *E. steelpoortensis* the spine-shields are often more or less elliptical and are sometimes short below the spines. Florally *E. steelpoortensis* differs from both the others by the fact that the cyathia are slightly more raised and the peduncle beneath the cyathium may elongate later to raise the capsule 2–5 mm above the branch.

Fig. 5.286. *Euphorbia steelpoortensis*, ± 1 m diam., among rocks, in very dry conditions but flowering profusely, *PVB*, near Penge, South Africa, 10 Sep. 2010 (© PVB).

Fig. 5.287. *Euphorbia steelpoortensis*, very rhizomatous growth, *PVB*, near Penge, South Africa, 10 Sep. 2010 (© PVB).

Fig. 5.288. *Euphorbia steelpoortensis*, ± 15 cm diam., tightly wedged among rocks, just beginning to flower, *PVB 13740*, near Haenertzburg, South Africa, 30 Jul. 2019 (© PVB).

5.1 Sect. Euphorbia

Fig. 5.289. *Euphorbia steelpoortensis*, female stage, *PVB 13555*, Burgersfort, South Africa, 1 Oct. 2019 (© PVB).

In the Olifants River Valley, several collections have been made where *Euphorbia steelpoortensis* grew with *E. clivicola*. Where they grow together, they flower at different times, with *E. clivicola* flowering strongly in July-August and *E. steelpoortensis* sterile, only flowering later, in September and October. However, in the Strydpoort Mountains south-west of Haenertzburg (where only *E. steelpoortensis*) occurs, it flowers in late August.

History

Plants now placed under *Euphorbia steelpoortensis* have been known for a long time, with the first recorded collection made in July 1912 by Marloth. The area where it occurs has long attracted attention for its wealth of succulents and it was also recorded in June 1936 by W.G. Barnard and in July 1937 by F.Z. van der Merwe. Most of these collections were considered to belong to *Euphorbia schinzii* (e.g. White et al. 1941: fig. 837) but some were also placed tentatively under *E. aeruginosa* (White et al. 1941: 742).

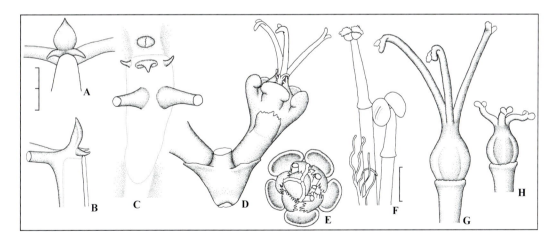

Fig. 5.290. *Euphorbia steelpoortensis*. **A**, young spines and leaf-rudiment from above (scale 2 mm, as for **B–E**). **B**, young spines and leaf-rudiment from side. **C**, spine-complex from above showing shape of spine-shield (dotted line). **D**, side view of part of cyme in female stage. **E**, face view of cyathium in late male stage. **F**, anthers and bracteole (scale 1 mm, as for **G, H**). **G, H**, female floret. Drawn from: **A–G**, *PVB 13555*, Burgersfort, Olifants River Valley, South Africa; **H**, SW of Haenertsburg, *P.Winter 5196* (NBG) (© PVB).

Euphorbia stellata Willd., *Sp. Pl.*, ed. 4, 2: 886 (1799). Lectotype (Bruyns 2012): F. le Vaillant, *Reise in das Innere von Afrika* 4: 245, t. 11 (1797).

Bisexual spiny dwarf glabrous succulent usually not more than 50 mm tall with rosette of 5–30 simple branches from apex of much swollen subterranean spindle-shaped stem from which fibrous roots arise. *Branches* prostrate to erect, 25–150 (200) × 8–15 mm and becoming slender and cylindrical near base, 2- to 5-angled, not or only indistinctly constricted into segments, smooth, green to dark purple-green or reddish green along angles usually mottled with paler green or cream between angles; *tubercles* fused into 2–5 wing-like angles with surface slightly to deeply concave between angles (when 2-angled, branches usually prostrate with upper surface concave and lower convex), conical and truncate, laterally flattened and projecting 2–7 mm from angles, with ± circular to elliptic spine-shield ± 2 mm diam. at tip of tubercle around spines and remaining separate from next below, bearing 2 spreading to slightly deflexed initially red-brown and later grey-brown fine spines 2–12 mm long; *leaf-rudiments* on tips of new tubercles towards apices of branches, 1–2 × 1–1.5 mm, erect, fleeting, narrowly ovate-lanceolate, sessile, with minute brown stipules < 0.3 mm long. *Synflorescences* 1–30 per branch mainly towards apex, each a solitary cyme in axil of tubercle, on short peduncle 1–2 × ± 2 mm, each cyme with 3 transversely disposed cyathia, central male, lateral 2 bisexual and developing later each on short peduncle 1–2 mm long and thick, with 1–2 ovate reddish bracts 1–2 × 1–1.5 mm subtending lateral cyathia; *cyathia* cupular, glabrous, 2.5–4.5 mm broad (2–2.5 mm long below insertion of glands), with 5 obovate red lobes with deeply incised margins, green sometimes suffused with red to red; *glands* 5, transversely elliptic and contiguous, 2–2.5 mm broad, deep green (sometimes with red margins) to red, spreading, inner margins flat to slightly raised, outer margins entire and spreading, surface between two margins dull; *stamens* glabrous, red with red to purple-red anthers, bracteoles enveloping groups of males, with finely divided tips, glabrous; *ovary* triangular in cross-section, glabrous, green to red, slightly exserted on erect pedicel ± 1.5 mm long and soon pushed out beyond cyathium; styles 2–2.5 mm long, green to red, branched to just below middle and often spreading horizontally. *Capsule* 5 mm diam., deeply 3-angled, glabrous, green to red, exserted on decurved and later erect pedicel 4–5 mm long.

Endemic to the Eastern Cape, *E. stellata* is recorded between Port Elizabeth and East London and further inland to near Jansenville and Cradock.

Within subg. *Euphorbia* in southern Africa, *E. stellata* is unusual for its almost geophytic habit. The stem is subterranean and usually swollen into an inflated and elongated, spindle-shaped tuber, narrowing both to its base and apex. Around its apex, which is usually located just beneath the surface of the ground, it bears a rosette of spreading to erect branches that project from the soil. Each branch is joined to the stem by a narrow, cylindrical neck. The branches are not known to rebranch, but are succulent and may survive for several years. Since the above-ground parts (i.e. the branches) are neither deciduous nor even particularly short-lived, the plant is not strictly a geophyte. Nevertheless, if the branches are severely scorched in a fire or are eaten off, new ones are readily produced from the apex of the stem. Generally, in *E. stellata* the spines are borne on fairly prominent tubercles and the spine-

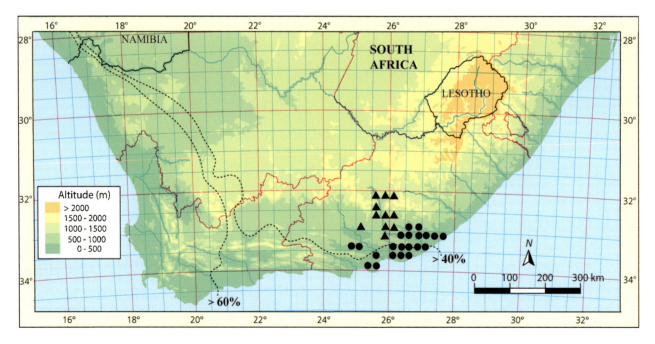

Fig. 5.291. Distribution of *Euphorbia stellata* (● = ssp. *stellata*, ▲ = ssp. *micracantha*) (© PVB).

5.1 Sect. Euphorbia

Fig. 5.292. *Euphorbia stellata* ssp. *stellata*, branches 2-angled and flat on the ground, here 8 to 10 cm long, west of Uitenhage, South Africa, 20 May 2012 (© PVB).

shield is small and restricted to the area immediately around the base of the spines at the apex of the tubercle. A relatively small, pointed leaf-rudiment subtended by two tiny stipular outgrowths is present at the tip of each new tubercle.

Euphorbia stellata shares with several other species (such as *E. grandidens*, *E. griseola*, *E. knuthii* and *E. mlanjeana*) the strongly triangular capsules that become well exserted from the cyathium just after maturing. It is to these species that *E. stellata* is related but its precise relationships within this group are not known. It can nevertheless be said that, among the species which occur close to it, only *E. grandidens* is fairly closely allied and, despite their very different respective habits (the one a large tree; the other a small near-geophyte), the two share as well the slender branches with very variable numbers of angles and comparatively prominent tubercles.

Dyer (1931) presented plenty of evidence of the variability of *E. stellata* and its close allies, *E. squarrosa* and *E. micracantha*, as they occur in the vicinity of Grahamstown. This suggested that it was unlikely that three distinct species were involved and that, for example while *E. micracantha* was easily distinguished from *E. stellata*, it was 'not so certain' that it could be separated from *E. squarrosa*. Nevertheless, White et al. (1941) continued to recognise the three species: *E. micracantha* (plants with mainly 4-angled, erect branches, low tubercles 2–4 mm long and relatively

Fig. 5.293. *Euphorbia stellata* ssp. *stellata*, west of Uitenhage, South Africa, 20 May 2012 (© PVB).

long spines), *E. squarrosa* (plants with mainly 3-angled, often spreading branches, particularly prominent tubercles 4–10 mm long and relatively short spines) and *E. stellata* (plants with mainly 2-angled, spreading branches usually pressed to the ground and with conspicuous mottling on the upper concave surface, relatively low tubercles 2–4 mm long and relatively short spines). However, they illustrated many plants which were intermediate between these three and expressed doubt that three species could be distinguished: 'And in the event that distinct species are involved, their limits can hardly be defined accurately' (p. 730). This arrangement of three species was followed in Bruyns et al. (2006) but, as it is quite often impossible to place a plant under one of these three names with any certainty, a broader view was taken in Bruyns (2012) and also here only a single species is recognised.

Fig. 5.294. *Euphorbia stellata* ssp. *stellata*, branches to ± 6 cm long and strikingly mottled, *PVB 10620*, NW of Uitenhage, South Africa, 4 Dec. 2006 (© PVB).

Following White et al. (1941), two subspecies are recognized here. These may be separated as follows:

1. Branches mainly 4-angled (but up to 6-angled, sometimes with spiralling angles), erect, cyathia red with red glands..subsp. **micracantha**
1. Branches mainly 2- to 3-angled, spreading and often pressed to ground, cyathia often green or suffused with red and with green glands (sometimes with red margins)..subsp. **stellata**

Euphorbia stellata subsp. stellata

Euphorbia procumbens Meerburg, *Pl. Rar.*: t. 55 (1789), *nom. illegit.*, *non* Mill. (1786).
Euphorbia uncinata DC., *Pl. Hist. Succ.* 27: t. 151 (1805). Lectotype (Bruyns 2012): figure by Redouté opposite p. 151.
Euphorbia radiata Thunb., *Prodr. Fl. Cap.* 2: 86 (1800). Type: South Africa, Cape, *Thunberg* (UPS-THUNB, holo.).
Euphorbia squarrosa Haw., *Philos. Mag. Ann. Chem.* 1: 276 (1827). Lectotype (Bruyns 2012): South Africa, Cape of Good Hope (no specimen preserved: painting number 295/423 by G. Bond at K, see Fig. 5.300).

Branches mainly 2- to 3-angled, spreading, often pressed to ground and often concave above. Cyathia green or suffused with red and with green glands (sometimes with red margins).

Distribution & Habitat

Euphorbia stellata subsp. *stellata* occurs in semi-arid parts of the Eastern Cape, where it has been recorded between Port Elizabeth, Grahamstown and East London.

Plants of subsp. *stellata* are usually found in flat areas or on gently sloping, shaly to alluvial hills, where they grow among stones in grassland or in gravel among low shrubs, usually in the company of many other small succulents and small tufts of fine grasses.

5.1 Sect. Euphorbia

Fig. 5.295. *Euphorbia stellata* ssp. *stellata*, some with branches 2-angled and flat on ground but not mottled, *PVB 12254*, NE of Peddie, South Africa, 5 Dec. 2012 (© PVB).

Fig. 5.296. *Euphorbia stellata* ssp. *stellata*, others with erect 3-angled branches, *PVB 12254*, NE of Peddie, South Africa, 5 Dec. 2012 (© PVB).

Diagnostic Features & Relationships

Even within the present concept of subsp. *stellata* there is significant, regional variation in several features of the branches. In the Port-Elizabeth-Uitenhage-Addo area, the branches are 2-angled and pressed to the ground, with the upper surface concave or channelled and the lower surface convex. In these, the upper surface is normally very attractively mottled with purple on dark green and the tubercles are relatively small, with short spines.

Outside the Port Elizabeth-Uitenhage-Addo area, the habit and shape of the branches varies greatly: the branches may be more erect, they are often not so distinctively marked and are much more variable in the number of angles. Mostly they have three to four angles but on some plants 2-angled branches are found among 3-, 4- and even 5-angled ones. In such plants one may find branches with two angles for the first few centimetres, with the number increasing as the branch becomes longer. There is particular variability in the number of angles on the branches in plants between

Fig. 5.297. *Euphorbia stellata* ssp. *stellata*, cyathia with green glands and reddish central area, with small sedge, Coega, Port Elizabeth, South Africa, 5 Dec. 2001 (© PVB).

Grahamstown, Fort Beaufort and Peddie. Here some plants have only prostrate, 2-angled branches, in others the lower parts of the branches are 2-angled and the upper parts 3-angled and in these the branches are usually ascending or erect. Also, the tubercles are particularly prominent, giving the branches a characteristically jagged shape.

Flowering in subsp. *stellata* mostly takes place between October and December, though specimens may flower at other times too. The small cyathia are borne in the axils of the tubercles near the tips of the branches and are mostly green, sometimes suffused with red or with red margins on the glands. When the styles become receptive the ovary is usually still just hidden within the cyathial lobes (though raised on a significant pedicel) but, as the male flowers in the same cyathium begin to mature the pedicel elongates and becomes decurved so that the capsule and the often still receptive styles are pushed right out of the cyathium and held alongside it, thus reducing the likelihood of self-fertilization.

History

The first collection of *Euphorbia stellata* appears to be that of C.P. Thunberg, made near Uitenhage probably around the middle of December 1773 in the company of Francis Masson. He named it *E. radiata* in 1800. The name *E. stellata* was based by Willdenow on a figure of the 'caterpillar euphorbia' that appeared in the German translation of the second book of travels of François le Vaillant (1795, vol. 2: t. XI). According to Le Vaillant's own account *Euphorbia stellata* was found near the Fish River in what is now Namibia. However, his field-notes

Fig. 5.298. *Euphorbia stellata* ssp. *stellata*, branches ascending but cyathia the same colour as those with branches flat on ground, *PVB 7480*, Kransdrift, Fish River valley, South Africa, 24 Nov. 2011 (© PVB).

5.1 Sect. Euphorbia

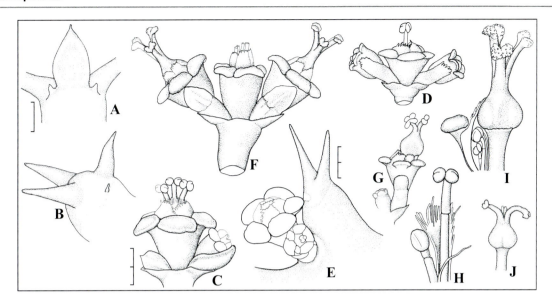

Fig. 5.299. *Euphorbia stellata* ssp. *stellata* (**A, B, C, E, F, H, I**) and ssp. *micracantha* (**D, G, J**). **A**, young spines and leaf-rudiment from above (scale 1 mm, as for **B, H–J**). **B**, young spines and leaf-rudiment from side. **C, D**, side view of cyme in first male stage (scale 2 mm, as for **D, F, G**). **E**, cyathium at beginning of female stage (scale 2 mm). **F, G**, cyme and cyathia in female stage. **H**, anthers and bracteole. **I, J**, female floret. Drawn from: **A, B, C, E, F, H, I**, *PVB 7480*, Kransdrift, Fish River valley, South Africa. **D, G, J**, *PVB 11920*, SE of Cradock, South Africa (© PVB).

may have been scanty and disorganized and his memory may have deceived him into confusing the Fish River of southern Namibia (where *E. stellata* does not occur) with the Great and Little Fish Rivers of the Eastern Cape, where *E. stellata* does occur (Jordaan 1973). It is even possible that Le Vaillant did not collect this species at all, for Dyer (1949: 14) pointed out that Le Vaillant's illustration is a completion of an unfinished sketch of Robert Gordon's among Paterson's collection of paintings (see the similar cases of *E. meloformis*, *E. cucumerina* and *E. avasmontana*). Another early collection of *E. stellata* was made by William Burchell on the 1st of October 1813 between Blue Krantz and the source of the Kasuga River, near Bathurst (*Burchell 3901*, K). He labelled this collection *Euphorbia stapelioides*.

The unusual growth-form and semi-geophytic habit of *E. stellata* has fascinated many botanists and it has been illustrated many times in the literature, starting with a very early figure by Meerburg (1789), that of Le Vaillant (1795) and another by Redouté in De Candolle (1805). One of the few cases where a clearly incorrect name was applied was that of Baker (1869), who provided the description accompanying an illustration of *E. stellata* in Saunders' *Refugium Botanicum* under the name of *E. tetragona* Haw.

Fig. 5.300. *Euphorbia stellata* ssp. *stellata*, South Africa, Cape of Good Hope. Watercolour 295/423 by G. Bond (lectotype of *E. squarrosa*, © RBG Kew).

Euphorbia stellata subsp. **micracantha** (Boiss.) Bruyns, *comb et stat. nov*.

Euphorbia micracantha Boiss., *Cent. Euphorb.*: 25 (1860). Lectotype (Bruyns 2012): South Africa, Cape, between Zuurberg and Klein Bruintjieshoogte, 2000–2500', Oct. *Drège 8206a* (K; MO, S, iso.).

Euphorbia gilbertii A.Berger, *Sukkul. Euphorb.*: 39 (1906). Type: South Africa, Cape, *Cooper* (missing).

Euphorbia lombardensis Nel, *Kakteenk.*: 194 (1933). Type: South Africa, Cape, Mortimer, 1200–1300 m, Dec. 1933, *M. Lombard sub SUG 1564* (NBG, holo.).

Branches 4- to 6-angled (angles sometimes spiralling), usually erect in a dense cluster. Cyathia red with red glands.

Distribution & Habitat

Euphorbia stellata subsp. *micracantha* is also endemic to the Eastern Cape, but occurs further inland than subsp. *stellata*, especially along the valley of the Fish River and its tributaries as far north as the vicinity of Cradock. It is also known towards Jansenville, to the west of the Fish River Valley.

Fig. 5.301. *Euphorbia stellata* ssp. *micracantha*, branches to 6 cm long, west of Cradock, South Africa, 2 May 2008 (© PVB).

Fig. 5.302. *Euphorbia stellata* ssp. *micracantha, PVB 11055*, west of Witmos, south of Cradock, South Africa, 30 Apr. 2008 (© PVB).

Plants of subsp. *micracantha* are also usually found in flat areas or on gently sloping, shaly hills, but are usually at least partly concealed among low shrubs or between stones. The plants are often heavily grazed and, even when they grow in the protection of a small, spiny shrub, the ends of the branches may be chewed off or whole branches pulled out.

Diagnostic Features & Relationships

Euphorbia stellata subsp. *micracantha* always produces short, erect, mainly 4- or 5-angled branches. Some are attractively mottled with dark green along the angles and paler between them, but this is not a constant feature. The spines are also considerably longer and may be quite close together, to provide some protection to the branches. Some of these plants are particularly inconspicuous, with tiny branches 20–40 mm long in a small cluster close to the ground.

Flowering in subsp. *micracantha* also occurs mainly in late spring and here the cyathia are entirely red or purplish red with red pedicels, dark red anthers, red ovaries and capsules.

History

Euphorbia stellata subsp. *micracantha* was first collected by J.F. Drège possibly together with his brother Carl in the first quarter of 1833 along the Fish River Valley, south of Somerset East. When he described *Euphorbia micracantha*, Boissier cited two specimens of Drège's, the other collected between the Fish River and Fort Beaufort. White et al. (1941) associated the collection *Burchell 3901* mentioned above with '*E. micracantha*', but it is more likely to belong under subsp. *stellata*.

Fig. 5.303. *Euphorbia stellata* ssp. *micracantha*, cyathia in first male stage, *PVB 11920*, SE of Cradock, South Africa, 16 Oct. 2012 (© PVB).

Fig. 5.304. *Euphorbia stellata* ssp. *micracantha*, cyathia in female stage, *PVB 11920*, SE of Cradock, South Africa, 15 Oct. 2015 (© PVB).

Fig. 5.305. *Euphorbia stellata* ssp. *micracantha,* capsules, *PVB 11920,* SE of Cradock, South Africa, 10 Nov. 2012 (© PVB).

Euphorbia subsalsa Hiern, Cat. Afr. Pl. 1 (4): 948 (1900). Type: Angola, Moçamedes distr., Pedra de Sal between Moçamedes and Bumbo, Rio Maiombo (Giraul), 3 Oct. 1859, *Welwitsch 642* (BM, holo.; K, LISU, iso.).

Leach (1976a) recognized two subspecies of *E. subsalsa,* subsp. *subsalsa* found in southern Angola and subsp. *fluvialis* from southern Angola and northern Namibia. Both were said to share features of branches < 15 mm thick and the suberect, concave cyathial glands.

Subsp. *subsalsa* is now known to be widely distributed in the coastal parts of southern Angola from Benguela southwards to near Pediva and eastwards to between Oncocua and Iona, close to the Cunene River. This is a somewhat wider distribution than Leach (1976a) gave. Here the plants branch mainly from the base, the often straggling, spreading 4-angled branches are 8–16 mm thick and marked with pale streaks between the angles, the spines are 6–19 mm long, with the stipular prickles 1.5–5 mm long and the spine-shields become very slender below the spines and taper off to leave a gap of 3–5 mm above the next axillary bud. Leach (1976a) considered that erect cyathial glands that are concave above are typical, since they were shown in a sketch made from the type (White et al. 1941: fig. 884). However, this was made from dried material. In living plants the glands are spreading. They vary between slightly concave, flat and slightly convex above and they are yellow to brownish yellow. The female floret is sessile.

Subsp. *fluvialis* is associated with rocky places on both the Angolan and Namibian sides of the valley of the Cunene River, between Ruacana and west of Epupa. It is separated from subsp. *subsalsa* by its more robust habit, with plants branching from the base and freely above the base, 4-angled branches that are erect, uniformly greyish green (unstriped) and 15–30 mm thick, the spines are 7–15 mm long, with the stipular prickles 1.5–6 mm long. The quite robust spine-shields are parallel-sided below the spines with a rounded base and initially have a slight gap (± 1 mm long) above the next axillary bud, but they later may fuse to form a ± continuous margin along the angles. Here the cyathial glands are erect, concave above and are suffused with pink, while the female floret is raised on a short pedicel around 1 mm long. Most of these features are quite closely matched in *E. kaokoensis* and *E. otjipembana.* Consequently subsp. *fluvialis* was changed to be a subspecies of *E. otjipembana,* with which it shares the thicker branches, longer and more robust spine-shields, the erect concave cyathial glands and the raised female floret

A further segregate of *E. subsalsa,* subsp. *otzenii* was described in 2018. This is found south of Kaoko Otavi in the Kaokoveld of Namibia. Here the 4-angled branches are uniformly greyish green and 7–10 mm thick, bearing spines 7–12 mm long and stipular prickles 1.5–3 mm long. The small spine-shields are slender and only 2–4 mm long below the spines, tapering off to leave a gap of at least 5 mm above the next axillary bud. The spreading, bright yellow cyathial glands are slightly concave above and the female floret is ± sessile. Subsp. *otzenii* is kept under *E. subsalsa* and it differs from subsp. *subsalsa* by the uniformly coloured branches and the smaller spine-shields.

Euphorbia subsalsa is closely related to *E. schinzii* from South Africa. Vegetatively they are hard to distinguish, with subsp. *otzenii* particularly similar to *E. schinzii.* Subsp. *subsalsa* may usually be separated from *E. schinzii* by its longer spines and longer stipular prickles but, in subsp. *otzenii*, it is mainly the much shorter spine-shields that distinguish it from *E. schinzii*. Their cyathia are also similar, though in subsp. *otzenii* the glands are slightly concave above while in *E. schinzii* they are always convex above.

Euphorbia subsalsa subsp. **otzenii** Bruyns, *Haseltonia* 25: 52 (2018). Type: Namibia, Kaokoveld, Otjomatemba, 18 Feb. 1993, *Bruyns 5551* (WIND, holo.; BOL, K, iso.).

Spiny succulent 0.05–0.25 × 0.2–0.3 m, much branched at ground level. *Branches* erect to spreading, 20–200 × 7–10 mm, 4-angled, not constricted into segments, smooth, grey-green; *tubercles* fused into 4 low angles with surface often slightly convex between angles, deltoid, laterally flattened and projecting 2–3 mm from angles, with spine-shield 3–6 mm long and 1–2 mm broad around apex and spreading slightly down (for 2–4 mm below spines) remaining well separated from next, bearing 2 spreading to slightly deflexed initially brown (later grey) spines 7–12 mm long; *leaf-rudiments* on tips of new tubercles towards apices of branches, ± 1 × 1 mm, erect, fleeting, ovate-deltoid, sessile, with spreading brown stipular prickles 1.5–3 mm long. *Synflorescences* 1–15 per branch towards apex, each a solitary cyme in axil of tubercle, sessile, each cyme with 3 transversely disposed cyathia, central usually male, lateral 2 bisexual and developing later each on short peduncle 1 mm long and 2 mm thick, with 2 lanceolate bracts 1–1.2 × ± 0.5 mm subtending lateral cyathia; *cyathia* shallowly conical-cupular, glabrous, 3–4.5 mm broad (2 mm long below insertion of glands), with 5 obovate lobes with deeply incised margins, bright greenish yellow; *glands* 5, semi-circular and slightly overlapping in lateral cyathia, ± 1.5 mm broad, greenish yellow to bright yellow, spreading, inner margins slightly raised, outer margins entire and spreading, surface between two margins slightly concave; *stamens* glabrous, bracteoles enveloping groups of males, with finely divided tips, glabrous; *ovary* obtusely 3-angled, glabrous, green, raised on short pedicel < 0.3 mm long; styles 3.5 mm long, branched to 1 mm from base. *Capsule* 3–4 mm diam., obtusely 3-angled, glabrous, sessile.

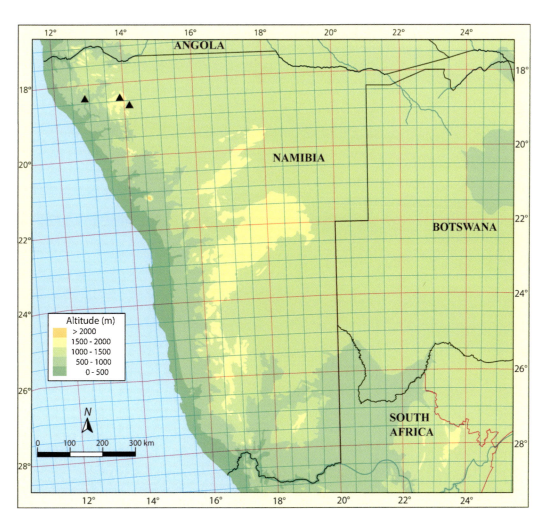

Fig. 5.306. Distribution of *Euphorbia* ssp. *otzenii* in southern Africa (© PVB).

Distribution & Habitat

Euphorbia subsalsa subsp. *otzenii* is endemic to the hilly country of the southern Kaokoveld, to the south of Kaoko Otavi.

Here it occurs in relatively flat areas with trees of *Commiphora*, *Dichrostachys cinerea*, mopane, *Sesamothamnus lugardii* and shrubs of *Catophractes alexandrii* as well as succulents such as *Aloe hereroensis*, *Euphorbia monteiroi* and *E. transvaalensis*, *Ceropegia oculatoides*, *Ceropegia schinzii* and *Ceropegia barklyana*. All of these grow on rocky ground among chunks of white calcrete and small amounts of grey to red soil. *Euphorbia subsalsa* subsp. *otzenii* usually grows wedged among these stones, unprotected by other shrubs.

Diagnostic Features & Relationships

Subsp. *otzenii* forms dense, small clumps, which consist of many, quite short branches arising from a central stem that is indistinguishable from the branches. The branches are more or less square in cross-section, usually around 10 mm thick and have a distinctive grey-green colour. They are armed with stout pairs of initially red then later whitish spines. Each of these is seated on a slightly swollen tubercle, with a small pair of upwardly curved, sharp stipular prickles just above it next to the minute, fleeting leaf.

Subsp. *otzenii* flowers mainly in October and November. When this happens, the upper 10–40 mm of the branches become covered with the bright yellow to greenish yellow

Fig. 5.307. *Euphorbia subsalsa* ssp. *otzenii*, shrublet ± 20 cm tall, among chunks of calcrete and scanty grasses, *PVB 5551*, north of Sesfontein, Namibia, 18 Feb. 1993 (© PVB).

Fig. 5.308. *Euphorbia subsalsa* ssp. *otzenii*, very sparsely flowering, *PVB 5551*, north of Sesfontein, Namibia, 18 Feb. 1993 (© PVB).

cyathia, with their relatively conspicuous, spreading, semicircular glands.

While the branches of subsp. *otzenii* are more slender than in any other spiny species in Namibia, they are closely matched in thickness in subsp. *subsalsa* from north of Namibe in Angola. However, in subsp. *subsalsa* the spines may be longer and stouter and they are more densely arranged around the branches. The stipular prickles in subsp. *otzenii* are also often much smaller than in subsp. *subsalsa*. Generally, in subsp. *otzenii* the spine-shields run down below the spines for less than 4 mm along quite stout and rounded angles, while they are decurrent for about 10 mm along much more slender angles in subsp. *subsalsa*. The cyathia of subsp. *otzenii* differ in their much narrower subtending bracts and they are bright yellow.

Fig. 5.309. *Euphorbia subsalsa* ssp. *otzenii*, bright yellow cyathia, all three in cyme maturing in quick succession, *PVB 5551*, north of Sesfontein, Namibia, 29 Nov. 2011 (© PVB).

Fig. 5.310. *Euphorbia subsalsa* ssp. *otzenii*, cyathia at female stage, *PVB 5551*, north of Sesfontein, Namibia, 1 Nov. 2014 (© PVB).

History

Euphorbia subsalsa subsp. *otzenii* was first collected by Max Otzen late in November 1939 (*Otzen 11*, PRE) at a spot called Osata. This material was shown in Figures 1095, 1096 of White et al. (1941: 981), as an undetermined species allied to *E. schinzii* but sometimes 'more branched and taller' and also 'somewhat related to *E. subsalsa*'. This collection of Otzen was missed by Leach (1976a) in his survey of the Namibian and Angolan members of *Euphorbia* 'sect. *Tetracanthae*' and was first recollected in February 1993.

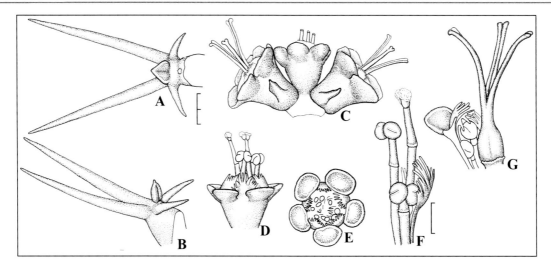

Fig. 5.311. *Euphorbia subsalsa* ssp. *otzenii*. **A**, young spines and leaf-rudiment from above (scale 2 mm, as for **B–E**). **B**, young spines and leaf-rudiment from side. **C**, side view of cyme in female stage. **D**, side view of central male cyathium. **E**, face view of central male cyathium. **F**, anthers and bracteole (scale 1 mm, as for **G**). **G**, female floret in dissected cyathium. Drawn from: *PVB 5551*, north of Sesfontein, Namibia (© PVB).

Euphorbia tetragona Haw., *Philos. Mag. Ann. Chem.* 1: 276 (1827). Lectotype (Bruyns 2012): South Africa, Cape of Good Hope, received 1823, *Bowie* (no specimen preserved: painting number 291/1060 by G. Bond at K).

Bisexual spiny glabrous succulent tree 3–15 m tall, with gradually deciduous branches forming often broad flattened bowl-shaped crown in upper part of cylindrical or faintly 6- to 8-angled erect solitary or sparingly forked trunk-like stem 3–15 m tall, trunk naked except for spines below crown, with many fibrous roots spreading from base. *Branches* spreading then ascending and often rebranching, older branches drying up and falling off stem after a few years, 0.1–1.5 m × 15–30 mm, (3-) 4- to 5-angled, occasionally slightly constricted into segments, smooth, pale to dark green; *tubercles* very evenly fused (in mature plants) into (3) 4–5 prominent angles with surface flat to slightly concave between angles, low-conical, laterally flattened and projecting 3–5 mm from angles, spine-shields separate, 2–3 mm broad along angles and projecting downwards for 2–8 mm and tapering rapidly below spines, bearing 2 spreading and widely diverging brown to grey spines 2–12 mm long (sometimes absent); *leaf-rudiments* on tips of new tubercles towards apices of branches, 1–2 × 2–4 mm, spreading, rapidly deciduous, deltoid and scale-like, sessile, mostly with small brown deltoid stipules ± 0.5 mm long. *Synflorescences* many per branch towards apex, with 1 cyme in axil of each tubercle, very shortly peduncled, each cyme with 3 transversely disposed cyathia, central male, outer two bisexual and developing later each on short peduncle 1–2 × 2–3 mm, with 2 small broadly ovate bracts 2–3 × ± 3 mm subtending lateral cyathia; *cyathia* shallowly conical, glabrous, 5–8 mm broad (1 mm long below insertion of glands), with 5 obovate lobes with toothed margins, yellow-green; *glands* 5, transversely narrowly rectangular to elliptic, contiguous, 2.5–3.5 mm broad, bright green turning yellow with age, slightly spreading, inner margins raised, outer margins entire and slightly spreading, surface between two margins pitted and wrinkled and concave; *stamens* with glabrous pedicels, anthers yellow, bracteoles enveloping groups of males, with finely divided tips, glabrous; *ovary* globose but slightly 3-angled, raised and partly exserted on pedicel ± 1.5 mm long, with slight calyx around base, glabrous, green; styles 2–2.5 mm long, widely spreading, branched to near base. *Capsule* 7–9 mm diam., very obtusely 3-angled, glabrous, red above and pale green below or wholly pale green, exserted on spreading to erect pedicel 2.5–6 mm long.

Distribution & Habitat

Euphorbia tetragona is a denizen of the Eastern Cape and southern Kwazulu-Natal of South Africa. It is especially associated with the valley of the Fish River and its tributaries, from around Grahamstown to just south of Cradock and in many of the smaller valleys leading off the Fish River even into the drier country near Pearston. The westernmost records are from Lootskloof near Jansenville and from the valley of the Sundays River and its tributaries between Pearston, Kirkwood and Jansenville. Further east and north-east, it is known in the valleys of the Keiskamma, Kei and Umzimvubu

5.1 Sect. Euphorbia

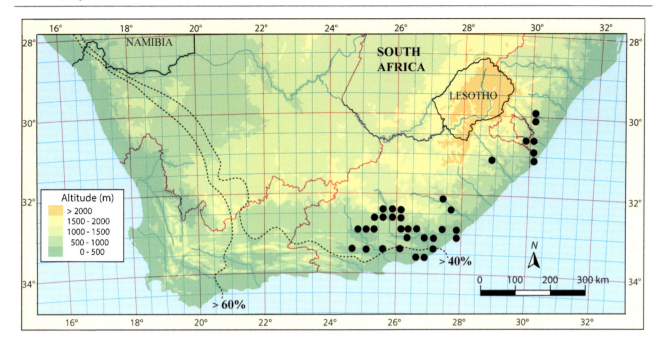

Fig. 5.312. Distribution of *Euphorbia tetragona* (© PVB).

Rivers of the Eastern Cape and there is the usual gap found across the former 'Transkei'. In Kwazulu-Natal it is recorded at least in the valley of the Umtamvuna and the Umkomaas Rivers, in the latter of which it reaches its northern-most stations, not far from the town of Richmond.

Large trees of this species form sparse to dense colonies on stony ground or among outcrops of rock in a variety of habitats, from steep, north- or west-facing slopes (often along the valleys of large rivers) to flat patches on the tops of hills. They are often found in dense, scrub-like and relatively impenetrable bush but, where this has been partly cleared or heavily grazed, these trees now occupy rather exposed situations. Many of the colonies show little regeneration and are badly damaged by foraging baboons, vervet monkeys and goats so that, in some parts of its distribution, it appears to be on the decline.

Diagnostic Features & Relationships

Specimens of *E. tetragona* develop into trees that often reach 10 m in height (though in the drier parts of its distribution they are mostly not more than 5 m tall), with a stout, cylindrical, initially spiny and later naked trunk. When young (in plants less than 0.5 m tall), the stem is 3- to 4-angled but the number of angles rapidly increases to around 8 (soon only visible at the apex of the stem, which is cylindrical lower down). Young plants have an attractive mottling of cream on green on the stem but this youthful colouring is soon lost and is replaced by a greyish corky bark as the trunk becomes cylindrical, which usually occurs before the plants reach 1 m tall. In some trees of *E. tetragona*, the trunk is forked but equally often it remains undivided. Towards its apex, the

Fig. 5.313. *Euphorbia tetragona*, young tree, *PVB 11914*, Trelawney, SW of Cradock, South Africa, 4 Aug. 2011 (© PVB).

Fig. 5.314. *Euphorbia tetragona*, very exposed mature tree with protecting bush grazed away, *PVB 12981*, Lootskloof, east of Jansenville, South Africa, 3 Sep. 2015 (© PVB).

Fig. 5.315. *Euphorbia tetragona*, remnant grove of old trees, near Carlisle Bridge, South Africa, 13 Jan. 2008 (© PVB).

trunk bears a dense, usually broad and bowl-shaped crown of slender, gradually ascending branches. Although occasional branches may be 3- or 5-angled, by far the majority is 4-angled and they are almost square in cross-section. In mature trees, the angles on the branches are very regular, with the tubercles hardly projecting from the surface (though in young plants the tubercles are fairly prominent). Spines, where present, are usually robust and broadly spreading, but they gradually become obsolete in the upper branches, though they may even be absent on quite young plants. The roughly elliptic spine-shields are relatively small and short and mainly surround the base of the spine, projecting a short way downwards towards the next tubercle. On older plants they often become somewhat corky. In *E. tetragona* the leaf-rudiments are very small, triangular, thick and scale-like and they rapidly disappear away from the growing apex of the branch.

Flowering in *E. tetragona* takes place mainly between mid-October and November (possibly also to early December). Trees in full flower have the upper parts of the branches covered with cyathia and they become particularly visible (the tree assumes a yellowish hue from the yellow on the tips of the branches) as they move into the final male stage, when the cyathial glands turn yellow and secrete nectar. They also then emit a slight, sweetish odour and, on a warm day, the whole tree may buzz with huge numbers of visiting flies, wasps and beetles. In the mature cyathia of *E. tetragona*, the glands are closely contiguous and their upper surface is very much broader than long, as well as distinctly concave. Initially bright green, they turn yellow with age. The female florets are borne on a significant pedicel so that the top of the ovary protrudes from among the bracteoles when the floret is mature. After fertilization the pedicel elongates somewhat, so that the capsule

5.1 Sect. Euphorbia

Fig. 5.316. *Euphorbia tetragona*, tree flowering profusely, *PVB 11915*, Witmos, south of Cradock, South Africa, 26 Oct. 2012 (© PVB).

Fig. 5.317. *Euphorbia tetragona*, cyathia just before female stage, *PVB 12169*, Perdepoort, east of Jansenville, South Africa, 25 Oct. 2012 (© PVB).

is later exserted a bit further. In *E. tetragona* the female florets have a slight calyx encasing their bases and the styles are short and particularly deeply divided. When mature, the capsule is very obtusely 3-angled and exserted from the remains of the cyathium for up to 6 mm. Many of them are red above and the trees may assume a reddish hue as the capsules mature towards the tips of their branches in years of good flowering. Seeds are usually released in late November and early December.

Euphorbia tetragona is of similar general appearance to *E. grandidens* and *E. triangularis* and may occur with at least one of them, though it often occurs where neither of the others grow. Analysis of DNA-data (Bruyns et al. 2011) showed that it is closely allied to *E. triangularis*, but is not especially close to *E. grandidens*, which it resembles far more closely vegetatively. *Euphorbia triangularis* is readily separated from *E. tetragona* by the much broader, mostly 3-angled branches which do not rebranch and the much larger leaf-rudiments on young growth. From *E. grandidens*, trees of *E. tetragona* are recognisable as distinct mainly by their quite broadly bowl-shaped crowns in which the branches are almost all square in cross-section and are sparingly rebranched. The leaf-rudiments of *E. tetragona* are also more triangular (more ovate in *E. grandidens*).

Florally the three species can be separated by various features. The cyathia in *E. tetragona* are broader than those of the other two, where they are the most slender in *E. grandidens*. In *E. tetragona* the upper surface of the glands is

Fig. 5.318. *Euphorbia tetragona*, capsules, *PVB 12169*, Perdepoort, east of Jansenville, South Africa, 25 Oct. 2012 (© PVB).

particularly short and broad (i.e. shorter radially than transversely) and is concave, while it is longer and less broad in both the others. It has a distinctive deep green colour when young in *E. tetragona* whereas in *E. grandidens* it is faintly suffused with red. The female florets in *E. tetragona* and *E. grandidens* have short styles divided nearly to their bases and lack the longer fused lower parts of the styles that are typical of *E. triangularis*. They are often more robust with thicker styles in *E. tetragona* than in *E. grandidens*, but this is not always the case. The anthers in *E. tetragona* are yellow while in *E. grandidens* they are yellow suffused with red. In *E. tetragona* the capsule is more obtusely 3-angled than in both *E. triangularis* and *E. grandidens*.

History

Euphorbia tetragona was brought to England by James Bowie in 1823 and the lectotype is a painting by George Bond, made from the material seen by Haworth (see Fig. 5.321). It is likely to have been collected in 1820 or 1821 (Gunn and Codd 1981). Brown (1915: 373) mentioned that the type plant, or part of it, was still flourishing in 1915 at Kew. It seems to have flowered rarely, for he mentioned that he had never seen flowers of it and for some reason he did not press any branches from this venerable individual, as he had done for the cultivated specimens collected by Bowie of *E. caerulescens* and *E. stellata*. It was also gathered by J.F. Drège and his brother Carl in the valley of the Great Fish River in October 1829 (S. records, *Drège 8209*), though material identified as *E. tetragona* by Boissier (1862) is *E. caerulescens* (S records, *Drège 8210*).

Fig. 5.319. *Euphorbia tetragona*, more brightly coloured capsules, *PVB 12169*, Perdepoort, east of Jansenville, South Africa, 25 Oct. 2012 (© PVB).

5.1 Sect. Euphorbia

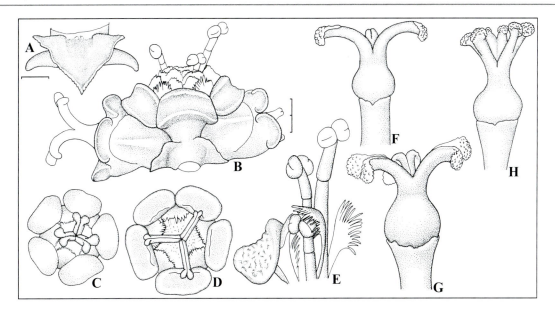

Fig. 5.320. *Euphorbia tetragona*. **A**, young spines and leaf-rudiment from above (scale 1 mm, as for **E–H**). **B**, side view of cyme in early female stage (scale 2 mm, as for **C**, **D**). **C**, **D**, face view of cyathium in female stage. **E**, anthers and bracteoles in dissected cyathium (scale 1 mm, as for **F–H**). **F–H**, female floret. Drawn from: **A–C, E, H**, *PVB 10968*, NE of Carlisle Bridge, South Africa. **D, F, G**, *PVB 12169*, Perdepoort, east of Jansenville, South Africa (© PVB).

Boissier mis-interpreted *E. tetragona*, since he included *E. canariensis* Thunb. in its synonymy, while this is today taken as equal to *E. caerulescens*.

Schimper maintained that some of his collections (e.g. *Schimper 1790*) from Ethiopia represented *E. tetragona* Haw. and some were given the name *E. tetragona* by A. Richard (since *E. tetragona* A. Rich. was described in 1851, it is illegitimate). This mistaken identity even persisted into the 'World Checklist of Euphorbiaceae' of 2000 and some of these specimens were still on the *JSTOR Plant Science* website in 2012 as types of *E. tetragona* Haw., despite the fact that they were quite clearly and correctly determined by M.G. Gilbert in 1988 as the widespread north-east African and Arabian species *Euphorbia polyacantha*. This shrub- to small tree-forming species with 4-angled branches is not closely related to *E. tetragona* Haw., but rather to others from north-east Africa (Bruyns et al. 2011).

Fig. 5.321. *Euphorbia tetragona*, South Africa, Cape of Good Hope, received 1823, *Bowie*. Watercolour 291/1060 by G. Bond (© RBG Kew).

Euphorbia tortirama R.A.Dyer, *Fl. Pl. South Africa* 17: t. 644 (1937). Type: South Africa, Transvaal, Bandolierskop, *Soll & S.W.Smith sub PRE 21371* (PRE, holo.; K, PRE, W, iso.).

Euphorbia groenewaldii R.A.Dyer, *Fl. Pl. South Africa* 18: t. 714 (1938). Type: South Africa, Transvaal, 10 miles NE of Pietersburg towards Mokeetsi, Nov. 1936, *B.H.Groenewald sub Van der Merwe 1186 (sub PRE 23397)* (PRE, holo.; K, PRE, iso.).

Bisexual spiny glabrous succulent 50–200 × 50–500 mm, with many branches from central carrot-shaped mainly subterranean stem up to 300 mm long and 150 mm thick tapering off below into fibrous roots. *Branches* ascending to spreading, 50–300 × 20–50 mm, with age obscurely constricted into ± elliptic segments, smooth, greyish green; *tubercles* fused into (2–) 3 (–4) prominent strongly and tightly spiralling wing-like angles, laterally flattened and projecting 6–20 mm from angles, with narrow spine-shield around apex spreading up along edge of angle to axil of tubercle (or extending slightly above spines sometimes as two narrow lines) and down for 3–10 mm below spines (sometimes continuous along edge of angles), at tips bearing 2 slightly to widely spreading yellow-brown to brown or grey spines 5–22 mm long; *leaf-rudiments* on tips of new tubercles towards apices of branches, 1–2 × 1–2 mm, erect, fleeting, ovate, sessile, with small lanceolate dark brown stipules < 1 mm long. *Synflorescences* 1–10 per branch towards apex, each a solitary cyme in axil of tubercle, on peduncle 2–6 × 3–6 mm, each cyme with 3 vertically to somewhat transversely disposed cyathia, central male, lateral 2 bisexual and developing later each on stout peduncle 3–5 × 3–5 mm, with 2 ovate often reddish bracts 1–3 × 3–4 mm subtending cyathia; *cyathia* cupular, glabrous, 4–8 mm broad (1–2 mm long below insertion of glands), with 5 lobes with deeply incised margins and sometimes prominent midrib, yellowish green sometimes suffused with red; *glands* 5, transversely oblong, very short, 3–4 mm broad, dull or dark yellowish green, erect to spreading, inner margins slightly raised, outer margins entire rounded and spreading to slightly incurved; *stamens* glabrous and sometimes suffused with red (anthers then also red), bracteoles enveloping groups of males, with finely divided tips, glabrous; *ovary* slightly 3-angled, glabrous, raised on very short pedicel ± 0.3 mm long; styles 1.5–4 mm long, branched in upper two thirds. *Capsule* 8–11 mm diam., obtusely and deeply 3-angled, glabrous, dull green with red along edges, sessile.

Distribution & Habitat

Euphorbia tortirama is found in the former Transvaal from around Bandolierskop (the type locality) and Pietersburg westwards to near the border with Botswana, west of Ellisras.

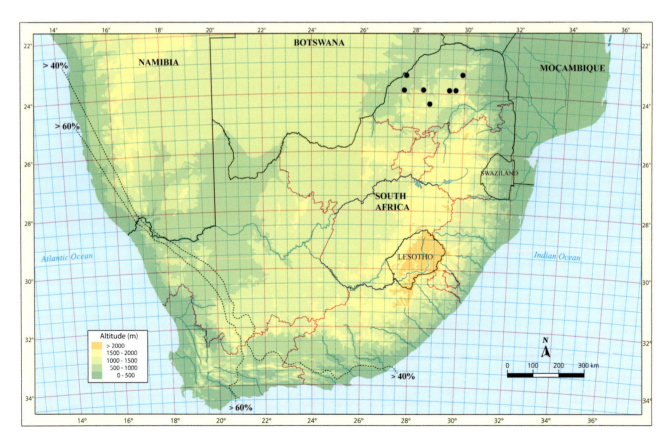

Fig. 5.322. Distribution of *Euphorbia tortirama* (© PVB).

5.1 Sect. Euphorbia

Fig. 5.323. *Euphorbia tortirama*, ± 30 cm diam., in sandy ground between rocks and scattered trees, *PVB 12425*, Riebeek-Wes, South Africa, 5 Jan. 2013 (© PVB).

Fig. 5.324. *Euphorbia tortirama*, two plants at the base of a rock, *PVB 12425*, Riebeek-Wes, South Africa, 5 Jan. 2013 (© PVB).

A record from Moçambique (Carter and Leach 2001: 405) is unlikely to belong to *E. tortirama* and more likely represents *E. clavigera*, which is known to occur there.

Plants occur in stony flat areas among low bushes and trees on substrates derived from sandstones or granites and may become very common. Many localities are heavily threatened by urban development and others have been overwhelmed by bush-encroachment.

Diagnostic Features & Relationships

Euphorbia tortirama forms low, mound-like plants usually not more than 30 cm in diameter. These consist of a thick, central, mostly subterranean stem (occasionally divided in older plants into several similar branches) covered with grey bark. Many short, stout branches that radiate from around the apex of the stem, some of them spreading on the ground and others nearly erect. Each branch is made up of a central core around which the long, laterally flattened tubercles are joined into (mostly) three, deep, strongly spiralling angles, giving the branches a twisted appearance (from where the name is derived), twisting either to the left or right as one looks along the branch towards the tip. The branches persist for several years (after which they die off and are shed) and they are constricted where growth for each year ceased and where the new growth in the next season began. In *E. tortirama* the tubercles are unusually long and the tip of each is ornamented with a pair of stout spines, around which the spine shield is broadest, continuing below in a very slender line and sometimes above also in a slender line to the axillary bud, where it is broader again.

Fig. 5.325. *Euphorbia tortirama*, plant in flower in gravelly ground among rocks and short grasses, ± 10 cm diam., *PVB 11887*, east of Polokwane (Pietersburg), South Africa, 13 Jan. 2011 (© PVB).

Fig. 5.326. *Euphorbia tortirama*, *PVB 11887*, east of Polokwane (Pietersburg), South Africa, 24 Nov. 2018 (© PVB).

In *E. tortirama*, flowering takes place over most of the warmer months, from September to April and in cultivation this sometimes continues until June. The sweetly scented cyathia are broad and short, seated on short, thick peduncles that are almost as broad as the cyathium itself. They have very short, broad and erect glands with a thick and rounded outer margin and some bear noticeably reddish anthers. As in many of the spine-paired species, the bracteoles are very variable in shape, some being roughly hand-shaped and erect, while others are much broader and envelop whole bunches of male flowers quite tightly. The comparatively large, deeply 3-angled capsules develop quickly and may be present over much of the summer around the tips of the branches, with attractively reddish edges on a dull green background.

Euphorbia tortirama is one of several closely related species, namely *E. clavigera*, *E. enormis*, *E. groenewaldii* and *E. vandermerwei*, which share a very similar habit. *Euphorbia clavigera* shares with *E. tortirama* the slender extensions of the hardened spine-shield above the spines for some distance to the axillary bud (in this respect differing from both *E. enormis* and *E. vandermerwei*). *Euphorbia clavigera* differs from *E. tortirama* usually by its considerably more brightly marked branches on which the angles

Fig. 5.327. *Euphorbia tortirama*, second male stage, *PVB 11887*, east of Polokwane (Pietersburg), South Africa, 25 Nov. 2014 (© PVB).

Fig. 5.328. *Euphorbia tortirama*, capsules, *PVB 6648*, south of Ellisras, South Africa, 14 Dec. 2012 (© PVB).

are not twisted along the length of the branch and the tubercles are closely fused into continuous and clear angles. The lack of twisting of the angles and their continuous nature also distinguish both *E. enormis* and *E. vandermerwei* from *E. tortirama*.

Euphorbia tortirama shares with *E. groenewaldii* the manner in which the tubercles are very loosely united (i.e. with the flesh quite deeply sculpted into gaps between successive tubercles) into relatively slender, tall angles which are then fairly tightly twisted into spirals along the length of the branches. The two were said to differ by the fewer branches that were not constricted into segments in *E. groenewaldii*, on account of the branches remaining on the plant for only 1 year and the lack of a continuous, horny margin on the angles from the spine-shields in *E. groenewaldii* (White et al. 1941). However, in plants seen east of Pietersburg (the area where 'typical' *E. groenewaldii* occurs) the branches may persist for at least three seasons and many of them were observed to be constricted at least once or twice. Generally, they were indeed fewer in number per plant and the stem was also much thicker and more deeply sunken into the ground (i.e. forming more of a tuber). The horny margin on the angles formed by the spine-shield is indeed usually not continuous but it has also been found not to be continuous in many plants of *E. tortirama* (as was first pointed out by Fourie 1988). The much smaller stature of the plants, their more geophytic appearance and shorter, more ephemeral branches may be associated with the much harsher environment in which *E. groenewaldii* occurs, with the plants usually tightly wedged into crevices in schist outcrops. Florally there is not much to separate the two. The cyathial glands are usually more spreading in *E. groenewaldii* and the anthers are often red (with more erect outer margins and yellow anthers in *E. tortirama*) but these features are also variable and so *E. groenewaldii* is reduced here to synonymy.

History

The first recorded collection of this species was made in March 1929 by I.B. Pole Evans to the north-west of Ellisras on the banks of the Palala River. Marloth believed these belonged to *E. uhligiana*, an East African species (PRE records). It no longer seems to occur in this area and I was also unable to locate it at another spot near Ellisras where I had seen it 20 years previously. The encroachment of dense bush and thick grasses between some of the trees seems to be at least partly responsible for the disappearance of some of the populations.

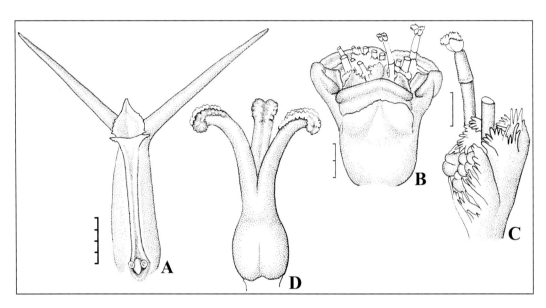

Fig. 5.329. *Euphorbia tortirama*. **A**, young spines and leaf-rudiment from above (scale 4 mm). **B**, side view of lateral cyathium in male stage (scale 2 mm). **C**, anthers and bracteoles (scale 1 mm, as for **D**). **D**, female floret. Drawn from: *PVB 6648*, south of Ellisras, South Africa (© PVB).

Euphorbia triangularis Desf. ex A.Berger, *Sukkul. Euphorb.*: 57 (1906). Lectotype (Dyer 1974a): South Africa, Cape, cultivated plant at Kew Gardens, pressed 30 Oct. 1913 by N.E. Brown (K).

Bisexual spiny glabrous succulent tree 3–18 m tall, with gradually deciduous branches forming rounded crown and arising in upper part of cylindrical or faintly 4- to 5-angled erect solitary or sparingly forked trunk-like stem 3–15 m tall, trunk naked except for spines below crown, with many fibrous roots spreading from base. *Branches* spreading then ascending and rarely rebranching, older branches drying up and falling off stem after a few years, 0.3–1.5 m × 35–60 (100) mm, 3- (to 5-) angled, shallowly constricted into segments 75–300 mm long, smooth, green; *tubercles* very evenly fused into 3 (–5) prominent wing-like angles with surface slightly to deeply concave between angles, low-conical, laterally flattened and projecting 3–5 mm from angles, spine-shields forming continuous hard grey to brown margin 2–3 mm broad (often expanding around axillary bud) along angles or projecting downwards 3–5 mm below spines and not continuous, bearing 2 spreading and widely diverging brown to grey spines 3–8 mm long; *leaf-rudiments* on tips of new tubercles towards apices of branches, 3–7 × 4–6 mm, spreading, deciduous, broadly ovate to orbicular, sessile, mostly with small irregular brown ridge-like stipules up to 1 mm long. *Synflorescences* many per branch towards apex, with (2) 3 (4) cymes in axil of each tubercle, shortly peduncled, each cyme with 3 usually vertically disposed cyathia, central male, outer two bisexual and developing later each on short peduncle 1.5–3 × 2 mm, with 2 small broadly ovate bracts 1–2 × 2 mm subtending lateral cyathia; *cyathia* conical-cupular, glabrous, 4–6 mm broad (2 mm long below insertion of glands), with 5 obovate lobes with toothed margins, yellow to yellow-green; *glands* 4–5, transversely elliptic, contiguous to slightly separated, 1.5–2 mm broad, yellow to green, slightly spreading, inner margins not raised, outer margins entire and slightly spreading, surface between two margins convex and often pitted and wrinkled; *stamens* with glabrous pedicels, anthers yellow, bracteoles enveloping groups of males, with finely divided tips, glabrous; *ovary* globose, glabrous, green, raised and partly exserted on erect pedicel 1.5–2.5 mm long; styles 2–4.5 mm long, widely spreading, divided to near middle. *Capsule* 6–8 mm diam., obtusely triangular, glabrous, mainly red above and becoming green near base, exserted from cyathium on decurved and later erect pedicel 1.5–9 mm long.

Distribution & Habitat

Euphorbia triangularis is widely distributed in southern Africa along the eastern side of the subcontinent, reaching its northernmost stations between Barberton, Nelspruit and Malelane in the former eastern Transvaal. From here it occurs somewhat sporadically in Swaziland and into

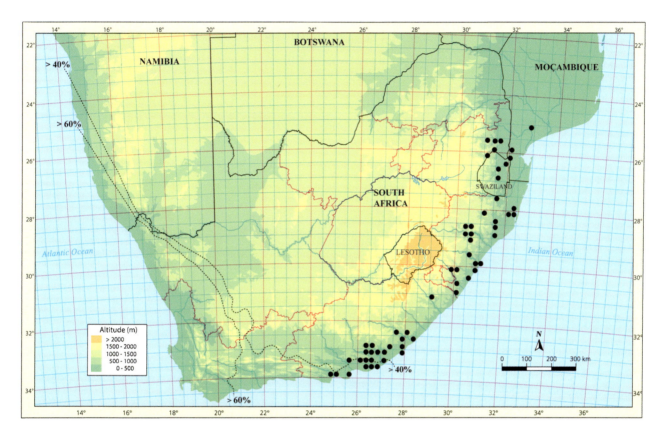

Fig. 5.330. Distribution of *Euphorbia triangularis* (© PVB).

Moçambique (near Namaacha and on Inhaca Island according to Carter and Leach 2001), as well as sporadically southwards through Natal and the former Transkei into the Eastern Cape. In the Eastern Cape it is plentiful from the lower reaches of the Kei River and around East London westwards past Port Elizabeth to the valley of the Gamtoos River around Patensie.

Euphorbia triangularis is mostly found on rocky, well-wooded slopes and is characteristic of steep places above or around cliffs of sandstone or shale along the valleys of rivers. In the northern areas, it has occasionally been recorded in coastal forest on sand. Plants usually occur in large and dense but often quite scattered colonies, where they dominate the vegetation, growing in a scanty soil that is often quite rich in humus and leaf-litter. *Euphorbia triangularis* occurs near or together with *E. grandidens* over a large proportion of their mutual distributions and, in the Eastern Cape, it also sometimes grows together with *E. tetragona*. However, it appears to be less drought-resistant than either of these species.

Fig. 5.331. *Euphorbia triangularis,* grove of trees in thick bush, *PVB 11874,* 10 km south of Matsamo, Swaziland, 8 Jan. 2011 (© PVB).

Fig. 5.332. *Euphorbia triangularis,* lone much-branched tree next to road, *PVB 12157,* between Kirkwood and Enon, South Africa, 21 May 2012 (© PVB).

Fig. 5.333. *Euphorbia triangularis,* tree ± 8 m tall projecting from bush, *PVB 12157,* above Umkomaas River, South Africa, 8 Jan. 2009 (© PVB).

Fig. 5.334. *Euphorbia triangularis*, tip of flowering branch in first male stage, *PVB 9407*, near Bathurst, South Africa, 24 Apr. 2016 (© PVB).

Fig. 5.335. *Euphorbia triangularis*, female stage but ovaries not exserted, *PVB 12157*, between Kirkwood and Enon, South Africa, 21 May 2012 (© PVB).

Diagnostic Features & Relationships

Euphorbia triangularis forms impressive and picturesque trees with a comparatively slender, occasionally forked trunk, from which the branches spread outwards and then ascend slightly to form the characteristic chandelier-like shape associated with the species. Although the trunk starts off 4-angled, it often becomes 5-angled, but this is only clear in larger specimens near its apex, as lower down it swells up between the angles to become almost cylindrical. In very small plants the trunk is green, without any cream mottling. In older plants only the apex of the trunk retains the green, photosynthetic surface and the rest becomes covered with a grey, corky bark. Most of the branches have three prominent, wing-like angles but on some trees 4-angled and even 5-angled branches are found. The branches, which very rarely rebranch, are regularly constricted into quite long segments and remain green for several seasons, after which they dry out and fall off. Spines are borne along the angles on tubercles that are initially quite well raised out of the angle but gradually merge with it to almost disappear. With time, too, the spines, which are never large or very hard, wear off and older branches are often spineless. The stem also generally becomes spineless with age. In *E. triangularis* the tubercles bear fairly conspicuous, broad, spreading and almost circular leaf-rudiments that can usually be seen on new growth from alongside small and medium-sized trees. At the base of the leaf, the stipules form a deltoid, scar-like swelling that is discrete from the leaf-rudiment. The spine-shields are fairly conspicuous and may be continuous along the angles or discrete (sometimes varying between these two conditions on one branch).

Euphorbia triangularis usually flowers between April and July (occasionally also continuing into early August). When flowering takes place, vast numbers of the bright yel-

Fig. 5.336. *Euphorbia triangularis,* second male stage with many ovaries exserted, *PVB*, between Kaapmuiden and Barberton, South Africa, 1 Aug. 2019 (© PVB).

Fig. 5.337. *Euphorbia triangularis,* capsules (some already partly dried out and about to dehisce), *PVB 9407*, south of Hankey, South Africa, 30 Oct. 2012 (© PVB).

low or yellow-green cyathia are produced near the tips of the younger branches, lending a quite different colour to the tree. The comparatively small cyathia are borne in groups of three, which are vertically disposed on each peduncle. Within them, the female floret is on a fairly long pedicel so that the ovary protrudes slightly from among the cyathial lobes. The female floret is also notable for the long, fused portion at the base of the styles. After flowering is over, the rounded, very slightly 3-angled and medium-sized, mostly red capsules develop in large numbers. They may be erect or decurved on their pedicels but are always at least slightly exserted from the cyathium, often on long pedicels. They seem to be produced on few trees in each population and are then usually visible as a reddish tinge towards the tips of the branches. They remain on the tree until late October–early November before the seeds are released.

Among the more slender-branched, tree-forming, spine-shield-bearing species of *Euphorbia* in the south-eastern parts of South Africa, *E. triangularis* is readily recognised, even from a distance, by the fact that its branches are much thicker than those of either *E. grandidens* or *E. tetragona* and rarely rebranch. Furthermore, the leaf-rudiments are different in both shape and size from those in the other two. While the stipules are discrete from the base of the leaf-rudiment in *E. triangularis*, they are partly fused to the base of the leaf-rudiment in the other two. The main floral difference between the three lies in the longer basal fused portion of the styles in *E. triangularis*. The leaf-rudiments are similar to those of the very local *E. keithii* and the differences between them are discussed under that species. *Euphorbia triangularis* is most closely related to *E. keithii* and *E. tetragona*. It is much less closely related to *E. grandidens*, with which it often grows. These two species may flower at the same time, but no hybrids between them have been recorded.

Euphorbia confinalis and *E. triangularis* look similar, but they are not closely related (Bruyns et al. 2011). In *E. confinalis* the young stems are distinctly but finely mottled with

5.1 Sect. Euphorbia

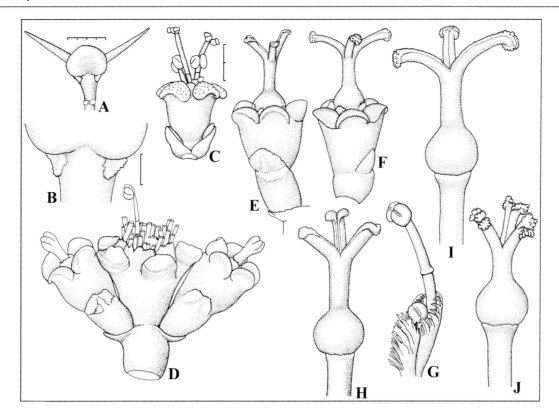

Fig. 5.338. *Euphorbia triangularis*. **A**, young spines and leaf-rudiment from above (scale 4 mm). **B**, detail of base of leaf-rudiment and stipular structures (scale 1 mm, as for **G–J**). **C**, side view of cyme in early first male stage (scale 2 mm, as for **D–F**). **D**, side view of cyme in late first male stage. **E, F**, cyathium in female stage from side. **G**, anthers and bracteole. **H–J**, female floret. Drawn from: **A, B, D–F, H, I**, *PVB 12157*, between Kirkwood and Enon, South Africa. **C, G, J**, *PVB 11909*, Royal Sheba Mine, Barberton, South Africa (© PVB).

cream on green and this mottling is generally lacking in *E. triangularis*. In trees of *E. triangularis* the branches are more closely-spaced along the trunk while they are more grouped into whorls which are more widely separated on the trunk in *E. confinalis*. The branches in *E. confinalis* are also stouter and have a more conspicuous, often continuous and somewhat wavy, horny margin provided by the spine-shields as well as smaller leaf-rudiments that are narrower and with more pointed tips. When flowering profusely, *E. confinalis* produces far more and more densely clustered cyathia than *E. triangularis*. At 6–8 mm in diameter, the cyathia are larger in *E. confinalis* than in *E. triangularis*, where they are more delicate and rarely reach 6 mm broad. On the more subtle level, the ovary and the capsules in *E. confinalis* are far more prominently 3-angled (with somewhat flattened sides), the fused part of the style is much shorter and the capsules become larger before dehiscence (while the ovary of *E. triangularis* is more spherical in shape and the capsules are also more rounded). Furthermore, the red colouring in the anthers in *E. confinalis* is lacking in *E. triangularis*. The cyathial glands have slightly upturned outer margins and are smooth above in *E. confinalis*.

History

The name *Euphorbia triangularis* first appeared in a catalogue of plants growing in the garden at the Natural History Museum in Paris, listed as a 'shrub' in the 'Caldarium' and possibly from the Cape (Desfontaines 1829, which is usually known as *Catalogus Plantarum Horti Regii Parisiensis*). It was unknown who had collected it and brought it to Paris. Desfontaines was said to have provided a description or diagnosis (e.g. White et al. 1941: 891), but actually none was present and consequently Desfontaines did not validly publish this name. The first description was provided by Alwin Berger in 1906.

One or more plants of *Euphorbia triangularis* were reputedly brought over from Paris to Kew in London in the early 1800's. Berger (1906) reported seeing several trees 3–5 m tall at Kew and they were still growing there in October 1913, when N.E. Brown made several sheets of herbarium specimens from them. On one of these sheets, Brown commented that he had observed a plant in the garden there for over 40 years, ever since he had started working there in February 1873. Later, however, they all appear to have died and disappeared without trace (Dyer 1974a, J.H. Ross in PRE archives).

Euphorbia vandermerwei R.A.Dyer, *Fl. Pl. South Africa* 17: t. 660 (1937). Type: South Africa, Transvaal, White River, Sept. 1936, *Van der Merwe sub PRE 22436* (PRE, holo.; K, P, SRGH, iso.).

Bisexual spiny glabrous succulent 50–200 × 0.1–1 m, with few to many branches from central swollen subterranean stem up to 150 × 75 mm from which fibrous roots arise, apically sometimes divided into 2–10 stem-like branches bearing aerial branches. *Branches* spreading and ascending towards tips, 50–300 × 15–30 mm, obscurely constricted into ± elliptic segments, smooth, dark green; *tubercles* fused into (3) 4–5 prominent sometimes slightly spiralling wing-like angles, laterally flattened and projecting 3–6 mm from angles, with narrow spine-shield around apex spreading up along edge of angle to axil of tubercle and down for 2–8 mm below spines (sometimes continuous along edge of angles), bearing 2 spreading red-brown to grey spines 3–10 mm long; *leaf-rudiments* on tips of new tubercles towards apices of branches, 1–2 × 1–2 mm, erect, fleeting, ovate-lanceolate and slightly concave above, sessile, with small irregularly lanceolate dark brown stipules < 1 mm long. *Synflorescences* 1–10 per branch towards apex, each a solitary bisexual cyathium (rarely with additional rudimentary lateral and vertically disposed cyathia) in axil of tubercle, on reddish green peduncle 1.5–4 × 3–5 mm, with 2 ovate dark red marginally ciliate bracts 3–4 × 2–4 mm clasping cyathium; *cyathia* initially cupular later broadly conical, glabrous, 4–8 mm broad (1–2 mm long below insertion of glands), with 5 dark red lobes with deeply incised margins, dark pink to red; *glands* 5–6, transversely oblong, very short, 2–3 mm broad, yellow to green above and red on outside, erect and later spreading, inner margins not raised, outer margins entire and rounded, surface between two margins shiny with secretion; *stamens* glabrous, bracteoles enveloping groups of males, with finely divided red tips, glabrous; *ovary* slightly 3-angled and red above, glabrous, raised on very short pedicel ± 0.3 mm long; styles 2–4 mm long, branched above middle, yellow. *Capsule* 5–8 mm diam., deeply 3-angled, glabrous and shiny, red above and green towards base or wholly dark red, sessile.

Distribution & Habitat

Euphorbia vandermerwei is of very restricted distribution in the hilly country at the foot of the escarpment between Nelspruit, Hazyview and Kaapmuiden. Here it grows on gently sloping, solid granitic outcrops, usually where these are in valleys around the bases of larger hills.

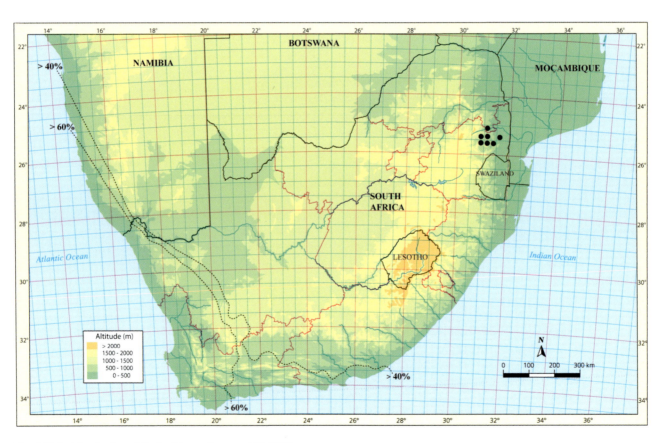

Fig. 5.339. Distribution of *Euphorbia vandermerwei* (© PVB).

Euphorbia vandermerwei usually grows in shallow mats of humous-rich soil on flat expanses of rock or in deeper crevices in the company of a scanty cover of short grasses, *Aloe petricola*, *Myrothamnus*, *Plectranthus*, *Selaginella dregei* and *Xerophyta*. It is not generally found in patches of denser grass where the frequent fires would destroy the plants.

Fig. 5.340. *Euphorbia vandermerwei*, growing on granitic slabs with tufts of *Coloechloa*, *PVB 11772*, south of Nelspruit, South Africa, 10 Sep. 2010 (© PVB).

Fig. 5.341. *Euphorbia vandermerwei*, nearly 0.5 m diam., branches mainly 4-angled, *PVB 11772*, south of Nelspruit, South Africa, 10 Sep. 2010 (© PVB).

Diagnostic Features & Relationships

Each plant of *Euphorbia vandermerwei* possesses a thickened stem, which rapidly tapers beneath the surface into spreading roots. Its apex is just above the surface of the ground and bears a fairly dense rosette of branches which hides it from view entirely. In older plants, the stem is itself sometimes forked near the apex and each fork bears a rosette of branches, so that the entire plant may be anything up to 1 m in diameter. The somewhat untidily arranged, often slightly entangled branches initially remain close to the ground, often for considerable distances and become erect towards their tips. They are usually dark green in colour and of very variable thickness. Towards their bases they are mostly 4-angled (but may be 3-angled) and of quite uniform thickness, with the angles straight or sometimes loosely spirally twisted. Once flowering begins, they are of more irregular thickness and the angles may also become more disorganised. The branches persist for several years, with constrictions of variable thickness and length distinguishing each season's growth. In many plants the spines are variable in length (3 mm long on some parts of a branch and up to 10 mm long elsewhere on the same branch) and they are usually accompanied by small spine-shields that only rarely spread continuously along the angles.

Flowering in *E. vandermerwei* takes place between August and December as new growth begins (though it may continue throughout the summer until April in some years), with the cyathia developing quickly in the axils of freshly developed tubercles just behind the tip of each branch. Each cyme consists of a solitary cyathium enclosed in a pair of red bracts, which form a sphere that opens up as the cyathium swells and the female floret matures. At this stage the cyathium is quite slender and the glands are held in an ascending position. After fertilization the capsule begins to grow, the cyathium becomes much broader (and relatively shallower), the glands become horizontal (and usually also become covered with nectar) and the

Fig. 5.342. *Euphorbia vandermerwei*, granitic slabs with *Selaginella* in very dry conditions, *PVB 11905*, NE of Nelspruit towards Numbi Gate, South Africa, 25 Jul. 2011 (© PVB).

Fig. 5.343. *Euphorbia vandermerwei*, branches mainly 5-angled, *PVB 11905*, NE of Nelspruit towards Numbi Gate, South Africa, 13 Dec. 2002 (© PVB).

male florets begin to mature in large numbers. A dark red colour pervades the bracts, the outside of the cyathium, the cyathial lobes, the tips of the bracteoles in the cyathium and the outside of the cyathial glands. The inner surface of the glands is yellow or green but often also suffused with red, while the anthers and the styles are yellow. Red is again found on the top of the ovary, which is red from its early stages onwards and this continues into the mature capsule, which is shiny and bright red, often darker red along the edges. The pedicel on which the ovary is borne is very short and it does not elongate as the capsule matures, so that the capsule ripens more or less inside the cyathium, which shrivels up and persists only around its base.

Euphorbia vandermerwei is closely related to *E. clavigera, E. enormis* and *E. tortirama*. Among these *E. vandermerwei* is unusual for the development of only a single cyathium per peduncle (though this is found also in the southernmost known populations of *E. clavigera*) and for the generally dark red colour of the cyathia. Other significant differences between them are discussed under *E. enormis* and *E. tortirama*.

Fig. 5.344. *Euphorbia vandermerwei,* dark cyathia mostly solitary per cyme, *PVB 11772,* south of Nelspruit, South Africa, 10 Sep. 2010 (© PVB).

Fig. 5.345. *Euphorbia vandermerwei,* cyathia, *PVB 11905*, NE of Nelspruit towards Numbi Gate, South Africa, 22 Jan. 2014 (© PVB).

Fig. 5.346. *Euphorbia vandermerwei*, cyathia with some capsules developing, *PVB 11905*, NE of Nelspruit towards Numbi Gate, South Africa, 13 Dec. 2002 (© PVB).

History

Euphorbia vandermerwei was first collected by I.B. Pole Evans in August 1917 and again by L.C.C. Liebenberg in March 1931, both near Nelspruit. The type was collected by F.Z. van der Merwe in September 1936 near White River, not far from Nelspruit either.

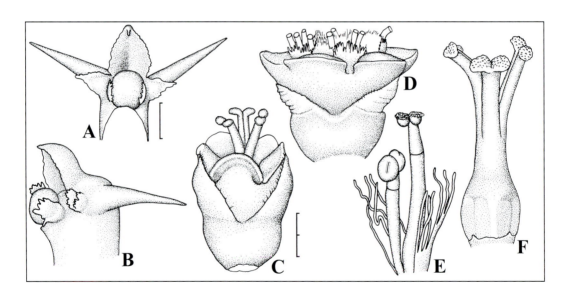

Fig. 5.347. *Euphorbia vandermerwei*. **A**, young spines and leaf-rudiment from above (scale 1 mm, as for **B**, **E**, **F**). **B**, young spines and leaf-rudiment from side. **C**, side view of cyathium in early female stage (scale 2 mm, as for **D**). **D**, side view of cyme in late male stage. **E**, anthers and bracteoles. **F**, female floret. Drawn from: *PVB 11772*, south of Nelspruit, South Africa (© PVB).

Euphorbia venteri L.C.Leach ex R.Archer & S.Carter, *Fl. Pl. Africa* 57: 86, t. 2176 (2001). Type: Botswana, near Tsessebe, c. 45 km north of Francistown, 12 Dec. 1991, *Venter & al. 174* (PRE, holo.; K, UNIN, iso.).

Bisexual spiny glabrous succulent 0.02–0.2 (0.3) × 0.15–0.5 m, with many branches from similar stem, stem and some branches often somewhat swollen underground with fibrous roots arising from them, densely branched at and slightly below ground level and frequently rebranching to form small clumps, often with some rhizomatous branches spreading underground from it and giving rise to further clumps. *Branches* erect to spreading, 20–200 × 4–8 mm, subcylindrical to obscurely 4-angled, not constricted into segments, smooth, pale grey-green to red, often slightly darker along angles and paler between them; *tubercles* in decussate pairs arranged into 4 obscure rows with surface flat or convex (indistinctly concave) between rows, conical and truncate, projecting 2–4 mm from angles, with spine-shields 2.5–6 mm long, ± 1 × 1–3 mm above spines and 2–4 mm long and very slender below spines and remaining well separated from next, bearing 2 spreading to slightly deflexed initially reddish brown (later dark brown or grey-brown) spines 2–8 (10) mm long; *leaf-rudiments* on tips of new tubercles towards apices of branches, 1–1.5 × 1–1.5 mm, erect, fleeting, ovate, sessile, with small brown stipular prickles 0.2–0.5 mm long. *Synflorescences* many per branch usually towards apex, each a solitary cyme in axil of tubercle, sessile, each cyme with 3 transversely disposed cyathia, central male, lateral 2 bisexual and developing later each on short peduncle 0.5–1 × 0.5–1 mm, with 2 linear bracts 1–2 × 1–1.5 mm subtending lateral cyathia; *cyathia* shallowly cupular, glabrous, 3–5.5 mm broad (2 mm long below insertion of glands), with 5 white obovate lobes with deeply incised margins, dull yellow-green; *glands* 5, transversely rectangular and contiguous, 2–2.5 mm broad, dark yellowish green to brownish green sometimes suffused with red towards margins, spreading, inner margins red and slightly raised to flat, outer margins entire and spreading, surface between two margins dull; *stamens* glabrous with reddish filamens and red anthers, bracteoles enveloping groups of males, with finely divided tips, glabrous; *ovary* 3-angled, glabrous, suffused with red, raised on short pedicel < 0.5 mm long; styles 2.5–3 mm long, branched to below middle. *Capsule* 3–4 mm diam., obtusely 3-angled, glabrous, sessile.

Distribution & Habitat

Only known in Botswana in the North-East District, *E. venteri* is common in the relatively flat, but high country

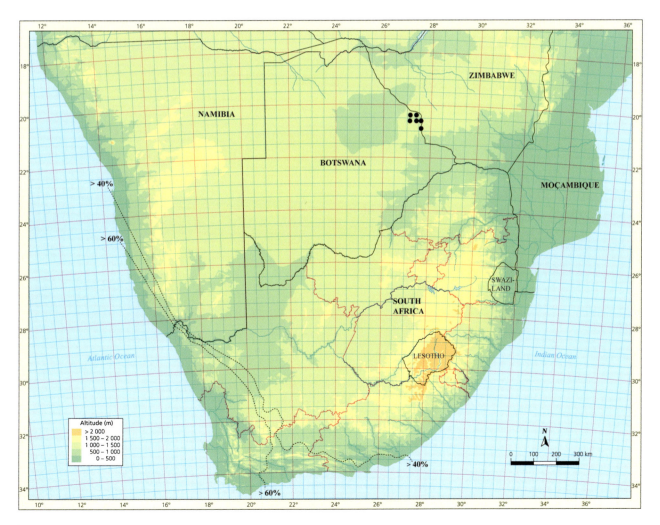

Fig. 5.348. Distribution of *Euphorbia venteri* (© PVB).

(1100–1300 m) between the villages of Tsessebe, Zwenhambe and Tutuma, to the north-east of Francistown. It is possible that it occurs in the neighbouring parts of Zimbabwe as well.

Euphorbia venteri generally occurs in pale ground of granitic origin, especially in eroded, gravelly and stony, slightly bare patches near to streams and dry river-beds, sometimes among low outcrops of rock. These areas are inhabited by scattered trees such as *Acacia tortilis*, *mopane* and several species of *Combretum*. There are often many other succulents together with it. This includes various stapeliads such as *Ceropegia* (*Duvalia*) *polita*, *Ceropegia* (*Huernia*) *verekeri*, *Ceropegia* (*Orbea*) *lutea*, *Ceropegia* (*Orbea*) *rogersii*, members of *Kalanchoe* and several species of *Aloe*, such as *A. esculenta*, *A. globuligemma* and *A. littoralis*.

Diagnostic Features & Relationships

The shape and size of the plant in *E. venteri* depends on the habitat. Some plants in exposed, shallow soils overlaying

Fig. 5.349. *Euphorbia venteri*, branches to ± 6 cm tall and very rhizomatous, in open between *mopane* trees growing in granitic gravel, *PVB 12353*, Tsesebe, near Francistown, Botswana, 27 Dec. 2012 (© PVB).

Fig. 5.350. *Euphorbia venteri*, denser-branched plant with branches to ± 15 cm tall, among *mopane* trees in fine pale granitic gravel, *PVB 12359*, west of Zwenhambe, near Francistown, Botswana, 28 Dec. 2012 (© PVB).

Fig. 5.351. *Euphorbia venteri*, densely branched plant almost 1 m diam. with branches to 20 cm tall, *PVB 12359*, west of Zwenhambe, near Francistown, Botswana, 28 Dec. 2012 (© PVB).

rocks have short branches (mostly only 20–60 mm long), that only slightly project from the ground and are connected beneath the ground over an area of up to 0.5 m by rhizomes. These branches are armed with short spines up to 3–5 mm long. In areas where the soil is deeper the plants are more compact, considerably more hemispherically clump-shaped, less rhizomatous (or not rhizomatous at all) and the branches are regularly between 0.1 and 0.3 m long. In these plants the spines may reach 10 mm long but are regularly at least 6 mm long. Generally, the branches have a distinctive, pale, grey-green colour with paler cream patches in the grooves between the angles and a darker grey-green on the angles and the tubercles. The relatively low to quite prominent tubercles are arranged into four rows but are not clearly fused into angles along the branches. They are most prominent at the position where the spines are attached and taper considerably towards the next pair below them, so that from the side the branch may appear to be made up of a series of wedge-shaped segments.

Cyathia appear in large numbers in spring from September to November, mainly towards the tips of the branches. Their glands are generally dark yellowish green to brownish green and may be suffused with red towards their margins and the anthers are red.

Euphorbia venteri differs from *E. schinzii* by lacking clearly 4-angled branches which are paler in colour and by the yellowish green or brownish green glands with red around their edges and the red colour of the anthers.

Fig. 5.352. *Euphorbia venteri*, grey branches with cyathia in first male stage, *PVB 12353*, Tsesebe, near Francistown, Botswana, 18 Oct. 2014 (© PVB).

Fig. 5.353. *Euphorbia venteri,* cyathia in second male stage, *PVB 12355,* west of Tsesebe, near Francistown, Botswana, 27 Sep. 2014 (© PVB).

Fig. 5.354. *Euphorbia venteri,* brownish glands, cyathia in second male stage, *PVB 12354,* west of Tsesebe, near Francistown, Botswana, 27 Sep. 2014 (© PVB).

Fig. 5.355. *Euphorbia venteri,* cyathia in female stage, *PVB 12363,* Tutuma, near Francistown, Botswana, 27 Sep. 2014 (© PVB).

History

Euphorbia venteri was first recorded in December 1991 and material from this collection was supplied to L.C. Leach, who recognised it as representing a new species but did not publish it. It was, until recently, known only from the type locality, where it was said to be rare and difficult to find (Archer and Carter 2001).

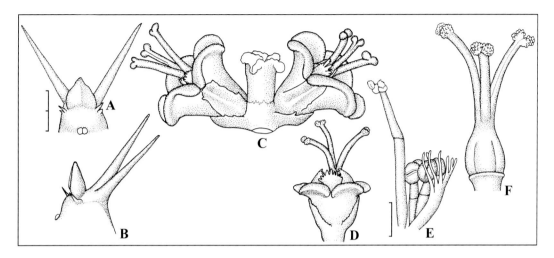

Fig. 5.356. *Euphorbia venteri*. **A**, young spines and leaf-rudiment from above (scale 2 mm, as for **B**–**D**). **B**, young spines and leaf-rudiment from side. **C**, side view of cyme in female stage. **D**, side view of female cyathium. **E**, anthers and bracteole (scale 1 mm, as for **F**). **F**, female floret. Drawn from: **A**, **B**, **D**, **F**, *PVB 12353*, Tsesebe, near Francistown, Botswana. **C**, *PVB 12354*, west of Tsesebe, near Francistown, Botswana. **E**, *PVB 12355*, west of Tsesebe, near Francistown, Botswana (© PVB).

Euphorbia virosa Willd., *Sp. Pl.*, ed. 4, 2: 882 (1799). Lectotype (Bruyns 2012): Paterson, *Reisen in das Land der Hottentotten und der Kaffern, während der Jahre 1777, 1778 und 1779*: 60, t. 9, 10 (1790).

Euphorbia bellica Hiern, *Cat. Welw. Afr. Pl.* 1: 945 (1900). Type: Angola, Moçamedes distr., frequent in sandy coastal hills from Giraul up to Cape Negro, Jul. 1859, *Welwitsch 643* (BM, holo.; LISU, iso.).

Euphorbia dinteri A.Berger, *Monatsschr. Kakteenk.* 16: 109 (1906). Type: Namibia, Khan River, received 1904, *Dinter* (NY, holo.).

Euphorbia virosa f. *caespitosa* H.Jacobsen, *Nat. Cact. Succ. J.* 10: 81 (1955). Type: none indicated.

Euphorbia virosa f. *striata* H.Jacobsen, *Nat. Cact. Succ. J.* 10: 81 (1955). Type: none indicated.

Euphorbia virosa subsp. *arenicola* L.C.Leach, Bol. Soc. Brot., sér. 2, 45: 355 (1971). Type: Angola, Moçamedes distr., *Leach & Cannell 14034* (LISC, holo.; BM, PRE, SRGH, iso.).

Bisexual spiny large glabrous succulent shrub 0.5–2.5 (3) × 0.7–4 (6) m with many branches at and just above ground level from short spirally-angled central stem to 0.3 (2) m tall tapering rapidly into a short taproot bearing woody and fibrous roots. *Branches* initially spreading then erect, mostly simple but occasionally branching again above middle, 0.2–2 m × 50–90 mm, (4) 5- to 8-angled, frequently but irregularly constricted into segments, smooth, grey-green sometimes with paler horizontal bands; *tubercles* fused into (4) 5–8 sinuate wing-like angles with initially V-shaped and soon rounded grooves 10–20 mm deep between angles, laterally flattened and projecting 4–10 mm from angles, with spine-shields united into continuous horny sinuate grey margin along angles 3–5 mm broad above spines and 3–5 mm broad below them (rarely separate), bearing 2 spreading and widely diverging initially glossy dark red and later grey spines 5–15 mm long; *leaf-rudiments* on tips of new tubercles towards apices of branches, 2–4 × 2–3.5 mm, erect, fleeting, lanceolate-ovate, sessile, with small obtuse reddish to brown irregular stipules to 0.5 mm long. *Synflorescences* many per branch towards apex, each a solitary cyme in axil of tubercle, on stout green peduncle 5–12 × 5–9 mm, each cyme usually with 3 usually vertically disposed cyathia (though up to 6 may develop around central cyathium), central male, lateral 2 bisexual and developing later each on short peduncle 2–5 × 4–6 mm, with 2 broadly ovate yellow bracts 2–4 × 5–9 mm subtending lateral cyathia; *cyathia* broadly conical-cupular, glabrous, 8–12 mm broad (3–4 mm long below insertion of glands), with (5) 7–12 obovate lobes with deeply incised margins, yellow-green; *glands* (5) 7–12, transversely elliptic and contiguous, 3–5 mm broad, yellow-green, ascending, usually contiguous, inner margins sometimes raised, outer margins entire, surface between two margins with some nectar; *stamens* glabrous, bracteoles enveloping groups of males, with finely divided tips, glabrous; *ovary* globose, glabrous, green, partly exserted from cyathium and tapering below into pedicel 0.5 mm long; styles 1.5–3.5 mm long, divided into 3–4 branches to near base. *Capsule* 10–24 mm diam., spherical to obscurely (2) 3–4 (6) angled, glabrous, (2) 3–4 (6)-locular, shiny pale green to bright red, ± sessile.

Distribution & Habitat

Euphorbia virosa is associated somewhat patchily with most of the length of the Namib Desert, with populations known from nearly 100 km north of Namibe in Angola to the valley of the Orange River on the border between Namibia and South Africa. Along the Orange River it is found east of Sendelings Drift, from near Marinkas Quellen (west of Noordoewer) to about 40 km west of the Aughrabies Falls (west of Kakamas). At about 350 km from the sea, this is by far the furthest inland that it is known to occur (records from further east, such as in Wilman 1946: 256, refer not to *E. virosa* but to *E. avasmontana*).

Throughout its range, *E. virosa* grows in places of exceptional aridity, often where the average annual rainfall is less than 100 mm. Plants are usually found growing in crevices between rocks on north-facing slopes of fairly bare, rocky to boulder-strewn slopes of schistose, gneissic or quartzitic hills or, more rarely, in stony or gravelly flats. Along the Orange River they often occur where the rocks are dark brown or nearly black and become exceedingly hot during the day in summer. Near Pella, along the Orange River near Pofadder in South Africa, it occurs on steep slopes among quartz-boulders and this is one of a few localities (others are known near Solitaire, Namibia) where it grows together with *E. avasmontana*. In most of its habitats along the Orange River the vegetation is sparse, where scattered specimens of *Portulacaria fruticulosa*, *P. namaquensis*, *Tylecodon hallii*, *T. rubrovenosus* and *Mesembryanthemum lignescens* are often the only other plants on the same slopes.

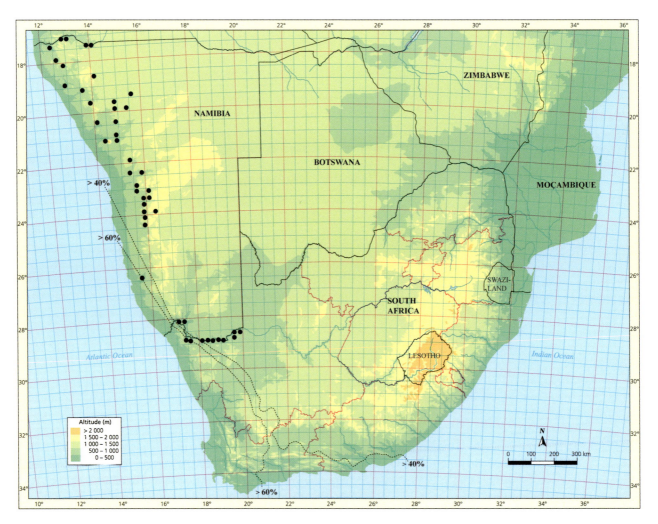

Fig. 5.357. Distribution of *Euphorbia virosa* in southern Africa (© PVB).

Fig. 5.358. *Euphorbia virosa*, inhospitable gneissic slopes near lower Orange River, *PVB 11365*, east of Onseepkans, South Africa, 8 Apr. 2009 (© PVB).

Fig. 5.359. *Euphorbia virosa*, occasionally growing in plains (here over 2 m tall, with Julian Bruyns), Namib Desert SE of Walvis Bay, Namibia, 19 Dec. 2018 (© PVB).

In Angola it is common on gentle schistose slopes or gravelly flats, usually a little beyond the low-lying, calcareous coastal plain, except just south of Namibe, where it occurs within 2 km of the coast. In the south of Angola, it will sometimes be found with vast populations of *Welwitschia*, more often with scattered *Acacia*, occasional specimens of *Commiphora*, *Euphorbia currorii* and *Sesamothamnus benguellensis*.

Diagnostic Features & Relationships

Euphorbia virosa can form immense, imposing shrubs and large specimens are frequently as much as 2 m tall and may be considerably broader. Each shrub consists of a thick and short stem that is rarely more than a third of a metre tall, from which many formidably spiky branches initially spread outwards and then become erect. The stem starts off 4-angled and prettily mottled horizontally with green on cream, with a

Fig. 5.360. *Euphorbia virosa*, sandy plains, plants rarely taller than 0.5 m, from the type locality of Leach's subsp. *arenicola*, Namib Desert 5 km south of Namibe, Angola, 12 Jan. 2006 (© PVB).

green line running up the middle of the groove between the angles. Before it is 50 mm tall, the number of angles increases to 7 or 8, though the youthful mottling of green on cream is more gradually lost and usually disappears completely only after the first branches appear. Except in the lower 50 mm or so, the angles along the stem are usually quite conspicuously spirally twisted. Branching begins after a height of about 10 cm is reached and all the branches generally remain alive on the plant for its entire life (i.e. they do not die and fall off as they age). On the branches the angles, which are also usually six or seven in number, are straight (i.e. not spiralling along the length of the branch) but are always distinctly sinuate. The branches are regularly constricted into short segments, each of which represents a season's growth and each segment is usually at most as long as the branch is thick. All the branches and the stem are armed with formidable spines. On the young, growing tip of a branch the soft spines are initially green, soon becoming shiny, dark red with a green base, after which they rapidly harden to a dark, almost black colour with the broad spine shield soon becoming similarly nearly black. Spines and spine-shields on the previous season's segment and lower down are pale grey, some of which is caused by a thick layer of wax. The spine-shields are comparatively broad and are typically continuous along the angles to form a hard, horny, pale margin along them, which is usually as broad above the spines as it is below them. The sinuate nature of the angles and consequently of this horny margin as well, places the spines at varying angles along the branch. This

Fig. 5.361. *Euphorbia virosa*, steep, well-vegetated granitic slopes (with many small *Myrothamnus* bushes and the author), SW of Kamanjab, Namibia, 6 Jan. 1990 (© PVB).

forms a formidable armour along the branch and makes the softer parts of the branch very difficult to reach. Nevertheless, these are not enough to protect the plants from determined attack and, on several occasions, I have seen specimens along the Orange River in which the upper parts of some of the branches had been eaten off, presumably by kudu or baboons. Figure 1.90 shows a plant almost completely demolished by black rhinos in Namibia.

Flowering in *E. virosa* takes place mainly between November and February and, if enough rain has been received in the previous season, the uppermost segments of many of the branches become covered with cyathia. A single cyme is produced in the axil of each tubercle. As is usual, a central functionally male cyathium first arises in each cyme after which two to six (possibly more) lateral bisexual cyathia develop. The comparatively large cyathia are almost circular in outline around their edge and this is brought about by the unusually large number of glands (there are usually at least seven and up to 12 of them per cyathium, though White et al. (1941: 793) gave 'five' per cyathium, which is incorrect) being closely pressed to each other laterally. Large numbers of male florets are densely packed into the cyathia. In the bisexual cyathia the plump ovary is slightly exserted by the time the styles are mature and it is seated on an unusually broad, flask-shaped base which merges with the thick, short pedicel. Ovaries may have three or four branches to the style. After fertilization, the ovary develops into an almost spherical, berry-like capsule with a variable number of locules. On one plant one will often find capsules with as few as two locules and others with the more common numbers of four to six locules. These strange, shiny, berry-like capsules are sometimes pale green, bright

Fig. 5.362. *Euphorbia virosa*, cyathia in first male stage, Kuiseb River, SE of Walvis Bay, Namibia, 19 Dec. 2014 (© PVB).

Fig. 5.363. *Euphorbia virosa*, cyathia in first male stage, east of Foz do Cunene, Angola, 11 Jan. 2007 (© PVB).

Fig. 5.364. *Euphorbia virosa*, cyathia in female stage (one precocious cyathium has already formed a capsule), west of Iona, Angola, 10 Jan. 2007 (© PVB).

green or even bright red, with faint paler lines indicating the edges of the locules. The capsules dry out on the plant and either disintegrate while still attached or fall apart after dropping to the ground. On drying out, they lose their rounded shape to become more clearly angled and the number of angles depends on the number of seeds inside. At this stage they are also quite thin-walled, somewhat parchment-like and easily broken by hand, with the seeds then falling out fairly easily. Berger (1906) referred to the seeds as 'the size of peas' and, at 6–13 mm in diameter, they are indeed exceptionally large. They are almost spherical in shape, quite smooth and hard and sometimes attractively marked (Fig. 1.56).

Euphorbia virosa is separated from most of the other spine-shield-bearing species by its robust and very large, shrubby habit with comparatively thick, grey-green, persistent branches and its hard spines borne on a continuous, tough, sinuate, horny margin along the angles. The almost circular outline of the large cyathia, with the closely contiguous and numerous glands are also distinctive. Another unusual feature is the short styles, which are united for a short distance and then abruptly join the much inflated ovary (in most others with similarly thick branches, the styles taper gradually into the relatively small ovary). As distinctive as the cyathia are the more or less spherical, usually 4-locular, berry-like capsules, which are among the largest in the genus. The fact that the capsules do not dehisce explosively but fall apart on the plant or on the ground below it is also diagnostic. The seeds are larger than any others in sect. *Euphorbia*.

The similar-looking *E. avasmontana* and *E. virosa* both inhabit the lower reaches of the Orange River and occur together on steep slopes west of Pella. The differences between them are discussed under *E. avasmontana*. However,

Fig. 5.365. *Euphorbia virosa*, pale capsules on furthest northern plant known, *PVB 13439*, 100 km north of Namibe towards Sao Nicolau, Angola, 18 Mar. 2017 (© PVB).

they are not closely related, with *E. virosa* closest to *E. otavibergensis* and to other large shrubby species from Angola.

History

Euphorbia virosa was first observed by William Paterson during his second journey from the Cape to Namaqualand and the Orange River. He found it on or just after the 7th of September 1778 with Robert Gordon, when they reached the Orange River near Goodhouse and explored its banks (Gunn and Codd 1981). In his account of this journey, he remarked on the barren appearance of the landscape here, with the mountains 'being in general naked rocks; though they are in some places adorned by a variety of succulent plant, and in particular Euphorbia, which grows to a height of 15 feet, and supplied the Hottentots with an ingredient for poisoning their arrows'. Two drawings were made of *E. virosa*, one of a large plant in habitat among rocks and another of a flowering branch. Paterson (1789, 1790) did not say who had made these illustrations, but it is possible that they were made by Robert Gordon (Dyer 1949). Both figures depict the characteristic features of the species unmistakably and, in the second, one can clearly see the considerable number of cyathial glands, which is typical of *E. virosa*. Below the figure of the plant, he added 'supposed to be the strongest Vegetable Poison in Africa'. *Euphorbia virosa* was observed again in this area by Le Vaillant, who travelled to the Orange River at Goodhouse and into the south-eastern part of Namibia between late 1783 and early 1784.

Fig. 5.366. *Euphorbia virosa*, brightly shiny berry-like capsules, northern foot of Stormberg, west of Noordoewer, Namibia, 16 Dec. 2018 (© PVB).

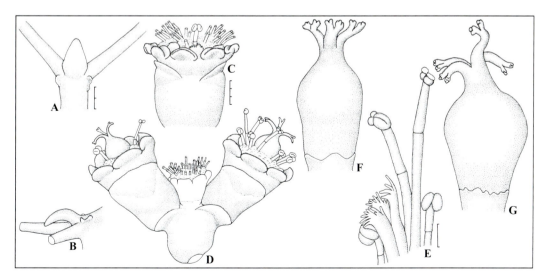

Fig. 5.367. *Euphorbia virosa*. **A**, young spines and leaf-rudiment from above (scale 2 mm, as for **B**). **B**, young spines and leaf-rudiment from side. **C**, side view of cyme in first male stage (scale 3 mm, as for **D**). **D**, side view of cyme in female stage. **E**, anthers and bracteoles (scale 1 mm, as for **F**, **G**). **F**, **G**, female floret. Drawn from: *PVB*, ± 20 km south of Vioolsdrift, South Africa (© PVB).

Taxonomically, much confusion has surrounded the identity of *E. virosa* and, for example, Boissier (1862) considered that *E. caerulescens* belonged to *E. virosa*. Further confusion arose with Berger's name *Euphorbia dinteri*. Marloth (1930) showed that this is none other than *E. virosa* and the photograph by Dinter (Frick 1936) supports this conclusion. N.E. Brown (1915) tried to sort this out but was hampered by the lack of herbarium material. He pointed out that he had only seen four specimens that could belong to *E. virosa* and, of these, three possessed cyathia with five glands and most probably belonged to other species. In this he was prophetic and all three are now included under the present concept of *E. avasmontana*. In fact, in the case of *E. virosa*, at the time of both Boissier's and N.E. Brown's revisions, nothing was known of *E. virosa* apart from the information supplied by Paterson, who did not see and therefore neither described nor illustrated the fruit. As pointed out by Marloth (1930), the earlier muddle mainly arose since living plants were not available to the various botanists who treated these species and the available herbarium material was scanty. Marloth (1930) did much to sort this out and both White et al. (1941) and Leach (1971) further clarified the identity of *E. virosa*. Nevertheless, incorrect identifications still occurred, such as those of Wilman (1946) for the Flora of Griqualand West, where *E. virosa* is listed, but does not occur.

The low-growing form described by Leach as subsp. *arenicola* was observed and photographed by H.H.W. Pearson on 23 May 1909 on the plains 4 miles SE of Moçamedes in Angola. This is the only area where this dwarf form occurs.

Euphorbia waterbergensis R.A.Dyer, *Fl. Pl. Africa* 28: t. 1095 (1951). Lectotype (Bruyns 2012): South Africa, Transvaal, 2.5 miles north of Elmerston P.O. towards Ellisras, 3300', Apr. 1948, *Codd & Erens 4018* (PRE; BOL, K, PRE, SRGH, iso.).

Bisexual spiny glabrous succulent shrub 0.1–1.5 × 0.3–1 m, with many branches from short erect unbranched stem (usually not more than 100 mm long) with many fibrous roots. *Branches* erect and occasionally rebranching above, 0.1–1.5 m × 15–25 mm, (4-) 5- to 6-angled, constricted into segments 20–200 mm long, smooth, dull dark grey-green; *tubercles* laterally flattened and fused into (4) 5–6 clear wing-like angles with surface slightly concave between angles, conical and truncate, projecting 1–3 mm from angles, spine-shields forming continuous hard grey to brown margin 1.5–2 mm broad along angles (initially separate but soon becoming continuous), bearing 2 spreading to slightly deflexed and widely diverging initially reddish black and later almost black spines 3–8 mm long; *leaf-rudiments* on tips of new tubercles towards apices of branches, 1 × 1.5 mm, erect, fleeting, shortly ovate and scale-like, sessile, sometimes with minute stipular prickles ± 0.5 mm long. *Synflorescences* many per branch towards apex, each a solitary cyme in axil of tubercle, on very short peduncle ± 1 mm long, each cyme with 3–5 cyathia, central male, 2 lateral cyathia transversely disposed and other 1–2 laterals (if present) vertically disposed, lateral cyathia bisexual and developing later each on short peduncle 1–2 × 1–2 mm, with 2–4 broadly ovate bracts 1 × 2–3 mm subtending lateral cyathia; *cyathia* conical-cupular, glabrous, 3–5 mm broad (2 mm long below insertion of glands), with 5 obovate lobes with deeply incised margins, yellow to green; *glands* 5, transversely rectangular and contiguous, 1.5–2.5 mm broad, yellow to green, spreading, inner margins not raised, outer margins entire and spreading, surface between two margins dull; *stamens* with glabrous pedicels, anthers pink with red pores, bracteoles enveloping groups of males, with finely divided tips, glabrous; *ovary* obtusely 3-angled, glabrous, green to slightly reddish green, raised on erect pedicel ± 1.5 mm long; styles 2–2.5 mm long, widely

Fig. 5.368. *Euphorbia waterbergensis*, shrub ± 2 m tall among many others, Waterval, near Ellisras, South Africa, c. 1986 (© S.P. Fourie).

spreading, divided to well below middle. *Capsule* 3–4 mm diam., obtusely 3-angled, glabrous, green to reddish, exserted and held erect on pedicel 1–2 mm long.

Distribution & Habitat

Euphorbia waterbergensis is restricted to the sandstone hills of the Waterberg, where it is known from a few scattered localities near Ellisras and considerably further to the northeast along the Palala River.

Habitats for *E. waterbergensis* are usually dry, terraced areas of red sandstone with shallow, sandy soils and a scanty cover of sparse grasses and some shrubs. In such spots *E. waterbergensis* usually occurs with various other succulents as well as the more ubiquitous *E. griseola* and *E. schinzii*.

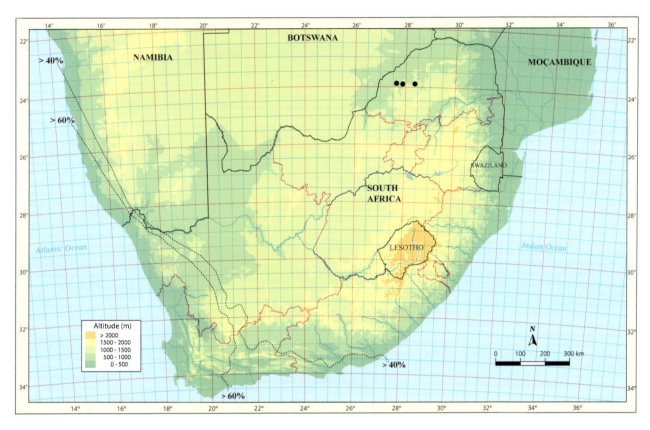

Fig. 5.369. Distribution of *Euphorbia waterbergensis* (© PVB).

Fig. 5.370. *Euphorbia waterbergensis,* shrub ± 1 m tall, *PVB 6539a,* turnoff to Strydom Dam, near Ellisras, South Africa, 1 Jan. 1996 (© PVB).

Diagnostic Features & Relationships

Euphorbia waterbergensis is a distinctive species because of its relatively sparsely branched habit in which the usually 5-angled branches do not rebranch significantly. This results in elegant and fairly open shrubs. The segmented shape and grey-green colour of the branches are also characteristic. On new growth one finds the very short and broad leaf-rudiments and the branches are fortified with quite short, but persistent spines. In *E. waterbergensis* the spines are not the usual red colour when they are formed but are only faintly reddish (closer to black) and the spine-shield between the two spines is thick and reddish black as well, forming a rather heavy boss over the tubercle. In most plants the spine-shields are continuous along the angles, only occasionally remaining separate. Stipular prickles in *E. waterbergensis* are minute and are well separated from the base of the leaf-rudiment, but a fine ridge (only visible in the young spine-shields) links them to it.

Euphorbia waterbergensis usually flowers between late March and mid-May and is the first of the autumn- to winter-flowering species to come into bloom. The comparatively small cyathia are produced on very short peduncles near the tips of the branches and are green with slightly darker green glands. They are unusual for the pink anthers each with a red

Fig. 5.371. *Euphorbia waterbergensis,* among reddish sandstone boulders, Marulafontein, near Ellisras, South Africa, 5 Jan. 2013 (© PVB).

Fig. 5.372. *Euphorbia waterbergensis,* cyathia in first male stage, showing the pinkish anthers, *PVB 6539a*, turnoff to Strydom Dam, near Ellisras, South Africa, 3 Apr. 2019 (© PVB).

Fig. 5.373. *Euphorbia waterbergensis,* central male cyathium sometimes surrounded by three other cyathia, *PVB 6539a*, turnoff to Strydom Dam, near Ellisras, South Africa, 3 Apr. 2019 (© PVB).

Fig. 5.374. *Euphorbia waterbergensis,* cyathia in second male stage, some anthers slightly pinkish, *PVB 8754*, Lapalala, Waterberg, South Africa, 3 Apr. 2019 (© PVB).

pore, but the pollen is the usual yellow colour. The female floret is borne on a noticeable pedicel, which just places the ovary at the level of the cyathial lobes. On fertilization, its pedicel elongates slightly and remains erect as the capsule matures.

Euphorbia waterbergensis frequently occurs together with *E. griseola* and *E. schinzii*. Of these two species, *E. griseola* is more closely related to *E. waterbergensis* and is vegetatively somewhat similar, with branches that are often 5-angled and of a similar thickness as well as continuous spine-shields along the angles. On the other hand, *E. schinzii* is more distantly related, as one sees from the smaller plants of much shorter and more slender, 4-angled branches, sessile ovaries and verrucose seeds. Despite their close relationship and many shared features, *E. griseola* and *E. waterbergensis* are quite easily separated. In *E. griseola* the branches are a darker green (frequently mottled with creamy patches) and they are more fiercely armed with longer spines. It flowers later than *E. waterbergensis* and the female floret protrudes much further from the cyathium than in *E. waterbergensis* and soon bends to one side. The manner in which the ovary is well raised above the base of the cyathium on a significant pedicel is a feature of all the species in the group to which the two belong. More recent results (Bruyns et al. 2011) have shown that *E. waterbergensis* is most closely related to *E. excelsa* and this unexpected fact is discussed under the latter species.

Fig. 5.375. *Euphorbia waterbergensis*, capsules, *PVB 6539a*, turnoff to Strydom Dam, near Ellisras, South Africa, 26 Aug. 2007 (© PVB).

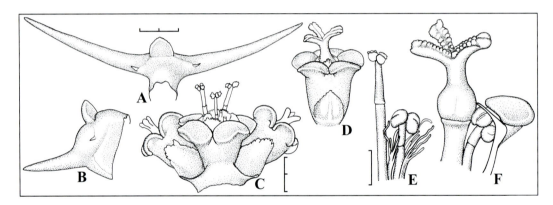

Fig. 5.376. *Euphorbia waterbergensis*. **A**, young spines and leaf-rudiment from above (scale 2 mm, as for **B**). **B**, young spines and leaf-rudiment from side. **C**, side view of cyme in late first male stage (scale 2 mm, as for **D**). **D**, side view of female cyathium. **E**, anthers and bracteole (scale 1 mm, as for **F**). **F**, female floret (with part of dissected cyathium). Drawn from: **A**, **B**, *PVB 8754*, Lapalala, Waterberg, South Africa. **C**–**F**, *PVB 6539a*, turnoff to Strydom Dam, near Ellisras, South Africa (© PVB).

History

The first record of *E. waterbergensis* was made by N. Allen Lever in January 1940 in rocky areas along the Palala River in the Waterberg.

Euphorbia zoutpansbergensis R.A.Dyer, *Fl. Pl. South Africa* 18: t. 715 (1938). Type: South Africa, Transvaal, Soutpansberg, Wylliespoort, Sept. 1937, *Dyer 3873 sub PRE 23393* (PRE, holo.; BM, E, K, MO, PRE, US, iso.).

Bisexual spiny glabrous succulent shrub to tree 1–6 m tall, with gradually deciduous branches forming almost spherical crown and arising in upper part of cylindrical and faintly (6- to) 8-angled erect solitary (rarely forked) trunk-like stem 1–5 m tall, trunk mostly naked except for spines just below crown, with many fibrous roots spreading from base. *Branches* spreading then ascending and rarely rebranching (except occasionally near tips), older branches drying up and falling off stem after a few years, 0.2–1.5 m × (20) 30–45 mm, (5-) 6- to 9-angled, constricted into segments 50–120 mm long, smooth, greyish green; *tubercles* very evenly fused into (5) 6–9 prominent wing-like angles with surface deeply concave between angles, low-conical, laterally flattened and projecting 3–5 mm from angles, spine-shields forming continuous (rarely interrupted) hard grey to brown margin 1–2 mm broad along angles, bearing 2 spreading and widely diverging initially red then brown to grey spines 4–12 mm long; *leaf-rudiments* on tips of new tubercles towards apices of branches, 2–3 × 2–3 mm, spreading and recurved near tip, deciduous, broadly ovate-deltate, sessile, mostly with small brown to black irregular ovate tooth-like stipules up to 1 mm long. *Synflorescences* many per branch towards apex, with 1 (–3) cymes in axil of each tubercle, with short peduncle 2–4 mm long, each cyme with 3 vertically disposed cyathia, central usually male, outer two bisexual and developing later each on short peduncle 2–3 × 2 mm, with 2 small broadly ovate bracts 2–2.5 × 2–2.5 mm subtending lateral cyathia; *cyathia* broadly cupular, glabrous, 5–6.5 mm broad (3 mm long below insertion of glands), with 5 short obovate lobes with deeply toothed margins, bright yellow to yellow-green; *glands* 5–6, transversely elliptic, contiguous to slightly separated, 2–2.5 mm broad, yellow, slightly spreading, inner margins not raised, outer margins entire and slightly spreading, surface between two margins dull; *stamens* with glabrous pedicels, anthers yellow, bracteoles enveloping groups of males, with finely divided tips, glabrous; *ovary* triangular in cross-section (3-angled with distinctly flattened sides), on pedicel ± 2.5 mm long, more or less entirely exserted and soon pushed out beyond cyathium, glabrous, green; styles ± 2 mm long, widely spreading, branched to lower third. *Capsule* 6–10 mm diam., triangular, glabrous, slightly shiny, bright red to pale green, exserted from cyathium on decurved and later erect pedicel 4–20 mm long.

Distribution & Habitat

Euphorbia zoutpansbergensis is only found in the Soutpansberg, the northernmost mountain range in South Africa. Here it occurs somewhat sporadically, mainly on the sides of ravines and river valleys along the northern slopes of the mountains from Wyllies Poort westwards to the slopes above Soutpan. Occasionally it also inhabits slopes to the south of the highest ridge of the range.

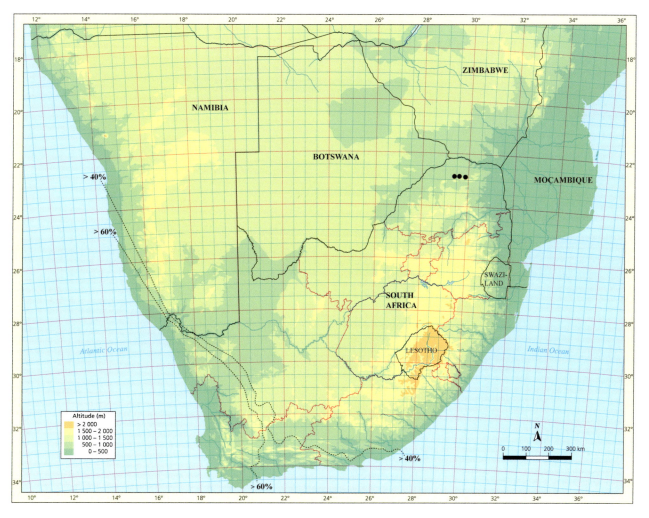

Fig. 5.377. Distribution of *Euphorbia zoutpansbergensis* (© PVB).

Fig. 5.378. *Euphorbia zoutpansbergensis,* tree ± 4 m tall, among sandstone rocks and bushes, *PVB 12053*, Leshaba Wilderness, west of Louis Trichardt, Soutpansberg, South Africa, 2 Nov. 2011 (© PVB).

Fig. 5.379. *Euphorbia zoutpansbergensis, PVB 12053*, Leshaba Wilderness, west of Louis Trichardt, Soutpansberg, South Africa, 2 Nov. 2011 (© PVB).

Plants form small colonies on steep, rocky hillsides, where they usually grow amongst other trees of a similar height in shallow soils derived from sandstones. They sometimes occur with or near to *Aloe angelica* and various other succulents including, occasionally, *E. ingens*.

Diagnostic Features & Relationships

Euphorbia zoutpansbergensis forms a handsome tree that is usually between 2 and 5 m tall, with a slender, almost cylindrical trunk and a fairly dense, broadly circular crown of branches. In young plants the trunk is dark green faintly mottled with cream and is armed with sharp spines but these wear off with age so that it is later smooth. The branches are slender and regularly, but fairly distantly constricted along their length. Most of them are 6- to 9-angled, with deep grooves between the relatively thin angles and their greyish green colour is not very prominent. The branches bear sharp spines along the angles that are joined up by a slender, but continuous, horny margin formed by the spine-shields. When young, each spine-pair is accompanied by a somewhat recurved, almost triangular leaf-rudiment and two almost leaf-like, spreading, blackish stipular outgrowths at its base.

Flowering in *E. zoutpansbergensis* usually takes place between early October and the middle of November (though the type was collected in flower in September). During this period on certain trees the last segment on many of the branches becomes covered with cyathia (in groups of up to

5.1 Sect. Euphorbia

Fig. 5.380. *Euphorbia zoutpansbergensis*, PVB 12053, Leshaba Wilderness, west of Louis Trichardt, Soutpansberg, South Africa, 2 Nov. 2011 (© PVB).

Fig. 5.381. *Euphorbia zoutpansbergensis*, cyathia in first male stage, PVB 7099, above Soutpan, Soutpansberg, South Africa, 23 Oct. 2012 (© PVB).

three cymes per leaf-axil) and these trees assume a yellowish hue. The fairly broadly cupular cyathia are usually bright yellow, with bright yellow glands. The lateral cyathia in each cyme are placed vertically (relative to the axis of the branch) and develop very gradually in the axils of the bracts of the first cyathium. They are often a slightly less bright yellow than the central one. In them, the female floret is raised on a substantial pedicel so that most of the ovary (with its quite short styles) is exserted from the cyathium. After fertilization, the pedicel bearing the ovary elongates greatly and may reach 20 mm long, initially decurved and then becoming erect shortly before the capsule explodes. In colour the immature capsules vary greatly: on some trees they are bright red while on others they are pale green. They are triangular in shape with relatively flat sides and fairly slender edges. Ripening takes place quickly after fertilization and the seeds are usually released by late November and early December.

Euphorbia zoutpansbergensis is closely allied to *E. eduardoi* and *E. vallaris* from Angola (Bruyns et al. 2011) and also shows some affinity to the Angolan endemic *E. candelabrum* Welw. and *E. confinalis*. It is not closely allied to any of the other tree-like species in South Africa and from them it is easily distinguished by the larger number of angles on

Fig. 5.382. *Euphorbia zoutpansbergensis,* second male stage with developing capsules, *PVB 12053*, Leshaba Wilderness, west of Louis Trichardt, Soutpansberg, South Africa, 2 Nov. 2011 (© PVB).

Fig. 5.383. *Euphorbia zoutpansbergensis,* branch with dark pink capsules similar in shape and colour to those of Welwitsch's *E. candelabrum* from Angola, *PVB 12053*, Leshaba Wilderness, west of Louis Trichardt, Soutpansberg, South Africa, 2 Nov. 2011 (© PVB).

Fig. 5.384. *Euphorbia zoutpansbergensis,* branch with triangular pale capsules, *PVB 12053*, Leshaba Wilderness, west of Louis Trichardt, Soutpansberg, South Africa, 2 Nov. 2011 (© PVB).

5.2 Sect. Monadenium

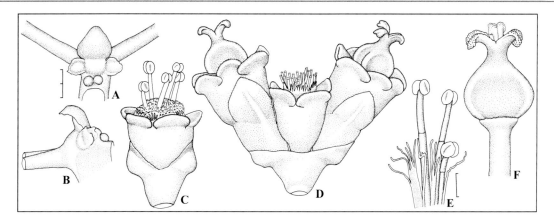

Fig. 5.385. *Euphorbia zoutpansbergensis.* **A**, young spines and leaf-rudiment from above (scale 2 mm, as for **B–D**). **B**, young spines and leaf-rudiment from side. **C**, side view of cyme in early first male stage. **D**, side view of cyme in late female stage. **E**, anthers and bracteoles (scale 1 mm, as for **F**). **F**, female floret. Drawn from: *PVB 7099*, above Soutpan, Soutpansberg, South Africa (© PVB).

the branches and the particularly far-exserted triangular capsules with flattish sides (somewhat similar to those of *E. candelabrum*). While far-exserted capsules are also found in such species as *E. grandidens* and *E. sekukuniensis*, in these the branches are far fewer-angled and much more slender and the capsules are differently shaped.

History

Euphorbia zoutpansbergensis was first observed by J. Hutchinson and Jan Gillett in August 1930 on a farm 13 miles west of Wyllies Poort in the Soutpansberg (*Hutchinson & Gillett 4454* (K)). Hutchinson (1946: 350) identified this as '*E. caerulescens* Harv.' It was again observed by Frederick Z. van der Merwe in May 1937 (photo at PRE) at the southern entrance to Wyllies Poort and a Mr & Mrs van der Vyver brought flowering material to R.A. Dyer in September 1937 from this spot. In the same month they and F.Z. van der Merwe took Dyer to this place to show him the new species in habitat and it was believed that this was 'the only known locality' for this species (Dyer 1938b). As Fourie (1985) indicated, it has subsequently been found at a few other localities in the Soutpansberg and all of these lie to the west of the type locality.

5.2 Sect. Monadenium

Euphorbia sect. **Monadenium** (Pax) Bruyns, *Taxon* 55: 411 (2006). *Monadenium* Pax, *Bot. Jahrb. Syst.* 19: 128 (1894). Type: *Monadenium coccineum* Pax (= *Euphorbia neococcinea* Bruyns).

This section is almost as diverse vegetatively as subg. *Athymalus* or even the rest of *Euphorbia*. Here one finds small hysteranthous geophytes, small shrubs with pencil-like branches, somewhat woody larger shrubs with cylindrical stems and shiny yellowish peeling bark and others (sometimes small trees) with green photosynthetic bark on the succulent branches, which may even be slightly angled after the manner of species like the American *E. pteroneura*. The leaves may be borne on tubercles and may be subtended by a thorn, sometimes with two stipular prickles or thorns flanking its base (for the diversity in thorns see Fig. 1.33). Amongst all this diversity, the leaves are always relatively prominent (Fig. 1.35) and somewhat fleshy but are deciduous. The cyathial glands are either fused into a cylinder (as in the species of the former *Synadenium*, which are nested within sect. *Monadenium*) or some of them are fused into a horse-shoe-like structure (typical of the former *Monadenium*).

Sect. *Monadenium* is most diverse in the tropics on the eastern side of Africa and is particularly well represented in East Africa. Only one of the 90 species occurs naturally in southern Africa.

Euphorbia lugardiae (N.E.Br.) Bruyns, *Taxon* 55: 413 (2006). *Monadenium lugardiae* N.E.Br., *Bull. Misc. Inform.* 1909: 138 (1909). Type: Botswana, foot of Kwebe Peak, Kwebe Hills, 3500', fl. 31 Aug. 1897 & leaves Feb. 1898, *Mrs Nell Lugard 22* (K).

Bisexual spineless glabrous succulent shrub 0.2–0.6 m tall with few to many branches from similar central stem

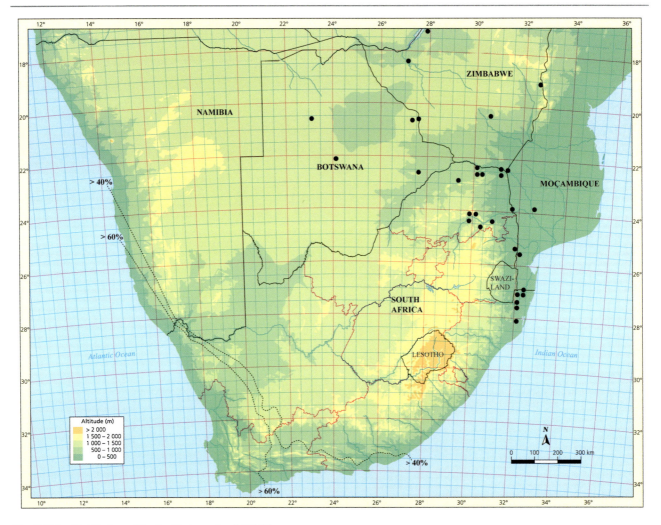

Fig. 5.386. Distribution of *Euphorbia lugardiae* (© PVB).

bearing fibrous roots, branching mainly from base. *Branches* ascending to erect (often sprawling when dry), 0.1–0.6 m × 15–30 mm, cylindrical, not constricted into segments, smooth, pale green becoming covered with corky bark and grey towards base; *tubercles* large and rhomboidal, neither fused together nor arranged into angles, with shallow groove around each, much flattened and projecting at most 1–2 mm from surface of branch, spine-shields absent; *leaves* on tips of new tubercles towards apices of stem and branches, 15–50 (90) × 5–20 (40) mm, spreading, deciduous, spathulate to obovate, acute or obtuse, subsessile to tapering into short cylindrical petiole, fleshy and with crenulate margins towards apex, puberulous, green, with small brown deciduous stipular prickles 0.5–1.5 mm long. *Synflorescences* several per branch towards apex, each a solitary slightly nodding cyme in axil of tubercle, on short stout peduncle 3–5 mm long, each cyme with 3 transversely disposed cyathia, central bisexual sessile and persistent, lateral 2 bisexual and developing later each on short peduncle 2–5 × 2–3 mm, with 2 ovate abaxially folded slightly adaxially thicker and fused grey-green ± veined and papillate-puberulous bracts 3–9 × 2–4 mm enveloping most of peduncle as well as initially enveloping lateral cyathia and slightly exceeding central cyathium; *cyathia* cupular, papillate-puberulous, 4.5–5 mm broad (3 mm long below insertion of glands), with 5 obovate lobes with deeply incised margins, pale green, open to about halfway down in front; *gland* 1, cupular and enclosing most of cyathium above except for gap in front

5.2 Sect. Monadenium

Fig. 5.387. *Euphorbia lugardiae,* large clump under trees with branches ± 0.5 m tall, Penge, west of Burgersfort, South Africa, 21 May 1980 (© PVB).

where capsule later exserted, 3 × 5 mm, pale green, erect, inner margins slightly raised against backs of lobes, outer margins erect, outer surface papillate-puberulous; *stamens* glabrous, bracteoles filiform and enveloping groups of males, sparsely pubescent; *ovary* ± acutely and deeply 3-angled, papillate-puberulous and with slight papillate wings along edges near apex, pale green suffused with red, raised on thick erect pedicel 2–3 mm long and rapidly becoming exserted horizontally or vertically through gap in gland as styles dry off; styles 2–2.5 mm long, branched to just above base and widely spreading. *Capsule* 3–6 mm diam., deeply 3-angled, green sometimes suffused with red, with finely papillate and slightly fleshy outer layer spreading and coarsely flap-like on angles and along base, exserted horizontally (later becoming erect) on pedicel 3–6 mm long and much swollen towards apex, dehiscing explosively.

Distribution & Habitat

Euphorbia lugardiae is known in Botswana, Moçambique, South Africa, Swaziland and Zimbabwe. There are only a few very scattered known localities in Botswana, Moçambique, Swaziland and Zimbabwe, while it has been more frequently recorded in the northern parts of South Africa, especially north of the Soutpansberg and in the area formerly known as Sekukuniland. Many records have been made also in the north of Natal, especially between Ndumu and Mkuzi.

Fig. 5.388. *Euphorbia lugardiae, PVB 11899*, south of Penge, west of Burgersfort, South Africa, 23 Jul. 2011 (© PVB).

Euphorbia lugardiae occurs in rocky habitats (some of these are outcrops of granite) among trees and scattered deciduous bushes along river valleys or stony hillsides, often in shallow soils overlying rocks with much loose leaf-litter on the ground. Plants may be sheltered or quite exposed and they often grow with many other succulents, including species of *Aloe*, *Ceropegia*, *Cynanchum gerrardii*, *C. viminale*, *Euphorbia cooperi*, *E. grandidens* and *Plectranthus*.

Diagnostic Features & Relationships
Euphorbia lugardiae forms quite large but often very diffuse and untidy clumps of thick, pale green branches which root readily wherever they lie on or touch the ground. The cylindrical branches are covered with large, flat tubercles, which are not arranged into rows or angles but are outlined clearly by a fine groove around their perimeter. At the tip of the branch, each tubercle bears a prominent, fleshy leaf and these together form a rosette around the apex of the branch. Initially bright and shiny green, the leaves are later similarly coloured to the branches, but have prominent paler veins. They last for most of the growing season, after which they shrivel and fall off, leaving a prominent scar a little above the middle of the tubercle. Around their bases there are two tiny stipular prickles but, apart from these, the plant is spineless.

Fig. 5.389. *Euphorbia lugardiae*, with *Aloe wickensii*, *Sansevieria* and *E. cooperi*, *PVB 11899*, south of Penge, west of Burgersfort, South Africa, 23 Jul. 2011 (© PVB).

Fig. 5.390. *Euphorbia lugardiae,* cymes with mature central cyathium, *PVB 11899*, south of Penge, west of Burgersfort, South Africa, 17 Jan. 2012 (© PVB).

5.2 Sect. Monadenium

Flowering in *E. lugardiae* takes place usually between October and January, with the cymes arising in the axils of the new leaves (one per leaf) at the beginning of the summer rains. Each cyme has two bracts whose margins are fused behind the cyme (i.e. closest to the branch). In this way they form a 'bract-cup' that partly surrounds the cyme, but is more open in front (i.e. away from the branch). The cyme consists of three bisexual cyathia of which the central one develops first and the outer two later. Here there is only one gland in each cyathium and its outer edge is erect to form a cupular structure with a broad opening at the front. If the cyathium is carefully dissected the gland is seen to have an inner edge too which is pressed up against the lobes. There are five lobes of which the three at the back (i.e. inside the high part of the gland) are mostly hidden and are fused much higher up to the gland just below its inner edge. The 'outer' two (at the open 'front' of the gland) are more prominent, fused lower down and enclose the male florets in the 'front' of the cyathium. As always, the female floret develops first and, as the styles shrivel up, it bends forward out of the cyathium through the gap in the 'front' in the gland, though with only slight lengthening of the pedicel. The ovary is covered (up to the base of the styles) with a reddish papillate layer which, when peeled away, reveals the pale green young capsule below. It would appear that this extra layer is the calyx. Once fertilized, the capsule is exserted a little further (often horizontally) on an often bright red pedicel which develops a very thick apex just beneath the capsule. The capsule is covered by a finely papillate, fleshy layer which is more coarsely papillate or toothed along the base and along each ridge. When the capsule is ripe, the pedicel turns to hold it facing upwards (sometimes 2–3 months after fertilization) and it soon dehisces explo-

Fig. 5.391. *Euphorbia lugardiae,* cyme with developing capsule and mature lateral cyathia, *PVB 11899*, south of Penge, west of Burgersfort, South Africa, 18 Jan. 2018 (© PVB).

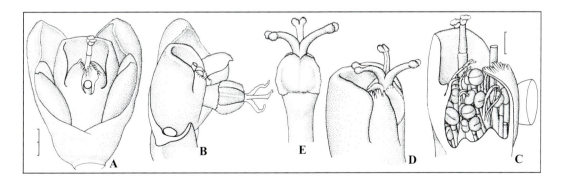

Fig. 5.392. *Euphorbia lugardiae*. **A**, cyme from side with young female floret removed (scale 2 mm, as for **B**). **B**, cyme from side with older female floret, with bract and one branch removed. **C**, dissected cyathium showing anthers and bracteoles (scale 1 mm, as for **D**, **E**, **F**). **D**, young female floret in cyathium. **E**, female floret. Drawn from: *PVB 7470*, Nzhelele Dam, SW of Messina, South Africa (© PVB).

sively. The outer fleshy layer dries out with the rest of the capsule but remains attached to the locular walls.

Euphorbia lugardiae is closely related to *E. kimberleyana* from south-eastern Zimbabwe and to *E. schubei* from East Africa (Dorsey et al. 2013). *Euphorbia kimberleyana* differs mainly by its larger size, by the possession of a single, small spine below each leaf and by the slightly more projecting centre of each tubercle (Williamson 1999). From the various cylindrical- and thick-branched species from further north (such as *E. schubei*), *E. lugardiae* differs quite clearly by the much flatter tubercles and the lack of more prominent spines around the base of each leaf.

History

According to Lugard (1938) he and his wife Charlotte Eleanor (Nell) Lugard discovered a single clump of this species on 31 August 1897 at the foot of Kwebe Peak in the Kwebe Hills, about 20 km south of Lake Ngami in western Botswana. Mrs Lugard made a colour painting of it on the same day, which was reproduced in black & white in Lugard (1938) and in White et al. (1941) and is here shown in colour (Fig. 5.393). It was also figured in Marloth (1925: fig. 80), but in this case under the name *Synadenium arborescens* Boiss.

Fig. 5.393. *Euphorbia lugardiae*, *Nell Lugard 22*, Kwebe Hills, Botswana, 18 May 1898 (watercolour by Nell Lugard, © RBG Kew).

5.3 Sect. Tirucalli

Euphorbia sect. **Tirucalli** Boiss. in DC., *Prodr.* 15 (2): 10, 94 (1862). *Euphorbia* subsect. *Tirucalli* (Boiss.) Benth. & Hook.f., *Gen. Pl.* 3 (1): 280 (1880). *Euphorbia* subg. *Tirucalli* (Boiss.) S.Carter, *Kew Bull.* 40: 823 (1985). Type: *Euphorbia tirucalli* L.

Arthrothamnus Klotzsch & Garcke, *Abh. Königl. Akad. Wiss. Berlin* 1859: 62 (1860). *Euphorbia* sect. *Arthrothamnus* (Klotzsch & Garcke) Boiss. in DC., *Prodr.* 15 (2): 10, 74 (1862). Type: *Euphorbia tirucalli* L.

Tirucalia Raf., *Fl. Tellur.* 4: 112 (1838). Type (designated here): *Euphorbia tirucalli* L.

Unisexual densely branched large tough succulent shrub becoming woody towards base. *Branches* grey-green often with white to tawny pubescence near tips, slender and terete, sometimes slightly vertically ridged. *Leaves* ephemeral, to 30 mm long, alternate, lanceolate with entire margins, pubescent, shortly petiolate or sessile, sometimes with small glandular stipules. *Synflorescences* on short axillary shoots near tips of branches (more in males than females), with 1 unisexual densely pubescent cyathium 3–14 mm diam. (broader in male than female) with 5 glands, subtended by minute bracts, glands entire, without appendages. *Capsule* obtusely 3- to 6-lobed, 7–30 mm diam., well exserted, densely pubescent, often drying up and falling to ground (dehiscing explosively). *Seeds* 3–6 per capsule, smooth, 4–8 mm long, ± cylindrical or obtusely 4-angled, grey to brown, with or without caruncle.

Sect. *Tirucalli* is one of the first African branches to diverge in subg. *Euphorbia*. Of the 24 species included in this section (Dorsey et al. 2013), 15 are found in Madagascar, seven in Africa, one in the Arabian Peninsula and one on Socotra (Fig. 1.70). Of the seven African species, one occurs in Somalia and the other six are endemic to south-western Africa. Here they occur from the hyper-arid southern end of the Namib Desert (around the lower reaches of the Orange River) past the northern end of this desert in coastal Angola to about 50 km north of Lobito. As with several such clades associated with the Namib Desert, the diversity is low, it has many close relatives in Madagascar and also some relatives in north-east Africa (cf. Didiereaceae, Bruyns et al. 2014, *Pachypodium* of the Apocynaceae and *Aloidendron* of the Asphodelaceae).

Although *E. tirucalli* is common along the eastern side of southern Africa, it is most likely that it is native to Madagascar (Dorsey et al. 2013). Consequently it is left out of this account of the naturally occurring species of sect. *Tirucalli* of southern Africa, though notes on it and some illustrations are included below.

5.3 Sect. Tirucalli

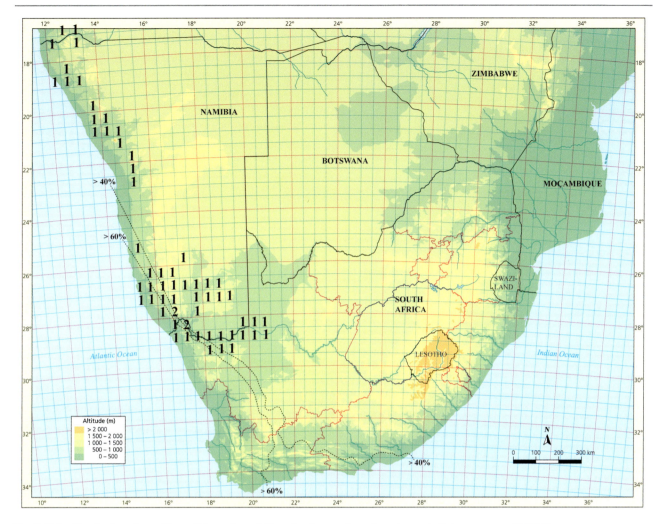

Fig. 5.394. Distribution of sect. *Tirucalli* in southern Africa, showing number of species per half-degree square (© PVB).

Plants of this section are typically large shrubs that are often quite woody towards the base. Their branches are succulent, but are tough and very fibrous except right at the tips when growing actively. The branches are cylindrical but may have slight irregular ridges that become more prominent when they are dry in *E. gummifera*.

Key to the species of sect. *Tirucalli* occurring naturally in southern Africa:

1. Divided parts of styles on slender cylindrical column above ovary, branches 3–6 mm thick, plant pale grey-green (felt at tips of branches white), capsule 7–8 mm diam., seeds 4–4.5 mm long, with caruncle..............................**E. congestiflora**
1. Divided parts of styles ± sessile ontop of ovary, branches 6–10 mm thick, plant brownish green (felt at tips of branches often slightly brownish or tawny), capsule 10–30 mm diam., seeds at least 5 mm long, without caruncle......................2.
2. Branches much rebranched above, often becoming 3-ribbed if slightly dry. leaf-rudiments 1–4 mm long (minute), capsule 10–12 mm diam., seeds 5.5–6 mm long...**E. gummifera**
2. Branches sparingly rebranched above (mainly branching from base), not becoming 3-ribbed if slightly dry, leaf-rudiments (2.5) 5–27 mm long (more conspicuous), capsule 18–30 mm diam., seeds 7–8 mm long..3.
3. Leaf-rudiments 2.5–11 mm long, ascending to recurved and channelled above, estipulate, capsule exserted on pedicel 10–25 mm long..**E. gregaria**
3. Leaf-rudiments (2.5) 5–27 mm long, erect and usually flat, with stipules, capsule exserted on pedicel 2–4 mm long.......
 ....**E. damarana**

Euphorbia congestiflora L.C.Leach, *Bol. Soc. Brot.*, sér. 2, 44: 197 (1970). *Tirucalia congestiflora* (L.C.Leach) P.V.Heath, *Calyx* 5: 89 (1996). Type: Angola, Namibe distr., between Cumilunga & Curoca Rivers, 11 Jan. 1956, *Mendes 1265* (LISC, holo.; BM, LUA, M, SRGH, iso.).

Many-branched, unisexual (possibly sometimes bisexual?), spineless, succulent shrub 1–2 × 0.5–1.5 m, arising from later woody stem with fibrous roots. *Branches* ascending to erect, subdensely to sparsely alternately rebranched towards tips, terete and not articulated at segments, 0.2–1 m × 3–6 mm, smooth and white-felty (sometimes pale brown) near growing tips becoming grey to grey-green with age, fairly rigid but tips not dying off to form slight spikes. *Leaf-rudiments* towards apex of young branch, alternate, narrowly ovate-lanceolate, 7–22 × 2–6 mm, very slightly fleshy, ascending, usually slightly longitudinally folded upwards to flat, fleeting, scantily pubescent, with entire margins, acute, tapering below into slightly swollen base but epetiolate, with small transparent to later dark brown spherical stipules on either side at base. *Synflorescences* terminal on branchlets, finely white-pubescent, each a cluster of 4–10 unisexual cyathia on short peduncles 1–3 mm long, each subtended by 2–4 small, scale-like, broadly ovate, pubescent, grey-green later red to brown bracts, 1–1.5 × 1–1.5 mm (often with small irregular brown stipules at base); *cyathia* conical-cylindrical to cupular, finely white-pubescent, 3–3.5 mm diam., ± 3 mm long below insertion of glands, with 5 small ± semicircular finely and densely pubescent lobes with densely pubescent margins, grey-green, finely pubescent over most of outside (except glands); *glands* 5, transversely elliptic, 1.5 mm broad, ascending, yellow-green to dark green, flat to convex above, widely spaced, outer

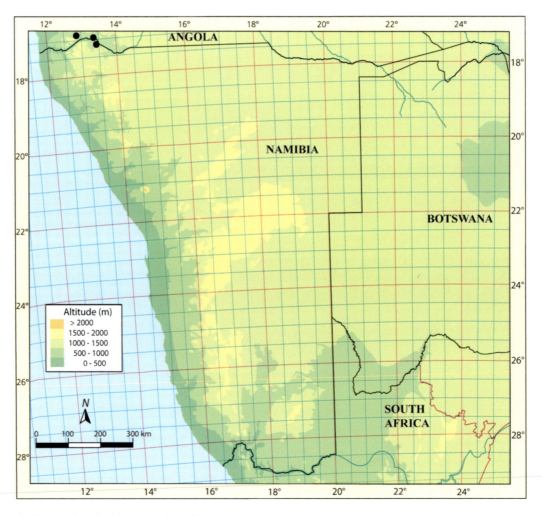

Fig. 5.395. Distribution of *Euphorbia congestiflora* (© PVB).

Fig. 5.396. *Euphorbia congestiflora*, large clump of shrubs (most ± 1.5 m tall), *PVB 5600*, east of Epupa Falls, Namibia, 25 Feb. 1993 (© PVB).

Fig. 5.397. *Euphorbia congestiflora*, *PVB 10373*, west of Oncocua, Angola, 15 Dec. 2016 (© PVB).

Fig. 5.398. *Euphorbia congestiflora*, *PVB 10373*, west of Oncocua, Angola, 8 Jan. 2006, with Patricia Craven (© PVB).

Fig. 5.399. *Euphorbia congestiflora*, male cyathia, *PVB 10373*, west of Oncocua, Angola, 15 Dec. 2016 (© PVB).

Fig. 5.400. *Euphorbia congestiflora*, female cyathia, *PVB 10373*, west of Oncocua, Angola, 15 Dec. 2016 (© PVB).

margin entire, inner margin not raised; *stamens* with glabrous pedicels, interspersed with filiform to linear apically pubescent bracteoles; *ovary* slightly ellipsoidal, finely and densely white-pubescent, surrounded tightly by many adpressed sterile male florets, raised on stout white-pubescent pedicel ± 1.5 mm long; styles 7–8 mm long, divided near apex, erect with divided parts spreading. *Capsule* deeply 3-angled, 7–8 mm diam., finely pubescent, exserted from cyathium on erect pubescent pedicel 7–8 mm long, yellow-brown; *seeds* ellipsoidal, 4–4.5 × 2.5 mm, smooth, pale to dark orange, with white apical caruncle.

Distribution & Habitat

Euphorbia congestiflora is found in Namibia along the Cunene River to the east of Epupa Falls and in the adjacent parts of southern Angola mainly between Oncocua and Iona.

Generally, *E. congestiflora* grows in relatively flat, stony, dry country, mostly avoiding the higher hills. It is usually

Fig. 5.401. *Euphorbia congestiflora*, capsules, *PVB 10373*, west of Oncocua, Angola, 15 Dec. 2016 (© PVB).

locally plentiful among other shrubs of almost equal height, with *mopane* and other scattered trees, but it does not tend to dominate the vegetation as other members of sect. *Tirucalli* often do.

Diagnostic Features & Relationships
Euphorbia congestiflora generally forms quite large shrubs, which are dense and made up of many branches, with a distinctive pale, grey-green colour. Although the branches are long and slender (rebranching comparatively rarely and then mainly towards their bases), they are fairly rigid (though not as much so as the other southern African species of sect. *Tirucalli*). Young branches are covered with a dense, white to pale brown, felt-like layer of hairs around their tips which soon wears off. Alternating, slender, green leaves (not covered with the same felty layer) are also present on new growth, but they soon fall off. Each of them has two small, spherical stipules at its base which are transparent to begin with and may later become dark brown.

Cymes are produced at the tips of some of the branches in small clusters, mainly in mid-summer (December to January) and the cyathia are similarly coloured to the branches. They are also mostly covered with a felt-like indumentum of fine, pale hairs, making them relatively inconspicuous. The female cyathia are comparatively long, with long styles branched only near their tips and a dense mat of sterile male florets pressed into a mass of felt-like hairs around their bases and pressed into the pedicel.

Analysis of DNA-data (Bruyns et al. 2011), showed that *E. congestiflora* belongs to a group of three species found in the northern part of the Namib Desert, consisting of *E. carunculifera*, *E. congestiflora* and *E. neochamaeclada* (of which the first and last are endemic to Angola). These three species all possess caruncles on their seeds (which are absent in the seeds of the other three members of sect. *Tirucalli* in southern Africa) and among these species *E. congestiflora* is most closely related to *E. neochamaeclada*. These two have slender branches with pale felt on young growth, while the branches of *E. carunculifera* are much thicker with darker felt on young growth. The stems in *E. neochamaeclada* are even more slender than in *E. congestiflora* and in *E. neochamaeclada* the plant has a low-growing, almost mat-like habit with much more branching towards the tips of the major branches. Leaves are generally smaller in *E. neochamaeclada* than in *E. congestiflora*. While the cyathia in both are of a similar breadth, they are slightly longer in *E. congestiflora* and the styles are more than double the length of those in *E. neochamaeclada*.

History
Euphorbia congestiflora was first recorded in Angola by Friedrich Welwitsch in August 1859 (mixed in with *E. negromontana* on the sheet *Welwitsch 632* at LISU). More recently it was collected in January 1950 in Angola by J. Brito Teixeira and was located in Namibia for the first time in July 1976 by O.A. Leistner and others. Leach described it in 1970, but he never saw living plants of it, as it mainly occurs in the relatively inaccessible, southernmost part of Angola. His description did not include details of the female cyathia and these are supplied here for the first time.

Fig. 5.402. *Euphorbia congestiflora*. **A**, tip of shoot with leaves (scale 2 mm). **B**, female cyathium (scale 2 mm). **C**, female floret (scale 1 mm). Drawn from: *PVB 10373*, west of Oncocua, Angola (© PVB).

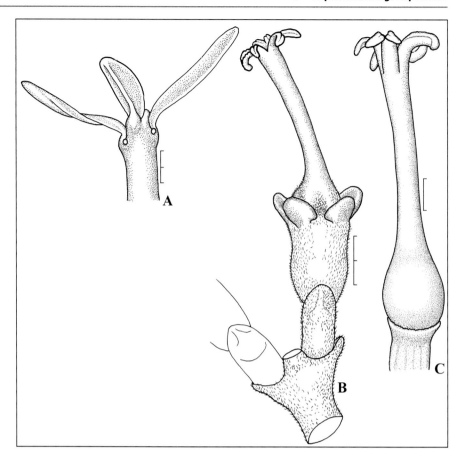

Euphorbia damarana L.C.Leach, *Bothalia* 11: 500 (1975). *Tirucalia damarana* (L.C.Leach) P.V.Heath, *Calyx* 5: 89 (1996). Type: Namibia, Damaraland, ± 64 km west of Khorixas, 27 July 1973, *Leach & Cannell 15064a* (LISC, holo.; K, M, PRE, SRGH, WIND, iso.).

Unisexual spineless succulent free-standing almost glabrous shrub 0.5–3 × 0.5–8 m, mainly branching from base of similar stem to 1 m tall, with slender hard woody rootstock and fibrous roots. *Branches* erect, occasionally alternately rebranching, terete and articulated at joints, 0.3–2 m × 6–12 mm, without distinct tubercles, smooth and finely tawny- to white-pubescent near growing tips, rigid, brownish green to green becoming grey-green with age; *leaf-rudiments* towards apices of branches, alternate, 2.5–27 × 2.5–8 mm, very slightly fleshy, erect, flat (rarely slightly longitudinally folded upwards), fleeting, narrowly to broadly obovate to elliptic, very sparsely tomentose (densely when still small), green above but purplish and faintly veined below, with entire margins, obtuse, tapering strongly into small swollen base but epetiolate, with small dark brown irregularly globular stipules on either side at base. *Synflorescences* terminal or lateral near apices of branches (rarely lower down), finely tomentose, each of 1–8 densely clustered sessile unisexual cyathia (1–5 and more scattered in females), each subtended by 2–4 small scale-like broadly ovate tomentose red to brown bracts 1–3 × 1.5–4 mm (often with small irregular brown stipules at base); *cyathia* obconical in males to almost cylindrical in females, densely tomentose, 6–14 mm broad (6–10 mm broad in females, ± 6 mm long below insertion of glands in males, ± 8 mm long in females), with 5 very short obovate lobes with truncate and dentate margins, reddish green, densely tomentose outside; *glands* 5, transversely elliptic and widely separated, 2–5 mm broad, usually slightly to strongly deflexed, dull red-brown, flat above in males and slightly swollen in females, outer margins usually somewhat undulating to crisped; *stamens* with finely tomentose pedicels, bracteoles filiform to hand-like to cuneate and membranous and woolly towards apices, densely intermingled with bracteoles and gradually forming raised woolly dome above lobes (but division into 5 fascicles often quite clear); *ovary* ellipsoidal, densely tomentose, surrounded by dense mat of tightly interwoven bracteoles somewhat embedded in tomentum, raised on stout tomentose pedicel < 1 mm long; styles 1–2 mm long, divided nearly to base, initially erect and later becoming horizontally spreading. *Capsule* 20–30 mm diam., obtusely 4- to 6-ribbed, slightly flattened-spherical, tawny-tomentose, yellow-brown, exserted from cyathium and ascending to slightly decurved (whole cyathium bending over) on thick slightly tomentose pedicel 2–4 mm long.

5.3 Sect. Tirucalli

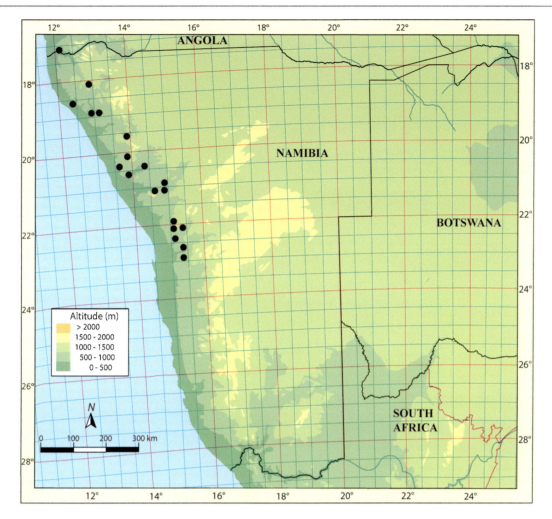

Fig. 5.403. Distribution of *Euphorbia damarana* (© PVB).

Fig. 5.404. *Euphorbia damarana*, plants in very flat area with almost no other vegetation, *PVB 11262*, Namib Desert, west of Usakos, Namibia, 6 Nov. 2008 (© PVB).

Distribution & Habitat

Primarily found in Namibia, *Euphorbia damarana* occurs in the tropical parts of the Namib Desert, from Tinkas Flats east of Walvis Bay to Orupembe and northwards, from where it just enters southern Angola. In Angola small numbers of plants occur on gentle slopes above the Cunene River, about 30 km east of Foz do Cunene, but this is the sole record for the country and they occur here with other species otherwise only known from Namibia, such as *Adenia pechuelii* and *Anacampseros albissima*.

Euphorbia damarana grows in gently sloping areas usually on substrates derived from igneous rocks (granites of various colours and dolerites in Damaraland), both on rocky hillsides and in the less stony ground in the flats and riverbeds. It is often the only succulent present and is frequently the dominant element in the scanty vegetation of the area. Plants are scattered in a regular manner over the countryside, apparently never crowding each other. Over their lifetime, organic material and fine sand accumulates under them so that the rootstock is surrounded by a layer of soil quite unlike that in the surroundings. Many small rodents, snakes, lizards (and even possibly small birds) find shelter around the bases of these plants, which enable them to escape partly from the harsh conditions of the area.

Diagnostic Features & Relationships

Euphorbia damarana forms enormous shrubs that often exceed 2 m in height and may spread over 5 m or more in diameter. Near the base of such a shrub the branches may reach 150 mm thick and become strong and woody. The

Fig. 5.405. *Euphorbia damarana*, stony plains, *PVB 12840*, north of Palmwag, Damaraland, Namibia, 23 Dec. 2014, with Cornelia Klak (© PVB).

Fig. 5.406. *Euphorbia damarana*, gentle stony slopes above Cunene River, *PVB 10698*, east of Foz do Cunene, Angola, 11 Jan. 2007 (© PVB).

younger branches that make up the bulk of the shrub are always cylindrical and rebranch mainly near their bases. When young they are covered with a dense, finely woolly tomentum of yellowish brown, curly hairs which gives them a faintly brownish colouring. These hairs wear off soon and are replaced with wax so that the branches are grey-green away from their tips. The inconspicuous, often slender leaves are borne near the tips of growing branches and soon fall off. They are not densely covered with hairs like the tips of the branches and the cyathia. Each leaf is dark green above and faintly purplish below and is accompanied by two small, somewhat irregularly shaped, dark brown stipules at their bases.

Flowering was observed in early November 2008 west of Usakos, but the species may well flower irregularly at any time in summer, whenever rains are received. Certainly, mature fruits were observed then too and mature fruits were also recorded by Leach (1975a) in June. Male cyathia are produced in clusters mainly at the tips of the branches. While the females are also often in clusters (though usually fewer in each cluster than in the males), they are often more scattered along the upper parts of the branches. The male cyathia are broader than the females and become covered above with a dense mound of hairs and old pedicels which projects somewhat above the glands and lobes. Within each cyathium, the division into five fascicles opposite the lobes by the larger bracteoles is usually quite clear. The whole structure is covered outside by a dense tomentum of whitish brown, curly hairs which extends onto the backs of the glands. From this mat of hairs, the glands protrude and are usually a dull red-brown above. Female cyathia are longer and slightly

Fig. 5.407. *Euphorbia damarana*, male cyathia, *PVB 11262*, Namib Desert, west of Usakos, Namibia, 6 Nov. 2008 (© PVB).

Fig. 5.408. *Euphorbia damarana*, female cyathia in different stages, *PVB 10698*, east of Foz do Cunene, Angola, 11 Jan. 2007 (© PVB).

narrower than the males and they are also covered outside with hairs. The ovary is much more densely covered with these hairs and the bracteoles around its base are somewhat matted into this tomentum. This tomentum extends onto the lower parts of the styles and only the upper, spreading and divided parts are more or less free of it. In each female cyathium there are between four and six styles and they are initially erect, spreading out as they mature. Fertilization of females rarely takes place in more than one per cluster of cyathia. A fertilized ovary rapidly expands into the slightly flattened-spherical, usually slightly ribbed capsules which may eventually exceed 25 mm in diameter. The dense tomentum covering the young ovary turns into a pale yellow-brown (tawny) mat of fine hairs on the mature capsule. Dehiscence of the capsule takes place by its falling apart on the bush, with the central stalk sometimes remaining on the plant. The seeds fall to the ground and appear to be dispersed by birds and rodents.

In their overall appearance, *E. damarana* and *E. gregaria* are very similar. *Euphorbia damarana* differs by its often larger size and the often much larger leaves. However, the most consistent differences are found in characters of the reproductive parts: both male and female cyathia are almost twice as broad in *E. damarana* as in *E. gregaria*; the ovaries are 4–6 locular with 4–6 seeds and there are 4–6 styles ontop of the ovary (mostly three locular and three styles in *E. gregaria*); in its younger stages (i.e. when the styles mature) the ovary itself is also far more massive than that of *E. gregaria*; when it is mature the capsule is subtended by a short and stout pedicel which holds it just beyond the remains

Fig. 5.409. *Euphorbia damarana*, capsules, *PVB 11262*, Namib Desert, west of Usakos, Namibia, 6 Nov. 2008 (© PVB).

Fig. 5.410. *Euphorbia damarana*, capsules, *PVB 10698*, east of Foz do Cunene, Angola, 11 Jan. 2007 (© PVB).

of the cyathium, while in *E. gregaria* the capsule is exserted from the cyathium for at least 10 mm.

History

Although only described in 1975, *E. damarana* has been known for a long time. The first recorded collection was made by M. Rautanen between Husab and Ubib in August 1892. It was also collected by Kurt Dinter in April 1913, at Jakkalswater (north of the Swakop River in Namibia) and it was known to Marloth from a collection by E. Reuning, made in September 1920 near Trekkopje, also near the Swakop River. The first collection from Angola was made by G. Brito Texeira, along the northern bank of the Cunene River and it was again gathered there in January 2007 by the present author.

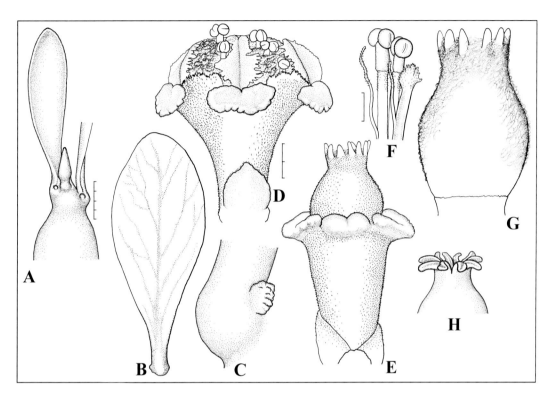

Fig. 5.411. *Euphorbia damarana*. **A**, tip of growing branch (scale 4 mm, as for **B**). **B**, leaf. **C**, base of leaf with stipular structure (scale of F = 0.5 mm). **D**, male cyathium (scale 2 mm, as for **E**, **H**). **E**, female cyathium. **F**, anthers and bracteoles (scale 1 mm, as for **G**). **G**, female floret with erect styles. **H**, top of female floret with diverging styles. Drawn from: *PVB 11262*, Namib Desert, west of Usakos, Namibia (© PVB).

Euphorbia gregaria Marloth, *Trans. Roy. Soc. South Africa* 2: 36 (1910). *Tirucalia gregaria* (Marloth) P.V.Heath, *Calyx* 5: 90 (1996). Type: Namibia, Kuibis, *Marloth 4683* (PRE, holo.; K, iso.).

Unisexual spineless succulent free-standing almost glabrous shrub 0.5–2.5 × 0.5–6 m, mainly branching from base of similar stem to 1 m tall, with slender hard woody rootstock and fibrous roots. *Branches* erect, occasionally alternately rebranching, terete and articulated at joints, 0.3–1.5 m × (4) 6–12 mm, without distinct tubercles, smooth and finely tawny- to white-pubescent near growing tips, rigid, purplish green to green becoming grey-green with age; *leaf-rudiments* towards apices of branches, alternate, 2.5–11 × 1.5–4 mm, fleshy, ascending to recurved, usually channelled above, fleeting, narrowly obovate to elliptic, sparsely tomentose, green to purple-green sometimes faintly veined below, with entire margins, obtuse, tapering often strongly into small swollen base but epetiolate, estipulate. *Synflorescences* terminal or lateral near apices of branches, finely tomentose, each of many densely clustered sessile unisexual cyathia (fewer and much more scattered in females), each subtended by 2–3 small scale-like broadly ovate tomentose red to brown bracts 1–4 × 1.5–2 mm; *cyathia* obconical to almost cylindrical in females, densely tomentose, 5–6.5 mm broad (± 3 mm long below insertion of glands), green suffused with red, with 4–5 obovate reddish green lobes densely tomentose outside and with truncate to dentate margins; *glands* 4–5, semi-circular to circular in males and transversely elliptic in females and widely separated in both sexes, 1.5–2 mm broad, spreading to slightly deflexed, dull red-brown, flat and slightly rugulose above in males to convex and green with red-streaked margins in females, outer margins flat and entire; *stamens* with finely tomentose pedicels, bracteoles filiform to hand-like to cuneate and membraneous and woolly towards apices; *ovary* ellipsoidal, densely tomentose, surrounded by dense mat of tightly interwoven bracteoles somewhat embedded in tomentum, soon exserted on stout tomentose pedicel ± 1 mm long; styles 1.5–2 mm long, divided nearly to base, often horizontally spreading. *Capsule* (10) 15–26 mm diam., obtusely 3 (–5)-ribbed, spherical to slightly obovoid, tawny-tomentose, greyish orange to red or brown, far exserted from cyathium and pendulous on densely tomentose decurved pedicel 10–25 mm long.

Distribution & Habitat

Euphorbia gregaria occurs over much of the very inhospitable territory of southern Namibia to the east of the Namib Desert, from north-east of Aus to the Fish River Valley and then eastwards past the Little and Great Karas Mountains to Ariamsvlei. It also occurs in the adjacent part of South Africa, where it is known in the very arid region from west of Vioolsdrift, via Goodhouse and the Naip Mountain eastwards to Aggeneys, Pofadder and further east to Kakamas.

Euphorbia gregaria mostly grows in barren, stony or gravelly plains and on the lower slopes of hills and mountains. In western Bushmanland at such mountains as Aggenys, Dabenoris and Naip it is found from around the base of the mountain to the summit, among white quartzitic rocks. Although it regularly occurs with succulents such as *Aloe dichotoma*, *Euphorbia avasmontana*, *E. gariepina*, *E. spinea*, *E. virosa* and *Kleinia longiflora*, in many of the areas where it is common it is the only succulent present. *Euphorbia gregaria* is often dominant, forming vast, diffuse colonies of scattered individuals over many kilometres of the countryside.

Diagnostic Features & Relationships

Euphorbia gregaria usually forms imposing, rounded shrubs that are often as much as 2 m tall and individuals may be much broader than this, since specimens up to 5 or 6 m in diameter are common. However, these shrubs can be rela-

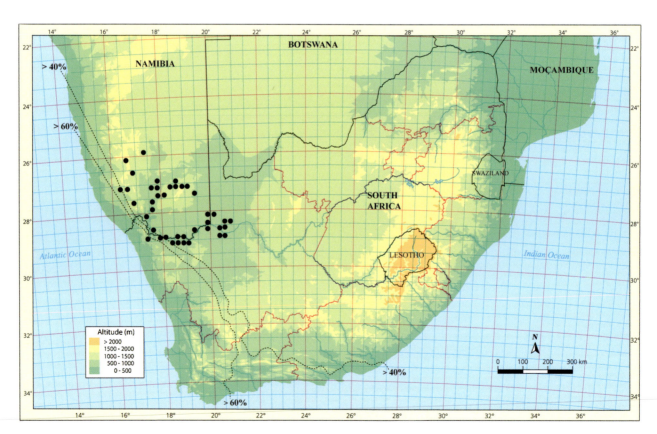

Fig. 5.412. Distribution of *Euphorbia gregaria* (© PVB).

5.3 Sect. Tirucalli

Fig. 5.413. *Euphorbia gregaria*, spaced out shrubs on stony gneissic hillside, *PVB 11119*, east of Jakkalswater, South Africa, 9 Jul. 2008 (© PVB).

Fig. 5.414. *Euphorbia gregaria*, shrub on gently sloping gravelly area, north of Onseepkans, Namibia, 10 Jan. 2000 (© PVB).

tively small (around 0.5 m tall) at maturity, when they grow on the tops of mountains, as they do around Aggenys, Dabenoris and Naip, in western Bushmanland and in the hills west of Vioolsdrift. Each shrub consists of many erect, very stiff, tough branches. In some areas, as for example around Keetmanshoop in Namibia, the branches rebranch sparingly or not at all. One will find, however, especially in the southern areas, such as between Vioolsdrift and Goodhouse much richer rebranching occurs (sometimes even with many short, horizontal branchlets, as in *E. gummifera*) and the shrubs are also shorter and broader. Older branches are grey-green from a thick covering of wax, while younger branches are purplish green to green and papillate, with a fine and dense covering of yellow-brown hairs at their tips and especially on very young growth, where the small leaf-rudiments (of somewhat variable shape) are also briefly present. In dry years the tips of the branches are often chewed off by various buck, though much of the fibrous interior often remains (as in Fig. 1.55 for *E. damarana*).

Flowering takes place between October and January and possibly later too, depending on rains. The cyathia are produced in large numbers, bringing with them a great deal of activity of insects (ants, flies and wasps) and the male cyathia, at least, give off a slight, sourish smell. They are not conspicuous, despite developing in dense clusters, and most of their colour is obscured by a dense covering of fine hairs that covers everything except the upper surfaces of the glands and the male florets. The female plants are remarkably difficult to locate in the early stages of flowering, since the female cyathia are often solitary (and only rarely in clusters of up to five, but even then they are still inconspicuous) and they can often be spotted easiest by the presence of a few old capsules or their remains. Once the capsules begin to develop, they become easier to find. The male cyathia are conical, with four or five flattish, often nearly circular glands that can be seen easily by their red-brown colour. In the female cyathium, the whole structure is nearly cylindrical, the glands are more swollen and more elliptic and are much less conspicuous

Fig. 5.415. *Euphorbia gregaria,* male cyathia, *PVB 11129,* towards Amam, west of Pofadder, South Africa, 18 Oct. 2008 (© PVB).

Fig. 5.416. *Euphorbia gregaria,* branches from male and female plants, 60 km south of Keetmanshoop, Namibia, 21 Jan. 2006 (© PVB).

(often green with red streaks on the outer margins), the lobes are much reduced and (inside the cyathium) the floret is surrounded by a densely packed carpet of bracteoles which are impossible to tease apart. Each female floret consists of a comparatively large ovary on the top of which are seated the three stumpy and deeply divided styles. The styles elongate slightly and spread out on the top of the ovary as the floret matures. The ovary itself is seated on a stout pedicel but the join of the pedicel to the ovary can only be seen if the dense tomentum that covers both of them is scraped away. Although the ovary is included within the cyathium in early stages and is still at least half-included in it when the styles mature, it gradually becomes further and further exserted until it is entirely outside the cyathium. Capsules are borne on quite stout, dependent pedicels and they dry up on the plant in this pendulous position, with the capsule tending to split slightly along the sutures of the locules as it dries. If knocked off, the capsule will often shatter on hitting the ground but some of them appear to remain intact and roll about among the stones and lower branches until blown away. Some of them even disintegrate on the plant and then the central column of the capsule may remain for a long time after the rest of the capsule has fallen off. Fragments of broken-up capsules and old, damaged seeds lie around the bases of female plants until blown away.

Of its various relatives, *E. damarana* is the most similar, but the plants in *E. gregaria* are generally smaller and more densely branched, with the branches considerably more rigid. The two differ also in their leaf-rudiments, which are shorter and more slender in *E. gregaria* than in *E. damarana*. The dense pubescence on the synflorescence is also differently coloured: whitish brown in *E. gregaria* and yellowish brown in *E. damarana*. Other differences are mentioned under *E. damarana*.

Another very similar species is *E. gummifera*. Just as *E. gregaria* is the dominant species over large portions of southern Namibia and the adjoining parts of South Africa, east of the winter-rainfall region (predominantly in the 'Nama

Fig. 5.417. *Euphorbia gregaria,* somewhat beaked capsules, 60 km south of Keetmanshoop, Namibia, 26 Dec. 2005 (© PVB).

Fig. 5.418. *Euphorbia gregaria,* stouter and rounder capsules, Steinfeld, Klein Karasberge, Namibia, 7 Jan. 2000 (© PVB).

Karoo'), so *E. gummifera* also forms characteristically vast, scattered colonies within the winter-rainfall region of the southern Namib Desert and the Orange River valley (predominantly in the 'Succulent Karoo'). The two species have been found growing together in a rocky valley east of the Fish River Mouth (*Bruyns 10024, 10025*, both at WIND), where they were separated altitudinally (*E. gregaria* occurring on the lower slopes at 400–600 m, *E. gummifera* higher up at 600–1000 m). The differences between them are discussed under *E. gummifera*.

Euphorbia gregaria is another of the species of *Euphorbia* whose roots are regularly parasitized by *Hydnora africana*.

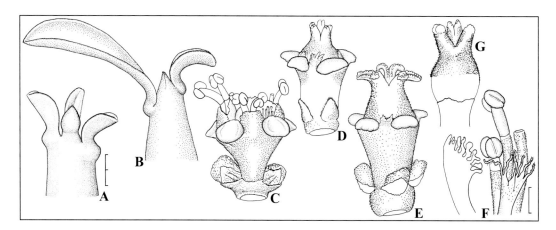

Fig. 5.419. *Euphorbia gregaria*. **A**, **B**, tip of growing branch (scale 2 mm, as for **B–E**). **C**, side view of male cyathium. **D**, **E**, side view of female cyathium. **F**, anthers and bracteoles (scale 1 mm, as for **G**). **G**, female floret (with indumentum scraped off lower down to show join to pedicel). Drawn from: *PVB 11129*, towards Amam, west of Pofadder, South Africa (© PVB).

As in several of the species from the arid regions and in contrast to others from the wetter parts, the germination of seedlings in *E. gregaria* does not appear to happen within the leaf-litter and humus that accumulates around the base of the plant. It appears that sometimes several seedlings germinate near one another, possibly all from a single capsule which has become buried as a whole during a heavy shower of rain. It is possible that some of the massive shrubs develop from these clusters, since it has been noted that not all the very big shrubs consist of a single specimen. However, once a plant is established, it is rare for young ones to develop within about 3 m of it. This gives rise to the very even spacing among individuals which is often seen in the vast colonies of this species.

History
Euphorbia gregaria was discovered by M.K. Dinter in November 1897. It was collected again by Rudolf Marloth in November 1908, a little east of Aus in Namibia. Shortly afterwards, in January 1909, it was collected by H.H.W. Pearson around Aggenys in Bushmanland, South Africa. Since then, records have gradually accumulated to show that it is a widespread and common species and Leach (1975a) summarized the differences between it and its close relatives, *E. damarana* and *E. gummifera*.

Euphorbia gummifera Boiss., *Cent. Euphorb.*: 26 (1860). *Tirucalia gummifera* (Boiss.) P.V.Heath, *Calyx* 5: 90 (1996). Type: South Africa, Cape, low-lying areas between Verleptpram and the mouth of the Orange River, Sept. 1830, *Drège 2944* (P, holo.; S, iso.).

Unisexual spineless succulent free-standing almost glabrous shrub 0.5–1.5 (1.8) × 0.5–2 m, branching and rebranching extensively from base of similar stem to 1 m tall and above, with slender hard woody rootstock and fibrous roots. *Branches* erect, repeatedly rebranching, cylindrical to obscurely and irregularly 3-ribbed and not articulated at joints, 30–300 × 7–10 mm, without distinct tubercles, smooth and finely tawny-pubescent near growing tips, rigid, grey-green; *leaf-rudiments* near apices of branches, alternate, 1–4 × 2–3 mm, fleshy, erect to spreading and often strongly recurved, channelled above, fleeting, elliptic to ± circular, finely pubescent to nearly glabrous above, dark red-green to dark green with red margins, with entire margins, obtuse, epetiolate, with very small dark brown ± spherical stipules on either side at base. *Synflorescences* terminal or lateral near apices of branches, finely tawny-pubescent, each of many densely clustered sessile unisexual cyathia (sometimes solitary, usually fewer and more scattered in females than males), each subtended by 2–4 small circular or scale-like red bracts 0.7–1.5 × 0.7–1 mm; *cyathia* obconical and shallowly cupular, finely white-pubescent becoming tomentose towards base, green usually suffused with red, 4–5.5 mm broad (± 2 mm long below insertion of glands), with 5 obovate red lobes pubescent outside and with deeply dentate margins; *glands* 4–6, transversely elliptic to circular, 1–1.5 mm broad, spreading to decurved, dark red to reddish green or yellow-green, outer margins flat and entire; *stamens* with glabrous pedicels, bracteoles filiform and sparsely pubescent; *ovary* ellipsoidal, sparsely pubescent, pale green suffused with red, on glabrous pedicel ± 0.8 mm long; styles ± 2 mm long, divided nearly to base, with branches often horizontally spreading. *Capsule* 10–12 mm diam., obtusely 3- to 4-lobed, finely pubescent, green to deep pink, very slightly exserted from cyathium on short erect pedicel ± 2 mm long.

Distribution & Habitat

Euphorbia gummifera is associated exclusively with the very arid part of the southern Namib Desert that receives its rainfall in winter. Here it is known mainly on the coastal plain from around Lüderitz and Hahlenberg to just south of the Orange River, where it occurs from east of Alexander Bay to near Kubus.

Plants are usually found on rocky lower slopes of hills or in gravelly flat plains. Exceptions to this are in the mountains east of Rosh Pinah, where it has been seen on some of the higher peaks, at altitudes from 600 to 1000 m, and especially on limestones.

Euphorbia gummifera generally occurs under conditions of exceptional aridity, where the annual rainfall rarely exceeds 50 mm per year, though this amount is usually supplemented by an unknown quantity of precipitation from fogs coming off the sea. Where it occurs in flat areas, *E. gummifera* is often the only larger shrub found.

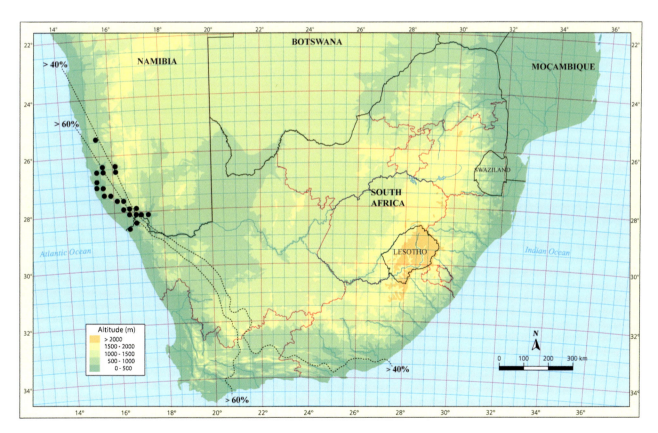

Fig. 5.420. Distribution of *Euphorbia gummifera* (© PVB).

Fig. 5.421. *Euphorbia gummifera,* low plants ± 0.8 m tall, east of Beesbank near Orange River, South Africa, 8 Jul. 2007, with Cornelia Klak and Julian Bruyns (© PVB).

Fig. 5.422. *Euphorbia gummifera,* densely branched plant ± 1.5 m tall, Namuskluft, Namibia, 8 Jul. 2013 (© PVB).

Diagnostic Features & Relationships

Euphorbia gummifera forms large shrubs that are often much broader than tall. Those occurring at higher altitudes are generally short and rarely exceed 0.5 m tall. Those occurring in flats and lower slopes often reach and exceed 1 m in height and 2 m in diameter. They are usually very extensively branched and rebranched so as to have quite a densely twiggy habit and the branches are remarkably rigid, even when quite young. This extensive branching begins in young plants in which the stem often seems to cease growth after 10–30 cm, after which growth is taken over by an extensive system of branches which again cease to elongate and start to rebranch often after only 10 cm or less. The branches are covered with a brownish pubescence near their tips and, if at all dry, they shrink somewhat to exhibit an extensive system of irregular, low, longitudinal ridges. Tiny, darkly coloured, usually only scantily pubescent leaf-rudiments are present near the tips of

5.3 Sect. Tirucalli

Fig. 5.423. *Euphorbia gummifera*, densely branched plant ± 1.5 m tall, *PVB 10026*, west of Gamkab River, Namibia, 12 Jul. 2005 (© PVB).

Fig. 5.424. *Euphorbia gummifera*, dark male cyathia in dense, almost spherical cluster, east of Beesbank near Orange River, South Africa, 8 Jul. 2007 (© PVB).

Fig. 5.425. *Euphorbia gummifera*, pale male cyathia, *PVB 10823a*, Annisfontein, near Orange River, South Africa, 8 Jul. 2007 (© PVB).

the branches and last for only a short while before drying up and falling off.

Flowering in *E. gummifera* usually takes place in the winter and spring months of June to September (though late summer rains may also bring on flowering earlier and fruiting by mid-July). Male cyathia are usually produced in dense clusters of up to around 10 (though in years of poor rainfall they may be solitary and then one can see that these clusters are formed from quick development of buds in the axils of the bracts below a terminal cyathium that develops slightly earlier – the development of these lateral cyathia being retarded in years of poor rainfall), while the females are generally solitary or in groups of up to three at the apex of the branches. During anthesis, the ovary protrudes somewhat from the cyathium to give the whole structure a pyramidal shape and the pedicel does not increase much in length once it is fertilised. The capsules are produced shortly afterwards and, before maturing, they sometimes assume a rather amazingly bright pink colour. They always remain erect and are exserted slightly from the old cyathium.

Although superficially similar to species such as *E. dregeana* and *E. mauritanica* (especially some forms of *E. mauritanica*, which can have thick, densely forked, grey branches with small and inconspicuous leaves, as for example at the top of Little Hellskloof, where they resemble *E. gummifera* quite closely), *E. gummifera* is a typical member of sect. *Tirucalli*. This one can easily see from the tawny-pubescence of young growth and the unisexual plants with clusters of cyathia around the tips of the stems, which are far more densely clustered in the male plants than in the females.

Euphorbia gummifera is closely related to *E. damarana* and *E. gregaria*. Vegetatively it is separated from them by the shrubs being mainly broader than tall, with their very much more extensively reforked branches and the manner in which the somewhat 3-ribbed branchlets run into the main branches without a definite articulation at their bases.

Fig. 5.426. *Euphorbia gummifera*, female cyathia, east of Beesbank near Orange River, South Africa, 8 Jul. 2007 (© PVB).

Fig. 5.427. *Euphorbia gummifera*, capsules, *PVB 10823a*, Annisfontein, near Orange River, South Africa, 8 Jul. 2007 (© PVB).

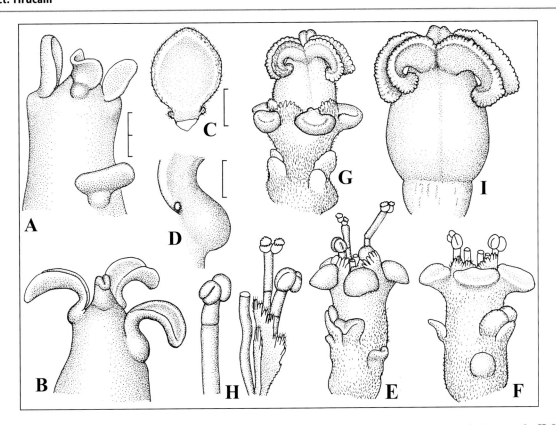

Fig. 5.428. *Euphorbia gummifera*. **A, B**, tip of growing branch (scale 2 mm, as for **B, E–G**). **C**, leaf-rudiment (scale 1 mm, as for **H, I**). **D**, base of leaf-rudiment with stipule (scale 0.5 mm). **E, F**, side view of male cyathium. **G**, side view of female cyathium. **H**, anthers and bracteole. **I**, female floret. Drawn from: *PVB 10823a,* Annisfontein, near Orange River, South Africa (© PVB).

The leaf-rudiments are also much shorter and broader than in the other two. Further differences are found in the floral parts. The cyathia in *E. gummifera* are somewhat shorter than in the others but, in particular, the dense pubescence that covers almost all parts of the cyathium in the others is considerably reduced in *E. gummifera*: the pedicels of the male florets are glabrous; instead of being densely covered with felt, the ovary is sparsely pubescent and as a consequence the capsule is almost glabrous and slightly shiny (nevertheless, as in the other two species, the pedicel of the female is surrounded with a dense packing of bracteoles which are not separable into individuals and which has to be scraped away to expose the pedicel); the lobes are sparsely pubescent and the outside of the cyathium is only tomentose near its base. Further differences are found in the much longer, often horizontally spreading branches of the style, the smaller, mostly 3-lobed capsules which are not exserted as far as in the others.

History
Among the members of sect. *Tirucalli* that occur in or around the Namib Desert, *E. gummifera* has been known by far the longest. It was discovered by J.F. Drège and his brother Carl sometime between the 16th and 29th of September 1830, along the lower parts of the Orange River, during their joint expedition to Namaqualand. Ernst Meyer had intended to give this collection the name *Euphorbia sessiliflora* but, as this had already been used, Boissier had to find another name for it.

White et al. (1941) showed photographs of plants of *E. gummifera* from north of Goodhouse, at Goodhouse and also from south of Lekkersing. Those around Goodhouse are specimens of *E. gregaria*, while the specimen at Lekkersing may be a very robust individual of *E. mauritanica*, since no members of sect. *Tirucalli* are known to grow so far to the south. Material included by Leach (1975a) under *E. gummifera* from between Vioolsdrift and Stinkfontein (such as *Werger 392*) belongs to *E. gregaria*.

Euphorbia tirucalli L., *Sp. Pl.* 1: 452 (1753). *Arthrothamnus tirucalli* (L.) Klotzsch & Garcke, *Abh. Königl. Akad. Wiss. Berlin* 1859: 62 (1860). *Tirucalia tirucalli* (L.) P.V.Heath, *Calyx* 5: 93 (1996). Lectotype (Leach 1973a): Ceylon, J. Commelijn, *Hort. Med. Amsterd. Rar. Pl.* 1: 27, t. 14 (1697).
Euphorbia rhipsalioides Welw. ex N.E.Br., *Fl. Trop. Afr.* 6 (1): 557 (1911), *nom. illegit., non* Lem. (1857). *Euphorbia tirucalli* var. *rhipsaloides* (Welw. ex N.E.Br.) Chev., *Rev.*

Fig. 5.429. *Euphorbia tirucalli*, ± 2 m tall, bushy growth, Jaglist, between Burgersfort and Chuniespoort, South Africa, 22 Jul. 2011 (© PVB).

Bot. Appl. Trop. 13: 547 (1933). Lectotype (Leach 1973a): Angola, Luanda, *Welwitsch 630* (BM; G, K, LISU, P, iso.).

Euphorbia laro Drake, *Bull. Mus. Hist. Nat. (Paris)* 1899: 307 (1899). Type: Madagascar.

Euphorbia media N.E.Br., *Fl. Trop. Afr.* 6 (1): 556 (1911). Syntypes: Tanzania, Malawi.

Euphorbia media var. *bagshawei* N.E.Br., *Fl. Trop. Afr.* 6 (1): 556 (1911). Syntypes: Uganda, *Bagshawe 898* (BM); *1196* (BM).

Euphorbia scoparia N.E.Br., *Fl. Trop. Afr.* 6 (1): 557 (1911). Syntypes: Eritrea, *Schweinfurth 345*; Sudan, *Muriel 67* (K); Ethiopia, *Schimper 896*.

Euphorbia tirucalli occurs widely in the tropical and subtropical parts of Africa, Madagascar, India and at least as far east as the Philippines and Taiwan. Linnaeus (1753) believed it to come from India (the name '*tirucalli*' appears to be of Tamil origin) and the lectotype selected by Leach came from cultivated plants in Sri Lanka. However, members of this section only occur as far east as Dhofar in Oman and do not occur naturally on the Indian subcontinent. So *E. tirucalli* is usually presumed to have originated in Africa (e.g. Leach 1973a). White et al. (1941: 102) believed it to be native to 'the southern half of Africa', while Leach (1973a) suggested eastern Africa or Angola. Croizat (1965) suggested that it may have originated in Madagascar and spread from there to its present, very wide distribution.

Molecular methods were used to show that the widely cultivated '*Caralluma subulata*' (= *Ceropegia adscendens* var *fimbriata*) is of Indian origin (Bruyns et al. 2010) and that the even more widely used *Aloe vera* originated in the

Fig. 5.430. *Euphorbia tirucalli*, tree ± 1.5 m tall in shallow soil on granitic slab, 12 km east of Nampula, Moçambique, 12 Nov. 2000 (© PVB).

5.3 Sect. Tirucalli

Fig. 5.431. *Euphorbia tirucalli*, isolated tree ± 4 m tall, near Muden, South Africa, 13 Jan. 2009 (© PVB).

Fig. 5.432. *Euphorbia tirucalli*, male cyathia in dense clusters, near Burgersfort, South Africa, 10 Sep. 2010 (© PVB).

Fig. 5.433. *Euphorbia tirucalli*, female tree with pubescent capsules, *E. Schmidt*, South Africa, 4 Oct. 2009 (© E. Schmidt).

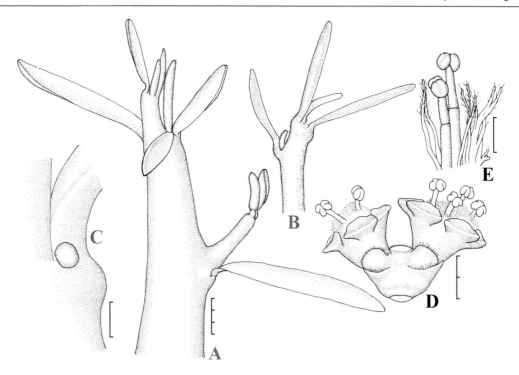

Fig. 5.434. *Euphorbia tirucalli*. **A**, **B**, tip of branch, smallest leaflets at top and top of branch finely furry (scale 3 mm, as for **B**). **C**, base of leaf with small ± spherical sparsely hairy stipule (scale 0.5 mm). **D**, cyme on male plant, with central cyathium already lost (scale 2 mm). **E**, anthers and bracteoles surrounding a minute sterile female floret (not shown) (scale 1 mm). Drawn from *PVB 6188a*, east of Morondava, Madagascar (© PVB).

Arabian Peninsula (Grace et al. 2016). In Bruyns et al. (2011) a collection of *E. tirucalli* from near Morondava, Madagascar was found to group with other species from Madagascar. Dorsey et al. (2013: fig. S1) investigated a total of 11 collections of *E. tirucalli* (2 from South Africa, 8 from near Tulear in Madagascar and one cultivated). According to the nuclear marker used, these grouped with other species in sect. *Tirucalli* from Madagascar. This suggests that Croizat was probably correct that *E. tirucalli* is a Madagascan species which has been spread widely by centuries of very successful cultivation, especially as a hedge.

Nevertheless, *E. tirucalli* is of such wide occurrence and importance in southern Africa that some figures of it are included here to assist with identification. Among all the species of sect. *Tirucalli* in Africa, *E. tirucalli* is distinctive for its tendency to develop into a tree with a distinct trunk (which is only exceeded in thickness and height by the Socotran *E. arbuscula*), for the finely longitudinally ridged branches and the small dark stipules at the bases of the leaves. It is also generally found in relatively moist areas. This is unlike the other southern African members of sect. *Tirucalli*, which are all associated with the Namib Desert on the western side of the subcontinent.

Addenda

6.1 Names of Uncertain Application or Excluded from Euphorbia and Naturally Occurring Hybrids

6.1.1 Names of Uncertain Application or Excluded from Euphorbia

Euphorbia aggregata A.Berger, *Sukkul. Euphorb.*: 92 (1906). Type: South Africa, Cape (missing).

No preserved material has been found of this species and it is difficult to be sure whether it falls under *E. ferox* or *E. pulvinata* or refers to the intermediates between them that occur widely over the eastern Karoo. It is therefore left among the names of uncertain application.

Euphorbia fleckii Pax, *Bull. Herb. Boiss.* 6: 738 (1898) Synypes: Namibia, Kuiseb, 1892, *Fleck 448a* (Z!), June 1893, *459a* (Z!), 1891, *466a* (Z!). Lectotype (designated here): Namibia, Kuiseb, 1891, *Fleck 466a* (Z 17073) = **Cynanchum viminale** (L.) Bassi ex L. (Apocynaceae).

Fleck 448a (Z 17074) is a mixture of *Cynanchum viminale* and *Kleinia longiflora*; *Fleck 448a* (Z 17075) is *Cynanchum viminale* (seen by the pairs of opposite branches and the fruit, as there are no flowers); *Fleck 459a* (Z 17076) is *Kleinia longiflora* (seen by the somewhat translucent branches and leaves, with several fine ridges below each leaf); *Fleck 466a* (Z 17073) is *Cynanchum viminale* with fruit. The last one is taken as the lectotype and so *E. fleckii* is another of the many synonyms of *C. viminale*.

Euphorbia parvimamma Boiss. in DC., *Prodr.* 15 (2): 86 (1862).

Euphorbia parvimamma was said to resemble *E. caput-medusae* but in all parts was smaller than *E. caput-medusae* (Berger 1899). However, the floral parts were not described by Boissier (1862) and no preserved material has been located. So it is impossible to be sure how this name should be applied.

Euphorbia viminalis L., *Sp. Pl.* 1: 452 (1753). Lectotype (Liede & Meve 1993): '*Felfel Tavil, Piper longum Aegyptium*' Veslingius in Alpinus, *De Plantis Aegypt.*: 190, t. 53 (1735). Epitype (Liede & Meve 1995: 47): *Bassi* in Herb Linn. 308.1(LINN). = **Cynanchum viminale** (L.). Bassi ex L. (Apocynaceae).

Several features of this illustration (such as the alternate branches and slender, alternate leaves) suggest that it is a poor representation of *Euphorbia tirucalli*, which would make *E. viminalis* another synonym of *E. tirucalli*, rather than a member of the Apocynaceae. However, the choice of an epitype ties it clearly to *C. viminale*.

Euphorbia viperina A.Berger, *Monatsschr. Kakteenk.* 12: 39 (1902). Type: South Africa, Cape of Good Hope?, collector unknown (missing).

The sinking of *E. viperina* under *E. inermis* (White et al. 1941) may not be correct, especially since Berger (1902) mentioned that his *E. viperina* stood between *E. caput-medusae* and *E. parvimamma* in respect of its flowers (though, as noted above under *E. inermis*, he later modified his description). The flowers of *E. parvimamma* are, however, unknown. No material of *E. viperina* seen by Berger has been traced, so its identity remains uncertain.

A small scrap of a branch labelled 'From the Type plant! Sent by Mr A. Berger, Aug. 1912' and filed at Kew under this name, was said to equal *E. inermis*. However, it is sterile, so there is no evidence to substantiate the statement that it belongs to *E. inermis*.

6.1.2 Naturally Occurring Hybrids

Croizat (1972: 102) stated that hybrids were plentiful among cultivated plants of *Euphorbia*. However, despite the fact that many species of *Euphorbia* grow socially in southern Africa, relatively few hybrids have been observed in nature. Hybrids are not known between species of the different subgenera and have only been recorded among members of sect. *Euphorbia* and of ser. *Meleuphorbia* of sect. *Anthacanthae* (subg. *Athymalus*).

(Subg. Athymalus) ser. Meleuphorbia

Euphorbia inconstantia R.A.Dyer, Rec. Albany Mus. 4: 93 (1931). Syntypes: Hellspoort, Oct. 1928, *Dyer 1076* (GRA); Grahamstown, Aug. 1927, *Dyer 1076* (GRA); 10 miles from Grahamstown on Queen's road, Nov. 1926, *Dyer 669* (GRA); Oct. 1927, *Dyer 1077* (GRA); Nov. 1926, *Dyer 669a* (GRA).

As mentioned under *E. polygona*, occasional hybrids are found between it and *E. heptagona* or between it and *E. pentagona*. Some of them are fertile, but they do not develop into breeding populations. A range of intermediates between these hybrids and their putative parents may also occur in the same habitats. Dyer (1931) believed that *E. inconstantia* was a naturally occurring hybrid between *E. polygona* and *E. pentagona* and he treated it as a species.

Other hybrids such as *E. obesa* ssp. *symmetrica* X *E. ferox* ssp. *ferox*, *E. ferox* ssp. *ferox* X *E. stellispina* and *E. heptagona* X *E. pseudoglobosa* ssp. *vlokii* have very occasionally been observed in nature (Figs. 6.2 and 6.3).

Fig. 6.2. Probably *Euphorbia ferox* ssp. *ferox* X *E. stellispina*, from among both parents, *PVB 11530*, near Rietbron, South Africa, 25 Dec. 2019 (© PVB).

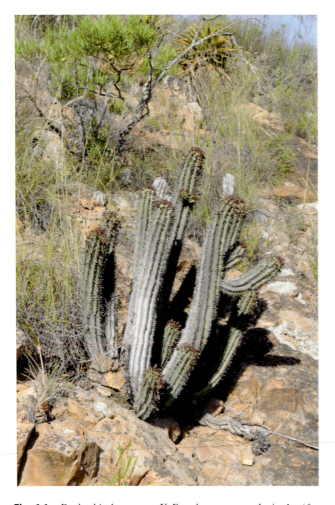

Fig. 6.1. *Euphorbia heptagona* X *E. polygona*, on rocks in dry 'fynbos', north of Joubertina, South Africa, 25 Oct. 2018 (with some pale *E. polygona* behind) (© PVB).

Fig. 6.3. Probably *Euphorbia obesa* ssp. *symmetrica* X *E. ferox* ssp. *ferox*, grown from seed collected in habitat on ssp. *symmetrica*, *PVB 3131a*, near Rietbron, South Africa, 25 Dec. 2019 (© PVB).

(Subg. Euphorbia) Sect. Euphorbia

Euphorbia anticaffra Lotsy & Goddijn, *Genetica* 10: 82 (1928).

Euphorbia bothae Lotsy & Goddijn, *Genetica* 10: 47 (1928).

In their study of Hybridization in the South African Flora, Lotsy & Goddijn (1928) wrote extensively about various species in the Eastern Cape and found that there was a range of intermediates between what they referred to as *E. ledienii* (*E. caerulescens* of the present work), *E. tetragona* and *E. triangularis*. Some of these intermediates were given names, as for the two above, though no types were designated. In some localities east and north-east of Grahamstown, there appear to be fertile hybrids between *E. caerulescens* and *E. triangularis*. The involvement of *E. tetragona* in these crosses has not been verified. These and back-crosses with both putative parents give rise to a bewildering variety of forms, but their origin has not been investigated by experimental crossing. This remains the only known area in southern Africa where such extensive hybridization appears to have taken place. The putative parents *E. caerulescens* and *E. triangularis* are known to be closely related and belong to the same major clade in sect. *Euphorbia* (Bruyns et al. 2011).

Fig. 6.4. *Euphorbia caerulescens* X *E. triangularis* 'curvirama', tree-like plant over 2 m tall, Fish River, west of Peddie, South Africa, 24 Oct. 2018. Here almost every stage between the two parents is found, some representing '*E. curvirama*' and others '*E. bothae*'. This tree probably represents '*E. curvirama*' (© PVB).

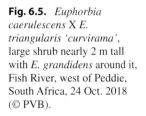

Fig. 6.5. *Euphorbia caerulescens* X *E. triangularis* 'curvirama', large shrub nearly 2 m tall with *E. grandidens* around it, Fish River, west of Peddie, South Africa, 24 Oct. 2018 (© PVB).

Euphorbia curvirama R.A.Dyer, *Rec. Albany Mus.* 4: 104 (1931). Type: South Africa, Cape, 28–30 miles from Grahamstown towards Peddie, Apr. 1928, *Dyer 1403* (GRA, holo.; K, PRE, iso.).

It appears that *E. curvirama* is a hybrid between *E. caerulescens* and *E. triangularis*. A wide range of intermediates between the two species can be observed on the slopes of the Fish River Valley, west of Peddie. These plants occur here over a particularly wide ecological range, from the low-lying, relatively arid, flat areas in the valley of the Fish River to the higher and steeper slopes.

Fig. 6.6. *Euphorbia caerulescens* X *E. triangularis* 'curvirama', Fish River, west of Peddie, South Africa, 24 Oct. 2018 (© PVB).

Fig. 6.7. *Euphorbia caerulescens* X *E. triangularis* 'bothae', capsules on shrub ± 1.5 m tall, Fish River, west of Peddie, South Africa, 24 Oct. 2018 (© PVB).

Euphorbia schinzii X *E. tortirama*

This unusual hybrid between a member of 'Tetracanthae' and a member of one of the other major clades in sect. *Euphorbia* (Bruyns et al. 2011) was noted growing among both parents by Fourie (1988). It is shown here in one of his photographs (Fig. 6.8).

A further hybrid in sect. *Euphorbia* was recorded by Venter (1987) and suspected to be *E. enormis* X *E. excelsa*.

Fig. 6.8. *Euphorbia schinzii* X *Euphorbia tortirama*, among both parents, near Ellisras, South Africa, c. 1986 (© S.P. Fourie).

6.2 The Species of Sect. *Euphorbia* and Sect. *Monadenium* in Moçambique

As a whole, Moçambique is mostly moist during the summer months, with an average annual rainfall that varies between 600 and 2000 mm (Jackson 1961) and dry during the winter. Despite this generally high rainfall, sect. *Euphorbia* is represented by 28 species in the country, of which 10 are endemic. They are found in two main habitats. One habitat is the relatively dry 'forest' of the coastal plain (as in Fig. 6.9). These forests have been much destroyed but remnants are found from the south around Maputo at least as far north as Beira. The other habitat where they are frequently encountered is in shallow pockets of soil on granitic domes. In Moçambique, granitic domes are mainly found north of the Zambezi River

Fig. 6.9. *Euphorbia bougheyi*, E. Schmidt, dense coastal bush near Mambone, Save River, Moçambique, 8 Jun. 2008 (© E. Schmidt).

and are often located in areas receiving an annual rainfall of well over 1000 mm. These domes vary from hillocks projecting 20–200 m or more from the surrounding bush (as in Fig. 6.10) to the granitic peaks of Mount Namuli, which rise to over 2400 m and are the highest mountains in the country. These granitic hills and mountains are sufficiently numerous and extensive that a unique and diverse flora, containing many endemic species, has developed on them. Molecular results indicate that, at least in *Euphorbia*, these species form a unique lineage that has arisen relatively recently.

Fig. 6.10. *Euphorbia corniculata*, several plants among short vegetation ontop of a granitic hill, with *Aloe mawii*, *Cynanchum viminale* and *Xerophyta*, *PVB 7714*, east of Nampula, Moçambique, 24 Dec. 1998 (© PVB).

In Moçambique the 'Tetracantheae' consists of *E. ambroseae*, *E. baylissii*, *E. contorta*, *E. corniculata*, *E. malevola*, *E. marrupana*, *E. namuliensis*, *E. plenispina* (incl. *E. stenocaulis*), *E. pseudocontorta*, *E. ramulosa*, *E. schinzii* and *E. unicornis*. Of these, *E. ambroseae*, *E. baylissii*, *E. malevola* and *E. schinzii* mainly inhabit coastal forest or inland scrub, with only *E. baylissii* endemic. *Euphorbia contorta*, *E. corniculata*, *E. marrupana*, *E. namuliensis*, *E. plenispina* (incl. *E. stenocaulis*), *E. pseudocontorta*, *E. ramulosa* and *E. unicornis* are all closely related to each other and they are exclusively found on granitic domes. All of them are endemic to Moçambique.

In Moçambique the remaining species are *E. angularis*, *E. bougheyi* (incl. *E. halipedicola*), *E. clavigera*, *E. confinalis*, *E. cooperi*, *E. decidua*, *E. grandicornis*, *E. grandidens*, *E. graniticola* (incl. *E. decliviticola*), *E. griseola*, *E. ingens*, *E. keithii*, *E. knuthii*, *E. lividiflora*, *E. mlanjeana* and *E. triangularis*. Of these, *E. cooperi*, *E. graniticola*, *E. griseola* and *E. mlangeana* are restricted to granitic domes while the rest mostly occur in coastal forest or on the slopes of the Lebombo Mountains. Most of them are found in the neighbouring countries of Malawi, South Africa, Swaziland, Tanzania and Zimbabwe and only *E. angularis* is possibly endemic to Moçambique.

Here, an account of those Moçambican taxa that are not found in the rest of southern Africa is given. It is possible that *E. persistentifolia* occurs along the steeper parts of the valley of the Zambezi River, but no specimens from Moçambique have been seen yet. Since keys were provided in Carter & Leach (2001) and in Burrows et al. (2018), they are not repeated here.

The Portuguese botanist António Rocha da Torre (Figs. 6.11 and 6.12) discovered several of these species. He made over 7000 collections in Angola and Moçambique and, although qualified as a pharmacist, he became one of the most important collectors of plants in these territories.

Fig. 6.11. António Rocha da Torre at the end of his days as a student in Coimbra, around 1930 (© J.A.R. de Paiva)

Fig. 6.12. (left to right) A Rocha da Torre, A. Salvador, M.F. Correia, J.A.R. de Paiva and M. Magalães, after a rainy day while collecting at Mepaluá near Mutáli, Moçambique, 25 Jan. 1964 (© J.A.R. de Paiva)

6.2.1 Sect. Euphorbia

Euphorbia ambroseae L.C.Leach, *Kirkia* 4: 15 (1964). Type: Moçambique, Sofala Prov., between Buzi and Gorongoza Rivers, ± 25', 29 Aug. 1960, *Leach 11238* (SRGH, holo.; CAH, G, K, LISC, LMA, PRE, iso.).

Euphorbia ambroseae var. *spinosa* L.C.Leach, *Kirkia* 10: 397 (1977). Type: Moçambique, Sofala Prov., ± 30 km west of Mambone, *Ambrose sub Leach 13159* (SRGH, holo.).

Bisexual weakly spiny glabrous succulent shrub 0.3–2 m tall with few branches around 4-angled erect stem (nearly cylindrical near base) 15–40 mm thick bearing fibrous roots at base. *Branches* ascending, 0.1–0.3 m × 10–20 mm, 4 (5)-angled, smooth, dull green sometimes paler between angles, sparingly rebranching; *tubercles* fused into 4 (5) low angles with branches slightly concave to flat between angles, laterally flattened and projecting ± 1 mm from angles, with spine-shields ± 1 mm broad, narrowly obovate, slightly separated or fused into slender continuous pale grey margin along angles, bearing 2 spreading and widely diverging grey spines 1–3 mm long; *leaf-rudiments* on tips of new tubercles towards apices of branches, 5–10 × 3–9 mm, erect, fleeting, deltate-ovate to nearly circular, acute, sessile, with small irregular brown stipular prickles to 2 mm long. *Synflorescences* towards apex of branch, each of 1 cyme in axil of tubercle, on peduncle 1–1.5 × ± 2 mm, with 3 horizontally disposed cyathia, central male or bisexual, lateral 2 bisexual and developing later each on peduncles ± 1.5 × 1.5 mm, with 2 broadly ovate bracts ± 2 × 2 mm subtending cyathia; *cyathia* broadly and shallowly funnel-shaped, glabrous, 5–7 mm broad (± 1 mm long below insertion of glands), with 5 broadly obovate lobes with fimbriate margins, pale greenish yellow suffused with red above; *glands* 5, transversely

Fig. 6.13. Distribution of *Euphorbia ambroseae* (© PVB).

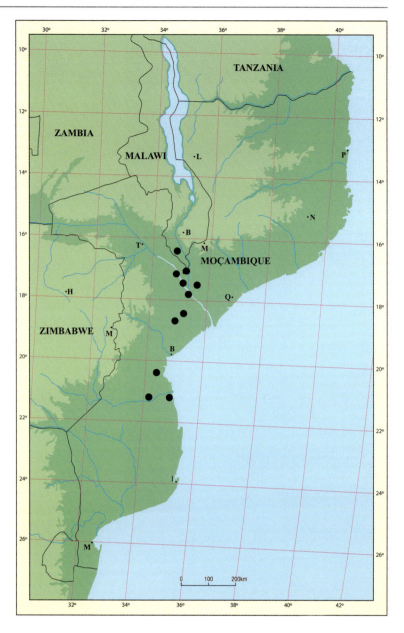

elliptic and closely contiguous, 3–4 mm broad, yellow-green with pinkish outer margin, spreading, inner margins flat, outer margins entire and spreading, surface between two margins ± flat; *stamens* glabrous, with bracteoles enveloping groups of males, with finely divided tips or filiform, glabrous; *ovary* globose and faintly 3-angled, glabrous, green, sessile to raised on pedicel ± 1.5 mm long; styles ± 2 mm long, branched from near base and widely spreading. *Capsule* 4–5 mm diam., obtusely 3-lobed, glabrous and pale red above, exserted ± 1.5 mm.

Distribution & Habitat
Euphorbia ambroseae is known in Malawi and Moçambique, where it mainly occurs in low-lying areas below 300 m.

Fig. 6.14. *Euphorbia ambroseae* (var. *spinosa*), *E. Schmidt*, Save River Bridge, Moçambique, Jan. 2016 (© E. Schmidt).

It usually inhabits dry, deciduous bush, growing in soft sand amongst leaf-litter.

Diagnostic Features & Relationships

Plants of *E. ambroseae* usually consist of an erect, 4-angled stem and a few spreading-erect branches. They may occasionally reach 2 m tall, when they are well sheltered, with the branches somewhat supported by the surrounding vegetation. However, they are usually much smaller, especially if exposed. The branches are usually around 10 mm thick. They are distinctive for the particularly slender spine-shields that typically form a fine, pale grey margin along the angles, contrasting strongly with the green of the branches. Many of the plants are almost spineless, while others have two small spines on the spine-shield and yet others have four small spines. The cyathia are relatively flat and, while not as broad as those of *E. quadrilatera*, they have a somewhat similar appearance.

Leach believed this species was close to *E. complexa*, but he also argued that *E. contorta* was closely related on account of their similar leaves, similar cyathia and capsules, though they differ markedly in the shape and colour of their branches. Unpublished molecular results show that *E. ambroseae* and *E. contorta* are indeed closely related.

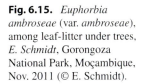

Fig. 6.15. *Euphorbia ambroseae* (var. *ambroseae*), among leaf-litter under trees, *E. Schmidt*, Gorongoza National Park, Moçambique, Nov. 2011 (© E. Schmidt).

Fig. 6.16. *Euphorbia ambroseae* (var. *ambroseae*), tip of branch with leaf-rudiments, *PVB 7699*, near Caia, Zambezi River, Moçambique, Apr. 2019 (© PVB).

Fig. 6.17. *Euphorbia ambroseae* (var. *ambroseae*), flowering branch in first male stage, *PVB 7699*, near Caia, Zambezi River, Moçambique, Oct. 2004 (© PVB).

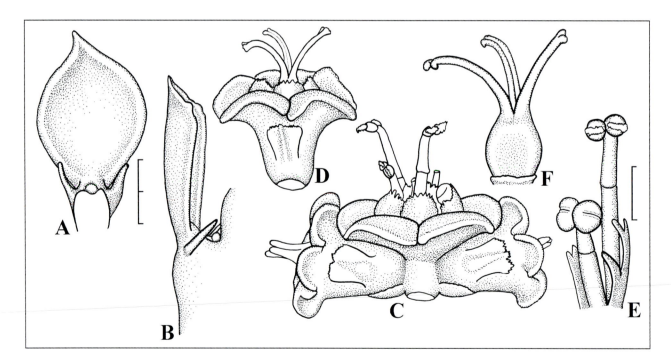

Fig. 6.18. *Euphorbia ambroseae* (var. *ambroseae*). **A**, young spines and leaf-rudiment from above (scale 2 mm, as for **B**–**D**). **B**, young spines and leaf-rudiment from side. **C**, side view of cyme in first male stage. **D**, side view of cyathium in female stage. **E**, anthers and bracteoles (scale 1 mm, as for **F**). **F**, female floret. Drawn from: *PVB 7699*, near Caia, Zambezi River, Moçambique (© PVB).

History

Euphorbia ambroseae was collected by Mrs Jean Ambrose in September 1960. She collected several succulents (including the beautifully flowered *Ceropegia* (*Orbea*) *halipedicola*) in the region around Mambone and seems to have first alerted Leach to the richness of the succulent flora of this area. However, it had been discovered already in July 1859 by John Kirk, who made a collection near the Zambezi River opposite the settlement of Sena. This was gathered when he accompanied Livingstone to the Zambezi River. He also collected *E. bougheyi* for the first time then and discovered *Aloe cryptopoda* in the same area.

Euphorbia angularis Klotzsch in W.Peters, *Naturw. Reise Mossambique*, Botanik, 1. Abt.: 92 (1861). Type: Moçambique, Nampula Prov., Moçambique distr., Goa Island, *Peters* (B, missing). Neotype (Leach 1970a): Goa Island, 5 m, 19 May 1961, *Leach 12361* (SRGH, neo.; K, LISC, MO, PRE, ZSS, iso.).
Euphorbia abyssinica var. *mozambicensis* Boiss. in DC., *Prodr.* 15 (2): 84 (1862). Type: Moçambique, *Peters* (B, missing).

Bisexual spiny glabrous densely branched succulent shrub 1–2 × 1–3 m with many branches around short initially 3- to 4-angled and later ± cylindrical and irregularly shaped erect stem bearing fibrous roots at base. *Branches* decumbent-erect, 0.1–1 m × 70–120 mm, 3 (4)-angled, deeply constricted into oblong deltoid segments 10–15 cm long, smooth, dull green, richly rebranching, branches prominently to only slightly winged; *tubercles* fused into 3 (4) often deeply wing-like angles sometimes with slightly undu-

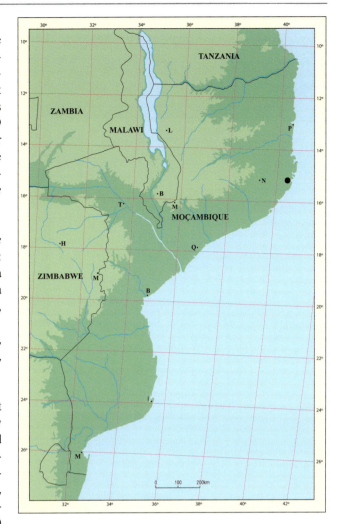

Fig. 6.19. Distribution of *Euphorbia angularis* (© PVB).

Fig. 6.20. *Euphorbia angularis*, habitat with vegetation around 1.5 m tall, *E. Schmidt*, Goa Island, Moçambique, Oct. 2005 (© E. Schmidt).

lating margins and flat to deeply concave between angles, laterally flattened and projecting 2–5 mm from angles, with spine-shields 2–5 mm broad, fused into continuous horny whitish grey margin along angles, bearing 2 spreading and widely diverging grey spines 6–12 mm long (sometimes additional small spines 1–3 mm long around leaf-axil; *leaf-rudiments* on tips of new tubercles towards apices of branches, ± 1.5 × 1.5 mm, erect to recurved, fleeting, ovate, acute, sessile, with small brown stipular prickles. *Synflorescences* many per branch towards apex, each of 1–3 cymes in axil of tubercle, each cyme on peduncle 1–3 × 4–6 mm, with 3 (or more) vertically disposed cyathia, central male, lateral 2 bisexual and developing later each on peduncles 3–5 × 4–6 mm, with 2 broadly ovate bracts ± 3 × 8 mm subtending cyathia; *cyathia* broadly cupular, glabrous, 8–10 mm broad (± 2 mm long below insertion of glands), with 5 obovate lobes with denticulate margins, yellowish green; *glands* 5, transversely elliptic and contiguous, 4.5–7 mm broad, yellow, spreading, inner and outer margins slightly raised, outer margins entire and spreading, surface between two margins slightly concave; *stamens* glabrous, with bracteoles enveloping groups of males, with finely divided tips or filiform, glabrous; *ovary* obtusely 3-lobed, glabrous, green, raised on short pedicel; styles 3.5–5 mm long, branched from near base to only in upper third. *Capsule* 14–18 mm diam., deeply 3-lobed, glabrous and bright red above, shortly exserted on slightly curved pedicel 5–7 mm long.

Fig. 6.21. *Euphorbia angularis*, plant around 0.8 m tall, *E. Schmidt*, Goa Island, Moçambique, Oct. 2005 (© E. Schmidt).

Fig. 6.22. *Euphorbia angularis*, flowering branch with broad tip and somewhat undulating angles, *E. Schmidt*, Goa Island, Moçambique, Oct. 2005 (© E. Schmidt).

Distribution & Habitat

Euphorbia angularis is only known on the small island of Goa, which lies at the entrance to Mossuril Bay near the town of Moçambique, in northern Moçambique. Here it is very common and dominates the vegetation, which grows on a mixture of broken-down remains of coral, sand and humus.

Diagnostic Features & Relationships

Euphorbia angularis is a shrubby species which reaches 2 m in height and may become considerably broader than tall. Its branches are mostly 3-angled and are constricted into segments of very variable shape and size, though many of them are not much longer than broad.

With its relatively broad yellow cyathia, *E. angularis* is somewhat similar to *E. cooperi* and *E. bougheyi*. It shares with the latter the slightly undulating margins of the branch-segments, as can be seen clearly in some of the young plants illustrated by Leach (1970a), though this is not found in branchlets on older plants. The deeply 3-lobed capsules are also similar, though they are smaller in *E. cooperi* and there the capsules are not always so deeply lobed. As Leach (1970a) said, *E. bougheyi* differs by the generally longer peduncles and the longer segments of the branches, but this could all be due to the very exposed and therefore considerably harsher and drier habitat of *E. angularis*.

History

Euphorbia angularis was one of the discoveries of the zoologist Wilhelm C.H. Peters, made during his expedition to Moçambique between June 1843 and August 1847. During most of this time he was based near Goa Island at Mossuril, so it is difficult to say exactly when he discovered *E. angularis* (Bauer et al. 1995, fig. 3).

It was first recollected by Leach in May 1961. A visit to Goa Island to see this species for himself and establish its identity had been a major goal for Leach. Two earlier attempts to get there were frustrated by illness and cartrouble, respectively, but his third attempt, together with R. Rutherford-Smith, was successful. Goa Island was also visited by Ernst Schmidt in October 2005 and he took the photographs of *E. angularis* that appear here.

Fig. 6.23. *Euphorbia angularis*, fruiting branch, *E. Schmidt*, Goa Island, Moçambique, Oct. 2005 (© E. Schmidt).

Euphorbia baylissii L.C.Leach, *J. S. African Bot.* 30: 213 (1964). Type: Moçambique, Inhambane Prov., Ponta Zavora, 3 Oct. 1963, *Leach & Bayliss 11796* (SRGH, holo.; BM, BRLU, COI, G, K, LISC, LMJ, MO, PRE, iso.)

Bisexual weakly spiny glabrous succulent shrub 0.5–1.5 (2) m tall with few branches around 4-angled erect stem (nearly cylindrical near base) 10–23 mm thick bearing fibrous roots at base. *Branches* ascending, 0.1–0.3 m × 6–20 mm, 4-angled, smooth, dull green sometimes paler (whitish) between angles, very sparingly rebranching; *tubercles* fused into 4 low angles with branches slightly concave between angles, laterally flattened and projecting 1–2 (4.5) mm from angles, with spine-shields ± 3–5 × 1 mm broad, narrowly obovate and acute below, widely separated, bearing 2 spreading and widely diverging fine grey spines 1.5–5 mm long; *leaf-rudiments* on tips of new tubercles towards apices of branches, ± 1.5 × 1 mm, erect, fleeting, ovate, acute, sessile, with small brown stipular prickles 1–3 mm long. *Synflorescences* towards apex of branch, each of 1 cyme in axil of tubercle, on peduncle ± 1.5 × 2 mm, with 3 horizontally disposed cyathia, central male or bisexual, lateral 2 bisexual and developing later each on peduncles ± 2.5 × 1.5 mm, with 2 broadly ovate bracts ± 1.5 × 2 mm subtending cyathia; *cyathia* broadly and shallowly funnel-shaped, glabrous, 5–6 mm broad (± 1.5 mm long below insertion of glands), with 5 broadly obovate lobes with fimbriate margins, pale yellow suffused with purple above; *glands* 5, transversely elliptic and closely contiguous, 3–4 mm broad, pale yellow sometimes with pinkish outer margin, spreading, inner margins flat, outer margins entire and spreading, surface between two margins ± flat; *stamens* glabrous, with bracteoles enveloping groups of males, with finely divided tips or filiform, glabrous; *ovary* globose and obtusely 3-angled, glabrous, pink above green below, raised on pedicel ± 2 mm long; styles ± 1.2 mm long, branched from base and widely spreading. *Capsule* 4–5.5 mm diam., obtusely 3-lobed, glabrous and purplish above and along sutures, exserted 2–3 mm.

Distribution & Habitat

Euphorbia baylissii is endemic to Moçambique, where it is found between Mambone, at the mouth of the Save River and Marracuene, just north of Maputo. It is a plant of coastal bush at altitudes of 50 m or less, growing on settled sand-dunes, where it flourishes among bushes in loose sand mixed with leaf-litter.

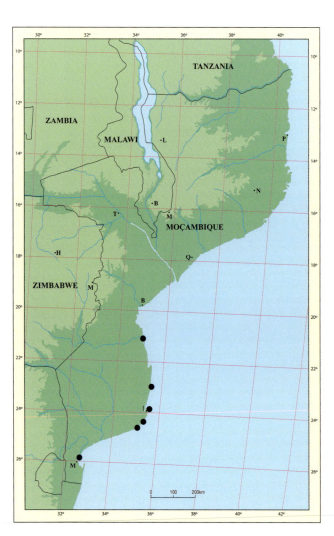

Fig. 6.24. Distribution of *Euphorbia baylissii* (© PVB).

Fig. 6.25. *Euphorbia baylissii*, plant in shade under trees, E. Schmidt, Ponta Zavora, Moçambique, Aug. 2006 (© E. Schmidt).

Fig. 6.26. *Euphorbia baylissii*, flowering branch, *PVB 8526*, Marracuene, Moçambique, Oct. 2004 (© PVB).

Fig. 6.27. *Euphorbia baylissii*, flowering branch, *PVB 7677*, west of Mambone, Save River, Moçambique, Apr. 2019 (© PVB).

Diagnostic Features & Relationships

A plant of somewhat spindly habit, *E. baylissii* usually consists of a slender stem and a few spreading branches, which are partly supported by the surrounding bushes. The colour of these branches is variable, from dark green-edged with whitish between the angles to uniformly dull green, but the spines are generally very fine and weak.

The cyathia of *E. baylissii* resemble those of *E. ambroseae* in being relatively broad and flat, but in *E. baylissii* there is less red or pink in them and the female floret is raised on a taller pedicel. The leaf-rudiments in *E. ambroseae* are also considerably larger (those in *E. baylissii* being very small) and the two species are not closely related, with *E. ambroseae* close to *E. contorta* and *E. baylissii* belonging to a group of Moçambican endemic species including *E. corniculata* and *E. ramulosa*.

History

Euphorbia baylissii was first collected by Margaret Moss in December 1935. She was curator of the Moss Herbarium at the University of the Witwatersrand, which was named after her husband, C.E. Moss. She regularly took groups of students from the University to coastal Moçambique (Gunn & Codd 1981), which is how she came to collect this species. It was also observed in the same area in 1939 by A.F. Gomes e Sousa and he made a sketch of this material that appeared as Fig. 1097 in White et al. (1941). Since it occurs right along the coast, it has been observed relatively often by visitors to the various coastal resorts in Moçambique. Leach named it for his regular travelling-companion Roy D.A. Bayliss, after whom he also named *Ceropegia* (*Tromotriche*) *baylissii*.

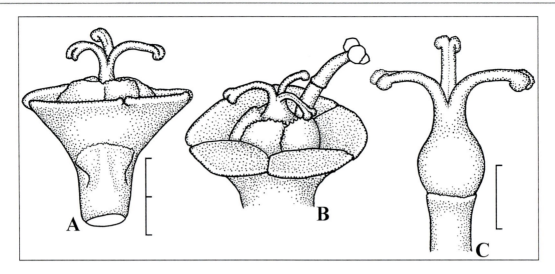

Fig. 6.28. *Euphorbia baylissii*. **A**, side view of lateral cyathium in female stage (scale 1 mm, as for **B**). **B**, lateral cyathium with female stage over. **C**, female floret (scale 1 mm). Drawn from: *PVB 7677*, west of Mambone, Save River, Moçambique (© PVB).

Euphorbia bougheyi L.C.Leach, *J. S. African Bot.* 30: 9 (1964). Type: Moçambique, Sofala Prov., Macuti Beach, near Beira, 1958, *A.S.Boughey 12206* (SRGH, holo.; CAH, COI, G, K, LISC, LMA, PRE, iso.).

Euphorbia halipedicola L.C.Leach, *J. S. African Bot.* 36: 42 (1970). Type: Moçambique, Sofala Prov., near Lake Gambue, *Leach & Wild 11130* (SRGH, holo.; K, LISC, PRE, iso.).

Bisexual spiny glabrous succulent shrub to small tree 1.5–5 (10) m tall with many branches around and often exceeding unbranched initially (3) 5- to 6 (9)-angled and later cylindrical erect stem 0.06–0.15 m thick bearing fibrous roots at base. *Branches* ascending-erect, 0.5–1 m × 50–200 mm, (2) 3- to 4-angled, deeply constricted into oblong deltoid or ovate segments 20–40 cm long, smooth, dull green, sparingly rebranching, branches deeply winged; *tubercles* fused into (2) 3–4 deeply and thinly (2–4 mm thick near edges) wing-like angles with markedly undulating margins and deep grooves between angles, laterally flattened and projecting 2–4 mm from angles, with spine-shields 2–4 mm broad, separate or fused into narrow continuous horny pale grey margin along angles, bearing 2 spreading and widely diverging grey spines 3–15 mm long and 2 smaller prickles 0.5–2 mm long alongside axillary buds; *leaf-rudiments* on tips of new tubercles towards apices of branches, 3–5 × 3–5 mm, erect, fleeting, ovate, acute, sessile, with small irregular brown stipular prickles ± 1 mm long. *Synflorescences* many per branch towards apices, each of 1–4 (6) cymes in axil of tubercle, each cyme on peduncle 2.5–12.5 × 4–5.5 mm, with 3 randomly to vertically disposed cyathia, central

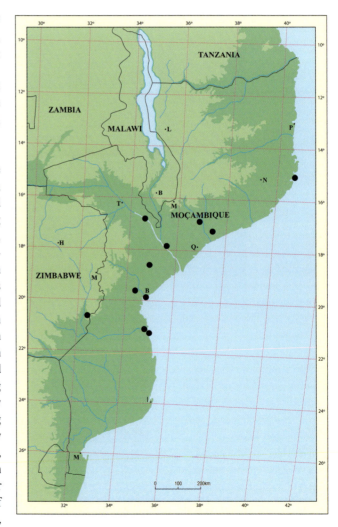

Fig. 6.29. Distribution of *Euphorbia bougheyi* (© PVB).

male, lateral 2 bisexual and developing later each on peduncles 3–12 × 2.5–4 mm, with 2 broadly ovate bracts 3.5–6 × 5.5–7 mm subtending cyathia; *cyathia* broadly cupular to funnel-shaped, glabrous, 8–10 mm broad (± 2 mm long below insertion of glands), with 5–6 obovate lobes with denticulate margins, yellowish green; *glands* 5–6, transversely elliptic and contiguous, 4–5.5 mm broad, yellow, spreading, inner margins slightly raised, outer margins entire and spreading, surface between two margins convex to concave; *stamens* glabrous, with bracteoles enveloping groups of males, with finely divided tips or filiform, glabrous; *ovary* globose, glabrous, green, ± sessile, with 3-lobed 'calyx' around ovary; styles 2.5–4.5 mm long, branched from near base to middle. *Capsule* 15–23 mm diam., deeply 3-lobed, glabrous and bright red above, slightly exserted on erect pedicel 3–8 mm long.

Distribution & Habitat
Euphorbia bougheyi is known in Malawi, Moçambique, Tanzania and Zimbabwe. In Moçambique it is mainly confined to coastal habitats, where it grows in dense and dry, deciduous bush along with other succulents. In the Gorongoza Park it occurs somewhat further from the coast, but there it is associated with similar vegetation.

Diagnostic Features & Relationships
Euphorbia bougheyi forms large plants of a shrubby nature or occasionally may develop into trees to 5 m tall or more. The stem is somewhat slender for so large a plant (perhaps since it is usually supported partly by the surrounding vegetation) and does not appear to form such a robust trunk as in *E. cooperi*. Nevertheless, with age it loses its angled shape to become cylindrical lower down. The branches are particularly broad and strongly winged. The thinness of these wings and their undulating or sinuate edges is one of the most characteristic features of the species. The spines on the branches are mostly relatively weak and rarely exceed 10 mm long.

Leach described two species here, *E. bougheyi* and *E. halipedicola*. The first he compared with *E. dawei* from Uganda and the second with *E. breviarticulata*, which is widespread in low-lying areas in East Africa, though it is not confined to the low-lying parts. However, they are very difficult to distinguish from each other and in Carter & Leach (2001: 394) this is achieved by separating them as follows: Spine-shields decurrent, very narrow = *E. bougheyi*; Spine-shields c. 5 mm in diameter, rounded to very obtusely triangular...= *E. halipedicola*. However, in *E. halipedicola* the spine-shields were said by Leach (1970a: 48) to provide the wings of the branches with a 'narrow continuous horny margin' and since the wings themselves were given as 'up to

Fig. 6.30. *Euphorbia bougheyi*, after rains, *PVB 7675*, east of Mambone, Save River, Moçambique, 14 Dec. 1998 (© PVB).

2 mm thick at the...margins' it is not possible for the spine-shields to reach 5 mm broad in this species. Furthermore, Leach recorded both species from the area between the Save and Buzi Rivers (in Manica e Sofala province, south of Beira: *Leach 11254* for *E. bougheyi*; *Leach 11130*, for example, for *E. halipedicola*) and it seems unlikely that two such similar species occur in this area. Consequently, both are placed here under the one name, *Euphorbia bougheyi*. This species is best separated from *E. breviarticulata* (where the wings along the angles may also be somewhat undulating and the fruits are similarly large) by the shorter spines (not exceeding 15 mm long, whereas the longest spines reach 40–80 mm in *E. breviarticulata*). From the somewhat similar *E. grandicornis* it is separated by the undulating margins of the angles, the considerably shorter spines and the considerably larger capsules.

Fig. 6.31. *Euphorbia bougheyi*, *M.Lötter*, Jays Camp, Moçambique, Aug. 2006 (© E. Schmidt).

Fig. 6.32. *Euphorbia bougheyi*, flowering branch, *E. Schmidt*, Gorongoza National Park, Moçambique, Jun. 2012 (© E. Schmidt).

Fig. 6.33. *Euphorbia bougheyi*, fruiting branch, *E. Schmidt*, Mambone, Save River, Moçambique, 8 Jun. 2008 (© E. Schmidt).

History

Euphorbia bougheyi was first collected by John Kirk in June 1859 along the lower Zambezi River. Almost 100 years later, in 1958, A.S. Boughey collected it again, this time near Beira. From these plants, material was eventually obtained for description.

Euphorbia contorta L.C.Leach, *Kirkia* 4: 17 (1964). Type: Moçambique, Nampula Prov., 30 miles west of Ribáuè, 2000', 16 May 1961, *Leach & Rutherford-Smith 10889* (SRGH, holo.; G, K, LISC, PRE, iso.).

Bisexual spiny glabrous succulent shrub 0.3–1 (3) m tall with few branches from erect or spreading usually 5-angled stem 25–30 mm thick bearing fibrous roots at base. *Branches* spreading, 0.05–0.3 m × 20–30 mm, (4) 5- to 7 (9)-angled, smooth, purplish grey to grey-green, sparingly rebranching; *tubercles* fused into (4) 5–7 (9) low often spiralling angles with branches slightly concave to flat between angles, laterally flattened and projecting ± 2 mm from angles, with spine-shields 5–10 × 2–3 mm, narrowly obovate and acute below, well separated along angles, bearing 2 spreading and widely diverging pink later grey spines 3–10 mm long; *leaf-*

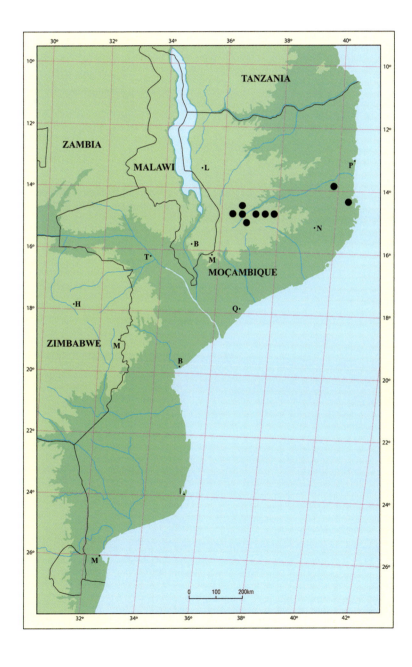

Fig. 6.34. Distribution of *Euphorbia contorta* (© PVB).

rudiments on tips of new tubercles towards apices of branches, 8–22 × 7–12 mm, erect, fleeting, elliptic-ovate, acute, sessile, with small stipular prickles to 3 mm long. *Synflorescences* towards apex of branch, each of 1 cyme in axil of tubercle, on peduncle 1–1.5 × ± 2 mm, with 3 horizontally disposed cyathia, central male, lateral 2 bisexual and developing later each on greenish peduncles ± 1.5 × 1.5 mm, with 2 ovate emarginate bracts ± 2 × 1–2 mm subtending cyathia; *cyathia* broadly and shallowly funnel-shaped, glabrous, 5–6 mm broad (± 2 mm long below insertion of glands), with 5 broadly obovate lobes with fimbriate margins, glaucous pale green to pinkish red; *glands* 5, transversely elliptic and closely contiguous, 2–3 mm broad, green suffused with red to yellow with red outer margin to wholly orange, spreading, inner margins flat, outer margins entire and spreading, surface between two margins ± flat; *stamens* glabrous, red, with bracteoles enveloping groups of males, with finely divided tips or filiform, glabrous; *ovary* globose and obtusely 3-angled, glabrous, pinkish red, raised on pedicel ± 1 mm long; styles ± 2 mm long, branched from near base and widely spreading, red. *Capsule* 3.5–4 mm diam., obtusely 3-lobed, glabrous and pale red above, exserted ± 1 mm.

Fig. 6.35. *Euphorbia contorta*, on low granitic outcrop, ± 1 m tall, *PVB 7732*, east of Cuamba, Moçambique, 26 Dec. 1998 (© PVB).

Fig. 6.36. *Euphorbia contorta*, on low granitic outcrop with *Xerophyta*, ± 1.5 m broad, *PVB 7732*, east of Cuamba, Moçambique, 26 Dec. 1998 (© PVB).

Fig. 6.37. *Euphorbia contorta*, ± 1.5 m diam., among bushes at base of granitic dome, *PVB 8551*, north of Alua, Moçambique, 12 Nov. 2000 (© PVB).

Distribution & Habitat

Euphorbia contorta is, at present, only known from Moçambique, where it is found between Cuamba, Lioma and Ribáuè at altitudes of 550–700 m. It always occurs on granitic 'whalebacks' but usually only near the foot of the larger hills, mostly in slightly loose soils and not in the very tight mats of soil found higher up on the bigger domes. Rather similar plants are found some 20–50 km from the coast near Pemba and are also included here. These occur among deciduous bushes in shallow, loose, humus-rich soil on granitic outcrops (Figs. 6.37 and 6.38).

Diagnostic Features & Relationships

Euphorbia contorta forms rather strange-looking and untidy plants usually not taller than 1 m, often with a relatively long stem that is initially erect then spreads horizontally, with a few similar, horizontally spreading branches. Plants in the eastern area (near Pemba) are more untidily branched (as in Fig. 6.37) and may develop into a dense and entangled mass of branches, though some remain small. All of this is armed with stout and fairly long spines. Both the stem and the branches have a greyish colour and the angles are often twisted spirally along the stem (less so along the branches). Those around Cuamba are uniformly grey-green and often 5- to 7-angled while those near Pemba are 4-angled and usually have a paler area between the angles. With their often much longer stems, these northern plants bear some resemblance to the Tanzanian endemic *E. quadrilatera*. In *E. contorta* the leaf-rudiments are fairly

Fig. 6.38. *Euphorbia contorta*, ± 2 m tall, *J. Burrows*, between Nacarroa and Namapa, Moçambique, Sep. 2009 (© J. Burrows).

Fig. 6.39. *Euphorbia contorta*, branch with leaf-rudiments, *PVB 7732*, east of Cuamba, Moçambique, 26 Dec. 1998 (© PVB).

Fig. 6.40. *Euphorbia contorta*, tip of branch with cyathia in first male stage, *PVB 7732*, east of Cuamba, Moçambique, Oct. 2007 (© PVB).

Fig. 6.41. *Euphorbia contorta*, tip of branch with cyathia in first male stage, *PVB 8600*, Mt Wesse, east of Malema, Moçambique, Nov. 2003 (© PVB).

6.2 Sect. *Euphorbia* and Sect. *Monadenium* in Moçambique

Fig. 6.42. *Euphorbia contorta*, flowering, *PVB 7708*, south of Nacala, Moçambique, Jul. 2001 (© PVB).

large and grey-green with a red edging. The spines also start off somewhat reddish and soon assume their characteristic grey colour.

As in several of the Moçambican endemics, in the floral parts there is a strong tendency towards a reddish colouring, with the cyathia, lobes, glands, stamens and ovary all at least suffused with red, sometimes quite brightly reddish (though this is not the case in all plants).

History

The first record of *E. contorta* was made in 1937 by António Rocha da Torre on the mountain near Lioma. His collections of this species from 1937 were grown in the botanical garden in Coimbra. It is possible that the sketch by A.F. Gomes e Sousa, published as Fig. 828 in White et al. (1941), may also be of *E. contorta*.

In the eastern area near Pemba, it was first observed in November 1998 and again in November 2000. A flowering branch of a plant from this area was shown under *E. lividiflora* in Burrows et al. (2018: 455).

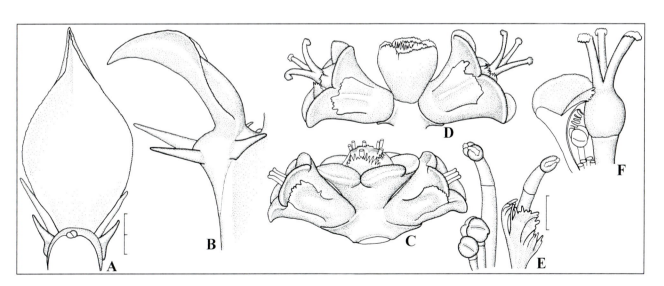

Fig. 6.43. *Euphorbia contorta*. **A**, young spine-complex and leaf-rudiment from above (scale 2 mm, as for **B–D**). **B**, young spine-complex and leaf-rudiment from side. **C**, side view of cyme at end of first male stage. **D**, side view of cyme in early female stage. **E**, anthers and bracteoles (scale 1 mm, as for **F**). **F**, female floret inside dissected cyathium. Drawn from: **A**, **B**, **C**, *PVB 8611*, Mitucue, Moçambique. **D**, **E**, **F**, *PVB 7732*, east of Cuamba, Moçambique (© PVB).

Euphorbia corniculata R.A.Dyer, *Fl. Pl. Africa* 27: t. 1076 (1949). Type: Moçambique, Nampula Prov., near Nampula, Mutivase, Aug. 1943, *Gomes e Sousa sub PRE 27271* (PRE).

Bisexual spiny glabrous densely branched dwarf cushion-forming succulent shrub 0.1–0.2 × 0.1–0.8 m with many branches around similar almost cylindrical erect stem bearing fibrous roots at base. *Branches* decumbent-erect, 50–150 × 10–15 mm, indistinctly 6- to 8-angled and almost cylindrical, not constricted into segments, smooth, green when young becoming dull metallic grey to brownish all over except in grooves between rows, rebranching; *tubercles* indistinctly arranged into 6–8 rows separated by slight grooves between rows, laterally flattened and projecting 2–4 mm, with spine-shields 3–5 mm broad covering tubercles and fused into continuous horny metallic grey covering separated only by narrow grooves between rows of tubercles, bearing 2 spreading and slightly diverging grey spines 4–8 mm long; *leaf-rudiments* on tips of new tubercles towards apices of branches, ± 1 × 1 mm, erect, fleeting, ovate, acute, sessile, with minute brown stipular prickles < 1 mm long. *Synflorescences* several per branch towards apex, each of 1 cyme in axil of tubercle, each cyme on peduncle 1–2 × 1–2 mm, with 3 horizontally disposed cyathia, central male, lateral 2 bisexual and developing later each on peduncles 2–4.5 × 1.5 mm, with 2 broadly ovate bracts ± 1.5 × 2 mm subtending cyathia; *cyathia* conical cupular, glabrous, 3.5–4.5 mm broad (± 1.5 mm long below insertion of glands), with 5 obovate lobes with den-

Fig. 6.44. Distribution of *Euphorbia corniculata* (© PVB).

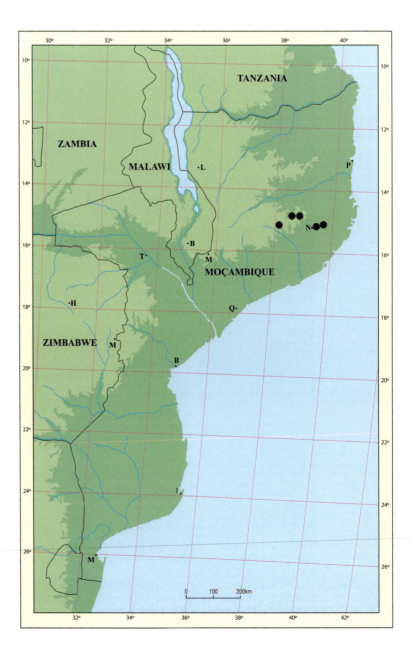

ticulate margins, purplish red; *glands* 5, transversely elliptic and slightly separated to contiguous, 1.5–3 mm broad, purplish red, spreading-erect, outer margins entire and spreading, surface between two margins slightly concave; *stamens* glabrous, with bracteoles enveloping groups of males, purplish red, with finely divided tips or filiform, glabrous; *ovary* obtusely 3-lobed, glabrous, green suffused with red, raised on short pedicel ± 0.5 mm long; styles 3–4 mm long, branched to 1 mm above base. *Capsule* 3–4 mm diam., obtusely 3-lobed, glabrous, dark purple-red above and green below, very shortly exserted on erect pedicel ± 1 mm long.

Distribution & Habitat
Euphorbia corniculata is endemic to northern Moçambique, where it is fairly widespread but has been rarely collected. Nevertheless, it is common on many of the bare granitic slabs around Nampula, though it is rarely so common elsewhere. Generally, it occurs at altitudes of 200–600 m and is confined to fairly barren spots with few grasses and sedges (as in Fig. 6.10). This prevents the plants from being burnt in the frequent fires of winter and spring.

Diagnostic Features & Relationships
Euphorbia corniculata forms relatively low dense shrubs that are rarely more than 15 cm tall, though they are often

Fig. 6.45. *Euphorbia corniculata*, large, dense plant nearly 1 m diam. ontop of granitic hill, *PVB 7714*, east of Nampula, Moçambique, 24 Dec. 1998 (© PVB).

Fig. 6.46. *Euphorbia corniculata*, medium-sized plant ± 0.3 m diam. with *Aloe mawii* and *Cynanchum viminale*, *PVB 7714*, east of Nampula, Moçambique, 24 Dec. 1998 (© PVB).

half a metre or more in diameter. With its almost cylindrical branches with indistinct rows of tubercles (not fused into angles) and small amounts of green tissue restricted to narrow grooves between the predominant metallic grey of most of the branches, this species has a most unusual appearance. At a first glance, the almost total lack of green suggests that the plants are dead, but this impression is quickly dispelled by the white sap that pours out of any damaged piece. When in flower, the purple-red of the cyathia contrasts somewhat against the dull colour of the branches. However, the cyathia are fairly small and are usually not produced in large numbers, so they do not render the plants any more conspicuous than before.

Dyer commented on the unusual appearance of this species but did not engage in any speculation on its affinities. Croizat (1972) and Leach (1976c) believed that this and *E. unicornis* were the closest relatives in Africa of species such as *E. milii* (i.e. the 'Christ thorns' of Madagascar). This was because of the lack of clear spine-shields on the branches and the way the branches are almost completely covered by a dark, metallic grey. Croizat (1972: 219) was so impressed with the unusual appearance of *E. corniculata* (which he knew from cuttings sent to him by Leach) that he even described a section especially for this and *E. unicornis* (namely sect. *Sterigmanthoides* Croizat, Webbia 27: 219 (1972). Type: *E. corniculata*). Analysis of molecular data showed that this interesting idea was improbable (Bruyns et al. 2006). These new investigations showed, for the first time, that *E. corniculata* and *E. unicornis* belong to a small group of relatively recently derived species that are endemic to the granitic mountains of northern

Fig. 6.47. *Euphorbia corniculata*, flowering, *PVB 7714*, east of Nampula, Moçambique, Dec. 2005 (© PVB).

Fig. 6.48. *Euphorbia corniculata*, with capsules, *PVB 7714*, east of Nampula, Moçambique, 24 Dec. 1998 (© PVB).

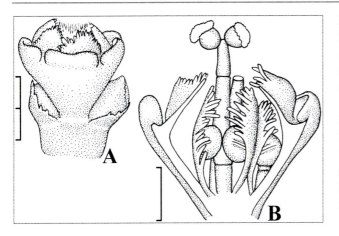

Fig. 6.49. *Euphorbia corniculata*. **A**, side view of cyme at end of first male stage (scale 2 mm). **B**, dissected side view of cyme in first male stage (scale 1 mm). Drawn from: *PVB 7714*, east of Nampula, Moçambique (© PVB).

Moçambique. Among these, *E. plenispina* has red or yellow cyathia but in the rest the cyathia are red to purple-red. All the others have separate spine-shields. In some the spines are paired, in *E. marrupana* they are either paired or solitary and in *E. unicornis* they are always solitary.

History

This species was first collected by António Rocha da Torre in 1934. He showed it to A.F. Gomes e Sousa, who later sent some plants to Dyer at the Botanical Research Institute, Pretoria. Dyer finally described it when these plants flowered in cultivation in Pretoria.

Euphorbia decidua P.R.O.Bally & L.C.Leach, *Kirkia* 10: 293 (1975) [described in *Candollea* 18: 347 (1963), but two syntypes were indicated and no holotype was selected, so this was invalid]. Type: Zambia, Mweru Wantipa, 10 Jan. 1944, *Bredo sub Bally E271* (K, holo.).

Bisexual spiny dwarf glabrous succulent usually not more than 120 mm tall with rosette of 5–30 mostly simple deciduous branches from apex of much swollen subterranean spindle-shaped stem 30–80 mm thick from which fibrous roots arise. *Branches* erect to sprawling, 50–200 × 5–8 mm and becoming slender and cylindrical near base, 3- to 4- (–6) angled, not constricted into segments, smooth, green to reddish or purplish green, usually simple (rebranching near base); *tubercles* fused into 3–4 (–6) wing-like angles with branches concave between angles, conical and truncate, laterally flattened and projecting 2–3 mm from angles, with small ± elliptic spine-shield 2–3 mm broad at tip of tubercle around spines and remaining separate from next below, bearing 2 often slightly deflexed initially red and later grey-brown slightly swollen spines 1.5–4.5 mm long; *leaf-rudiments* on tips of new tubercles towards apices of branches, 2–3 × 1–1.5 mm, erect, fleeting, lanceolate, sessile, with minute brown stipular prickles < 1 mm long (ascending to spreading *leaves* 30–50 × 10 mm, lanceolate and tapering into petiole 10–20 mm long sometimes arising in rosette before branches from apex of stem). *Synflorescences* of many cymes from apex of stem when branches absent, cymes 1- to few-branched, on ± terete peduncles 20–40 × 3–4 mm, each cyme with 2–6 cyathia on peduncles 5–30 × ± 1.5 mm (rarely few ± sessile solitary cymes near tips of branches in axils of tubercles, each cyme with 3 transversely disposed

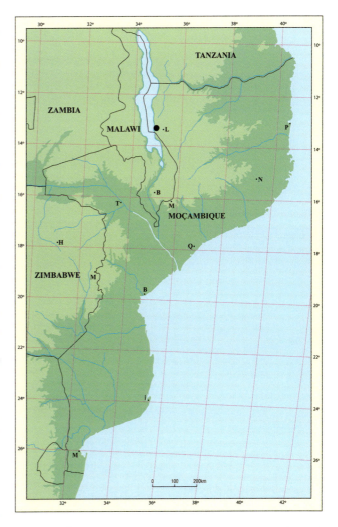

Fig. 6.50. Distribution of *Euphorbia decidua* in Moçambique (© PVB).

cyathia, central male, lateral 2 bisexual, with subtending ovate bracts 2–3 × 1.5–2 mm); *cyathia* cupular, glabrous, 3.5–4 mm broad, with 5 obovate lobes with incised margins, green; *glands* 5, transversely elliptic and contiguous, 1.5–2.5 mm broad, green to slightly yellow, spreading, upper surface slightly concave; *stamens* glabrous, bracteoles enveloping groups of males, with finely divided tips, glabrous; *ovary* obtusely 3-lobed, glabrous, soon exserted beyond cyathium on curved pedicel 3–6 mm long; styles 1.5–2 mm long, branched to near base and often spreading horizontally. *Capsule* 4.5–5 mm diam., deeply 3-lobed, glabrous, green to red, exserted on decurved and later erect pedicel ± 6 mm long.

Distribution & Habitat

Euphorbia decidua is widely distributed across south tropical Africa in D.R. Congo, Malawi, Tanzania, Zambia and Zimbabwe, with a single record from northern Moçambique. This was made near Meponda along the shores of Lake Malawi. In this area it was common on flat, pebbly ground among scattered *Brachystegia* trees and shrubs of *Xerophyta*.

Fig. 6.51. *Euphorbia decidua*, two plants among gravel and small tufts of grass, *PVB 7733*, east of Meponda, Moçambique, 27 Dec. 1998 (© PVB).

Fig. 6.52. *Euphorbia decidua*, *PVB 7733*, east of Meponda, Moçambique, 27 Dec. 1998 (© PVB).

Fig. 6.53. *Euphorbia decidua*, fleshy leaves and spiny branches just above tuber in cultivation, *PVB 7733*, east of Meponda, Moçambique, Feb. 2005 (© PVB).

Diagnostic Features & Relationships

Euphorbia decidua is one of the few geophytic species in sect. *Euphorbia* that occur in Africa south of the equator. Plants are relatively small, rarely more than 8–10 cm tall, with a rosette of erect, branches that wither and die off during the dry season. Very occasionally (as in Fig. 6.53), a few quite large fleshy leaves may develop first after the dry season, after which the normal branches emerge. The branches are relatively slender and soft, with weak spines.

Flowering may occasionally take place near the tips of the branches during the growing season. More usually the cyathia are borne on short, nearly terete peduncles (as in Fig. 6.54), that arise in September or October from the apex of the stem before the branches begin to grow. These peduncles branch and rebranch from the axils of bracts beneath cyathia to produce further cyathia in a unique manner within sect. *Euphorbia*. However, the cyathia are not unusual. They have the exserted female floret and an exserted quite deeply 3-lobed capsule held on an initially decurved pedicel (and smooth seeds) that are typical of many species such as *E. griseola*, *E. knuthii* and *E. stellata*. *Euphorbia decidua* has been found to be closer to these than to any of the other small, shrubby, spiny species.

History

Despite its wide distribution, *E. decidua* was recorded for the first time only in January 1944, perhaps because much of the collecting done earlier in these regions took place in the dry season when it is not visible. It was first recorded in Moçambique in December 1998 and this remains the only known gathering from the country.

Fig. 6.54. *Euphorbia decidua*, cyathia developing just above tuber on short, cylindrical, branched peduncles, *PVB 7733*, east of Meponda, Moçambique, 30 Oct. 2007 (© PVB).

Euphorbia grandicornis subsp. **sejuncta** L.C.Leach, *J. S, African Bot.* 36: 39 (1970). Type: Moçambique, ± 6 miles east of Nampula, ± 1250', 23 Jul. 1962, *Leach & Schelpe 11437* (SRGH, holo.; K, LISC, MO, PRE, iso.).

This subspecies of *E. grandicornis* is found in a small area of northern Moçambique, between Ribáuè and Nampula. Here it is usually common in shallow soils on low granitic domes, with a range of other succulents, which may include *E. corniculata*.

Subsp. *sejuncta* differs from subsp. *grandicornis* by its smaller stature, rarely exceeding 1 m in height. It also differs by the frequent occurrence of 2-angled (i.e. flat) branches, though 3- or even 4-angled branches are also found. The branches may be attractively mottled with cream, though this is not found on all plants.

Fig. 6.55. Distribution of *Euphorbia grandicornis* subsp. *sejuncta* (© PVB).

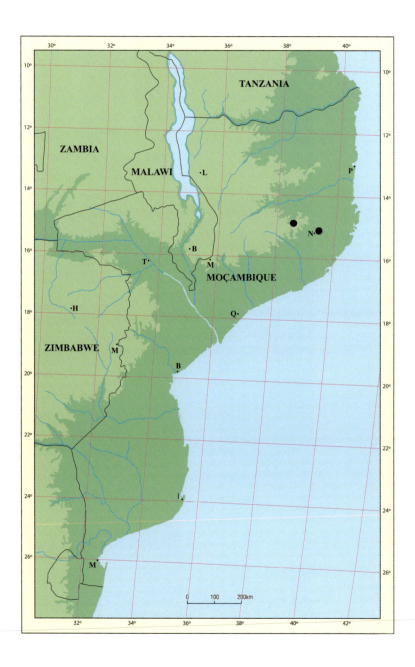

Fig. 6.56. *Euphorbia grandicornis* subsp. *sejuncta*, on low granitic outcrop with *Cynanchum viminale* and *E. tirucalli*, *PVB 8543*, east of Nampula, Moçambique, 12 Nov. 2000 (© PVB).

Fig. 6.57. *Euphorbia grandicornis* subsp. *sejuncta*, attractively mottled 2-angled branch in flower, *PVB 8543*, east of Nampula, Moçambique, Mar. 2005 (© PVB).

Fig. 6.58. *Euphorbia grandicornis* subsp. *sejuncta*, in fruit, *PVB 8568*, east of Ribáuè, Moçambique, 17 Nov. 2000 (© PVB).

Euphorbia graniticola L.C.Leach, *Kirkia* 4: 18 (1964). Type: Moçambique, Manica Prov., 10 miles SSW of Vila Pery (Chimoio), ± 2000', cult. Oct. 1962, *Leach 5103* (SRGH, holo.; G, K, LISC, LM, PRE, iso.)

Euphorbia decliviticola L.C.Leach, *J. S. African Bot*. 39: 13 (1973). Type: Moçambique, Nampula Prov., ± 2.5 km south of Posta Agricola, Ribáuè, cult. Nov. 1969, *Leach & Schelpe 11427* (SRGH, holo.; K, LISC, PRE, iso.).

Bisexual spiny glabrous succulent shrub to tree 0.5–4 × 1–2 m with many branches in dense crown at or above ground level from usually unbranched weakly 5- to 9- angled and later cylindrical erect stem 0.2–3 m tall bearing fibrous roots at base. *Branches* spreading then ascending, simple, 0.2–1 m × 30–65 mm, 4- to 6-angled, slightly constricted into ellipsoidal to oblong segments, gradually dying off and wearing away, smooth, greyish green; *tubercles* fused into 4–6 thin wing-like angles with deep grooves between angles, laterally flattened and projecting 1–5 mm from angles, with spine-shields slightly separated or later united into continuous horny grey margin 1.5–3 mm broad along angles, bearing 2 spreading and widely diverging grey spines 4–8 mm long; *leaf-rudiments* on tips of new tubercles towards apices of branches, 5–10 (35) × 2–4 mm, erect, fleeting, linear-ovate, acute, sessile, with small irregular brown tooth- to

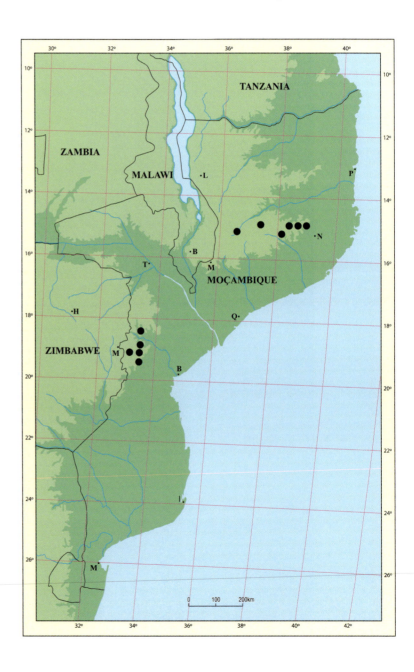

Fig. 6.59. Distribution of *Euphorbia graniticola* in Moçambique (© PVB).

Fig. 6.60. *Euphorbia graniticola*, about 0.8 m tall and almost stemless, on granitic outcrop among *Coleochloa* and *Myrothamnus*, *PVB 8744*, Mt Manga, north of Chimoio, Moçambique, 31 Dec. 2000 (© PVB).

scale-like stipules < 1 mm long. *Synflorescences* many per branch towards apex, each of (1–) 3 cymes in axil of tubercle, each cyme on peduncle ± 1 mm long, with 3 (horizontally to) vertically disposed cyathia, central male, lateral 2 bisexual and developing later each on short peduncle ± 1 × 1.5 mm, with 2 ovate to deltate bracts ± 2 × 1.5 mm subtending lateral cyathia; *cyathia* conical-cupular, glabrous, 3–4 (5) mm broad (2–3 mm long below insertion of glands), with 5 obovate lobes with deeply incised margins, greenish yellow; *glands* 5, transversely elliptic and contiguous, 1.5–2 mm broad, yellow to orange-yellow, spreading, inner margins slightly raised, outer margins entire and spreading, surface between two margins ± flat; *stamens* glabrous, bracteoles enveloping groups of males, with finely divided tips or filiform, glabrous; *ovary* subglobose, glabrous, green, soon exserted and projecting sideways on pedicel 4–6 mm long; styles 1–2 mm long, branched ± to middle. *Capsule* 5–6 mm diam., obtusely 3-angled, glabrous and green to red, exserted on decurved and later erect pedicel 5–9 mm long.

Distribution & Habitat
Euphorbia graniticola is known in Malawi in the hills east of Mangochi at the southern end of Lake Malawi. In Moçambique it is more widespread, from the centre around Chimoio and further north between Lioma and Nampula.

This species always grows in shallow pockets of soil on solid slabs of granite, usually where these are relatively gently sloping but occasionally on steep slopes too. As usual, it occurs amongst tufts of *Coloechloa*, often with other succulents and it can be very common in some localities. Its thick stems also provide it with some protection against fires.

Fig. 6.61. *Euphorbia graniticola*, about 3 m tall and festooned with several species of orchids, on low granitic outcrop, *PVB 8747*, north of Vanduzi, Moçambique, 1 Jan. 2001 (© PVB).

Fig. 6.62. *Euphorbia graniticola*, about 1.5 m tall and in fruit, *PVB 8570*, east of Ribáuè, Moçambique, 17 Nov. 2000 (© PVB).

Fig. 6.63. *Euphorbia graniticola*, to about 4 m tall, on steep and slippery, severely burnt granitic slope with Cornelia Klak, *PVB 8744*, Mitucue, Moçambique, 21 Nov. 2000 (© PVB).

Diagnostic Features & Relationships

Euphorbia graniticola forms robust shrubs to small trees to 2 m tall in the south of its distribution, but further north it is generally more tree-like and can reach 3 m in height. Nevertheless, even in this area there is often a mixture of shrubs and trees in the populations. The branches are typically grey-green and are shed below the crown so that the stem is bare of branches lower down.

Flowering takes place from September until February and here there are usually three cymes per axil, with the cyathia usually disposed vertically, though this is variable. The cyathia are fairly small, but they are produced in large numbers so that flowering trees become yellowish from them.

Leach distinguished two species here, *E. graniticola* in central Moçambique and *E. decliviticola* further north. The second was distinguished by its broader and shorter cyathia with broader bracts, with the cyathia more randomly disposed on the cymes and by its more rounded and obtusely 3-angled capsules on a slightly shorter pedicel. He also distinguished them on the size of the plant, with *E. graniticola* primarily an acaulescent shrub and *E. decliviticola* caulescent and tree-forming. However, in the descriptions he gave *E. graniticola* as shrubby or reaching 2 m as a small tree and *E. decliviticola* as shrub-like or tree-like to 3 m, so there is no clear difference here. Plants from near Lioma in the area where *E. decliviticola* occurs were found to have capsules that closely resembled those of *E. graniticola* and had similarly narrow cyathia, so the differences in the cyathia break down in some areas. Carter & Leach (2001: 394) separated them on the height of the plant, length of the branch-segments and length of the spines, which is not at all what Leach intended and does not work in the material seen between

6.2 Sect. *Euphorbia* and Sect. *Monadenium* in Moçambique

Fig. 6.64. *Euphorbia graniticola*, in flower, *PVB 8570*, east of Ribáuè, Moçambique, 17 Nov. 2000 (© PVB).

Fig. 6.65. *Euphorbia graniticola*, in flower, *PVB 7796*, SW of Lioma, Moçambique, Dec. 2017 (© PVB).

Fig. 6.66. *Euphorbia graniticola*, in fruit, *PVB 8570*, east of Ribáuè, Moçambique, 17 Nov. 2000 (© PVB).

1998 and 2004. Consequently, only one species is recognized here.

The relationships of *E. graniticola* appear to lie with South African species such as *E. tetragona* and *E. triangularis* and it is not related closely to *E. mlanjeana*, as Leach (1973) believed it to be. He also suggested *E. memoralis* as a close relative, but this has not been investigated.

History

This species was known to Leach by June 1959 and perhaps well before this too, though there are no earlier records.

Euphorbia lividiflora L.C.Leach, *Kirkia* 4: 20 (1964). Type: Moçambique, Sofala Prov., 18 miles south of Muda, SE of Lake Gambue, 28 Aug. 1961, *Leach 11129A* (SRGH, holo.; COI, G, K, PRE, iso.).

Bisexual spiny glabrous succulent tree 1.5–4 (10) m tall with many branches in broad crown from woody unbranched initially 5- or 6-angled later cylindrical erect trunk 0.05–0.1 (0.25) m thick bearing fibrous roots at base. *Branches* ascending and later spreading, 0.5–2 (3) m × 30–50 mm, 5-angled, slightly constricted into ellipsoidal segments, smooth, greyish green later becoming woody towards bases, repeatedly rebranching in widely separated whorls to near tips, branchlets 3- to 4-angled, slightly segmented, 12–25 mm thick, grey-green; *tubercles* fused into 3–5 wing-like angles with branches slightly concave between angles, laterally flattened and projecting 1–2 mm from angles, with small well separated spine-shields ± 5–7 × 3 mm, bearing 2 spreading and widely diverging grey spines 3–5 mm long (sometimes absent); *leaf-rudiments* on tips of new tubercles towards apices of

Fig. 6.67. Distribution of *Euphorbia lividiflora* (© PVB).

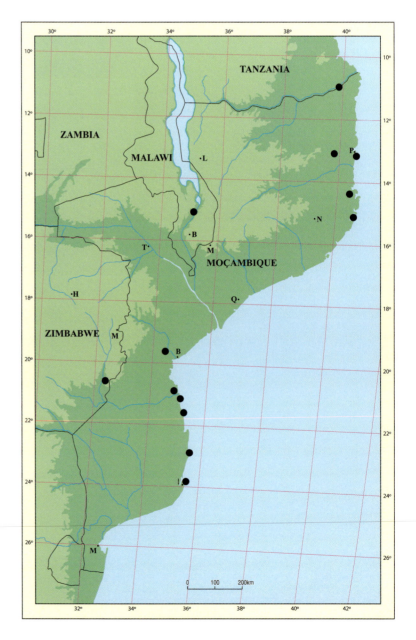

branches, 4–13 × 4–5 mm, erect, fleeting, ovate, acute, sessile, with small irregular brown stipular prickles ± 0.5 mm long. *Synflorescences* several per branch towards apex, each of 1 cyme in axil of tubercle, each cyme on purple-red peduncle 3–6 mm long, with 3 horizontally disposed cyathia (sometimes more cyathia developing), central male or bisexual, lateral 2 (or more) bisexual and developing later each on peduncles 3–9 × 2.5–4 mm, with 2 broadly ovate bracts subtending lateral cyathia; cyathia shallowly conical, glabrous, 6–9 mm broad (± 1 mm long below insertion of glands), with 5 broadly ovate deep pinkish red lobes with finely denticulate margins, reddish purple; *glands* 5 (6), transversely elliptic and usually separate, 4–4.5 mm broad, dark purple to maroon above (pink-red below), spreading, inner margins slightly raised, outer margins entire and spreading, surface between two margins ± convex; *stamens* glabrous, with dark purple to maroon anthers on paler maroon filaments, with many pale cream bracteoles enveloping groups of males with finely divided tips or filiform, glabrous; *ovary* slenderly 3-lobed and

Fig. 6.68. *Euphorbia lividiflora*, about 5 m tall in dense bush, *E. Schmidt*, Inhassoro, Moçambique, 21 Jan. 2016 (© E. Schmidt).

Fig. 6.69. *Euphorbia lividiflora*, about 1 m tall growing in open, *E. Schmidt*, Guludo, Moçambique, 6 Sep. 2014 (© E. Schmidt).

tapering gradually into styles, glabrous, red, erect on maroon pedicel 0.5–1 mm long with 3-lobed 'calyx' around ovary; styles 2–3 mm long, branched nearly to base, red. *Capsule* 15–25 mm diam., very deeply 3-lobed, glabrous and red, slightly exserted on erect pedicel 1–4 mm long.

Distribution & Habitat

Euphorbia lividiflora is known in Malawi, Moçambique, Tanzania and Zimbabwe. In Moçambique it only grows in coastal areas from Inhambane northwards, whereas in Malawi and Zimbabwe it grows much further from the coast, though still in low-lying areas.

Generally, *Euphorbia lividiflora* occurs in fairly dry places, in dense, deciduous bush. In Moçambique, such habitats are only found close to the sea. Sometimes (as around the mouth of the Save River) they support a wealth of other succulents, including *Aloe*, various species of *Ceropegia* and *E. baylissii*, *E. bougheyi*, *E. clavigera* and *E. knuthii*.

Diagnostic Features & Relationships

Euphorbia lividiflora forms quite large trees, but these are diffuse and are not easily seen within the thickets where they grow. The stem is fairly slender and supports itself on the surrounding bushes until it is older and sturdier, when it may ultimately reach 25 cm thick. The ultimate branchlets are always slender and grey-green, which makes them inconspicuous and the crown is never made up of a dense mass of them either.

Flowering takes place over much of the summer. The dark purple-red cyathia are fairly large and relatively flat,

Fig. 6.70. *Euphorbia lividiflora*, *PVB 7751*, Liwonde National Park, Malawi, Dec. 2015 (© PVB).

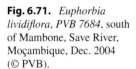

Fig. 6.71. *Euphorbia lividiflora*, *PVB 7684*, south of Mambone, Save River, Moçambique, Dec. 2004 (© PVB).

Fig. 6.72. *Euphorbia lividiflora*, pale capsules, *E. Schmidt*, De la Crusse (1440AD), Moçambique, 19 Mar. 2009 (© E. Schmidt).

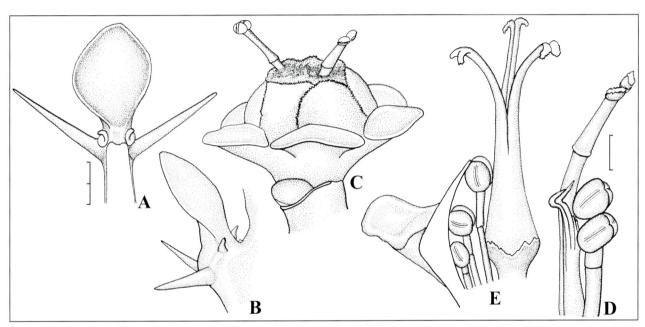

Fig. 6.73. *Euphorbia lividiflora.* **A**, young spine-complex and leaf-rudiment from above (scale 2 mm, as for **B–C**). **B**, young spines and leaf-rudiment from side. **C**, side view of cyathium at end of first male stage. **D**, anthers and bracteoles (scale 1 mm, as for **E**). **E**, female floret inside dissected cyathium. Drawn from: *PVB 7684*, south of Mambone, Save River, Moçambique (© PVB).

with unusually large lobes (which are particularly thick at their middle) and masses of bracteoles filling up the inside. They are also unusual for the slender ovary which gradually merges with the styles and the significant 'calyx' that surrounds its base. The capsules are also peculiar, with their three prominent and very much compressed lobes.

As Leach stated when he described *Euphorbia lividiflora*, it is close to *E. robecchi* from East and north-east Africa and also to *E. qarad* from the southern part of the Arabian Peninsula. *Euphorbia robecchi* is a much larger tree that regularly reaches and may exceed 10 m tall, with the main branches reaching 5 m long and forming a roughly hemispherical crown. It also differs by its yellow cyathial glands and the ± cylindrical branches without spines in mature trees. Furthermore, the capsules are never purplish or pinkish as they often are in *E. lividiflora*.

History

The first record of this species was made by A.O.D. Mogg, who found it on Bazaruto Island in October 1958. Leach first noticed it in May 1961 and in August 1961 he gathered the material from which he described it in 1964.

Euphorbia malevola L.C.Leach, *J. S. African Bot.* 30: 1 (1964). Type: Zimbabwe, Nuanetsi distr., 6 miles south of Lundi River on Salisbury-Beitbridge road, ± 2000', 27 Aug. 1962, *Leach 5083* (SRGH, holo.; K, LISC, PRE, iso).

Bisexual spiny glabrous succulent shrub 0.3–1.5 × 0.3–2 m, with many branches from above base of similar central stem from which fibrous roots arise. *Branches* erect to spreading, 0.1–0.5 m × 10–20 mm, 4 (5)-angled, not constricted into segments, smooth, pale grey-green sometimes becoming creamy-green between angles; *tubercles* fused into 4 (5) sinuate angles with flat to slightly concave area between angles, laterally flattened and projecting 2–3 mm from angles, with spine-shields 6–12 mm long, 2–3 × 2 mm above spines and 4–9 × 1–2 mm below spines and usually separate (rarely almost continuous), bearing 2 spreading and widely diverging red to brown and later grey spines 5–10 mm long; *leaf-rudiments* on tips of new tubercles towards apices of branches, 1.5–2 × 0.8–1.2 mm, recurved, fleeting, ovate-acuminate, sessile, with grey spreading stipular prickles 0.5–2 mm long. *Synflorescences* many per branch towards apex, each a solitary cyme in axil of tubercle, very shortly peduncled, each cyme with 3 transversely disposed cyathia, central male sessile and often deciduous, lateral 2 bisexual and developing later each on short peduncle 1–2 × 2–3 mm, with 2 broadly ovate denticulate green bracts 1.5–2 × 1.5–2 mm subtending lateral cyathia; *cyathia* funnel-shaped, glabrous, 5–6 mm broad (± 2 mm long below insertion of glands), with 5 obovate

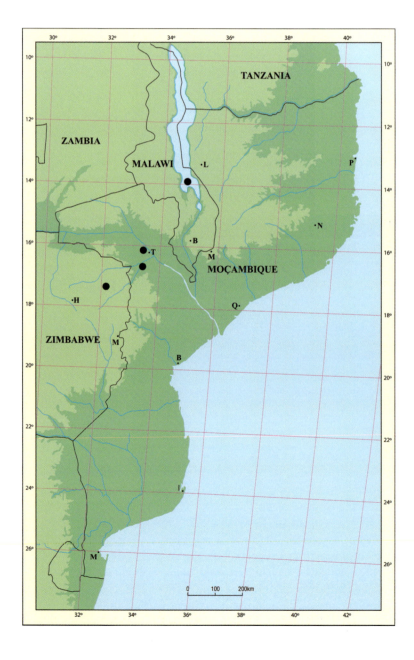

Fig. 6.74. Eastern side of the distribution of *Euphorbia malevola* (© PVB).

lobes pink or white streaked with pink with deeply incised margins, pale pink becoming grey-green below; *glands* 5, transversely elliptic, contiguous to separate, 2–3 mm broad, yellow with red to purple outer margins, spreading, with slightly raised inner margins, outer margins entire, surface between two margins flat; *stamens* glabrous, bracteoles enveloping groups of males, with finely divided tips, glabrous; *ovary* obtusely 3-angled, glabrous, green, very slightly raised on short thick pedicel less than 0.5 mm long; styles 3–4 mm long, white, branched to near base. *Capsule* 3.5–4 mm diam., obtusely 3-angled, green marked with purple, glabrous, ± sessile.

Distribution & Habitat
Euphorbia malevola is widespread in the valley of the Zambezi River in Zambia and Zimbabwe but is also known more widely in Zimbabwe and in Malawi. In Moçambique it occurs between Guro, Tete and Cabora Bassa near the Zambezi River valley. In this area it is usually found in locally dry places, sometimes on low granitic domes, but also in gravelly ground among *mopane* and *baobab* with *Aloe chabaudii*, *Cissus cactiformis*, *C. quadrangularis* and *Fockea multiflora*. It may become very common in denuded and overgrazed areas.

Fig. 6.75. *Euphorbia malevola*, about 1 m broad, near Sinazongwe, Zambia, 2 Dec. 2003 (© PVB).

Diagnostic Features & Relationships
Euphorbia malevola forms shrubs to 1 m tall, but usually around 0.5 m tall and often to 1 m broad or more. They are generally branched from the base and above, with many of the branches spreading horizontally towards their tips. Most of the branches are 4-angled and they are often slightly paler green or cream between the angles. The spines are relatively fine.

Flowering usually takes place between June and August (though up to November) and here the cyathia are suffused with red, with red margins to the yellow glands. The styles are fairly long and white and the ovary is more or less sessile.

Euphorbia malevola is part of the *E. schinzii*-complex and is not easily separated from some members of it. One of them, later called *E. limpopoana* (and now known as *E. schinzii* subsp. *bechuanica*), was first described by Leach as *E. malevola* subsp. *bechuanica*. It differs from other members of the *E. schinzii*-complex by forming larger plants with thicker branches (most of which are at least 10 mm thick), the slightly longer peduncles under the cyathia, the somewhat longer cyathia and the pink and red colour in the cyathia, especially the reddish margins of the glands.

History
The first collection of *E. malevola* was made by M. Wilman in July 1920 near Wankie, Zimbabwe and I. Pole Evans discovered it in Zambia in July 1930. In Moçambique the first record was made by António Rocha da Torre and Manuel Fernandes Correia in December 1965.

Fig. 6.76. *Euphorbia malevola*, cyathia in female stage, *PVB 7757*, towards Cabora Bassa, Moçambique, Oct. 2004 (© PVB).

Fig. 6.77. *Euphorbia malevola*, cyathia in second male stage, *PVB 7396*, south of Tete, Moçambique, Nov. 2006 (© PVB).

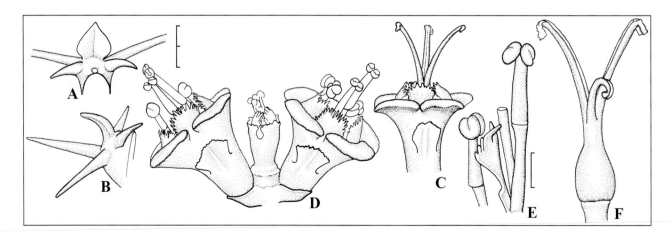

Fig. 6.78. *Euphorbia malevola*. **A**, young spine-complex and leaf-rudiment from above (scale 2 mm, as for **B–D**). **B**, young spine-complex and leaf-rudiment from side. **C**, side view of cyathium in female stage. **D**, side view of cyme in second male stage. **E**, anthers and bracteoles (scale 1 mm, as for **F**). **F**, female floret. Drawn from: *PVB 7757*, near Cabora Bassa, Moçambique (© PVB).

Euphorbia marrupana Bruyns, *Novon* 16: 456 (2006). Type: Moçambique, Niassa Prov., 30 km SE of Marrupa, 600 m, 2 Jan. 2004, *Bruyns 9708* (BOL, holo.).

Bisexual spiny glabrous densely branched succulent shrub 0.2–1 × 0.2–1.5 m with many branches in whorls above base of similar almost cylindrical erect stem bearing fibrous roots at base. *Branches* spreading to erect, 50–500 × 10–20 mm, 4- to 7-angled, square to almost cylindrical, becoming cylindrical towards bases, not constricted into segments, smooth, grey-green often mottled with cream on young growth, rebranching; *tubercles* indistinctly fused into 4–7 low angles with branches flat to concave between angles, laterally flattened and projecting 1–2 mm, with spine-shields 5–8 × 2–3 mm, elongated-rectangular to deltate, remaining separate, initially brown later hardening to dull grey, bearing 1 spreading red-brown later grey or black spine 6–12 mm long often forked above base; *leaf-rudiments* on tips of new tubercles towards apices of branches, 1–2 × 1–2 mm, erect, grey-green, fleeting, ovate, acute, sessile, with small brown stipular prickles 1–2 mm long. *Synflorescences* several per branch towards apex, each of 1 cyme in axil of tubercle, each cyme on peduncle ± 1 × 2 mm, with 3 horizontally disposed cyathia, central male, lateral 2 bisexual and developing later each on peduncles ± 2 × 2.5 mm, with 2 ovate pinkish bracts ± 1 × 1.5 mm subtending cyathia; *cyathia* broadly funnel-shaped, glabrous, 3–4 mm broad (± 1.5 mm long below insertion of glands),

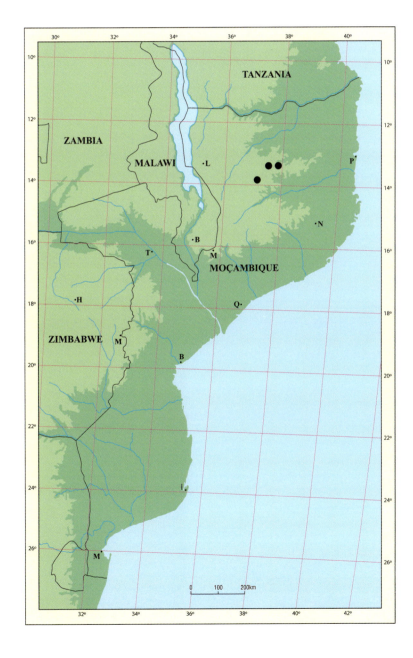

Fig. 6.79. Distribution of *Euphorbia marrupana* (© PVB).

with 5 obovate pinkish lobes with denticulate margins, pinkish yellow; *glands* 5, transversely elliptic, separate to contiguous, ± 2 mm broad, yellow with pinkish-red outer margins, spreading, outer margins entire and spreading, surface between two margins slightly convex; *stamens* glabrous, with red pedicels, with bracteoles enveloping groups of males, mostly pinkish, with finely divided tips or filiform, glabrous; *ovary* obtusely 3-lobed, glabrous, green, raised on short pedicel 1–1.5 mm long; styles 1–2 mm long, branched ± to base, red. *Capsule* 3–4 mm diam., obtusely 3-lobed, glabrous, pink-red above and maroon along sutures, slightly raised on short pedicel 1–1.5 mm long.

Distribution & Habitat

Euphorbia marrupana occurs in northern Moçambique, where it is known between the towns of Nungo and Maua. The rainfall in this area is high and probably exceeds 1500 mm per year, though it is confined to summer. It grows here on the steep slopes of tall granitic mountains. Where it occurs, it is fairly plentiful, but plants are usually quite scattered across the slopes. *Euphorbia marrupana* grows in shallow but very congested soils in dense mats of the fine-leaved, grass-like sedge *Coleochloa*. Fires are frequently lit on these mountains during the dry season. These damage the plants (and any other succulents) greatly and this is

Fig. 6.80. *Euphorbia marrupana*, densely branched old plant ± 1 × 1.5 m on steep granitic slope among tufts of *Coleochloa*, *PVB 9708*, east of Marrupa, Moçambique, 2 Jan. 2004 (© PVB).

Fig. 6.81. *Euphorbia marrupana*, younger plant ± 1 m tall on steep granitic slope among dense tufts of *Coleochloa*, *PVB 9708*, east of Marrupa, Moçambique, 2 Jan. 2004 (© PVB).

Fig. 6.82. *Euphorbia marrupana*, branch with both simple and forked spines, *PVB 9708*, east of Marrupa, Moçambique, 2 Jan. 2004 (© PVB).

probably the reason for their very scattered occurrence. Other succulents occurring with it are *Aloe mawii*, *Kalanchoe elizae* and *Cynanchum viminale*, but it is the only species of *Euphorbia* found on these slopes.

Diagnostic Features & Relationships

Plants of *E. marrupana* may reach 1 m tall. They are usually densely and irregularly branched, so as to form a fairly impenetrable shrub. The branches are peculiar for their variability from square and 4-angled to nearly cylindrical with 6 or 7 low angles. There is also unusual variability in the spines, which are always solitary on each tubercle, but are simple to deeply forked. In small seedlings the first spines that appear are four nearly equal prickles ± 1 mm long surrounding each leaf-rudiment (two below the leaf and two adjacent as stipular prickles). The stipular spines remain small while the others are modified later into a single spine, which is sometimes forked above.

In some respects, such as the often cylindrical stems and the single spines, *E. marrupana* resembles *E. unicornis*. However, *E. unicornis* branches more from lower down on the stem and tends to produce longer branches which rebranch less. Consequently, the plants are more rounded and symmetrical in appearance (as in Figs. 6.113 and 6.114). In *E. marrupana* the spine-shields are not continuous along the angles, but are much shorter than the internodes, so that much more of the grey-green of the branches is visible than in *E. unicornis*. The cyathia in *E. unicornis* are always deep purple-red and the glands form a narrow rim around the cyathium. In *E. marrupana* the cyathia are pale pinkish, while the glands are yellow with pink edges and are somewhat broader and flatter.

Fig. 6.83. *Euphorbia marrupana*, flowering branch in female stage, with simple spines, *PVB 9707*, east of Nungo, Moçambique, Oct. 2004 (© PVB).

Fig. 6.84. *Euphorbia marrupana*, flowering branch just after female stage, with simple spines, *PVB 9707*, east of Nungo, Moçambique, Nov. 2012 (© PVB).

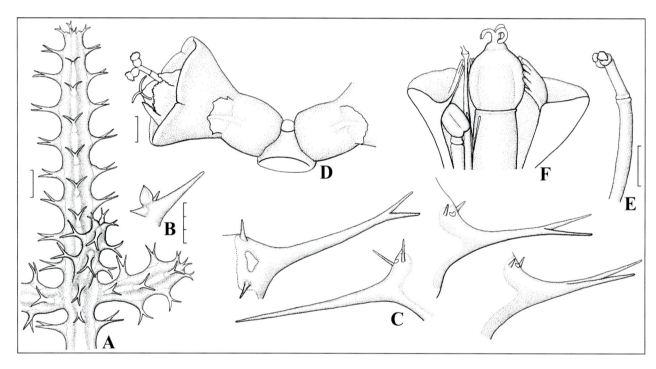

Fig. 6.85. *Euphorbia marrupana*. **A**, portion of branch (scale 10 mm). **B**, young spine-complex with leaf-rudiment from side (scale 3 mm, as for **C**). **C**, mature spine-complexes with different shapes of spines, from simple to deeply forked. **D**, side view of part of cyme in second male stage (scale 1 mm, as for **E**). **E**, anther (scale 1 mm, as for **F**). **F**, side view of dissected cyathium in late male stage (styles partly shrivelled). Drawn from: *PVB 9707*, east of Nungo, Moçambique (© PVB).

History

Euphorbia marrupana was first gathered in January 2004 and there do not appear to be other records of it.

Euphorbia mlanjeana L.C.Leach, *J. S. African Bot.* 39: 3 (1973). Type: Malawi, Mulanje Mountain, steep slopes above Little Malosa River, 15 Aug. 1971, *Leach & al. 14805* (SRGH, holo.; K, LISC, PRE, iso.).

Bisexual spiny glabrous succulent shrub 0.15–1 × 0.2–1.2 m with many branches at and above ground level from erect ± cylindrical rarely branched stem 0.1–0.5 m tall bearing fibrous roots at base. *Branches* spreading then ascending, simple, 0.2–0.6 m × 20–55 mm, (3) 4- to 5 (6)-angled, often slightly constricted into ellipsoidal segments (becoming evenly narrower towards base and top), soon dying off and rapidly wearing away, smooth, yellow-

Fig. 6.86. Distribution of *Euphorbia mlanjeana* (© PVB).

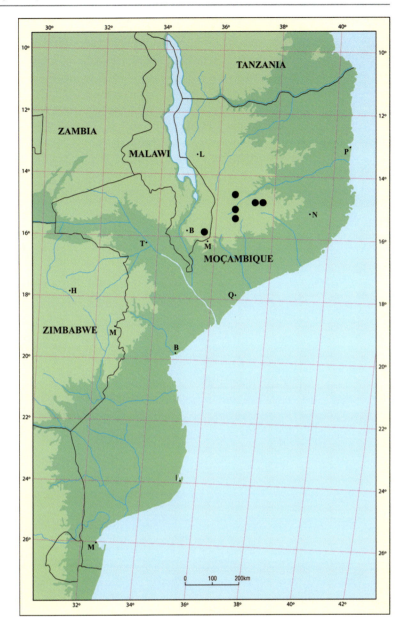

ish to bright green; *tubercles* fused into (3)4–5 (6) thin wing-like angles with deep grooves between angles, laterally flattened and projecting 1–5 mm from angles, with spine-shields slightly separated or later united into continuous horny grey margin ± 2 mm broad along angles, bearing 2 spreading and widely diverging grey spines 3–6 mm long; *leaf-rudiments* on tips of new tubercles towards apices of branches, 2–4 × 1–1.5 mm, erect, fleeting, linear-ovate, acute, sessile, with small irregular tooth-like brown stipules ± 0.5 mm long. *Synflorescences* many per branch towards apex, each a solitary cyme in axil of tubercle, on peduncle ± 1 mm long, with 3 horizontally disposed cyathia, central male, lateral 2 bisexual and developing later each on short peduncle ± 1 × 1.5 mm, with 2 ovate bracts ± 2 × 1.5 mm subtending lateral cyathia; *cyathia* conical-cupular, glabrous, 4–5 mm broad (2 mm long below insertion of glands), with 5 obovate lobes with deeply incised margins, yellow to greenish yellow; *glands* 5, transversely elliptic and contiguous, 1.5–2.5 mm broad, yellow, spreading, inner margins slightly raised, outer margins entire and spreading, surface between two margins covered with copious nectar; *stamens* glabrous, bracteoles enveloping groups of males, with finely divided tips or filiform, glabrous; *ovary* obtusely 3-angled, glabrous, green, exserted and projecting upwards or downwards on pedicel ± 1.5 mm long; styles 2–3 mm long, branched to above middle. *Capsule* 5–7 mm diam., acutely 3-angled, glabrous and green to shiny red, exserted on decurved and later erect pedicel 5–8 mm long.

Distribution & Habitat

In Malawi, *Euphorbia mlanjeana* is known at altitudes between 1000 and 2000 m on the slopes of Mount Mulanje, which is the highest mountain in the country and lies in its south-eastern corner. In Moçambique it is found some 150 km east of Mt Mulanje on the slopes of several of the higher peaks between Cuamba, Ribáuè and Errego, at altitudes between 700 and 1600 m.

Just as on Mt Mulanje, in Moçambique it occurs on steep slopes in shallow patches of soil on huge granitic outcrops. These patches of soil are dense, almost impenetrable mats of roots, especially of the grass-like sedge *Coleochloa* and can be exceedingly wet during the rainy season. Other succulents such as *Aloe munchii*, *A. torrei*, *Cynanchum mulanjense*, *C. oresbium*, *Ceropegia* (*Huernia*) *erectiloba*, *E. graniticola* and *Kalanchoe elizae* may occur together with *E. mlanjeana*. Its stems are often festooned with lichens, mosses and sometimes small epiphytic orchids and even plants of *Cynanchum oresbium* grow epiphytically on it. Its thick stems and ephemeral branches enable it to survive the frequent fires that regularly consume the grass-like vegetation on some of these slopes.

Fig. 6.87. *Euphorbia mlanjeana*, steep slopes with tall *Aloe mawii*, tufts of *Coloechloa* and *Xerophyta*, *PVB 8602*, SE of Malema, Moçambique, 19 Nov. 2000 (© PVB).

Fig. 6.88. *Euphorbia mlanjeana*, steep slopes among tufts of *Coloechloa* and *Xerophyta*, *PVB 8599*, Mt Wesse, east of Malema, Moçambique, 18 Nov. 2000 (© PVB).

Diagnostic Features & Relationships

Euphorbia mlanjeana is usually a shrub with a single rather bare and only rarely branched stem that is often heavily scorched by fires. In some places (such as Mt Namuli) the plants do not exceed 20 cm tall and here usually no stem is visible. In others (such as the type locality and several places in Moçambique) plants reach 1 m tall, with a quite tall and relatively stout stem. Its branches, though reaching 0.5 m long in larger plants, are ephemeral and only last a few seasons after which they die off or are scorched by fire and then die off. They are typically bright green, with relatively small spines.

In *E. mlanjeana* the cymes are solitary in each axil and the cyathia in each cyme are horizontally disposed. When the styles mature the ovary is on a pedicel around 1.5 mm long. Afterwards, the pedicel elongates considerably as the capsule expands so that it is pushed well out of the cyathium to one side. The shape of the leaf-rudiments combined with the solitary cymes per axil, the much exserted, deeply 3-angled capsules and smooth ± spherical seeds suggest a strong relationship between *E. mlanjeana* and *E. griseola* and its allies (such as *E. jubata* and *E. persistentifolia*).

The two species *E. graniticola* and *E. mlanjeana* may occur together. Small plants of the first and large ones of the second are similar and one needs to examine them closely to distinguish them. Branches of *E. mlanjeana* have a greener colour (often faintly lined and also slightly translucent), never the opaque greyish green of the other, while *E. graniti-*

Fig. 6.89. *Euphorbia mlanjeana*, green branch with darker mottling, *PVB 8610*, Mitucue, Moçambique, Mar. 2001 (© PVB).

Fig. 6.90. *Euphorbia mlanjeana*, flowering branch, *PVB 9739*, Mt Namuli, Moçambique, Nov. 2010 (© PVB).

Fig. 6.91. *Euphorbia mlanjeana*, capsules, *PVB 8602*, SE of Malema, Moçambique, Dec. 2006 (© PVB).

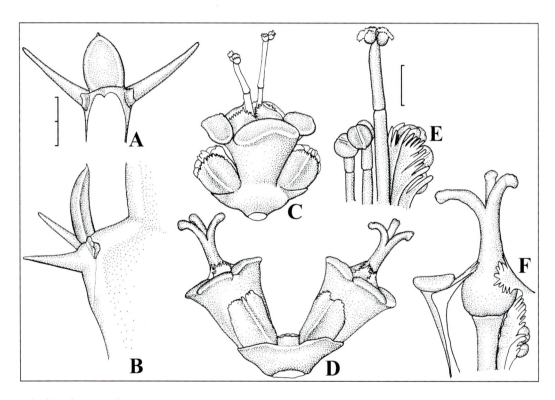

Fig. 6.92. *Euphorbia mlanjeana*. **A**, young spine-complex and leaf-rudiment from above (scale 2 mm, as for **B**–**D**). **B**, young spine-complex and leaf-rudiment from side. **C**, side view of cyathium at end of first male stage. **D**, side view of cyme at female stage. **E**, anthers and bracteoles (scale 1 mm, as for **F**). **E**, female floret inside dissected cyathium. Drawn from: *PVB 9739*, Mt Namuli, Moçambique (© PVB).

cola has much larger leaf-rudiments. Also, the cymes are solitary in each axil and the cyathia are horizontally disposed in *E. mlanjeana*, but usually three per axil with vertically disposed cyathia in *E. graniticola*. In both species the capsule is exserted once mature, but it is much more deeply 3-lobed in *E. mlanjeana*, whereas it is more shallowly and obtusely 3-lobed in *E. graniticola*.

History

Euphorbia mlanjeana was first observed on Mt Mulanje by the American mammologist Harald E. Anthony and the Australian botanist Leonard J. Brass, during the Vernay Nyasaland Expedition of May to October 1946 to Malawi. On this occasion, they spent some three weeks on the mountain during June 1946 (Anthony 1949). It was again collected by James D. Chapman on Mt Mulanje around 1958 and he referred to it as a small candelabriform *Euphorbia* in his account of the flora of the mountain (Chapman 1962). According to Leach (1973b) and Carter & Leach (2001), it is endemic to Mt Mulanje in Malawi and it was not recorded among the trees and shrubs of Moçambique by Burrows et al. (2018). The first plants were seen in Moçambique by the present author late in December 1998, just SW of Lioma and it was recorded at several more localities in Moçambique in 2000 and 2004.

Euphorbia namuliensis Bruyns, *Novon* 16: 454 (2006). Type: Moçambique, Zambézia Prov., northern slopes of Mt Namuli, 800–1500 m, 4 Jan. 2004, *Bruyns 9723* (BOL, holo.; E, MO, iso.).

Bisexual spiny glabrous shortly rhizomatous succulent shrub 0.2–0.4 × 0.2–0.5 m with few branches from base of similar erect stem bearing fibrous roots at base. *Branches* spreading underground for 0.05–0.25 m then erect, 50–400 × 12–20 mm, (5) 6-angled, mostly not constricted into segments, smooth, dark green, rarely rebranching, dying off after 2 or 3 seasons and replaced by new branches; *tubercles* fused into 6 angles with branches concave between angles, laterally flattened and projecting 1–2 mm, with spine-shields 1–5 × 1–2 mm, narrowly elliptical, remaining separate, initially brown later hardening to dull grey, bearing 4 brown later grey spines 4–6 mm long (2 below leaf spreading horizontally and 2 stipular prickles next to it ascending); *leaf-rudiments* on tips of new tubercles towards apices of branches, 2–3 × 2 mm, erect, grey-green, fleeting, narrowly ovate, acute, sessile, with grey stipular prickles 4–6 m long indistinguishable from spines. *Synflorescences* several per branch towards apex, each of 1 cyme in axil of tubercle, each cyme on peduncle 1–2 × 2 mm, with 3 horizontally disposed cyathia, central male, lateral 2 bisexual and developing later each on peduncles ± 2 × 2.5 mm, with 2 ovate reddish bracts ± 1 × 1.5 mm subtending cyathia; *cyathia* broadly funnel-shaped, glabrous, 3–4 mm broad (± 1.5 mm long below insertion of glands), with 5 obovate pinkish lobes with denticulate margins, pinkish yellow; *glands* 5, transversely elliptic-rectangular and contiguous, ± 2 mm broad, yellow to purplish green with pinkish to red to dark maroon outer margins, spreading, outer margins entire and spreading, surface between two margins slightly convex; *stamens* glabrous, with red pedicels, with bracteoles enveloping groups of males, mostly pinkish, with finely divided tips or filiform, glabrous; *ovary* obtusely 3-lobed, glabrous, green, raised on short pedicel ± 1 mm long; styles ± 1 mm long, branched nearly to base, red. *Capsule* 4–5 mm diam., obtusely 3-lobed, glabrous, red above and green below, slightly raised on short pedicel ± 1 mm long.

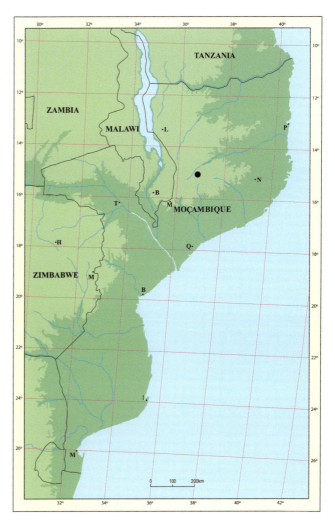

Fig. 6.93. Distribution of *Euphorbia namuliensis* (© PVB).

Fig. 6.94. *Euphorbia namuliensis*, ± 40 cm tall among tufts of *Coloechloa*, *PVB 9723*, Mt Namuli, Moçambique, 4 Jan. 2004 (© PVB).

Fig. 6.95. *Euphorbia namuliensis*, rhizomatous branches to 40 cm tall among tufts of *Coloechloa*, *PVB 9723*, Mt Namuli, Moçambique, 4 Jan. 2004 (© PVB).

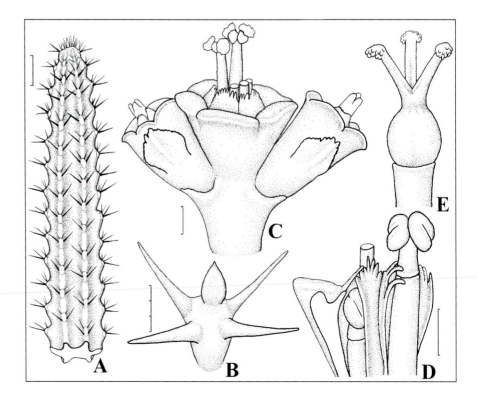

Fig. 6.96. *Euphorbia namuliensis*. **A**, portion of branch (scale 10 mm). **B**, young spine-complex and leaf-rudiment from below (scale 3 mm). **C**, side view of cyme in first male stage (scale 1 mm). **D**, anthers and bracteoles in dissected cyathium (scale 1 mm, as for **E**). **E**, female floret. Drawn from: *PVB 9723*, Mt Namuli, Moçambique (© PVB).

Fig. 6.97. *Euphorbia namuliensis*, flowering branch, *PVB 9723*, Mt Namuli, Moçambique, Dec. 2014 (© PVB).

Distribution & Habitat
Euphorbia namuliensis is common on the northern slopes of Mount Namuli, but this is its only known locality. On the mountain it was found to have a wide altitudinal range and plants were seen from some of the hotter outcrops near the base of the mountain at about 800 m to some of the cooler, higher slopes at 1500 m. It was always found among dense clumps of *Coleochloa* and various species of *Xerophyta* in shallow patches of densely matted soil on granitic domes. Other succulents sharing the habitat were *Aloe chabaudii*, *A. mawii*, *Kalanchoe elizae*, *K. humilis* and *Cynanchum viminale*. Although small forms of *E. mlanjeana* were observed further southeast on Mt Namuli, no other species of *Euphorbia* was seen growing near it. The rainfall in this area probably exceeds 1000 mm per annum (mainly in summer) and deeper soils on the mountain support an open forest of *Brachystegia*, while orchids and other mesophytic monocots were quite common in less exposed spots.

Diagnostic Features & Relationships
With their rhizomatous habit, plants of *Euphorbia namuliensis* form diffuse clumps with relatively few branches. Within a clump, the branches are connected underground by horizontal runners up to 20 cm long or more, that arise near the bases of older branches. These runners are able to burrow through the extremely tough, dense, fibrous masses of roots of *Coleochloa* and *Xerophyta* that make up the soil. The above-ground portions of the mainly 6-angled, almost cylindrical branches are erect, unbranched and dark green. Individual branches last for two or three seasons after which they die off and are replaced by new ones arising from beneath the ground. In *E. namuliensis* there are four spines of more or less identical size on each spine-shield. Two of these are the upwardly oriented stipular prickles flanking the leaf-rudiment and the other two spreading spines are located just below the leaf-rudiment. The spine-shield itself is inconspicuous and does not extend far downwards towards the next leaf-rudiment below.

Euphorbia namuliensis resembles *E. whellanii* from northern Zambia, in that they share a somewhat rhizomatous habit, more than 4-angled branches with stipular prickles and spines almost identical. However, it is actually closely related to *E. corniculata*, *E. marrupana* and *E. unicornis* and its cyathia are particularly similar to those of *E. marrupana*. The 6-angled simple branches and their rhizomatous habit are unlike anything found among these species, so that it is easily recognized by these vegetative features alone.

History
This species was discovered in January 2004 and it is not known from any other collections.

Euphorbia plenispina S.Carter, *Kew Bull.* 54: 964 (2000). Type: Moçambique, Manica e Sofala, between Machaze and Matindire, 28 Jul. 1968, *M.F.de Carvalho 1019* (LISC, holo.).

Euphorbia stenocaulis Bruyns, *Novon* 16: 457 (2006). Type: Moçambique, Zambézia Prov., Mocuba, 250 m, 10 Nov. 2000, *Bruyns 8534* (BOL, holo.; E, MO, iso.).

Bisexual spiny glabrous succulent shrublet $0.1–0.3 \times 0.2–0.4$ m with many branches from base of similar erect to creeping stem bearing fibrous roots at base. *Branches* occasionally shortly rhizomatous then erect to creeping, $30–300 \times 4–7$ mm, 4-angled, rarely slightly constricted into segments, smooth, green suffused with red, rebranching; *tubercles* fused into 4 low angles with branches ± flat between angles, laterally flattened and projecting < 1 mm (almost obsolete), with spine-shields $3–4 \times ± 1$ mm, narrowly ellipti-

Fig. 6.98. Distribution of *Euphorbia plenispina* (© PVB).

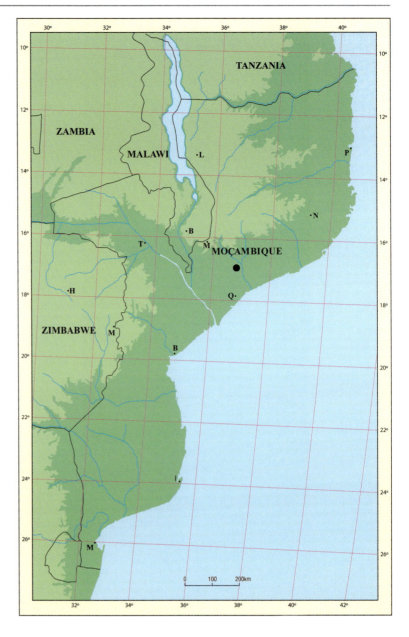

cal-linear, remaining separate, initially brown later hardening to dull grey, bearing 4 brown later grey spines 2–4 mm long in an X-like arrangement (2 below leaf descending and 2 stipular prickles next to it ascending); *leaf-rudiments* on tips of new tubercles towards apices of branches, 2–3 × 1 mm, erect, grey-green, fleeting, subulate, acute, sessile, with grey stipular prickles 2–4 mm long indistinguishable from spines. *Synflorescences* several per branch towards apex, each of 1 cyme in axil of tubercle, each cyme on peduncle 1–2.5 × 1.5 mm, with 3 horizontally disposed cyathia, central male, lateral 2 bisexual and developing later each on peduncle ± 1.5–2.5 × 1.5 mm, with 2 ovate green bracts with red midrib and base ± 1.5 × 1 mm subtending cyathia; *cyathia* broadly funnel-shaped, glabrous, (3) 4–5 mm broad (± 1.5 mm long below insertion of glands), with 5 obovate pale green to white lobes with denticulate margins, yellow-green; *glands* 5, transversely elliptic and closely contiguous to separate, ± 2.5 mm broad, yellow with paler margin or red, ascending-spreading, outer margins entire and ascending, surface between two margins ± flat; *stamens* glabrous, yellow, with bracteoles enveloping groups of males, with finely divided tips, glabrous; *ovary* obtusely 3-lobed, glabrous, green, raised on short pedicel ± 1.5 mm long; styles ± 2 mm long, branched nearly to base, pale yellow. *Capsule* not seen.

Distribution & Habitat

Euphorbia plenispina was common but local between Machaze and Matindire (not plotted on Fig. 6.98 since the locality is not precisely stated on the sheet) in Manica e Sofala Province, in stony ground with pebbles and much

Fig. 6.99. *Euphorbia plenispina*, flowering, *PVB 8534*, Monte Rue, Mocuba, Moçambique, Oct. 2014 (© PVB).

lichen on the vegetation (according to the type-sheet). *Euphorbia stenocaulis* is known on a small series of granitic hills near Mocuba in the southern part of Zambézia Province. Here it occurs among dense clumps of *Coleochloa* in shallow dense mats of roots and soil on gently sloping, solid granitic whale-backs. Where it was found, it was extremely common and occurred together with *Ceropegia* (*Huernia*) *erectiloba*, *Cynanchum viminale* and various species of *Xerophyta*.

Diagnostic Features & Relationships

In this slender-stemmed species the branches creep and may become somewhat rhizomatous. They are bright green with faint red markings. On each spine-shield there are four fine weak spines. These consist of two spines below the leaf-rudiment and another two more or less equally long stipular prickles alongside the bases of the margins of the leaf.

Euphorbia plenispina was described from a single, rather incomplete herbarium specimen. Nevertheless it is clear that, in respect of the slender branches and in the fine spines and prickles of ± equal length, it is extremely similar to *E. stenocaulis*. *Euphorbia stenocaulis* was said to differ from *E. plenispina* in the yellow cyathia and in the glands being continuous around the top of the cyathium, while they are red and separate in *E. plenispina*. While the glands are indeed yellow in *E. stenocaulis*, they are not continuous but separate, though they are pressed so closely together laterally that this was mis-interpreted as a continuous rim to the cyathium. They therefore differ only in the colour of their glands. Since this is variable in several other related species, these two are considered to be synonymous here.

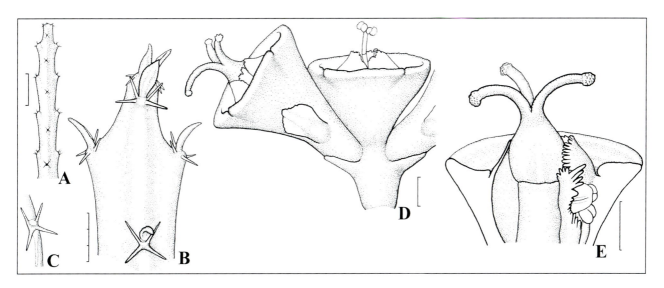

Fig. 6.100. *Euphorbia plenispina*. **A**, portion of branch (scale 10 mm). **B**, tip of branch with young spine-complexes and leaf-rudiments (scale 3 mm, as for **C**). **C**, mature spine-complex from side. **D**, side view of part of cyme in female stage (scale 1 mm). **E**, female floret inside dissected cyathium (scale 1 mm). Drawn from: *PVB 8534*, Monte Rue, Mocuba, Moçambique (© PVB).

History

Euphorbia plenispina is known from only two places, with the original collection made in 1968 by Manuel Fidalgo de Carvalho and the other one in November 2000.

Euphorbia pseudocontorta Bruyns, *Phytotaxa* 433: 299 (2020). Type: Moçambique, Zambézia Prov.: 60 km north-east of Molócuè, ± 500 m, 22 Dec. 1998, *Bruyns 7795* (BOL, holo.; NBG, iso.).

Bisexual spiny glabrous succulent 0.15–0.3 × 0.3–0.5 m with many branches from similar stem, densely branched at ground level, with fibrous roots arising from base. *Branches* ascending to ± prostrate, 100–200 × 15–25 mm, 3- to 4-angled, not constricted into segments, smooth, pale grey-green with paler stripes between angles; *tubercles* fused into 3 or 4 angles, laterally flattened and truncate, projecting ± 5 mm from branch, with spine-shields 5–7 mm long, 3–5 mm long and rounded below spines but remaining well

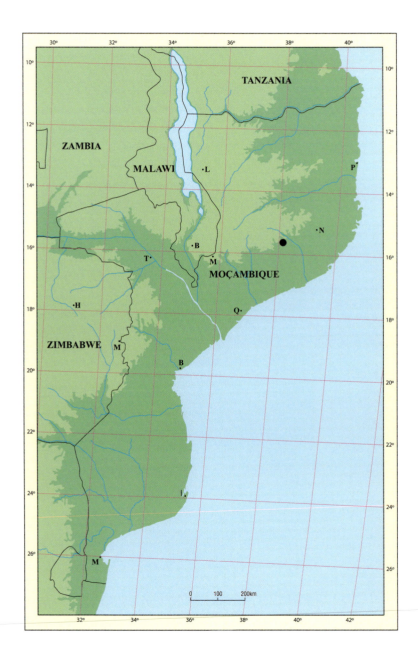

Fig. 6.101. Distribution of *Euphorbia pseudocontorta* (© PVB).

separated from next, bearing 2 spreading initially purple-red (later grey) spines ± 10 mm long; *leaf-rudiments* on tips of new tubercles towards apices of branches, 8–10 × 5–6 mm, erect, fleeting, ovate-lanceolate, sessile, with small stipular prickles ± 1.5 mm long. *Inflorescences* many per branch usually towards apex, each a solitary cyme in axil of tubercle, sessile, each cyme with 3 transversely disposed cyathia, central male, lateral 2 bisexual and developing later each on short peduncle 2–3 × ± 1 mm, with 2 ovate bracts ± 2 × 1–2 mm subtending lateral cyathia; *cyathia* shallowly funnel-shaped, glabrous, ± 5 mm broad (3 mm long below insertion of glands), with 5 pinkish red obovate lobes with deeply incised margins, grey-green variably suffused with red; *glands* 5, transversely rectangular and contiguous, 2.5–3 mm broad, dark yellow with red outer margins, spreading, inner margins slightly raised, outer margins entire and spreading; *stamens* glabrous, red, bracteoles enveloping groups of males, with finely divided red tips, glabrous; *ovary* 3-angled, glabrous, green, raised on pedicel ± 1 mm long; styles ± 2.5 mm long, pinkish, branched to near base. *Capsule* ± 4 mm diam., obtusely 3-angled, reddish green, glabrous, nearly sessile.

Fig. 6.102. *Euphorbia pseudocontorta*, ± 40 cm diam., with small *Selaginella* on flattish granitic outcrop, with 3- and 4-angled branches, *PVB 7795*, NE of Molocue, Moçambique, 22 Dec. 1998 (© PVB).

Fig. 6.103. *Euphorbia pseudocontorta*, female stage, *PVB 7795*, NE of Molocue, Moçambique, Oct. 2007 (© PVB).

Fig. 6.104. *Euphorbia pseudocontorta*, first male stage, *PVB 8538*, south of Ligonha, Moçambique, Oct. 2007 (© PVB).

Fig. 6.105. *Euphorbia pseudocontorta*, capsules, *E. Schmidt*, Ligonha, Moçambique, 18 Oct. 2005 (© E. Schmidt).

Distribution & Habitat

Euphorbia pseudocontorta is only known in the central province of Zambézia, where it is found in shallow pockets of soil on low granitic domes at altitudes of 500–600 m. Plants are common in shallow soils on otherwise fairly bare rocks.

Diagnostic Features & Relationships

Euphorbia pseudocontorta is similar to *E. contorta*, which occurs in shallow soil on granitic outcrops around the bases of larger hills some 150 km further to the north-west at altitudes of 550–700 m. In *E. contorta* the plant is sparingly branched, with the stem usually rising rather obliquely to 1 m or more and with a few horizontal branches spreading randomly from it along its length (as in Figs. 6.35 and 6.36). In contrast to this, *E. pseudocontorta* is densely branched from the base of the plant and so the stem is not visible among the branches. In *E. contorta* the stem and branches are uniformly dull grey and are mostly 5- to 7-angled (rarely 4-angled), while in *E. pseudocontorta* each branch has a paler longitudinal patch between the angles and the branches are 3- to 4-angled. Florally the two differ in the longer peduncles in *E. pseudocontorta*, both in the peduncle beneath the cyme and in the peduncles beneath each of the lateral cyathia in the cymes.

History

This species was first observed in December 1998 and was seen again in the same area in December 2000.

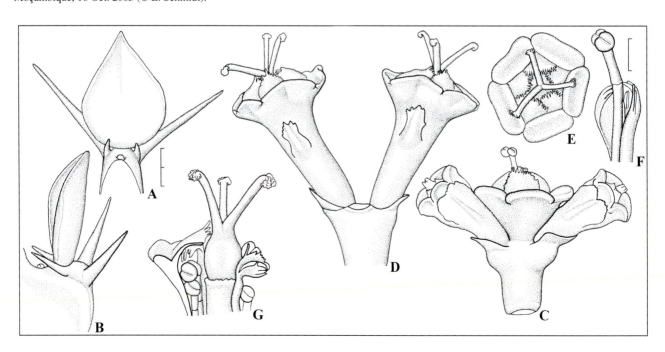

Fig. 6.106. *Euphorbia pseudocontorta*. **A**, young spine-complex and leaf-rudiment from above (scale 2 mm, as for **B–E**). **B**, young spine-complex and leaf-rudiment from side. **C**, side view of cyme in first male stage. **D**, side view of cyme in female stage. **E**, face view of cyathium. **F**, anther and bracteoles (scale 1 mm, as for **G**). **G**, female floret inside dissected cyathium. Drawn from: **A**, **B**, **C**, **E**, *PVB 8538*, south of Ligonha, Moçambique. **D**, **F**, **G**, *PVB 7795*, NE of Molocue, Moçambique (© PVB).

Euphorbia ramulosa L.C.Leach, *J. S. African Bot.* 32: 176 (1966). Type: Moçambique, Nampula Prov., ± 60 km west of Nampula, ± 800 m, *Leach & Schelpe 11440* (SRGH, holo.; K, LISC, PRE, iso.).

Bisexual spiny glabrous densely branched dwarf cushion-forming succulent shrub 0.1–0.2 (0.4) × 0.1–0.5 m with many branches around similar 4-angled erect stem bearing fibrous roots at base. *Branches* decumbent-erect, 50–180 × 8–12 mm, 4-angled becoming terete near bases, not constricted into segments, smooth, shiny dark green, rebranching; *tubercles* fused into 4 angles and slightly grooved and concave between angles, laterally flattened and projecting 2–4 mm from angles, with spine-shields ± 1.5 mm broad around spines and narrower below, fused into continuous narrow horny whitish grey margin along angles, bearing 2 spreading and widely diverging grey spines 4–8 mm long; *leaf-rudiments* on tips of new tubercles towards apices of branches, ± 1 × 1 mm, erect, fleeting, ovate, acute, sessile, with minute brown stipular prickles < 1 mm long. *Synflorescences* many per branch towards apex, each of 1 cyme in axil of tubercle, each cyme on peduncle 1–3 × 1–3 mm, with 3 horizontally disposed cyathia, central male, lateral 2 bisexual and developing later each on peduncle 1.5–4.5 × 1.5 mm, with 2 broadly ovate bracts ± 1.5 × 2 mm subtending cyathia; *cyathia* broadly funnel-shaped, glabrous, 3.5–5.5 mm broad (± 2 mm long below insertion of glands), with 5 obovate lobes with den-

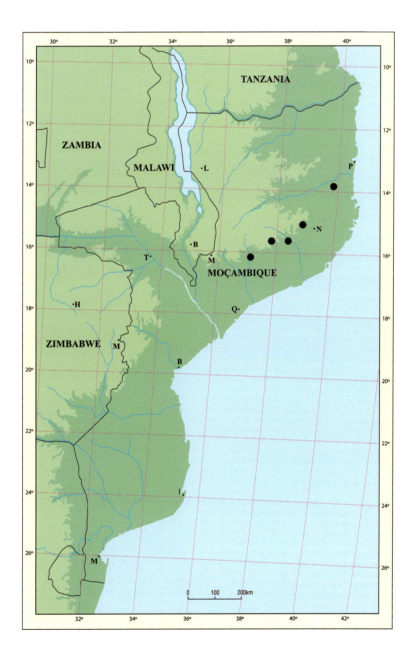

Fig. 6.107. Distribution of *Euphorbia ramulosa* (© PVB).

ticulate margins, usually purplish red; *glands* 5, transversely elliptic and contiguous, 1.5–3 mm broad, purplish red (greenish yellow), spreading, outer margins entire and spreading, surface between two margins flat to slightly concave; *stamens* glabrous, with bracteoles enveloping groups of males, red to dark purplish red, with finely divided tips or filiform, glabrous; *ovary* obtusely 3-lobed, glabrous, green suffused with red, raised on short pedicel ± 1.5 mm long; styles 1–1.5 mm long, branched to near base. *Capsule* 3–4 mm diam., obtusely 3-lobed, glabrous, dark purple-red above and green below, shortly exserted on erect pedicel ± 1.5 mm long.

Distribution & Habitat

Euphorbia ramulosa is endemic to Moçambique and is relatively widely distributed in the northern part, from near Alto Ligonha at least as far north as Alua, at altitudes between 350 and 800 m. It usually grows in shallow patches of gritty ground on relatively bare granitic slabs in places with little other vegetation, where the frequent grass-fires are not able

Fig. 6.108. *Euphorbia ramulosa*, ± 12 cm tall, several plants among rocks on edge of large expanse of bare rock on low granitic dome with *Coleochloa* and *Xerophyta*, *PVB 8540*, north of Ligonha, Moçambique, 11 Nov. 2000 (© PVB).

Fig. 6.109. *Euphorbia ramulosa*, ± 20 cm tall, densely branched ± hemispherical plant among rocks ontop of low granitic dome, *PVB 8540*, north of Ligonha, Moçambique, 11 Nov. 2000 (© PVB).

Fig. 6.110. *Euphorbia ramulosa*, rather diffuse plant in flower, *PVB 8550*, north of Alua, Moçambique, 12 Nov. 2000 (© PVB).

Fig. 6.111. *Euphorbia ramulosa*, fruiting, *PVB 8540*, north of Ligonha, Moçambique, 11 Nov. 2000 (© PVB).

to reach it. In some localities it occurs together with *E. corniculata*.

Diagnostic Features & Relationships

Euphorbia ramulosa forms low, cushion-like shrublets that rarely exceed 20 cm tall and are densely branched. The branches have a distinctive slightly shiny dark green colour and are typically around 10 mm thick, with short but tough spines.

Flowers appear during late spring and early summer. Although Leach (1966) gave the glands as greenish yellow, the cyathia seen were always purplish red with dark purple-red glands, as shown here. With its habit of growth, slightly exserted capsule and verrucose seeds, *E. ramulosa* looks like a typical member of the *E. schinzii*-complex, though the purple-red cyathia and relatively long peduncles of the cyathia make it easily recognizable. It is not closely related to *E. schinzii*, but rather to the rather different-looking *E. corniculata* and *E. unicornis*.

History

The first collection of *Euphorbia ramulosa* was made by Leach and Rutherford-Smith in May 1961 to the west of Nampula.

Euphorbia unicornis R.A.Dyer, *Bothalia* 6: 225 (1951). Type: Moçambique, Niassa Prov., Quissanga distr., Cuero Mt, 400 m, 14 Feb. 1949, *Pedro & Pedrogão 5091* (PRE).

Bisexual spiny glabrous succulent shrub 0.2–1 × 0.2–1 m with many branches (often nearly whorled) usually well above base around similar almost cylindrical erect stem bearing fibrous roots at base. *Branches* ascending-erect, 150–500 × 8–18 mm, indistinctly 6- to 8-angled and almost cylindrical, becoming cylindrical and spineless towards bases, not constricted into segments, smooth, brownish green when young becoming dull silvery all over except grey-green in narrow grooves between angles, rebranching sparingly; *tubercles* indistinctly fused into 6–8 very low angles separated by slight grooves between angles, laterally flattened and projecting 1–2 mm, with spine-shields 3–5 mm broad, elongated-rectangular, covering tubercles and decurrent to just above next axil below later often joining into continuous margin, initially brown later hardening to dull silvery, bearing 1 spreading red-brown later grey spine 4–10 mm long; *leaf-rudiments* on tips of new tubercles towards apices of branches, ± 2 × 2 mm, erect and pinkish, fleeting, ovate, acute, sessile, with small brown stipular prickles 1–2 mm long. *Synflorescences* several per branch towards apex, each of 1 cyme in axil of tubercle, each cyme on peduncle 1–2 × ± 2 mm, with 3 horizontally disposed cyathia, central male, lateral 2 bisexual and developing later each on peduncle 2–3 × 1.5 mm, with 2 ovate bracts ± 1 × 1.5 mm subtending cyathia; *cyathia* conical cupular, glabrous, 3–4 mm broad (± 1.5 mm long below insertion of glands), with 5 obovate lobes with denticulate margins, purplish red; *glands* 5, transversely elliptic and contiguous, 1.5–2 mm broad, purplish red,

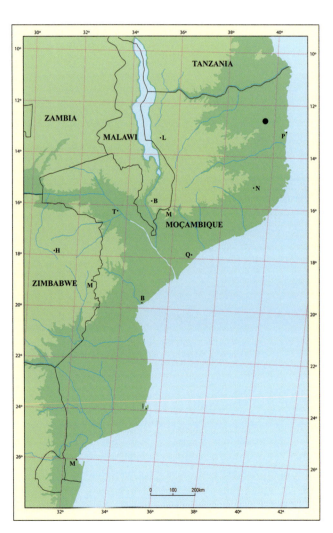

Fig. 6.112. Distribution of *Euphorbia unicornis* (© PVB).

Fig. 6.113. *Euphorbia unicornis*, ± 0.3 m tall, with green *Xerophyta* and very dry *Coleochloa*, *PVB 8554*, Cuero Mt, Meluco, Moçambique, 14 Nov. 2000 (© PVB).

ascending, outer margins entire and slightly spreading, surface between two margins slightly concave; *stamens* glabrous, with bracteoles enveloping groups of males, mostly red, with finely divided tips or filiform, glabrous; *ovary* obtusely 3-lobed, glabrous, green with red top, sessile or raised on very short pedicel ± 0.25 mm long; styles 2–2.5 mm long, branched to ± 0.5 mm above base, purple-red. *Capsule* 3–4 mm diam., obtusely 3-lobed, glabrous, purple-red above and green below, ± sessile.

Distribution & Habitat

Also endemic to Moçambique, *E. unicornis* is recorded from Cuero Mountain at Meluco and on many of the hills within a 20 km radius of Meluco. No further records are known. It is extremely common on these tall relatively bare granitic hills, at Cuero Mountain from the base at 300 m to the summit at just under 750 m. Here it grows with lots of *Aloe chabaudii* and an acaulescent form of *A. mawii*, *Ceropegia* (*Huernia*) *erectiloba*, *Cynanchum viminale*, *Kalanchoe humilis*, *Sansevieria*, *Euphorbia tirucalli* and *E. torrei*. The soil is very dense and hard (from roots and other fibres) and is often covered with a mat-forming *Selaginella*.

Diagnostic Features & Relationships

Plants vary very much in shape, from tall with tall branches hardly spreading away from the stem to specimens that are much broader than tall and from densely to quite laxly branched as well. The spine-shields are relatively broad and cover most of the tubercles. Initially they are reddish brown and are separate, with a slight gap between successive shields just above the axillary bud. Later they turn a silvery grey and this gap usually closes up around the axillary bud so that they form a continuous hard margin along the very low angles. Between adjacent angles there is usually a narrow groove between the spine-shields, where the pale grey-green of the epidermis is visible. This remains visible until the branch finally becomes ± terete near its base. Each spine-shield bears a single spine at the apex of the tubercle behind the leaf-rudiment. The colour of the spine follows that of the spine-shield from red-brown to silvery and the spines persist for many years.

Flowering takes place in *E. unicornis* between October and December, when the tops of many of the branches become covered with the attractive purplish red cyathia. Similarly-coloured capsules are soon formed.

The relationships of this remarkable species are discussed under *E. corniculata*, to which Dyer (1951) considered it to be close.

History

Plants of this species were sent to Pretoria in February 1949 by the two agronomists José Gomes Pedro (after whom the Portuguese endemic species *Euphorbia pedroi* was later named) and José Pedrogão de Jesus, from the type locality in the north of Moçambique. It was recollected there in November 2000 and is still very common on Cuero Mountain and some of the surrounding hills. No other records appear to exist.

Fig. 6.114. *Euphorbia unicornis*, flowering despite the dry conditions, *PVB 8554*, Cuero Mt, Meluco, Moçambique, 14 Nov. 2000 (© PVB).

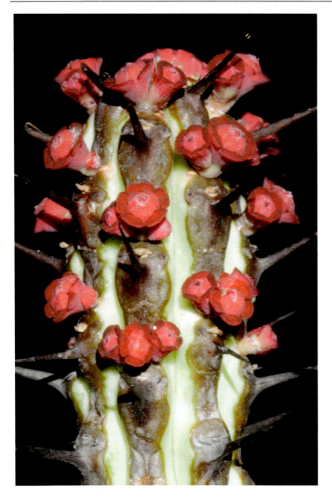

Fig. 6.115. *Euphorbia unicornis*, tip of flowering branch before first male stage, *PVB 8554*, Cuero Mt, Meluco, Moçambique, Nov. 2011 (© PVB).

Fig. 6.116. *Euphorbia unicornis*, flowering branch in female stage, *PVB 8554*, Cuero Mt, Meluco, Moçambique, Dec. 2016 (© PVB).

Fig. 6.117. *Euphorbia unicornis*, fruiting despite the dry conditions, *PVB 8554*, Cuero Mt, Meluco, Moçambique, 14 Nov. 2000 (© PVB).

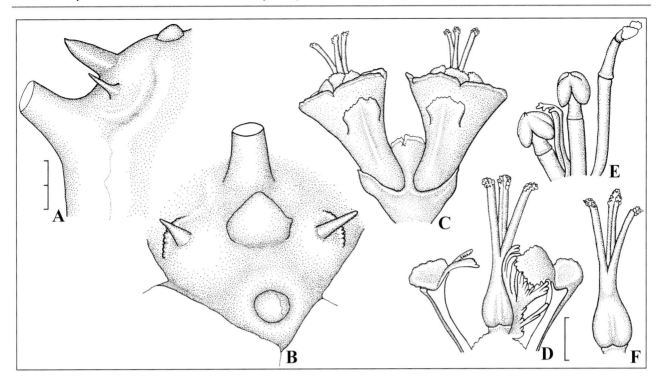

Fig. 6.118. *Euphorbia unicornis*. **A**, young spine-complex and leaf-rudiment from side (scale 2 mm, as for **B**, **C**). **B**, young spine-complex and leaf-rudiment from above. **C**, side view of cyme in female stage. **D**, side view of dissected cyathium (scale 1 mm, as for **E**, **F**). **E**, anthers and bracteole. **F**, female floret. Drawn from: *PVB 8554*, Cuero Mt, Meluco, Moçambique (© PVB).

6.2.2 Sect. Monadenium

For sect. *Monadenium*, six species are recorded from Moçambique, namely *E. crenata*, *E. lugardiae*, *E. neopedunculata*, *E. neorugosa*, *E. spinulosa* and *E. torrei*. Information is presented here on *E. neopedunculata* and *E. torrei* only.

Euphorbia neopedunculata (S.Carter) Bruyns, *Taxon* 55: 414 (2006). *Monadenium pedunculatum* S.Carter, *Kew Bull.* 42: 903 (1987). Type: Tanzania, Mpanda distr., Milumba Plain, *Richards 7053* (K, holo.).

Bisexual spineless glabrous succulent geophyte 0.05–0.2 m tall with 1 (4) short underground stems from slenderly turnip-shaped tuber 15–60 mm thick bearing fibrous roots. *Branches* 1–4 above ground, annual, erect, rarely rebranching, 0.05–0.2 m × 3–6 mm, cylindrical and longitudinally ridged, smooth or slightly papillate, grey-green; *tubercles* ± absent, spine-shields absent; *leaves* towards apex of branch, 10–90 × 3–15 (20) mm, ascending, deciduous, linear to lanceolate (obovate), acute, sessile, slightly fleshy and with entire undulating margins, grey-green, with minute glandular stipules. *Synflorescences* several towards apex of stem, each a solitary cyme in axil of leaf, on slender ascending pale green peduncle 15–80 × 1 mm, each cyme with 0–2 branches to 10 mm long, with 1–3 cyathia developing in succession, each cyathium surrounded with 2 ovate pale green bracts ± 3 × 2 mm; *cyathia* ± cylindrical widening in middle, held facing ± horizontally, 3–5 × 3 mm, with 5 oblong white lobes with deeply toothed margins; *gland* 1, erect, enclosing most of cyathium above except for gap in front, ± 3 mm long, white later pink to green; *ovary* obtusely 3-angled, becoming exserted horizontally through gap in gland as styles dry off; styles ± 1 mm long, branched to just above base. *Capsule* ± 5.5 mm diam., obtusely 3-angled, exserted horizontally (later becoming erect) on pedicel 4–12 mm long, dull green.

Distribution & Habitat

Euphorbia neopedunculata is widely distributed in D.R. Congo, Malawi, Tanzania and Zambia, as well as in the northern parts of Moçambique. It is usually a plant of open patches in areas covered by dry deciduous bush to

Fig. 6.119. Distribution of *Euphorbia neopedunculata* (© PVB).

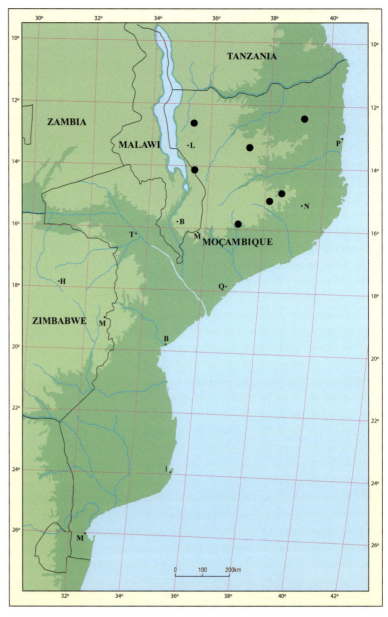

Fig. 6.120. *Euphorbia neopedunculata*, flowering, ± 8 cm tall, *PVB 8622*, south of Macaloge, Moçambique, 23 Nov. 2000 (© PVB).

Fig. 6.121. *Euphorbia neopedunculata*, fruiting, ± 8 cm tall, *PVB 8622*, south of Macaloge, Moçambique, 23 Nov. 2000 (© PVB).

Brachystegia-woodland. In *Brachystegia*-woodland it grows on pale soils with other small geophytes, but in Tanzania it may occur in fairly dry deciduous bush, in hard, loamy ground with many other succulents.

Diagnostic Features & Relationships
Euphorbia neopedunculata is a geophyte, where the few above-ground branches die off and the plant retreats to the tuber during the dry season. The tuber is usually shaped like an elongated turnip, but of variable thickness, depending on the harshness of the conditions.

In Moçambique, where it is one of several geophytic species of this section, plants are mostly shorter than 150 mm, with fairly inconspicuous grey-green leaves and small yellowish cyathia.

History
Bally (1961) placed some collections of this species under *E. chevalieri*, but it was later separated off as a distinct species, which is now known to occur widely in Africa south of the equator. It was first collected in Moçambique near Ribáuè in November 1940 by F.A. Mendonça.

Euphorbia torrei (L.C.Leach) Bruyns, *Taxon* 55: 415 (2006). *Monadenium torrei* L.C.Leach, *Garcia de Orta*, Sér. Bot. 1: 37 (1973). Type: Moçambique, Cabo Delgado, Montepuez distr., base of Mt Matuta, 5 km south of M'salo River near Nantulo, ± 350 m, 9 Apr. 1964, *Torre & Paiva 11790* (LISC, holo.; COI, K, LMU, PRE, SRGH, iso.)

Bisexual spiny mostly glabrous succulent shrub 0.5–3 m tall with few to many branches from similar central stem (to 35 mm thick) bearing fibrous roots, branching mainly well above base. *Branches* ascending to erect, 0.1–0.6 m × 8–15 mm, cylindrical, not constricted into segments, smooth, dull brown-green later covered with peeling pale brown bark towards base; *tubercles* initially prominent later indistinct, neither fused together nor arranged into angles, with fine groove around each, much flattened and projecting at most 1–2 mm from surface of branch and produced into spine 4–8 mm long just beneath leaf, spine-shields absent; *leaves* on tips of new tubercles towards apex of stem and branches, 40–120 × 10–35 mm, spreading, deciduous, lanceolate to obovate, acute or obtuse, sessile, slightly fleshy and with dentate margins, glabrous, grey-green, with minute brown deciduous stipular prickles 0.5–1 mm long. *Synflorescences* few per branch towards apex, each a solitary cyme in axil of tubercle, on slender pale green longitudinally ridged erect peduncle 60–100 × 3–4 mm glabrous except for few hairs at apex, each cyme with 3–6 longitudinally ridged erect branches to 50 mm long bent horizontally at tips, with several cyathia developing in succession, each cyathium surrounded with 2 ovate inwardly folded cream (pale green-veined) and puberulous bracts forming bract-cup ± 10 × 15 mm notched almost to base in middle; *cyathia* ± cylindrical widening in middle, held facing slightly downwards, pubescent inside, 7.5–10 × 3.5–5 mm, with 5 oblong sparsely pubescent lobes with deeply toothed margins, pale green becoming white near tips, open to about 2/3 down in front; *gland* 1,

Fig. 6.122. Distribution of *Euphorbia torrei* (© PVB).

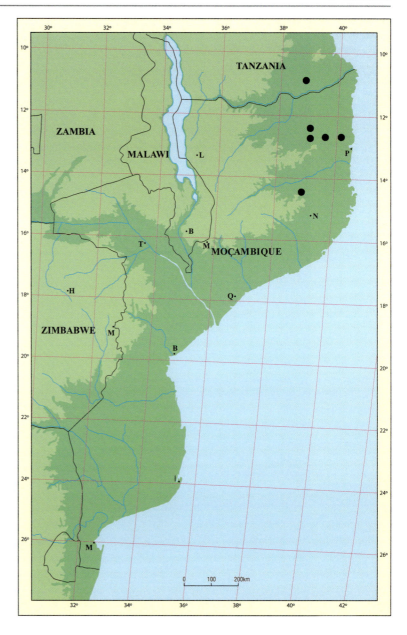

erect and bent inwards near middle, enclosing most of cyathium above except for gap in front, ± 6 mm long, green below becoming white above with bright pink edging at tip, inner margin slightly raised against backs of lobes, outer margin slightly reflexed, outer surface smooth; *stamens* glabrous, bracteoles filiform and enveloping groups of males, sparsely pubescent; *ovary* obtusely 3-angled, sparsely pubescent, pale green, raised on erect pedicel 2–3 mm long and rapidly becoming decurved through gap in gland as styles dry off; styles 3–4 mm long, branched to just above base and slightly spreading, pale yellow. *Capsule* ± 4.5 mm diam., obtusely 3-angled, sparsely pubescent, exserted horizontally (later becoming erect) on pedicel 3–6 mm long.

Distribution & Habitat

Euphorbia torrei is known in south-eastern Tanzania and northern Moçambique, where it occurs at altitudes of 170–600 m between Macomia and Nantulo. Usually it is found in shallow loose soils and leaf-litter on granitic outcrops in partial shade under deciduous trees.

Diagnostic Features & Relationships

Euphorbia torrei forms sparingly branched shrubs that regularly reach 2 m tall. The branches are mostly around 10 mm thick, covered with a dull surface from which some of the bark later peels off in brownish strips. A stout spine develops on each tubercle just below the position of the leaf and these persist to near the base of the stem (Fig 1.33 B).

Fig. 6.123. *Euphorbia torrei*, ± 0.8 m tall, in leaf-litter among dry bushes on rocks near base of tall granitic dome, with larger plants in background and *Aloe mawii*, *PVB 8553*, Cuero Mt, Meluco, Moçambique, 14 Nov. 2000 (© PVB).

Fig. 6.124. *Euphorbia torrei*, ± 1 m tall, in lush condition among rocks and bush, *E. Schmidt*, Moçambique, Mar. 2009 (© E. Schmidt).

Fig. 6.125. *Euphorbia torrei*, flowering branch, *PVB 8553*, Cuero Mt, Meluco, Moçambique, May 2019 (© PVB).

Leaves appear on new tubercles during the summer months and are notable for their toothed margins and variable shapes. They are quite thin in texture (not as fleshy as in *E. lugardiae*), so that the veins stand out on their lower surface.

Flowering in *E. torrei* takes place in autumn or early winter, usually as the leaves are in the process of falling off. The cyathia are produced on pale longitudinally ridged peduncles that often exceed 50 mm long. Each cyathium-bearing peduncle is bent horizontally at its tip so that the cyathia face horizontally or slightly downwards. Only the upturned lip of the gland (its outer margin) is usually edged with bright pink, otherwise most of the floral parts are pale green, yellow or cream.

Leach believed that *E. torrei*, along with the Angolan *E. neocannellii*, was closest to *Euphorbia neospinescens* (*Monadenium spinescens*) from Tanzania, but both seem to be closer to another Tanzanian species, *E. biselegans* (*M. elegans*). *Euphorbia neocannellii* and *E. torrei* both have a single spine beneath each leaf, while in *E. biselegans* there are three prominent spines around the leaf (Fig. 1.33 C). Two of these three spines correspond to the two minute prickles in *E. torrei*. *Euphorbia biselegans* shares the cylindrical branches with peeling bark lower down and the toothed margins of the only slightly fleshy leaves with *E. torrei*. In *E. neocannellii* the leaves are much larger and more succulent and have smooth margins (Fig. 1.35).

Fig. 6.126. *Euphorbia torrei*, close-up of cymes, *PVB 8553*, Cuero Mt, Meluco, Moçambique, Jun 2018 (© PVB).

Fig. 6.127. *Euphorbia torrei*, close-up of cymes, *PVB 9701*, north of Nantulo, Moçambique, Jun 2018 (© PVB).

History

The first record of *E. torrei* was made by J. Stocks in 1907 in northern Moçambique. It was again collected by Jacob Gerstner (of *Aloe gerstneri* and *Euphorbia gerstneriana* fame) in August 1949, this time near Lupaso in SE Tanzania. This specimen was sterile and could only be placed near *E. neospinescens* (Bally 1961: 101), though Bally later visited this area and collected it himself. The plants from which Leach described it were collected in flower in April 1964 by António Rocha da Torre and Jorge A.R. de Paiva. Living plants were not seen by Leach, who did not explore the northernmost parts of Moçambique.

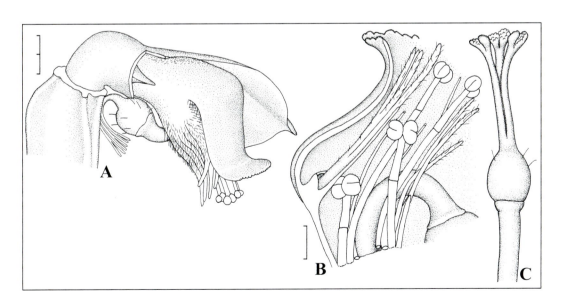

Fig. 6.128. *Euphorbia torrei*. **A**, side view of cyathium with one bract removed (scale 2 mm). **B**, side view of dissected cyathium (scale 1 mm, as for **C**). **C**, female floret. Drawn from: *PVB 8553*, Mt Cuero, Meluco, Moçambique (© PVB).

References

Aiton, W. (1789). *Hortus Kewensis*. Ed. 1, 2. London: G. Nicol.

Alpini, P. (1735). *Historiae naturalis Aegypti pars secunda, sive, de Plantis Aegypti Liber. Cum Observationibus & Notis Joannis Veslingii*. Leiden: G. Potvliet.

Anthony, H. J. (1949). A succulent enthusiast in Nyasaland. *Journal of the New York Botanical Garden, 50*, 97–103.

APG IV. (2016). An update of the Angiosperm Phylogeny Group classification for the orders and families of flowering plants: APG IV. *Botanical Journal of the Linnean Society, 181*, 1–20.

Arakaki, M., Christin, P.-A., Nyffeler, R., Lendel, A., Eggli, U., Ogburn, R. M., Spriggs, E., Moore, M. J., & Edwards, E. J. (2011). Contemporaneous and recent radiations of the world's major succulent plant lineages. *Proceedings of the National Academy of Sciences, 108*, 8379–8384.

Archer, R. H. (1998). *Euphorbia leistneri* (Euphorbiaceae), a new species from the Kaokoveld (Namibia). *South African Journal of Botany, 64*, 258–260.

Archer, R. H., & Carter, S. (2001). *Euphorbia venteri. Flowering Plants of Africa, 57*, 86–90.

Axelrod, D. J. (1983). Biogeography of oaks in the Arcto-Tertiary Province. *Annals of the Missouri Botanical Garden, 70*, 629–657.

Axelrod, D. J., & Raven, P. H. (1978). Late Cretaceous and Tertiary vegetation history of Africa. In M. J. A. Werger (Ed.), *Biogeography & ecology of Southern Africa* (pp. 77–130). The Hague: W. Junk.

Baker, J. G. (1869). Euphorbia tetragona, Haw. In W.W. Saunders (Ed.). *Refugium Botanicum, 1*, t. 39.

Bally, P. R. O. (1961). *THE GENUS MONADENIUM. With descriptions of 18 new species 32 plates in natural size and 88 figures*. Berne: Benteli Publishers.

Bally, P. R. O. (1963). Miscellaneous notes on the flora of Tropical East Africa including description of new taxa. 8–15. *Candollea, 18*, 335–357.

Bauer, A. M., Günther, R., & Klippel, M. (1995). *The herpetological contributions of Wilhelm Peters (1815–1883)*. Ithaca: Society for the Study of Amphibians and Reptiles.

Berger, A. (1899). Zwei verwechselte Euphorbien. *Monatsschr Kakteenk, 9*, 88–92.

Berger, A. (1902). Eine neue Euphorbia. *Monatsschr Kakteenk, 12*, 39–40.

Berger, A. (1906). *Sukkulente Euphorbien*. Stuttgart: E. Ulmer.

Berger, A. (1910). Einige neue afrikanische Sukkulenten. *Botanische Jahrbücher für Systematik, 45*, 223–233.

Bergius, P. J. (1767). *Descriptiones Plantarum ex Capite Bonae Spei*. Stockholm: Laurentius Salvius.

Boissier, P. E. (1860). *Centuria Euphorbiarum*. Paris: B. Hermann, Leipzig & J.-B. Baillière.

Boissier, E. (1862). Euphorbieae. In A. P. De Candolle (Ed.), *Prodromus* 15 (2): 3–188. Paris: V. Masson & Sons.

Brown, R. (1814). General remarks, geographical and systematical, on the botany of Terra Australis. In M. Flinders (Ed.), *A voyage to Terra Australis: Undertaken for the purpose of completing the discovery of that vast country* (Appendix III) (Vol. 2, pp. 533–613). London: G. W. Bulmer.

Brown, N. E. (1897). Euphorbia grandicornis, Goebel. *Hooker's Icones Plantarum, 26*, t. 2531, 2532.

Brown, N. E. (1911–12). *Euphorbia*. In W. T. Thiselton-Dyer (Ed.), *Flora Tropical Africa* 5 (2): 470–603. London: L. Reeve & Co.

Brown, N. E. (1915). *Euphorbia*. In W. T. Thiselton-Dyer (Ed.), *Flora Capensis* 5 (2): 222–375. London: L. Reeve & Co.

Bruyns, P. V. (1992a). Notes on *Euphorbia monteiroi. Aloe, 29*, 36–38.

Bruyns, P. V. (1992b). Notes on African plants. Euphorbiaceae. Notes on *Euphorbia* species from the Northwestern Cape. *Bothalia, 22*, 37–51.

Bruyns, P. V. (2000). *Euphorbia*. In P. Goldblatt & J. Manning (Eds.), *Cape Plants. Strelitzia* 9: 455–458. Pretoria: South African National Biodiversity Institute.

Bruyns, P. V. (2012). Nomenclature and typification of southern African species of *Euphorbia. Bothalia, 42*, 217–245.

Bruyns, P. V. (2013). Euphorbiaceae. In D. A. Snijman (Ed.), *Plants of the Greater Cape Floristic Region. 2: The Extra Cape Flora. Strelitzia* 30: 371–379. Pretoria: South African National Biodiversity Institute.

Bruyns, P. V. (2018). New taxa in *Euphorbia* (Euphorbiaceae) in southern Africa. *Haseltonia, 25*, 30–56.

Bruyns, P. V., & Berry, P. E. (2019). The nomenclature and application of the names *Euphorbia candelabrum* Welw. and *Euphorbia ingens* E. Mey ex Boiss. in tropical Africa. *Taxon, 68*, 828–838.

Bruyns, P. V., Mapaya, R. J., & Hedderson, T. (2006). A new subgeneric classification for *Euphorbia* (Euphorbiaceae) in southern Africa based on ITS and *psb*A-*trn*H sequence data. *Taxon, 55*, 397–420.

Bruyns, P. V., Al-Farsi, A., & Hedderson, T. (2010). Phylogenetic relationships of *Caralluma* R. Br. (Apocynaceae). *Taxon, 59*, 1031–1043.

Bruyns, P. V., Klak, C., & Hanáček, P. (2011). Age and diversity in Old World succulent species of *Euphorbia* (Euphorbiaceae). *Taxon, 60*, 1717–1733.

Bruyns, P. V., Oliveira-Neto, M., Melo-de-Pinna, F., & Klak, C. (2014). Phylogenetic relationships in the Didiereaceae with special reference to subfamily Portulacarioideae. *Taxon, 63*, 1053–1064.

Burman, J. (1737). *Thesaurus Zeylanicus*. Amsterdam: Janssonius-Waesberg. & S. Schouten.

Burman, J. (1738–9). *Rariorum Africanarum Plantarum*. Amsterdam: Boussière.

Burrows, J. E., Burrows, S., Schmidt, E., Lötter, M., & Wilson, E. O. (2018). *Trees and shrubs Mozambique*. Cape Town: Print Matters Heritage.

Carter, S. (1988). *Euphorbia* in Euphorbiaceae (Part 2). In R. M. Polhill (Ed.), *Flora of tropical East Africa* (pp. 409–533). Kew: Royal Botanic Gardens.

Carter, S. (1993). Euphorbia, In: M. Thulin (Ed.), *Flora of Somalia* 1: 306–337. Kew: Royal Botanic Gardens.

Carter, S. (1994). A preliminary classification of *Euphorbia* subgenus *Euphorbia*. *Annals of the Missouri Botanical Garden, 81*, 368–379.

Carter, S. (1999). New *Euphorbia* taxa in the Flora Zambesiaca area. *Kew Bulletin, 54*, 959–965.

Carter, S. (2002). *Euphorbia*. In U. Eggli (Ed.), *Illustrated handbook of succulent plants: Dicotyledons* (pp. 102–203). Berlin: Springer.

Carter, S. (2014). Typification of *Euphorbia xylophylloides* and *Euphorbia polygona*. *Euphorbia World, 10*, 18–20.

Carter, S., & Leach, L. C. (2001). Tribe Euphorbieae. In G. V. Pope (Ed.), *Flora Zambesiaca* 9 (5): 339–465. Kew: Royal Botanic Gardens.

Chapman, J. D. (1962). *The vegetation of Mlanje Mountains, Nyasaland*. Zomba: The Government Press.

Coe, M. J., & Skinner, J. D. (1993). Connections, disjunctions and endemism in the eastern and southern African mammal faunas. *Transactions of the Royal Society of South Africa, 48*, 233–255.

Coetzee, J. A. (1980). Tertiary environmental changes along the south-western African coast. *Palaeontologia Africana, 23*, 197–203.

Commelijn, C. (1697). *Horti Medici Amstelodamensis* 1. Amsterdam: A. van Someren.

Commelijn, C. (1703). *Praeludia botanica. Lugduni Batavorum* (Leiden). Leiden: F. Haringh.

Compton, R. H. (1929). The vegetation of the Karoo. *Journal of the Botanical Society of South Africa, 15*, 13–21.

Compton, R. H. (1931). The Flora of the Whitehill district. *Transactions of the Royal Society of South Africa, 19*, 269–329.

Compton, R. H. (1976). *The Flora of Swaziland* (Journal of South African Botany, Supplementary Volume 11). Kirstenbosch: Trustees of National Botanc Gardens.

Corner, E. J. H. (1976). *The Seeds of Dicotyledons*. Cambridge: Cambridge University Press.

Court, D. (1988). Euphorbia polycephala at Cranemere. *Euphorbia Journal, 5*, 39–42.

Croizat, L. (1932). A list of annotated observations on the remarks of Dr. K. von Poellnitz concerning A. Berger's classification of succulent Euphorbiae. *Cactus and Succulent Journal (US), 4*, 291–293.

Croizat, L. (1933). A list of annotated observations on the remarks of Dr. K. von Poellnitz concerning A. Berger's classification of succulent Euphorbiae. Part III. *Cactus and Succulent Journal (US), 4*, 330–334.

Croizat, L. (1934). *De Euphorbio Antiquorum Atque Officinarum*. A study of succulent Euphorbiae long known in cultivation. New York.

Croizat, L. (1937). On the classification of *Euphorbia*. II. How should the cyathium be interpreted? *Bulletin of the Torrey Botanical Club, 64*, 523–536.

Croizat, L. (1945). "*Euphorbia Esula*" in North America. *American Midland Naturalist, 33*, 231–243.

Croizat, L. (1965). An introduction to the subgeneric classification of *Euphorbia* L., with stress on the South African and Malagasy species. I. *Webbia, 20*, 473–706.

Croizat, L. (1967). Two new semi-succulent Euphorbias from Venezuela: Euphorbia lutzenbergeriana and E. lagunillarum. *Cactus and Succulent Journal (US), 39*, 142–144.

Croizat, L. (1972). An introduction to the subgeneric classification of *Euphorbia* L., with stress on the South African and Malagasy species III. *Webbia, 27*, 1–221.

Cronquist, A. (1988). *The evolution and classification of flowering plants* (2nd ed.). New York: New York Botanical Garden.

De Candolle, A. P. (1805). *Plantarum Historia Succulentarum*, Fasc. 27. Paris: Garnery.

D'Isnard, A.-T. D. (1720). Etablissement d'un Genre de Plante appellé Euphorbe. *Hist. Acad. Roy. Sci. Mém. Math. Phys. (Paris, 4°), 1720*, 384–399.

Desfontaines, R. L. (1829). *Tableau de l'École de Botanique du Muséum d'Histoire Naturelle* (3rd ed.). Paris: J.S. Chaudé.

Dillenius, J. J. (1732). *Hortus elthamensis* 2. London.

Dinter, M. K. (1921). Botanische Reisen in Deutsch-Südwest-Afrika. *Feddes Repert. Spec. Nov. Regni Veg., Beih.* Band 3.

Dinter, M. K. (1923). Succulentenforschung in Südwestafrika. I. Teil. *Feddes Repert. Spec. Nov. Regni Veg., Beih.* Band 23.

Dinter, M. K. (1928). Sukkulentenforschung in Südwestafrika. II. Teil. *Feddes Repert. Spec. Nov. Regni Veg., Beih.* Band 53.

Dorsey, B. L., Haevermans, T., Aubriot, X., Morawetz, J. J., Riina, R., Steinmann, V. W., & Berry, P. E. (2013). Phylogenetics, morphological evolution, and classification of *Euphorbia* subgenus *Euphorbia*. *Taxon, 62*, 291–315.

Drège, J. F. (1843). *Zwei planzengeographische Documente*. Besondere Beigabe zur Flora 1843, Band II.

Dressler, R. L. (1957). The genus *Pedilanthus* (Euphorbiaceae). *Contributions from the Gray Herbarium, 182*, 1–188.

Dudley, C. O. (1997). The candelabra tree (*Euphorbia ingens*): A source of water for black rhinoceros in Liwonde National Park, Malawi. *Koedoe, 40*, 57–62.

Dyer, R. A. (1931). Notes on *Euphorbia* species of the Eastern Cape Province with descriptions of three new species. *Records of the Albany Museum, 4*, 64–110.

Dyer, R. A. (1934a). Miscellaneous new species. *Bulletin of Miscellaneous Information, 1934*, 265–270.

Dyer, R. A. (1934b). Euphorbia duseimata. *Flowering Plants of South Africa, 14*, t. 530.

Dyer, R. A. (1935a). The seed germination of certain species of *Euphorbia*. *South African Journal of Science, 32*, 313–319.

Dyer, R. A. (1935b). Note on the meloformia group of Euphorbia. *Records of the Albany Museum, 4*, 254–255.

Dyer, R. A. (1937a). Euphorbia complexa. *Flowering Plants of South Africa, 17*, t. 643.

Dyer, R. A. (1937b). Euphorbia tubiglans. *Flowering Plants of South Africa, 17*, t. 646.

Dyer, R. A. (1938a). *Euphorbia fasciculata* Thunb. and *E. schoenlandii* Pax. *South African Journal of Science, 35*, 298–299.

Dyer, R. A. (1938b). Euphorbia zoutpansbergensis. *Flowering Plants of South Africa, 18*, t. 715.

Dyer, R. A. (1940a). *Euphorbia schoenlandii*. *Flowering Plants of South Africa, 20*, t. 772.

Dyer, R. A. (1940b). *Euphorbia obesa*. *Flowering Plants of South Africa, 20*, t. 788.

Dyer, R. A. (1949). Col. Robert Jacob Gordon's contribution to South African Botany. *South African Biological Society Pamphlet, 14*, 1–20.

Dyer, R. A. (1951). Newly described plants. Euphorbiaceae. *Bothalia, 6*, 221–227.

Dyer, R. A. (1953). A new succulent Euphorbia from the Western Cape Province. *Journal of South African Botany, 19*, 135–136.

Dyer, R. A. (1957). The classification of species of *Euphorbia* with stipular spines. *Bull. Jard. Bot. Bruxelles, 27*, 487–493.

Dyer, R. A. (1974a). *Euphorbia triangularis*. *Flowering Plants of Africa, 43*, t. 1687.

Dyer, R. A. (1974b). Notes on African plants. Euphorbiaceae. A new species of *Euphorbia*. *Bothalia, 11*, 278–280.

Dyer, R. A. (1979). Looking back. *Cactus and Succulent Journal (US), 51*, 116–117, 119.

Flores, G. (1970). Suggested origin of the Mozambique channel. *Transactions of the Geological Society of South Africa, 73*, 1–16.

Fourie, S. P. (1987). An introduction to the succulent Euphorbias of the Transvaal, Part 2: Shrubs. *Euphorbia Journal, 4*, 48–68.

Fourie, S. P. (1988). An introduction to the succulent Euphorbias of the Transvaal, Part 3: Dwarf shrubs I. *Euphorbia Journal, 5*, 83–93.

Fourie, S. P. (1989). An introduction to the succulent Euphorbias of the Transvaal, Part 3: Dwarf Shrubs II. *Euphorbia Journal, 6*, 113–125.

Frick, G. A. (1931). Euphorbia polycephala. *Cactus and Succulent Journal (US), 3*, 97–98.

Frick, G. A. (1936). Euphorbia Dinteri Berger. *Euphorbia Review, 2*, 2–3.

Frick, G. A. (1937). Euphorbia halleri Dtr. *Cactus and Succulent Journal (US), 9*, 39.

Frick, G. A. (1938). Habitat of Euphorbia pseudocactus. *Cactus and Succulent Journal (US), 9*, 143–144.

Genç, Z., & Rauh, W. (1984). Vergleichend-anatomische Untersuchungen an den Honigdrüsen einiger *Euphorbia*-Arten. In W. Rauh (Ed.), Anatomisch-Biochemische Untersuchungen an Euphorbien Teil 1. *Trop. subtrop. Pflanzenw.* 45: 9–54.

Gilbert, M. G. (1987). Two new geophytic species of *Euphorbia* with comments on the subgeneric grouping of its African members. *Kew Bulletin, 42*, 231–244.

Gilbert, M. G. (1995). Euphorbiaceae. In S. Edwards, M. Tadesse, & I. Hedberg (Eds.), *Flora of Ethiopia and Eritrea* 2 (2): 265–380. Addis Ababa: National Herbarium.

Glen, H. F., & Germishuizen, G. (2010). *Botanical Exploration of southern Africa* (Strelitzia 26) (2nd ed.). Pretoria: South African National Biodiversity Institute.

Goebel, K. (1889). *Pflanzenbiologische Schilderungen* 1. Marburg: N.G. Elwert'sche Verlagsbuchhandlung.

Grace, O. M., Buerki, S., Symonds, M. R. E., Forrest, F., Van Wyk, A. E., Smith, G. F., Klopper, R. R., Bjora, C. S., Neale, S., Demissew, S., Simmonds, M. S. J., & Rønstead, N. (2016). Evolutionary history and leaf succulence as explanations for medicinal use in aloes and the global popularity of Aloe vera. *Evolutionary Biology, 2016*, 16–29.

Grove, A. T. (1969). Landform and climatic change in the Kalahari and Ngamiland. *Geographical Journal, 135*, 191–212.

Gschwendt, M., & Hecker, E. (1973). Über die Wirkstoffe der Euphorbiaceen I. *Z. Krebsforsch., 80*, 335–350.

Gschwendt, M., & Hecker, E. (1974). Über die Wirkstoffe der Euphorbiaceen II. *Z. Krebsforsch., 81*, 193–210.

Gunn, M., & Codd, L. E. (1981). *Botanical exploration of Southern Africa*. Cape Town: A.A. Balkema.

Hall, H. (1949). Euphorbia fasciculata rediscovered. *Cactus and Succulent Journal (US), 21*, 140–141.

Hammer, S. A. (1993). *The genus Conophytum*. Pretoria: Succulent Plant Publications.

Hans, A. S. (1973). Chromosomal conspectus of the Euphorbiaceae. *Taxon, 22*, 591–636.

Hargreaves, B. J. (1976). Killing and curing. *Cactus and Succulent Journal (US), 48*, 190–196.

Hargreaves, B. J. (1984). Is "Chikhawo" Euphorbia decidua? *Euphorbia Journal, 2*, 91–94.

Hargreaves, B. J. (1992a). The succulent Euphorbias of Botswana. *Euphorbia Journal, 8*, 45–50.

Hargreaves, B. J. (1992b). Volcanos & spurges. *Euphorbia Journal, 8*, 103–110.

Hargreaves, B. J. (1992c). The other spurges of Lesotho (at least allegedly). *Euphorbia Journal, 8*, 126–132.

Hargreaves, B. J. (1994). The Euphorbia schinzii complex. *Euphorbia Journal, 9*, 147–153.

Hässler, A. (1931). Verwantschaftliche Gliederung der afrikanischen Euphorbien aus der Sektionen Trichadenia Pax und Rhizanthium Boiss. *Bot. Not., 1931*, 317–338.

Haworth, A. H. (1812). *Synopsis Plantarum Succulentarum*. London: R. Taylor.

Haworth, A. H. (1827). LVI. Description of new succulent plants. *Philosophical Magazine Annals of Chemistry, 1*, 271–277.

Hecker, E. (1981). Cocarcinogenesis and tumor promoters of the diterpene ester type as possible carcinogenic risk factors. *Journal of Cancer Research and Clinical Oncology, 99*, 103–124.

Hemsley, W. B. (1904). Euphorbia viperina. *Curtis's Botanical Magazine, 130*, t. 7971.

Hendey, Q. B. (1982). *Langebaanweg. A record of past life*. Cape Town: South African Museum.

Herre, H. (1936). Euphorbia Stapelioides Boiss. *Euphorbia Review, 2*, 18–19.

Herre, H. (1950). Euphorbia superans Nel spec. nov. *Desert Plants Life, 22*, 15.

Herre, H. (1973). Notes on South African succulents. *Cactus and Succulent Journal (US), 45*, 238–239.

Holmes, Carter, S. (1993). *Euphorbia*. In M. Thulin (Ed.), *Flora of Somalia* 1: 323–337. Kew: Royal Botanic Gardens.

Hooker, J. D. (1865). Euphorbia monteiri. *Curtis's Botanical Magazine, 91*, t. 5534.

Hooker, J. D. (1903). Euphorbia obesa. *Curtis's Botanical Magazine, 129*, t. 7888.

Hoppe, J. R. (1985). Die Morphogenese der Cyathiendrüsen und ihrer Anhänge, ihre blattypologische Deutung und Bedeutung. *Bot. Jahrb. Syst., 105*, 497–581.

Hoppe, J. R., & Uhlarz, H. (1982). Morphogenese und typologische Interpretation des Cyathiums von *Euphorbia*-Arten. *Beitr. Biol. Pflanzen, 56*, 63–98.

Horn, J. W., van Ee, B. W., Morawetz, J. J., Riina, R., Steinmann, V. W., Berry, P. E., & Wurdack, K. J. (2012). Phylogenetics and the evolution of major structural characters in the giant genus *Euphorbia* L. (Euphorbiaceae). *Molecular Phylogenetics and Evolution, 63*, 305–326.

Hutchinson, J. (1946). *A botanist in Southern Africa*. London: P.R. Gawthorn.

Ihlenfeld, H.-D., & Gerbaulet, M. (1990). Untersuchungen zum Merkmalsbestand und zur Taxonomie der Gattungen *Apatesia* N.E. Br., *Carpanthea* N.E. Br., *Conicosia* N.E. Br., *Herrea* Schwantes und *Hymenogyne* Haw. (Mesembryanthemaceae Fenzl). *Bot. Jahrb. Syst., 111*, 457–498.

Jackson, S. P. (1961). *Climatological Atlas of Africa*. Lagos: CCTA/CSA.

Jacobsen, N. H. G. (1988). Euphorbias of the Skeleton Coast National Park: Namibia and adjacent areas. *Euphorbia Journal, 5*, 59–67.

Jacot Guillarmod, A. (1971). *Flora of Lesotho (Basutoland)*. Lehre: J. Cramer.

Jones, K., & Smith, J. B. (1969). The chromosome identity of *Monadenium* Pax and *Synadenium* Pax (Euphorbiaceae). *Kew Bulletin, 23*, 491–498.

Jordaan, P. G. (1973). Le Vaillant's botanical paintings. In J. C. Quinton, A. M. Lewin Robinson, & P. W. M. Sellicks (Eds.), *François le Vaillant: Traveller in South Africa, and his collection of 165 Water-Colour Paintings*, 1781–1784, 2: 43–81. Cape Town: Library of Parliament.

Kimberley, M. J. (1982). Dr R.F. Rand. *Excelsa, 10*, 85–87.

Klak, C., & Linder, H. P. (1998). Systematics of *Psilocaulon* N.E. Br. & *Caulipsolon* Klak, gen. nov. (Mesembryanthemoideae, Aizoaceae). *Bot Jahrb. Syst., 120*, 301–375.

Klotzsch, J. F. (1860a). Linne's natürliche Pflanzenklasse *Tricoccae* des Berliner Herbarium's im Allgemeinen und die natürliche Ordnung *Euphorbiaceae* insbesondere. *Monatsber. Königl. Akad. Wiss. Berlin, 1859*, 236–254.

Klotzsch, J. F. (1860b). Linne's natürliche Pflanzenklasse *Tricoccae* des Berliner Herbarium's im Allgemeinen und die natürliche Ordnung *Euphorbiaceae* insbesondere. *Abh. Königl. Akad. Wiss. Berlin, 1859*, 1–108.

Koutnik, D. L. (1984a). A Brief Taxonomy of the Euphorbia clavaloricata complex (Treisia). *Euphorbia Journal, 2*, 38–50.

Koutnik, D. L. (1984b). *Chamaesyce* (Euphorbiaceae) – A newly recognized genus in southern Africa. *South African Journal of Botany, 3*, 262–264.

Koutnik, D. L. (1987). A taxonomic revision of the Hawaiian species of the genus *Chamaesyce* (Euphorbiaceae). *Allertonia, 4*, 331–388.

Krauss, C. F. F. (1844). Pflanzen des Cap- und Natal-landes, gesammelt und zusammengestellt von Dr. Ferdinand Krauss (Fortsetzung). *Flora, 27*, 277–307.

Krenzelok, E. P., Jacobsen, T. D., & Aronis, J. M. (1996). Poinsettia exposures have good outcomes…Just as we thought. *American Journal of Emergency Medicine, 14*, 671–674.

Kunkel, G. (1978). *Flora de Gran Canaria 3. Las plantas suculentas*. Barcelona: Seix y Barral Hnos.

Lamarck, J.-B. (1786). *Encyclopédie Méthodique. Botanique* 2. Paris/Liege: Panckoucke/Plomteux.

Lawant, P., & Van Veldhuisen, R. (2014a). *Euphorbia leachii* Lawant & van Veldhuisen sp. nov., a new species from South Africa. *Euphorbia World, 10*, 5–8.

Lawant, P., & Van Veldhuisen, R. (2014b). Some dubious interpretations of *Euphorbia patula* Mill. *Euphorbia World, 10*, 12–15.

Leach, L. C. (1963). Euphorbia johnsonii. Its rediscovery and an amplified description. *Kirkia, 3*, 34–36.

Leach, L. C. (1964). Euphorbia species from the Flora Zambesiaca area III. *Journal of South African Botany, 30*, 209–217.

Leach, L. C. (1966). Euphorbia species from the Flora Zambesiaca area: V. *Journal of South African Botany, 33*, 173–182.

Leach, L. C. (1967). Euphorbia species from the Flora Zambesiaca area: VII. *Journal of South African Botany, 33*, 247–262.

Leach, L. C. (1968a). Euphorbia species from the Flora Zambesiaca area: VI. *Kirkia, 6*, 133–143.

Leach, L. C. (1968b). Euphorbiae succulentae angolenses–I. *Bol. Soc. Brot., Sér. 2, 42*, 161–179.

Leach, L. C. (1969). Euphorbiae succulentae angolenses: II. *Bol. Soc. Brot., Sér. 2, 43*, 163–182.

Leach, L. C. (1970a). Euphorbia species from the Flora Zambesiaca Area: IX. *Journal of South African Botany, 36*, 13–52.

Leach, L. C. (1970b). Euphorbiae Succulentae Angolensis: III. *Bol. Soc. Brot., 44*, 185–206.

Leach, L. C. (1971). Euphorbia virosa Willd. with a new synonym and two new taxa from Angola. *Bol. Soc. Brot., Sér. 2, 45*, 349–362.

Leach, L. C. (1973a). *Euphorbia tirucalli* L, its typification, synonymy and relationships, with notes on 'Almeidina' and 'Cassoneira'. *Kirkia, 9*, 69–86.

Leach, L. C. (1973b). Euphorbia species from the Flora Zambesiaca Area: 10. *Journal of South African Botany, 39*, 3–22.

Leach, L. C. (1974). Euphorbia succulentae Angolenses: IV. *Garcia de Orta, Sér. Bot., 2*, 31–54.

Leach, L. C. (1975a). Notes on *Euphorbia mauritanica*, *E. gossypina* and some related species with an amplified description of *E. berotica*. *Bothalia, 11*, 505–510.

Leach, L. C. (1975b). Euphorbia succulentae Angolenses: V. *Garcia de Orta, Sér. Bot., 2*, 111–116.

Leach, L. C. (1976a). Euphorbia (*Tetracanthae*) in Angola and northern Kaokoland. *Dinteria, 12*, 1–35.

Leach, L. C. (1976b). A review of the *Euphorbia brevis-imitata-decidua* complex with descriptions of five new species. *Bull. Jard. Bot. Nat. Belg., 46*, 241–263.

Leach, L. C. (1976c). Distributional and morphological studies of the tribe Euphorbieae (Euphorbiaceae) and their relevance to its classification and possible evolution. *Excelsa, 6*, 3–18.

Leach, L. C. (1977). Euphorbiae succulentae Angolenses: VI. *Garcia da Orta, Sér. Bot., 3*, 99–102.

Leach, L. C. (1980). A taxonomic review of *Euphorbia gariepina* Boiss., its distribution, variability and synonymy. *Excelsa Taxon Series, 2*, 74–81.

Leach, L. C. (1981). A new species of *Euphorbia* from the Karoo. *Journal of South African Botany, 47*, 103–107.

Leach, L. C. (1982). A new species of *Euphorbia* from the Namib, South West Africa. *Dinteria, 16*, 27–31.

Leach, L. C. (1983). A new *Euphorbia* from South West Africa. *Journal of South African Botany, 49*, 189–192.

Leach, L. C. (1984a). A new *Euphorbia* from South Africa. *Journal of South African Botany, 50*, 341–345.

Leach, L. C. (1984b). A new *Euphorbia* from the Richtersveld. *Journal of South African Botany, 50*, 563–568.

Leach, L. C. (1985). A new species of *Euphorbia* (Euphorbiaceae) from the Cape Province. *South African Journal of Botany, 51*, 281–283.

Leach, L. C. (1986a). *Euphorbia filiflora*. *Flowering Plants of Africa, 49*, t. 1927.

Leach, L. C. (1986b). A new *Euphorbia* (Euphorbiaceae) from the western Cape Province. *South African Journal of Botany, 52*, 10–12.

Leach, L. C. (1986c). A new *Euphorbia* from the western Knersvlakte, Cape Province. *South African Journal of Botany, 52*, 369–371.

Leach, L. C. (1988a). A new species of *Euphorbia* (Euphorbiaceae) from the south-western Cape. *South African Journal of Botany, 54*, 501–503.

Leach, L. C. (1988b). The *Euphorbia juttae-gentilis* complex, with a new species and a new subspecies. *South African Journal of Botany, 54*, 534–538.

Leach, L. C. (1991). Euphorbia griseola Pax: Its subspecies & relationships. *The Euphorbia Journal, 7*, 131–137.

Leach, L. C. (1992). Euphorbia knuthii Pax: Its distribution and other matters. *Euphorbia Journal, 8*, 72–73.

Leach, L. C., & Williamson, G. (1990). The identities of two confused species of *Euphorbia* (Euphorbiaceae) with descriptions of two closely related new species from Namaqualand. *South African Journal of Botany, 56*, 71–78.

Lemaire, C. (1855). Observations diagnostico-nomenclaturales sur les Euphorbes charnues du Cap. *Ill. Hort.* 2: misc. 65–69.

Lemaire, C. (1858). Nouvelles Euphorbes. *Ill. Hort.* 5: misc. 63–64.

Le Vaillant, F. (1790). *Voyage de M. le Vaillant dans l'intérier de l/Afrique par Le Cap de Bonne Espérence, dans les années 1783, 84 & 85*. Paris: Leroy.

Le Vaillant, F. (1795). *Second voyage dans l'intérieur de l'Afrique: par le Cap de Bonne-Espérance, dans les années 1783, 84, et 85*. Paris: H.J. Jansen.

Le Vaillant, F. (1797). *Reise in das Innere con Afrika, vom Vorgebürge der guten Hoffnung aus, in den Jahren 1780 bis 85* (Vols. 3, 4). Frankfurt: P.H. Guilhauman.

Liede, S., & Meve, U. (1993). Towards an understanding of the *Sarcostemma viminale* (Asclepiadaceae) complex. *Botanical Journal of the Linnean Society, 112*, 1–15.

Linneaus, C. (1753). *Species Plantarum*. Stockholm: Laurentius Salvius.

Löffler, E., & Uhlarz, H. (1969). Observations on some Euphorbias from Tropical West-Africa. *Cactus and Succulent Journal (US), 41*, 210–220.

Loffler, L., & Loffler, P. (2005). *Swaziland tree Atlas, including selected shrubs and climbers*. Pretoria: SABONET.

Lotsy, J. P., & Goddijn, W. A. (1928). Hybrisization in the native flora. *Genetica, 10*, 1–129.

Loutit, B. D., Louw, G. N., & Seely, M. K. (1987). First approximation of food preferences and the chemical composition of the diet of the desert-dwelling black rhinoceros, *Diceros bicornis* L. *Madoqua, 15*, 35–54.

Lugard, E. (1938). Monadenium lugardae N.E.Br. *Cactus and Succulent Journal (US), 9*, 178–179.

Macnae, M. M., & Davidson, L. E. (1969). The volume "*Icones Plantarum et Animalium*" in the Africana Museum, Johannesburg, and its relationship to the *Codex Witsenii* quoted by Jan Burman in his "*Decades Rariorum Africanarum Plantarum*". *Journal of South African Botany, 35*, 65–81.

Manning, J., & Goldblatt, P. (Eds.). (2012). *Plants of the Greater Cape Floristic Region*. 1: The Core Cape Flora. (Ed.). *Strelitzia* 29. Pretoria: South African National Biodiversity Institute.

Marloth, R. (1908). *Das Kapland, insonderheit das Reich der Kapflora, das Waldgebiet und die Karroo, pflanzengeographisch dargestellt.* Jena: G. Fischer.

Marloth, R. (1910). Some new South African succulents. Part III. *Transactions of the Royal Society of South Africa, 2,* 33–39.

Marloth, R. (1925). *The flora of South Africa.* 2 (2). Cambridge: Cambridge University Press.

Marloth, R. (1928). The meloformia group of Euphorbias. *South African Gardening & Country Life, 28,* 45–46.

Marloth, R. (1930). A revision of the group *Virosae* of the genus *Euphorbia* as far as represented in South Africa. *South African Journal of Science, 27,* 331–340.

Marloth, R. (1931). Euphorbias III. *South African Gardening & Country Life, 21*(127), 133.

Marx, J. G. (1987). *Euphorbia albipollinifera* Leach. A recently described species from the South-Eastern Cape Province. *Aloe, 24,* 18–19.

Marx, J. G. (1988). Euphorbia meloformis Aiton and Euphorbia valida N.E. Brown: Some observations in habitat. *Euphorbia Journal, 5,* 94–103.

Marx, J. G. (1992). The succulent Euphorbias of the southeastern Cape Province, Part 1: Dwarf species and smaller shrubs. *Euphorbia Journal, 8,* 74–102.

Marx, J. G. (1993). The subglobose euphorbias and relatives. *Aloe, 30,* 103–107.

Marx, J. G. (1994). Euphorbia brevirama N.E. Brown: The quest continues. *Euphorbia Journal, 9,* 221–222.

Marx, J. G. (1999a). The South African Melon-shaped Euphorbias: The full picture as known to date. *Euphorbiaceae Study Group Bulletin, 12,* 13–34.

Marx, J. G. (1999b). *Euphorbia suppressa* J.G. Marx and *Euphorbia gamkensis* J.G. Marx, two hitherto-unnamed species from the Western Cape Province, South Africa. *Cactus and Succulent Journal (US), 71,* 33–40.

Marx, J. G. (2015). Accumulation of errors regarding *Euphorbia crassipes* Marloth. *Euphorbia World, 11,* 20–26.

Mauseth, J. D. (2004a). The structure of photosynthetic succulent stems in plants other than cacti. *International Journal of Plant Sciences, 165,* 1–9.

Mauseth, J. D. (2004b). Cacti and other succulents: Stem anatomy of 'other succulents' has little in common with that of cact. *Bradleya, 22,* 131–140.

Meerburg, N. (1789). *Plantae rariores vivis coloribus depictae.* Leiden: J. Meerburg.

Meyer, P. G. (1967). Euphorbiaceae. In H. Merxmüller (Ed.), *Prodromus einer Flora von Südwestafrika* (p. 67). Lehre: J. Cramer.

Miller, P. (1768). *The gardeners dictionary* (8th ed., p. London).

Millspaugh, C. H. (1909). Praenunciae bahamensis II. Publ. Field Columb. Mus. *Bot Ser, 2,* 298–322.

Monteiro, J. J. (1875). *Angola and the River Congo.* 2 Vols. London: MacMillan & Co.

Nel, G. C. (1935). *Jahrb. Deutsch. Kakteen-Ges.* 1: 29–32, 42–43.

Nordenstam, B. (1974). The flora of the Brandberg. *Dinteria, 11,* 1–67.

Oudemans, C. A. J. A. (Ed.). (1865). *Neerland's Plantentuin* 1. Groningen: J.B. Wolters.

Palmer, E. (1966). *The plains of Camdeboo.* London: Collins.

Panhuysen, L. (2011). *Een Nederlander in de Wildernis. De Ontdekkingsreizen van Robert Jakob Gordon (1743–1795) in Zuid-Afrika.* Rijksmuseum, Amsterdam (Derde, Herziene Druk).

Park, K.-R. (1996). Phylogeny of New World subtribe Euphorbiinae (Euphorbiaceae). *Korean Journal of Plant Taxonomy, 26,* 235–256.

Park, K.-R., & Elisens, W. J. (2000). A phylogenetic study of tribe Euphorbieae (Euphorbiaceae). *International Journal of Plant Sciences, 161,* 425–434.

Park, K.-R., & Jansen, R. K. (2007). A phylogeny of Euphorbieae subtribe Euphorbiinae (Euphorbiaceae) based on molecular data. *Journal of Plant Biology, 50,* 644–649.

Partridge, T. C., & Maud, R. R. (2000). *Macro-scale geomorphic evolution of southern Africa.* In T. C. Partridge & R. R. Maud (Eds.), *The Cenozoic of southern Africa* (pp. 3–18). Oxford: Oxford University Press.

Paterson, W. (1789). *A narrative of four journeys into the country of the Hottentots, and Caffraria: In the years one thousand seven hundred and seventy-seven, eight, and nine.* London: J. Johnson.

Paterson, W. (1790). *Reisen in das Land der Hottentotten und der Kaffern während der Jahre 1777, 1778 und 1779.* C.F. Voss, Berlin [translated from the English by J.R. Forster].

Pax, F. (1894). Euphorbiaceae africanae. II. *Bot. Jahrb. Syst., 19,* 76–127.

Pax, F. (1900). Euphorbiaceae. In Zahlbruckner, A. (Ed.), Plantae Pentherianae. *Ann. K. Naturhist. Hofmus.* 15: 48–51.

Pax, F. (1904). Monographische Übersicht über die afrikanische Arten aus der Sektion *Diacanthium* der Gattung *Euphorbia. Bot. Jahrb. Syst., 34,* 61–85.

Pax, F. (1905). XXIII. Euphorbia schoenlandii Pax... *Feddes Repert. Spec. Nov Regni Veg.* 1: 59.

Pax, F. (1908). Euphorbiaceae. In Schinz, H. (Ed.), Beiträge zur Kenntnis der Afrikanischen-Flora 21. *Bull. Herb. Boiss.,* Sér. 2, 8: 634–637.

Pax, F. (1909). Euphorbiaceae africanae. IX. *Bot. Jahrb. Syst., 43,* 75–90.

Pax, F. (1921). Euphorbiaceae. In A. Engler & O. Drude (Eds.): 1–168. *Die Vegetation der Erde,* 9, *Die Pflanzenwelt Afrikas, insbesondere seiner tropischen Gebiete* 3 (2). Leipzig: Engelmann.

Pearson, H. H. W. (1914). Observations on the internal temperatures of *Euphorbia virosa* and *Aloe dichotoma. Annals of the Bolus Herbarium, 1,* 41–66.

Peirson, J. A., Bruyns, P. V., Riina, R., Morawetz, J. J., & Berry, P. E. (2013). A molecular phylogeny and classification of the largely succulent and mainly African *Euphorbia* subg. *Athymalus* (Euphorbiaceae). *Taxon, 62,* 1178–1199.

Petiver, J. (1709–11). *Catalogus Classicus & Topicus Omnium Rerum Figuratum in V. Decadibus, Seu primo [-secundo] volumine Gazophylacii Naturae & Artis.* London: C. Bateman.

Phillips, E. P. (1924). Euphorbia cooperi. *Flowering Plants of South Africa, 4,* t. 157.

Phillips, E. P. (1925). Euphorbia tridentata. *Flowering Plants of South Africa, 5,* t. 197.

Phillips, E. P. (1926). Euphorbia monteiri. *Flowering Plants of South Africa, 6,* t. 218.

Phillips, E. P. (1928a). Euphorbia bubalina. *Flowering Plants of South Africa, 8,* t. 285.

Phillips, E. P. (1928b). Euphorbia trichadenia. *Flowering Plants of South Africa, 8,* t. 288.

Phillips, E. P. (1929a). Euphorbia stellaespina. *Flowering Plants of South Africa, 9,* t. 344.

Phillips, E. P. (1929b). Euphorbia kunthii. *Flowering Plants of South Africa, 9,* t. 348.

Phillips, E. P. (1929c). Euphorbia fusca. *Flowering Plants of South Africa, 9,* t. 350.

Phillips, E. P. (1930). Euphorbia heptagona. *Flowering Plants of South Africa, 10,* t. 390.

Phillips, E. P. (1934). Euphorbia valida. *Flowering Plants of South Africa, 14,* t. 526, t. 527.

Pickford, M., & Senut, B. (2000). Geology and palaeobiology of the central and southern Namib Desert, southwestern Africa, Vol. 1. Geology and history of study. *Memoir Geological Survey of Namibia, 18,* 1–155.

Plukenet, L. (1692). *Phytographia seu plantae quamplurimae novae…* 3. London.

Poiret, J. L. M. (1812). *Encyclopédie Méthodique, Botanique* (Lamarck), Supplément, 2 (2). Paris: H. Agasse.

Porembski, S., & Barthlott, W. (2000). Granitic and gneissic outcrops (inselbergs) as centers of diversity for dessication-tolerant vascular plants. *Plant Ecology, 151,* 19–28.

Prain, D. (1909). *Euphorbia ledienii. Curtis's Botanical Magazine, 135,* t. 8275.

Prenner, G., & Rudall, P. J. (2007). Comparative Ontogeny of the cyathium in *Euphorbia* (Euphorbiaceae) and its allies: Exploring the organ-flower-inflorescence boundary. *American Journal of Botany, 94,* 1612–1629.

Prenner, G., Hopper, S. D., & Rudall, P. J. (2008). Pseudanthium development in *Calycopeplus paucifolius*, with particular reference to the evolution of the cyathium in Euphorbieae (Euphorbiaceae-Malpighiales). *Australian Systematic Botany, 21,* 153–161.

Rabinowitz, P. D., Coffin, M. F., & Falvey, D. (1983). The separation of Madagascar and Africa. *Science, 220,* 67–69.

Raper, P. E., & Boucher, M. (Eds.). (1988). *Robert Jacob Gordon: Cape Travels 1777–1786.* Johannesburg: Brenthurst Press.

Rauh, W. (1967). *Die grossartige Welt der Sukkulenten.* Hamburg: Paul Parey.

Raven, P. H. (1975). The bases of angiosperm phylogeny: Cytology. *Annals of the Missouri Botanical Garden, 62,* 724–764.

Raven, P. H., & Axelrod, D. J. (1974). Angiosperm biogeography and past continental movements. *Annals of the Missouri Botanical Garden, 61,* 539–657.

Renner, S. (2004). Plant dispersal across the tropical Atlantic by wind and sea currents. *International Journal of Plant Sciences, 165*(Suppl), 523–533.

Reynolds, G. W. (1966). *The Aloes of tropical Africa and Madagascar.* Mbabane: The Aloes Book Fund.

Riina, R., Peirson, J. A., Geltman, D. V., Molero, J., Frajman, B., Pahlevani, A., Barres, L., Morawetz, J. J., Salmaki, Y., Zarre, S., Kryukov, A., Bruyns, P. V., & Berry, P. E. (2013). A worldwide molecular phylogeny and classification of the leafy spurges, *Euphorbia* subgenus *Esula* (Euphorbiaceae). *Taxon, 62,* 316–342.

Röper, J. A. C. (1824). *Enumeratio Euphorbiarum quae in Germannia et Pannonia gignuntur.* Göttingen: C.E. Rosenbusch.

Rolfe, R. A. (1889). *Matabele Land and the Victoria Falls: A naturalist's wanderings in the interior of South Africa. From the letters & journals of the late Frank Oates, F.R.G.S.* (C. G. Oates, Ed.). 2nd ed. Appendix V. Botany: 390–413. London: Kegan Paul, Trench & Co.

Rutherford, M. C., & Westfall, R. H. (1986). Biomes of southern Africa – An objective categorisation. *Memoirs of the Botanical Survey of South Africa, 54,* 1–98.

Schnabel, D. H. (2013). A morphology based taxonomic revision of the *Euphorbia polygona* species complex. *Euphorbia World, 9,* 5–25.

Schnabel, D. H. (2018). *A monograph of Euphorbia polygona from the perspective of an amateur botanist.* Haan: D.H. Schnabel.

Schnepf, E., & Deichgräber, G. (1984). Electron microscopical studies of nectaries of some Euphorbia species. In W. Rauh (Ed.), Anatomisch-Biochemische Untersuchungen an Euphorbien Teil 1. *Trop. subtrop. Pflanzenw.* 45: 55–93.

Schweickerdt, H. G. (1935). In H. G. Schweickerdt & I. C. Verdoorn (Eds.), Notes on the Flora of southern Africa VI. *Bulletin of Miscellaneous Information* 1935: 204–209.

Seely, M. K., & Jacobson, K. M. (1994). Desertification and Namibia: A perspective. *Journal of African Zoology, 108,* 21–36.

Shah, J. J., & Jani, P. M. (1964). Shoot apex of *Euphorbia neriifolia* L. In *Proceedings of the National Academy of Sciences, India* 30, B, no. 2: 81–91.

Siesser, W. G. (1980). Late Miocene origin of the Benguela upwelling system off northern Namibia. *Science, 208,* 283–285.

Sloane, B. L. (1952). Alain White 1880–1951. *Cactus and Succulent Journal (US), 24,* 70–71.

Snijman, D. A. (Ed.). (2013). *Plants of the Greater Cape Floristic Region.* 2: The Extra Cape Flora. *Strelitzia* 30: 371–379. Pretoria: South African National Biodiversity Institute.

Srivastava, S., Narain, P., & Srivastava, G. S. (1987). A new basic chromosome number in *Euphorbia* based on male meiosis of *E. clandestina* Jacquin. *Hort. Cytologica, 52,* 627–630.

Steinmann, V. W. (2003). The submersion of *Pedilanthus* into *Euphorbia* (Euphorbiaceae). *Acta Botánica Mexicana, 65,* 45–50.

Steinmann, V. W., & Porter, J. M. (2002). Phylogenetic relationships in Euphorbieae (Euphorbiaceae) based on ITS and *ndh*F sequence data. *Annals of the Missouri Botanical Garden, 89,* 453–490.

Steinmann, V. W., Van Ee, B., Berry, P. E., & Gutiérrez, J. (2007). The systematic position of *Cubanthus* and other shrubby endemic species of *Euphorbia* (Euphorbiaceae) in Cuba. *Anales del Jardín Botánico de Madrid, 64,* 123–133.

Struck, M. (1992). Pollination ecology in the arid winter rainfall region of southern Africa: A case study. *Mitt. Inst. Allg. Bot. Hamburg, 24,* 61–90.

Swanepoel, W. (2009a). *Euphorbia ohiva* (Euphorbiaceae), a new species from Namibia and Angola. *South African Journal of Botany, 75,* 249–255.

Swanepoel, W. (2009b). *Euphorbia otjingandu* (Euphorbiaceae), a new species from the Kaokoveld, Namibia. *South African Journal of Botany, 75,* 497–504.

Swanepoel, W. (2013). *Euphorbia otjipembana* subsp. *okakoraensis* (Euphorbiaceae), a new subspecies from the Kaokoveld, Namibia, with an amplified description of *Euphorbia otjipembana*. *Phytotaxa, 117,* 51–57.

Thellung, A. (1916). Euphorbiaceae. In H. Schinz (Ed.), Beiträge zur Kenntnis der afrikanischen Flora. (XXVII). *Vierteljahrsschr. Naturforsch. Gesellsch. Zürich* 61: 431–461.

Thunberg, C. P. (1800). *Prodomus Plantarum Capensium* 2. Uppsala: J.F. Edman.

Thunberg, C. P. (1823). In J. A. Schultes (Ed.), *Flora Capensis, sistens plantas promontorii bonae spei Africes* 2. Stuttgart: J.G. Cotta.

Tian, X., Wang, Q., & Zhou, Y. (2018). *Euphorbia* section *Hainanensis* (Euphorbiaceae), a new section endemic to the Hainan Island of China from biogeographical, karyological, and phenotypical evidence. *Frontiers in Plant Science, 9,* 1–9.

Troll, W. (1935–7). *Vergleichende Morphologie der höheren Pflanzen* 1 (1). Berlin: Gebrüder Borntraeger.

Uhlarz, H. (1974). Entwicklungsgeschichtliche Untersuchungen zur Morphologie der basalen Blatteffigurationen sukkulenter Euphorbien aus den Subsektionen Diacanthium Boiss. und Goniostema Baill. *Trop. subtrop. Pflanzenw., 9,* 1–69.

Uhlarz, H. (1975a). Über die strittige Homologie sogenannter Stipulardrüsen bei einiger *Euphorbia*-Arten der Sektion *Euphorbium. Plant Syst. Evol., 124,* 229–250.

Uhlarz, H. (1975b). Über die Dorsalstacheln der Podarien von *Monadenium guentheri* Pax. *Beitr. Biol. Pflanzen, 51,* 335–352.

Uhlarz, H. (1978). Über die Stipularorgane der Euphorbiaceae, unter besonderer Berücksichtigung ihrer Rudimentation. *Trop. subtrop. Pflanzenw., 23,* 1–65.

Venter, K. (1987). A new field hybrid from the Transvaal?: A suspected hybrid between *Euphorbia excelsa* & *Euphorbia enormis*. *Euphorbia Journal, 4,* 69–70.

Vesling, J. (1638). *Ioannis Veslingii Mindani…De Plantis Aegyptis observationes et notae as Proseperum Alpinum.* Leiden: P. Frambottum.

Visser, J. (1981). *South African parasitic flowering plants.* Cape Town: Juta.

Vlok, J., & Schutte-Vlok, A. L. (2010). *Plants of the Klein Karoo.* Hatfield: Umdaus Press.

Vogel, S. (1954). Blütenbiologische Typen als Elemente der Sippengliederung. Dargestellt anhand der Flora Südafrikas. *Bot. Studien, 1*, 1–338.

Von Willert, D. J., Eller, B. M., Werger, M. J. A., Brinckmann, E., & Ihlenfeldt, H.-D. (1992). *Life strategies of succulents in deserts with special reference to the Namib Desert*. Cambridge: Cambridge University Press.

Vosa, C. G., & Bassi, P. (1991). Cromosome (sic) studies in the Southern African flora. 95–102. The basic caryotype of eight species of succulent *Euphorbia* L. *Caryologia, 44*, 27–33.

Ward, J. D., & Corbett, I. (1990). Towards an age for the Namib. In M. K. Seely (Ed.), *Namib ecology: 25 years of Namib research* (Transvaal Museum Monograph) (Vol. 7, pp. 17–26). Pretoria: Transvaal Museum.

Ward, J. D., Seely, M. K., & Lancaster, N. (1983). On the antiquity of the Namib. *South African Journal of Science, 79*, 175–183.

Watt, J. M., & Breyer-Brandwijk, M. G. (1962). *The medicinal and poisonous plants of Southern and Eastern Africa* (2nd ed.). Edinburgh/London: E. & S. Livingstone.

Webster, G. L. (1967). The genera of Euphorbiaceae in the southeastern United States. *Journal of the Arnold Arboretum, 48*, 363–430.

Webster, G. L. (1975). Conspectus of a new classification of the Euphorbiaceae. *Taxon, 24*, 593–601.

Webster, G. L. (1994). Synopsis of the genera and suprageneric taxa of Euphorbiaceae. *Annals of the Missouri Botanical Garden, 81*, 33–144.

Weiss, J. E. (1893). Empfehlenswerte Kakteen. *Neubert's Deutsch. Garten-Magazin, 46*, 286–292.

Wheeler, L. C. (1939). A miscellany of New World Euphorbiaceae 2. *Contributions from the Gray Herbarium, 127*, 48–78.

Wheeler, L. C. (1941). Euphorbia subg. Chamaesyce in Canada and the Unites States exclusive of southern Florida. *Rhodora, 43*, 97–154.

Wheeler, L. C. (1943). The genera of the living Euphorbieae. *American Midland Naturalist, 30*, 456–503.

White, A. C., Dyer, R. A., & Sloane, B. L. (1941). *The Succulent Euphorbieae (Southern Africa)*. Pasadena: Abbey Garden Press.

Wijnands, D. O. (1983). *The botany of the Commelins*. Rotterdam: A.A.Balkema.

Wijnands, D. O., Wilson, M. L., & Toussant van Hove, T. (Eds.). (1996). *Jan Commelin's monograph on Cape Flora*. Cape Town: The Editors.

Williamson, G. (1995). *Euphorbia versicolores*, a new species from the northwestern Cape, South Africa. *Cactus and Succulent Journal (US), 67*, 284–287.

Williamson, G. (1999). A new species of *Monadenium* (Euphorbiaceae) from south-east Zimbabwe. *Excelsa, 19*, 56–58.

Williamson, G. (2003). A new variety of *Euphorbia filiflora* Marloth from the Northern Cape Province of South Africa. *Bradleya, 21*, 49–52.

Williamson, G. (2007). Notes on *Euphorbia multiramosa* Nel (Euphorbiaceae) and related species. *Euphorbia World, 3*, 8–18.

Wilman, M. (1946). *Preliminary check list of the flowering plants and ferns of Griqualand West (Southern Africa)*. Cambridge: Deighton, Bell & Co.

Wilson, M. L., Toussant van Hove-Exalto, T., & van Rijssen, W. J. J. (Eds.). (2002). *Codex Witsenii*. Cape Town/Amsterdam: Iziko Museums of Cape Town/Davidii Media.

Worsdell, W. C. (1914). On some points in the stem-anatomy of *Euphorbia virosa* and *Aloe dichotoma*. *Annals of the Bolus Herbarium, 1*, 67–71.

Yang, Y., Riina, R., Morawetz, J. J., Haevermans, T., Aubriot, X., & Berry, P. E. (2012). Molecular phylogenetics and classification of *Euphorbia* subgenus *Chamaesyce* (Euphorbiaceae). *Taxon, 61*, 764–789.

Zimmermann, N. F. A., Ritz, C. M., & Hellwig, F. H. (2010). Further support for the phylogenetic relationships within *Euphorbia* L. (Euphorbiaceae) from nrITS and *trnL-trnF* IGS sequence data. *Plant Systematics and Evolution, 286*, 39–58.

Name Index

A
Acocks, J. P. H. 376
Alpinus, P. 488, 489
Ambrose, J. 903
Anderberg, A. 2
Anderson, C. 2
Anderson, T. 493
Anthony, H. E. 943
Archer, J. 344
Audissou, J. -A. 258
Auge, J.A. 617

B
Baines, J. T. 450, 797
Bally, P. R. O. 722, 963
Barnard, K. H. 336, 768
Barnard, W. G. 659, 701, 804, 809
Bayer, A. W. 217
Bayer, F. 677
Bayer, M. B. 2, 77
Bayer, W. F. 656, 742
Bayliss, R. D. A. 907
Becker, J. & O. 2
Becker, R. 2
Bell, J. 186
Bell, S. 2
Berger, A. 183, 372, 396, 664, 686, 722, 776, 837, 852, 854, 893
Bergius, C. H. 207
Bergius, P. J. 617
Bernhardi, J. J. 621, 628
Berry, P. E. 2, 6–8
Blanc, A. 707
Boissier, P. E. 97, 109, 134, 277, 394, 398, 437, 504, 554, 595, 641, 817, 827, 854, 889
Bolus, H. 111, 497, 516
Bond, G. 2, 151, 664, 665, 815, 826, 827
Boos, F. 97, 221
Boughey, A. S. 911
Bowie, J. 3, 68, 139, 147, 152, 207, 663–665, 712, 826
Brass, J. 943
Brauns, J. C. E. H. J. 263, 296, 407
Brito Teixeira, J. 873, 879
Britten, G. V. 408
Britten, L. 281
Brown, N. E. 1, 8, 86, 91, 122, 135, 145, 151, 155, 174, 216, 247, 263, 271, 284, 286, 292, 296, 314, 325, 348, 398, 400, 404, 407, 437, 467, 489, 532, 538, 580, 594, 605, 631, 663–665, 686, 693, 704, 722, 793, 797, 800, 837, 854
Bruyns, J. J. 849, 886
Bruyns, P. V. 759, 850

Burchell, W. 202, 207, 216, 304, 538, 554, 618, 621, 628, 631, 815
Burgers, C. J. 624
Burke, J. 87, 91, 247, 800
Burman, J. 69, 211, 225
Burns, S. 2
Burtt-Davy, J. 183, 291, 292, 304, 396, 670

C
Cannell, I. C. 594, 765
Carter, S. 6, 396
Challen, G. 2
Chapman, J. 450
Chapman, J. D. 943
Claudius, H. 69, 102, 188, 436, 601
Codd, L. E. 564, 681, 785, 789
Commelijn, C. 134
Commelijn, J. 188
Compton, R. H. 344, 605, 619
Cooper, T. 216, 325, 686
Correia, M. F. 899, 933
Court, D. 83
Craven, P. 871
Croizat, L. 892
Cronwright, W. E. 776
Crundall, A. H. 785
Cumming, D. 2
Curror, A. B. 467

D
de Carvalho, M. F. 948
de Jesus, J. P. 955
de Paiva, J. A. R. 2, 899, 963
de Wijn, G. J. 748
de Winter, B. 564, 594, 720, 785
Dean, S. & R. 2
Dekenah, J. 75
Desfontaines, R. L. 837
Dillenius, J. J. 488, 601
Dinter, J. 516
Dinter, M. K. 5, 306, 385, 445, 508, 538, 575, 654, 755, 854, 879, 884
D'Isnard, A.-T. D. 69
Drège, C. 2, 97, 109, 152, 202, 216, 233, 277, 296, 357, 398, 427, 432, 437, 493, 510, 544, 588, 613, 663, 722, 817, 826, 889
Drège, I. L. 233
Drège, J. F. 2, 97, 109, 152, 165, 202, 216, 247, 249, 258, 277, 296, 357, 398, 427, 432, 437, 493, 510, 544, 588, 613, 621, 631, 663, 722, 817, 826, 889
Duncanson, T. 2, 68, 187, 188, 207, 208, 664, 712

Dyer, R. A. 1, 3, 4, 8, 83, 116, 147, 166, 174, 197, 198, 229, 242, 247, 281, 299, 330, 344, 348, 408, 421, 495, 544, 564, 654, 670, 674, 681, 692, 716, 730, 734, 750, 772, 776, 781, 785, 789, 793, 863, 918, 919

E
Een, T. J. 450
Erb, E. 2

F
Fehr, J. W. 87
Fischer, G. A. 411, 563
Fleck, E. 446
Foord, J. 684
Fourie, S. P. 2, 665, 737, 793, 797, 832, 863, 896
Frick, G. A. 83, 432, 776
Friedrich, M. 334
Friedrich, W. 2

G
Galpin, E. E. 183, 538, 575, 797
Geldenhuys, J. 2
Geltman, D. V. 2
Gerrard, W. T. 216, 580
Gerstner, J. 340, 670, 963
Gilbert, M. G. 827
Giess, W. 5, 693
Gildenhuys, S. 2
Gillett, J. 797, 863
Goebel, K. I. E. 707
Gomes e Sousa, A. de F. 907, 915, 919
Gordon, R. J. 3, 97, 655, 815, 853
Gossweiler, J. 563, 693
Govaerts, R. 2
Grieve, G. & K. 2, 613
Grubb, M. 617
Gueinzius, W. 580
Gürich, G. 569

H
Hahn, C. H. 3, 692, 725
Hall, H. 344, 421, 768
Hanáček, P. 149
Haworth, A. H. 2, 3, 7, 58, 68, 151, 152, 165, 187, 188, 207, 212, 395, 436, 663–665, 712, 826
Hermann, P. 211
Herre, H. 105, 147, 197, 349, 367, 389, 459, 510, 544, 546, 547
Hill, J. 212
Hooker, J. D. 450, 451
Hunter, 654
Hutchinson, J. 797, 863
Hutton, C. 348

J
Jacobsen, N. H. G. 2, 524, 751
James, H. R. 2, 249
Jankowitz, W. 306
Joubert, J. E. V. 726

K
Keith, D. R. 730
Kensit, L. 174
Kirk, J. 291
Kirk, J. W. C. 291, 903, 911
Klak, C. 2, 566, 611, 685, 876, 886, 926
Knobel, J. C. J. 734
Knuth, P. 739
Köpfner, 584
Kolbe, F. C. 544
Krapohl, J. H. C. 319
Krauss, C. F. F. 53, 310, 621, 628
Kroon, M. & J. 2
Krüger, H. 166, 782

L
Labuscagne, F. & D. 2
Lavranos, J. 421
le Roux, J. A. v. Z. 198
le Vaillant, F. 166, 655, 814, 815, 853
Leach, L. C. 3–5, 9, 49, 206, 421, 455, 459, 497, 504, 507, 544, 547, 594, 641, 693, 716, 739, 742, 765, 768, 797, 847, 850, 854, 873, 890, 901, 903, 905, 907, 909, 926, 928, 931, 933, 953, 962, 963
Ledien, F. 664
Leistner, O. A. 572, 594, 873
Lemaire, C. 58
Lempp, K. F. 453
Lever, N. A. 858
Levyns, M. R. 376
Liebenberg, L. C. C. 842
Lochner, M., C. & M. 2
Loffler, P. & L. 2
Loutit, R. 524
Louw, W. J. 734, 742
Lugard, C.E. (Nell) 2, 450, 451, 461, 462, 868

M
MacOwan, P. 145, 605
Maire, L. 532
Marais, W. 564
Marloth, H. W. R. 3–5, 83, 103, 107, 122, 147, 174, 183, 193, 197, 198, 229, 281, 287, 294, 310, 318, 319, 340, 348, 354, 367, 372, 385, 389, 392, 421, 445, 558, 605, 654–656, 670, 697, 701, 716, 781, 804, 809, 832, 879, 884
Masson, F. 3, 97, 139, 165, 225, 314, 521, 532, 663, 814
McClean, A. P. D. 776
McKen, M. J. 580
McMurtry, D. 2, 804
Mendes, E. J. 693, 768
Mendonça, F. A. 959
Merxmüller, H. 693
Meyer, E. 398, 631, 889
Meyer, G. 111
Meyer, J. J. 2
Meyer, P. G. 693, 765
Mogg, A. O. D. 701, 931
Möller, A. 2
Moll, E. J. 217
Moninckx, J. 134, 394, 601
Moninckx, M. 270

Monteiro, J. J. 450
Moss, C. E. 907
Moss, H. 524
Moss, M. 907
Muir, J. 172, 183, 193, 252, 392
Müller, M. A. N. 524
Mund, J. L. L. 532

N
Nel, G. C. 319
Nesemann, A. 174
Neto, M. 149
Nordenstam, B. 454

O
Oates, C. G. 474
Oates, F. 474
Oates, W. E. 474
Obermeyer, A. A. 649, 797
Oliver, E. G. H. 459, 572
Oosthuizen, C. P. 299, 330
Otzen, M. 3, 5, 107, 166, 692, 693, 725, 758, 768, 782, 821

P
Pagan, H. H. 78
Palmer, C. N. 396
Palmer, S. 83
Paterson, W. 815, 853
Pax, F. 292, 403, 470, 641, 642, 739, 793
Pearson, H. H. W. 263, 287, 334, 380, 385, 446, 521, 538, 541, 605, 654, 854, 884
Peckover, R. 2
Pedro, J. G. 955
Pehlemann, I. 421
Penther, A. 470
Peters, W. C. H. 905
Pillans, C. E. 103
Pillans, N. S. 174, 372
Plowes, D. C. H. 674
Pogge, P. 91
Pole Evans, I. B. 197, 776, 832, 842, 933
Prain, D. 664
Purcell, W. F. 174

R
Rand, R. F. 474
Range, P. 385, 445, 558
Rautanen, M. 879
Rehmann, A. 800
Reuning, E. 879
Reynolds, G. W. 54
Rogers, F. A. 455
Rocha da Torre, A. 898, 899, 915, 919, 933, 955
Rolfe, R. A. 473, 474
Ross Frames, P. 544, 804
Ross, J. H. 837
Rowland Jones, M. 789
Rutherford-Smith, R. 905, 953

S
Salter, T. M. 344
Schimper, W. 827
Schinz, H. 793, 800
Schlechter, F. R. R. 257, 258, 404, 497, 580, 739
Schmidt, E. 2, 412, 905
Schnabel, D. H. 164
Scholl, G. 97, 221
Schönland, S. 4, 83, 122, 247, 263, 281, 296, 297, 348, 403, 404, 407
Schumacher, J. 655
Schweickerdt, H. G. K. 649
Seydel, R. H. W. 508
Shoesmith, H. W. 53, 83
Sloane, B. L. 1, 4, 730
Smith, A. 216, 722
Smith, C. A. 247, 363, 407, 624
Smith, G. G. 105, 242
Smith, J. 174
Smith, M. 296
Sonnerat, P. 618
Steenkamp, P. 572
Stephens, E. L. 124
Stocks, J. 963
Stone, J. B. 706
Strey, R. G. 217
Sutherland, P. C. 538, 580

T
Taljaard, G. 198
Taylor, H. C. 624
Theron, A. S., A. & P. 2
Thunberg, C. P. 3, 139, 207, 225, 314, 398, 437, 488, 521, 618, 628, 635, 663, 814
Tölken, H. R. 459
Tribe, G. 2, 690, 758

V
van Balen, J. C. 748
van Breda, P. A. 74
van der Merwe, F. Z. 3, 681, 701, 742, 750, 751, 772, 793, 804, 809, 842, 863
van der Stel, S. 69, 102, 188, 436, 601
van Heerden 493
van Jaarsveld, E. J. 493
van Vuuren, J. & R. 2
van Wyk, E. 2
Venter, F. 459, 677
Venter, P. I. 572
Verdoorn, I. C. 649
Vlok, J. 177
Volkmann, M. 755
Vorster, P. J. 572

W
Watermeyer, E. B. 229
Weiss, J. E. 707
Welwitsch, F. 91, 524, 594, 693, 873
White, A. C. 1, 4
Whittal, G. & A. 2

Willdenow, C. L. 814
Williamson, F. 459
Williamson, G. 321, 459
Wilman, M. 78, 716, 933
Wiman, J. J. F. 3
Winter, P. J. D. 2, 772

Wiss, H. J. 453
Wisura, W. 2, 421

Z
Zeyher, C. 3, 87, 91, 139, 202, 233, 357, 396, 497, 504

Subject Index

A

Acacia 359, 656, 704
Acacia tortilis 795, 797, 844
Acrodon 44, 200
Adenia pechuelii 876
Aizoaceae 9, 43, 44, 62, 65, 312, 322, 386, 391, 396, 423, 438
Aloe 679, 682, 695
Aloe africana 661
Aloe angelica 860
Aloe broomii 185
Aloe castanea 656, 705, 783
Aloe chabaudii 933, 945, 955
Aloe claviflora 142, 538
Aloe cryptopoda 903
Aloe dichotoma 423, 538, 880
Aloe dolomitica 783
Aloe esculenta 844
Aloe ferox 158, 218, 231
Aloe gerstneri 963
Aloe globuligemma 685, 844
Aloe greenii 145
Aloe hereroensis 538
Aloe littoralis 844
Aloe marlothii 704
Aloe mawii 898, 917, 937, 940, 945, 955, 961
Aloe microstigma 124
Aloe munchii 940
Aloe pillansii 423
Aloe torrei 940
Aloe vera 890
Aloe wickensii 656, 866
Aloidendron 868
Anacampseros albissima 876
Anacampseros sect. Avonia 43
Androstachys johnsonii 679, 788
Anthacantha desmetiana 125, 129
Apocynaceae 18, 43, 47, 62
Arthrothamnus 55, 639, 868
Arthrothamnus bergii 484
Arthrothamnus brachiatus 525
Arthrothamnus burmanni 484, 549
Arthrothamnus cymosus 549
Arthrothamnus densiflorus 525
Arthrothamnus ecklonii 549
Arthrothamnus scopiformis 549
Arthrothamnus tirucalli 889
Asteraceae 46
Astraloba 142
Astrophytum asterias 143

B

Bergeranthus 71
Brachystegia 920, 945

C

Cactaceae 18
Caralluma subulata 890
Carissa bispinosa 176
Ceropegia 63, 74, 866, 930
Ceropegia barklyana 820
Ceropegia (Caralluma) furta 145
Ceropegia (Duvalia) caespitosa 62
Ceropegia (Duvalia) modesta 71
Ceropegia (Duvalia) polita 694, 844
Ceropegia (Hoodia) gordonii 538
Ceropegia (Hoodia) pilifera subsp. annulata 158
Ceropegia (Huernia) erectiloba 940, 947, 955
Ceropegia (Huernia) nouhuysii 647
Ceropegia (Huernia) stapelioides 694
Ceropegia (Huernia) verekeri 844
Ceropegia (Huernia) whitesloaneana 647
Ceropegia mixta (Orbea variegata) 9
Ceropegia oculatoides 820
Ceropegia (Orbea) halipedicola 903
Ceropegia (Orbea) lutea 9, 844
Ceropegia (Orbea) rogersii 844
Ceropegia (Piaranthus) geminata 62
Ceropegia sect. Duvalia 72
Ceropegia sect. Piaranthus 72
Ceropegia (Stapelia) gigantea 694
Ceropegia (Stapelia) schinzii 820
Ceropegia (Tromotriche) baylissii 907
Ceropegia zeyheri 137
Chamaesyce 55, 475, 477
Chamaesyce glanduligera 582
Chasmatophyllum 71
Cissus 679
Cissus cactiformis 933
Cissus quadrangularis 933
Coleochloa 47, 925, 936, 940, 945, 947, 952
Commiphora 469, 470, 724, 849
Corpuscularia 65
Cotyledon barbeyi 715
Cotyledon woodii 158
Crassula 44, 74, 176, 200, 715
Crassulaceae 65
Crassula ericoides 231
Crassula perforata 231

Crassula rubricaulis 231
Croton tiglium 55
Cubanthus 6, 7
Cucurbitaceae 18
Cynanchum gerrardii 866
Cynanchum mulanjense 940
Cynanchum oresbium 940
Cynanchum pearsonianum 538, 598
Cynanchum viminale 124, 218, 538, 598, **676,** 866, 893, 898, 917, 937, 945, 947, 955
Cyphostemma currorii 453
Cyphostemma uter 571, 761

D
Dactylanthes 55, 57, 60
Dactylanthes anacantha 60, 74
Dactylanthes globosa 64, 68
Dactylanthes hamata 432, 436
Dactylanthes patula 69
Delosperma echinatum 65
Dichrostachys cinerea 795
Didiereaceae 43

E
Elaeophorbia 6, 7
Endadenium 6, 7
Eremophyton 6
Eriocephalus 226
Esula 6, 7, 55, 589, 607
Euclea 359
Euphorbia 55
Euphorbia abdelkuri 41
Euphorbia abyssinica 54, 721
Euphorbia abyssinica var. *mozambicensis* 903
Euphorbia acaulis 1, 12, 14, 22
Euphorbia acervata 791
Euphorbia adjurana 34
Euphorbia aequoris 516
Euphorbia aeruginosa 641, **645–649,** 742, 795, 797, 807, 809
Euphorbia aggregata 119, 122, 183, 893
Euphorbia albanica 629–631
Euphorbia albertensis 287, 293, 296
Euphorbia albipollinifera 64, 237, **238–242,** 245, 294, 393, 394
Euphorbia amarifontana 525, 532
Euphorbia ambroseae 898–903, 907
    var. *ambroseae* 901, 902
    var. *spinosa* 899, 901
Euphorbia ammak 721
Euphorbia ampliphylla 1, 22, 45, 54, 721, 722
Euphorbia anacantha 60, 74
Euphorbia angrae 525
Euphorbia angularis 898, **903–905**
Euphorbia anomala 582
Euphorbia anoplia 157
Euphorbia anticaffra 895
Euphorbia antiquorum 7, 17, 54, 55, 641
Euphorbia antisyphilitica 6, 26
Euphorbia antso 29, 34
Euphorbia appariciana 6
Euphorbia arceuthobioides 554
Euphorbia arida 234, 237, **242–249,** 292, 296
    subsp. **arida 243–247**
    subsp. **camdebooensis** 237, **247–249**
Euphorbia armata 97

Euphorbia artifolia 614
Euphorbia aspericaulis 517
Euphorbia astrispina 183
Euphorbia astrophora 293, 295
Euphorbia atrispina 8, 125, 129
    var. *viridis* 125
Euphorbia audissoui 237, **249–254**
Euphorbia austro-occidentalis 479
Euphorbia avasmontana 16, 28, 36, 37, 46, 48, 50, 52, 53, 644, **650–656,** 753–755, 789, 848, 852, 854, 880
Euphorbia bachmannii 624
Euphorbia baliola 282, 287
Euphorbia ballyi 705
Euphorbia balsamea 432
Euphorbia balsamifera 35
Euphorbia barnardii 21, 644, **656–659,** 696, 784
Euphorbia basutica 271, 277
Euphorbia baumii 448
Euphorbia bayeri 525
Euphorbia baylissii 898, **905–908,** 930
Euphorbia bellica 847
Euphorbia benguelensis vii, 86, 87
Euphorbia benthamii 461
Euphorbia bergeriana 427
Euphorbia bergii 303, 304
Euphorbia berotica 4, 591, **592–595**
Euphorbia biglandulosa 484
Euphorbia biselegans 22, 962
Euphorbia bisellenbeckii 22
Euphorbia bolusii 264, 271
Euphorbia bothae 895
Euphorbia bougheyi vii, 704, 705, 897, 898, 903, 905, **908–911,** 930
Euphorbia brachiata 511, 525
Euphorbia brakdamensis 237, **254–258,** 270, 318, 320, 388
Euphorbia braunsii 25, 234–236, **258–263,** 275, 276, 286, 303, 304, 318, 330, 352, 369
Euphorbia breviarticulata 704, 705, 707, 909
Euphorbia brevirama 242, 293, 296
    var. *supraterra* 293
Euphorbia bruynsii 62–64
Euphorbia bubalina 15, 27, 48, **213–217,** 221, 225
Euphorbia bupleurifolia 93–97
Euphorbia burmanni 6, 8, 23, 34, 52, 172, 250, 438, 482, **484–489,** 595
Euphorbia cactus 704, 705
Euphorbia caducifolia 22
Euphorbia caerulescens vii, 37, 113, 114, 345, 641, 644, **659–665,** 730, 781, 826, 827, 854, 863, 895, 896
    E. caerulescens X E. triangularis 'bothae,' 896
    E. caerulescens X E. triangularis 'curvirama,' 895–896
Euphorbia canaliculata 221
Euphorbia canariensis 7, 41, 641, 663
Euphorbia candelabrum 4, 681, 689, 693, 721, 861–863
Euphorbia caperonioides 461
Euphorbia captiosa 119, 122, 174
Euphorbia caput-medusae 3, 9, 48, 53, 69, 221, 235, 237, 254–256, 258, **263–271,** 270, 271, 287, 344, 357, 388, 394, 495, 544, 622, 893
    var. *geminata* 263
    var. *major* 264
    var. *minor* 264
Euphorbia carunculifera 32, 52, 54, 873
Euphorbia cataractarum 21
Euphorbia caterviflora 525
Euphorbiaceae 18
Euphorbia celata 8, 43, **414–422,** 430, 432, 445, 459

Euphorbia cereiformis 122, 130, 134, 174
    var. *echinata* 130
Euphorbia cervicornis 433
Euphorbia chamaesycoides 477, 481
Euphorbia chersina 525, 532
Euphorbia chevalieri 959
Euphorbia cibdela 533
Euphorbia ciliolata 572
Euphorbia clandestina 16, 35, 95, 213, 214, **216–221**, 225, 228, 250, 270
Euphorbia clava 65, 95, 214, 216, **221–225**, 232, 233, 270, 355
Euphorbia clavarioides 7, 10, 11, 37, 38, 52, 234, 236, **271–277**, 323, 406, 407, 546
    var. *truncata* 271
Euphorbia clavata 221
Euphorbia clavigera 10, 47, 338, 643, **665–670**, 696, 697, 829, 830, 841, 898, 930
Euphorbia claytonioides 581, 587
Euphorbia clivicola 338, 641, 645, **670–677**, 694, 769, 783, 809
    subsp. *calcritica* **675–677**
    subsp. *clivicola* **671–675**
Euphorbia colliculina 36, 237, 252, **277–282**, 310, 328
Euphorbia commelinii 264
Euphorbia commiphoroides 564
Euphorbia complexa 791, 793, 901
Euphorbia confinalis 644, **677–682**, 678, 692, 701, 836, 837, 861, 898
    subsp. *confinalis* 24, 25, 678
    subsp. *rhodesica* 678
Euphorbia confluens 264
Euphorbia conformis 523
Euphorbia congestiflora 33, **869–874**
Euphorbia conspicua 693
Euphorbia contorta 898, 901, 907, **911–915**, 950
Euphorbia cooperi 47, 54, 643, 665–666, **682–687**, 698, 719, 753, 784, 802, 803, 866, 898, 905, 909
    var. *calidicola* 682
    var. *cooperi* 47, **682–687**
    var. *ussanguensis* 682
Euphorbia corneliae 607, **608–611**, 622
Euphorbia corniculata 898, 907, **916–919**, 945, 953, 955
Euphorbia coronata 221
Euphorbia corymbosa 484
Euphorbia crassipes 12, 46, 48, 238, 247, **282–288**, 292, 296, 298, 299, 330, 538
Euphorbia crebrifolia 607, 624, 627
Euphorbia crispa 208, 212
Euphorbia cucumerina vii, 139, 157, 165–166
Euphorbia cumulata 112–117, 128, 129, 345
Euphorbia cuneata 473
Euphorbia curocana 440, 445
Euphorbia currorii 463–467, 849
Euphorbia curvirama 896
Euphorbia cussonioides 1
Euphorbia cylindrica 213, 216, 221, **225–229**
Euphorbia damarana 29, 32, 49, 52, **874–879**, 882–884, 888
Euphorbia davyi 234, 237, **288–292**, 300, 361, 362, 396
Euphorbia dawei 909
Euphorbia decepta 35, 45, 48, 146, 237, 240, 245, 246, 286, 287, **292–299**, 391, 393, 394
    subsp. *decepta* **293–297**
    subsp. *gamkaensis* **297–299**
Euphorbia decidua 22, 711, 898, **919–921**
Euphorbia decliviticola vii, 898, 924, 926
Euphorbia decussata 525
Euphorbia dinteri 755, 847, 854
Euphorbia discreta 321, 325

Euphorbia dispersa 754
Euphorbia dregeana 6, 15, 414, **415**, **422–427**, 432, 651, 888
Euphorbia dumosoides 26, 607, **611–613**
Euphorbia duseimata 234, 237, 291, **299–301**, 306, 361, 362
    subsp. *duseimata* **301–304**
    subsp. *pseudoduseimata* **304–306**
Euphorbia echinata 130
Euphorbia ecklonii 14, **199–203**, 206, 207, 211
Euphorbia eduardoi 5, 17, 27, 34–36, 643, **687–693**, 724, 757, 861
Euphorbia eendornensis 282
Euphorbia einensis 525
    var. *anemoarenicola* 526
Euphorbia elastica 422
Euphorbia elliotii 22
Euphorbia elliptica 203
    var. *undulata* 204, 212
Euphorbia engleriana 440
Euphorbia enneagona 130
Euphorbia enopla vii, 8, 125, 129
    var. *viridis* 125
Euphorbia enormis 643, 670, 684, **693–697**, 830, 831, 841
    E. enormis X *E. excelsa* 896
Euphorbia ephedroides 484, **489–490**, 498, 544
    subsp. *ephedroides* 489, **490–493**
    subsp. *gamsbergensis* 489, **493–494**
    subsp. *imminuta* 489, **495**
Euphorbia epicyparissias 624
    var. *puberula* 624
    var. *wahlbergii* 624
Euphorbia epiphylloides 41
Euphorbia ericoides 613, 617, 627
Euphorbia ernestii 321, 325, 407
Euphorbia erosa 130
Euphorbia erubescens 621, 628
Euphorbia erythrina 613, 616
    var. *burchellii* 614
    var. *meyeri* 613
Euphorbia esculenta 36, 46, 52, 53, 234, 238, 280, 281, 286, 292, **306–311**, 328, 344, 354, 357
Euphorbia espinosa 12, 19, **560–564**
Euphorbia eustacei 97, 101
Euphorbia evansii 707, 712
Euphorbia excelsa 27, 47, 643, 694, **697–701**, 857
Euphorbia exilis 23, 33, 43, 482, **496–499**
Euphorbia eylesii 477, 481
Euphorbia falsa 135
Euphorbia fasciculata 29, 43, 48, 234–236, **311–315**, 398, 402, 404
Euphorbia faucicola 684, 755
Euphorbia ferox 32, 35, 52, 112, **117–125**, 119–125, 128, 129, 133, 142, 143, 146, 158, 181, 182, 306, 345, 893
    E. ferox ssp. *ferox* X *E. stellispina* 894
    subsp. *calitzdorpensis* **122–125**, 176
    subsp. *ferox* 20, 33, 36, 119, 120, **122–124**
Euphorbia filiflora 236, 256–258, **315–321**, 344
    var. *nana* 315
Euphorbia fimbriata 130, 174
Euphorbia flanaganii 37, 236, 237, 276, **321–326**, 339, 340, 349, 390, 392, 393
Euphorbia fleckii vii, 893
Euphorbia foliosa 614, 616, 617
Euphorbia forskalii 479
Euphorbia fortuita 238, 252, 281, 310, **326–330**
Euphorbia fragiliramulosa 523
Euphorbia francescae 455, 459
Euphorbia franksiae 321, 325, 340
    var. *zuluensis* 325

Euphorbia frickiana 168
Euphorbia friedrichiae 238, **330–336**, 380, 381
   subsp. *friedrichiae* **331–334**, 336
   subsp. *pofadderensis* 330, **334–336**
Euphorbia fructus-pini 263
   var. *geminata* 264
Euphorbia frutescens 564
Euphorbia fusca 247, 282, 284, 287, 288
Euphorbia galpinii 572
Euphorbia gamkensis 299
Euphorbia gariepina 27, 64, 413, 414, 422, **427–432**, 445, 538, 880
   subsp. *balsamea* 432
   subsp. *gariepina* **427–432**
Euphorbia gatbergensis 321, 325
Euphorbia genistoides 33, 47, 607, **613–618**, 622, 627, 628, 630, 631
   var. *corifolia* 613
   var. *leiocarpa* vii, 613
   var. *major* vii, 613
   var. *puberula* 613
Euphorbia gentilis 52, 484, **500–504**, 515, 541, 546
   subsp. *tanquana* 500
Euphorbia gerstneriana 35, 234, 236, 325, **336–340**, 339, 340, 963
Euphorbia giessii 37, 484, **504–508**
Euphorbia gilbertii 816
Euphorbia glandularis 9, 496, 498
Euphorbia glanduligera 581, **582–584**, 588
Euphorbia glaucella 582
Euphorbia globosa 27, 54, **64–68**, 69, 82, 86, 355
Euphorbia glochidiata 20
Euphorbia glomerata 64
Euphorbia goetzei 572
Euphorbia gorgonis 389, 396
Euphorbia gossweileri 87
Euphorbia gracilicaulis 654
Euphorbia grandialata vii, 702, 705, 706
Euphorbia grandicornis vii, 21, 43, 644, 658, 684, 696, **702–707**, 737, 775, 776, 898, 909
   subsp. *grandicornis* 24, **702–707**
   subsp. *sejuncta* 703, **922–923**
Euphorbia grandidens 17, 33, 37, 47, 643, 700, **707–712**, 804, 811, 825, 833, 836, 863, 866, 898
Euphorbia graniticola vii, 898, **924–928**, 940, 941–942
Euphorbia grantii 451
Euphorbia graveolens 396, 400
Euphorbia gregaria 6, 23, 29, 30, 32, 52, 538, 651, 721, **878–884**, 888, 889
Euphorbia griseola 642, 645, 647, **713–717**, 739, 855, 857, 898, 921, 941
   subsp. *griseola* **713–717**
   subsp. *mashonica* 714
   subsp. *zambiensis* 714
Euphorbia groenewaldii 828, 830, 832
Euphorbia gueinzii 12, 37, 459, 474
   var. *albovillosa* 577
Euphorbia guentheri 21
Euphorbia guerichiana 10, 19, 26, 33, **564–569**
Euphorbia gummifera 29, 32, 38, 881, **883–889**
Euphorbia gymnocalycioides 1, 10, 11, 641
Euphorbia gynophora 560
Euphorbia halipedicola vii, 683, 898, 908, 909
Euphorbia hallii 8, 12, 33, 235, 236, **340–344**
Euphorbia hamata 415, 420–422, 426, 430, **432–440**, 445, 459, 546
   var. **hamata 433–437**
   var. **pedemontana 437–440**
Euphorbia hastisquama 525
Euphorbia haworthii 221

Euphorbia helioscopia 636
Euphorbia heptagona 8, 31, 47, 49, 111, ***112***, 117, 118, **125–130**, 133, 150, 151, 154, 155, 158, 162–164, 174, 176, 191, 197, 209, 250, 306, 894
   E. heptagona X *E. pseudoglobosa* ssp. *vlokii* 894
   var. *dentata* 125
   var. *fulvispina* 125
   var. *ramosa* 125
   var. *subsessilis* 125
   var. *viridis* 125
Euphorbia herrei 31, 37, 484, **508–511**, 546
Euphorbia heterochroma 716
Euphorbia hirta 477
Euphorbia hopetownensis 282, 288
Euphorbia horrida 156, 157, 164, 165
   var. *major* 157
   var. *noorsveldensis* 157
   var. *striata* 157
Euphorbia hottentota 650, 654–656
Euphorbia huttoniae 37, 71, 114, 238, 310, **345–349**, 356, 357
Euphorbia hydnorae 596
Euphorbia hypogaea 15, 235, 236, 275, 276, **349–354**, 352
Euphorbia hystrix 97
Euphorbia inaequilatera 35, 477, 479–481
Euphorbia inconstantia 164, 894
Euphorbia indecora 525
Euphorbia indurescens 413, 445
Euphorbia inelegans 467
Euphorbia inermis 12, 14, 238, 252, 308, 310, 347–349, **354–358**, 893
   var. *laniglans* 306
Euphorbia infausta 135
Euphorbia ingens 1, 2, 22, 25, 27, 32, 34, 43, 48, 52, 54, 643, 679, 684, 686, 694, 711, **718–722**, 860, 898
Euphorbia ingenticapsa 754
Euphorbia inornata 282, 288
Euphorbia insarmentosa 461
Euphorbia involucrata 624
   var. *megastegia* vii, 624
Euphorbia jaegeriana 467
Euphorbia jansenvillensis 193, 197, 198
Euphorbia jubata 716, 739, 941
Euphorbia juglans 168
Euphorbia juttae 46, 146, 422, 484, **511–517**
Euphorbia kalaharica 650, 655
Euphorbia kaokoensis 33, 641, 645, **723–726**, 763, 765, 767, 818
Euphorbia karroensis 8, 484, 546
Euphorbia keithii 644, 701, **727–730**, 836, 898
Euphorbia kimberleyana 868
Euphorbia knobelii 21, 644, **730–735**
Euphorbia knuthii 10, 29, 37, 639, 642–644, 704, **735–739**, 793, 898, 921, 930
   subsp. *johnsonii* 737
   subsp. **knuthii** 11, 737
Euphorbia kraussiana 607, **618–621**
   var. *erubescens* 618
Euphorbia kwebensis 582
Euphorbia lacei 41
Euphorbia lamarckii 601
Euphorbia laro 890
Euphorbia latimammillaris, vii, 172, 174, 175
Euphorbia laxiflora 214
Euphorbia leachii 70, 74
Euphorbia ledienii 659, 664, 665, 895
   var. *dregei* 659
Euphorbia leistneri 570–572

Euphorbia lignosa 15, 27, 37, 49, 413, 414, 415, 430, **440–446**, 441, 442, 445
Euphorbia limpopoana 794, 798, 933
Euphorbia livida 477, 478, 481
Euphorbia lividiflora 25, 898, 915, **928–931**
Euphorbia lombardensis 816
Euphorbia longibracteata 448
Euphorbia loricata 48, 92, 95, **97–103**, 105, 107, 438
Euphorbia louwii 641, 645, 649, **739–743**
Euphorbia lugardiae 21, 28, 30, 33, 35, **863–868**, 962
Euphorbia lumbricalis 8, 544, 547
Euphorbia lupatensis 477, 481, 482
Euphorbia lydenburgensis 10, 641, 645, **743–751**, 770, 772, 793, 807
    var. ***lydenburgensis*** **745–748**
    var. ***minor*** **749–751**
Euphorbia macella 484
Euphorbia macowanii 264
Euphorbia macra 581
Euphorbia maculata 479, 480
Euphorbia magnifica 22
Euphorbia maleolens 46, 234, 237, 291, 300, **358–363**, 694
Euphorbia malevola 744, 791, 797, 798, 898, **932–934**
    subsp. *bechuanica* 794, 933
Euphorbia mammillaris 112, **116–118**, 124, 128–135, 174, 250
    var. *spinosior* 130
Euphorbia marientalii 258
Euphorbia marlothiana 9, 264, 268
Euphorbia marlothii 448
Euphorbia marrupana 20, 21, 37, 898, **935–938**, 945
Euphorbia matabelensis 463, 466, **467–470**, 468–470, 473
Euphorbia mauritanica 6, 7, 32, 33, 35, 36, 37, 49, 50, 52, 146, 180, 250, 306, 314, 527, 544, 591, 594, **595–601**, 605, 888
    var. *corallothamnus* 596
    var. *foetens* 596
    var. *lignosa* 596
    var. *minor* 596
    var. *namaquensis* 596
Euphorbia media 890
    var. *bagshawei* 890
Euphorbia medusa 437
Euphorbia medusae 264
Euphorbia melanohydrata 238, 330, **363–368**, 384
    subsp. *conica* 363
Euphorbia melanosticta vii, 595
Euphorbia meloformis 3, 10, 11, 54, 65, 71, 111, 112, **135–140**, 143, 145, 166, 345, 355, 391
    subsp. *meloformis* f. *falsa* 135
    subsp. *meloformis* f. *magna* 135
    subsp. *valida* 135
    var. *pomiformis* 135
    var. *prolifera* 135
Euphorbia memoralis 928
Euphorbia meyeri 319
Euphorbia micracantha 800, 811, 816, 817
Euphorbia milii 54, 918
Euphorbia minuscula 607, **621–624**, 622
Euphorbia mira 9, 204, 206
Euphorbia miscella 8, 9, 415, 420, 421
Euphorbia mlanjeana 30, 37, 739, 898, 928, **938–943**, 940, 945
Euphorbia monteiroi 27, 344, 413, 414, 415, 430, **446–455**, 820
    subsp. ***brandbergensis*** **451–454**
    subsp. ***monteiroi*** **448–451**
    subsp. ***ramosa*** 26, **454–455**
Euphorbia morinii 125
Euphorbia mosaica 10
Euphorbia mossambicensis 477, 481

Euphorbia muirii 9, 264
Euphorbia multiceps 235, 238, 330, 352, **368–376**
    subsp. ***multiceps*** **369–372**
    subsp. ***tanquana*** 236, 262, **373–376**
Euphorbia multifolia 92, 102, **103–107**
Euphorbia multiramosa 376, 380, 381
Euphorbia mundii 525
Euphorbia muraltioides 629, 630
Euphorbia muricata 18, 19, 483, **517–521**
Euphorbia myrtifolia 631
Euphorbia namaquensis 238, 330, 366, **376–381**
Euphorbia namibensis 237, 330, 366, 367, **381–385**
Euphorbia namuliensis 10, 37, 898, **943–945**
Euphorbia namuskluftensis 8, 9, 415, 420–422
Euphorbia natalensis 37, 46, 607, 611, 613, 617, **624–628**, 722
Euphorbia negromontana 484, **522–525**, 873
    subsp. ***rimireptans*** **523–525**
Euphorbia nelii 315, 319
Euphorbia neoarborescens 22
Euphorbia neocannellii 23, 962
Euphorbia neochamaeclada 32, 33, 873
Euphorbia neopedunculata **957–959**
Euphorbia neopolycnemoides 477, 478, 481
Euphorbia neospinescens 21, 962, 963
Euphorbia neriifolia 16, 17
Euphorbia nivulia 22, 54
Euphorbia nodosa vii, 560
Euphorbia oatesii 43, **471–474**
Euphorbia obesa 1, 10, 27, 32, 35, 54, 111, 112, 139, **140–143**, 894
    subsp. ***obesa*** **141–147**
    subsp. ***symmetrica*** **145–147**
Euphorbia obtusifolia 601
Euphorbia odontophylla 130
Euphorbia officinarum 134, 188
Euphorbia ohiva 463, 466
Euphorbia ornithopus **70**, 74
Euphorbia otavibergensis 644, **751–755**, 758, 789, 852–853
Euphorbia otavimontana 751
Euphorbia otjingandu 644, 754, **755–759**
Euphorbia otjipembana 641, 642, 645, 725, **760–769**, 818
    subsp. *fluvialis* **765–769**
    subsp. *okakoraensis* 761
    subsp. *otjipembana* **760–769**
Euphorbia ovata 629–631
Euphorbia oxystegia 12, 33, 48, 92, 102, **107–111**, 459
Euphorbia parifolia 581
Euphorbia parvimamma 893
Euphorbia passa 321, 325
Euphorbia patula 10, 37, 48, 61, 64, 67, **69–79**, 77, 82, 86, 248, 345, 391, 419, 473
    subsp. ***anacantha*** **74–75**
    subsp. ***brucebayeri*** **75–77**
    subsp. ***patula*** 11, 68, **70–74**, 270
    subsp. ***wilmaniae*** **77–79**
Euphorbia paxiana 596
Euphorbia pedroi 955
Euphorbia peltigera 433, 437
Euphorbia pentagona 112, 114, 116, 128, **147–152**, 163–164, 181, 187, 345, 894
Euphorbia pentops 237, 254, 268, **385–389**
Euphorbia peplus 636
Euphorbia perangusta 730, 734
Euphorbia pergracilis 477, 481
Euphorbia perpera 498, 525
Euphorbia persistens 665, 670
Euphorbia persistentifolia 716, 898, 941

Euphorbia pfeilii 582
Euphorbia phylloclada 8, 37, 581, 582, **585–588**
Euphorbia phymatoclada 595
Euphorbia pillansii 112, 128, **152–156**, 173, 187
 var. *albovirens* 152
 var. *ramosissima* 152
Euphorbia pisima 641, 645, 744, 749, **769–773**
Euphorbia pistiifolia 199, 202
Euphorbia planiceps 77, 78
Euphorbia platycephala 31, 43, **409–412**
Euphorbia platymammillaris vii, 130
Euphorbia plenispina, vii, 898, **945–948**
Euphorbia polyacantha 827
Euphorbia polycephala 10, 26, 27, 37, 53–54, 62, 64, 67, 69, **79–83**
Euphorbia polygona 7, 10, 16, 18, 20, 47, 49–51, 111, 112, 127–129, 143, 146, **156–166**, 187, 894
 var. *alba* 157
 var. *ambigua* 157
 var. *anoplia* 157
 var. *exilis* 157
 var. *hebdomadalis* 157
 var. *horrida* 156, 157, 165
 var. *major* 157, 165
 var. *minor* 157, 165
 var. *nivea* 157
 var. *noorsveldensis* 157, 165
 var. *polygona* 165
 var. *striata* 157
Euphorbia polygonata vii, 156
Euphorbia pomiformis 135, 139
Euphorbia procumbens 8, 37, 235, 237, 240, 241, 245, 270, 292, 294, 324, **389–396**, 406, 812
Euphorbia prostrata 1, 477, 479
Euphorbia proteifolia vii, 93
Euphorbia pseudocactus 21, 644, 705, 730, **773–777**, 789
Euphorbia pseudocontorta 898, **948–950**
Euphorbia pseudoglobosa 34, 35, 111, 112, 128, **166–172**, 191, 198, 352, 622
 subsp. **nesemannii** 128, 166–168, **172–175**
 subsp. **pseudoglobosa** 168–172, 176, 191–193, 197
 subsp. **vlokii** 168, **175–177**
 var. *dysselsdorpensis* 175
 var. *oshoekensis* 168
Euphorbia pseudohypogaea 304, 306
Euphorbia pseudotuberosa 61, 62, **83–87**, 89–91, 459
Euphorbia pteroneura 863
Euphorbia pubiglans 213, 221, 225, **230–234**
Euphorbia pugniformis 325, 389, 394
Euphorbia pulcherrima 6, 7, 52, 54
Euphorbia pulvinata 32, 52, 111, 112, 119, 122, 163, **177–183**, 783, 893
Euphorbia pyriformis 135
Euphorbia qarad 931
Euphorbia quadrata 12, 26, 415, **455–460**
Euphorbia quadrilatera 901, 913
Euphorbia racemosa 533
Euphorbia radiata 396, 812, 814
Euphorbia radyeri 23, 25, 36, 37, 45, 46, 49, 53, 158, 194, 306, 644, 660, **777–782**
Euphorbia ramiglans 9, 264
Euphorbia ramulosa 898, 907, **951–953**
Euphorbia rangeana 258
Euphorbia rectirama 533, 535
Euphorbia restituta 18, 48, 234–236, 314, 315, **396–400**

Euphorbia restricta 21, 644, 658, **782–785**
Euphorbia rhipsalioides 889
Euphorbia rhombifolia 8, 9, 37, 42, 52, 53, 65, 71, 172, 377, 438, 445, 484, 495, 497, 510, 515, 518, 519, 521, **525–532**, 535, 540, 541, 544, 622
 var. *laxa* 533
 var. *triceps* 533
Euphorbia richardsiae 739
Euphorbia rimireptans 523, 524
Euphorbia robecchii 54, 931
Euphorbia rowlandii 644, **785–789**
Euphorbia rubella 1, 12, 14
Euphorbia rubriflora 477, 481
Euphorbia rudis 258, 263, 303
Euphorbia rudolfii 498, 525
Euphorbia ruscifolia 629–631
Euphorbia sagittaria 650, 655
Euphorbia sarcostemmatoides 596
Euphorbia sarcostemmoides 6
Euphorbia schaeferi 427
Euphorbia scheffleri 29, 30
Euphorbia schinzii 47, 641, 645, 648, 666, 715, 716, 737, 742, 744, 750, 772, **789–801**, 797, 809, 818, 819, 845, 855, 857, 898, 933, 953
 E. schinzii X *E. tortirama* **896–897**
 subsp. **bechuanica** 648, **794–798**, 800, 933
 subsp. **schinzii** 24, **791–794**
 subsp. **schinzioides** **798–801**
Euphorbia schizacantha 10, 12
Euphorbia schlechteri 477, 481
Euphorbia schoenlandii 33, 34, 234–236, 314, 315, 344, 398, **400–404**
Euphorbia schubei 868
Euphorbia sclerophylla 607, 617, **628–631**
 var. *myrtifolia* 629
 var. *puberula* 629
 var. *ruscifolia* 629
Euphorbia scoparia 890
Euphorbia scopoliana 130
Euphorbia sect. *Aculeatae* 641
Euphorbia sect. **Anisophyllum** 7, 28, 37, **477–482**
Euphorbia sect. **Anthacanthae** 15, **58–460**
Euphorbia sect. **Aphyllis** **591–607**
Euphorbia sect. *Arthrothamnus* 868
Euphorbia sect. **Articulofruticosae** 18, 23, 37, 41–42, 43, **482–558**
Euphorbia sect. *Crepidaria* 26, 54
Euphorbia sect. **Crotonoides** 58, **460–462**
Euphorbia sect. *Diacanthae* 641
Euphorbia sect. *Diacanthium* 641
Euphorbia sect. **Espinosae** **559–569**
Euphorbia sect. **Esula** **607–637**
Euphorbia sect. **Euphorbia** 16, 18, 22, 26, 35, 37, 38, 40–43, 48, 55, 639, **641–863**, 895–897, **899–957**
Euphorbia sect. *Euphorbium* 60
Euphorbia sect. *Florispinae* 35
Euphorbia sect. **Frondosae** **569–577**
Euphorbia sect. **Gueinziae** **577–580**
Euphorbia sect. *Intermediae* 641
Euphorbia sect. **Lyciopsis** 43, 58, **462–474**
Euphorbia sect. *Monacanthae* 641
Euphorbia sect. **Monadenium** 12, 21, 22, 26, 28, 33, 35, 55, **863–868**, **957–963**
Euphorbia sect. *Pachysanthae* 29
Euphorbia sect. *Pseudomedusea* 234
Euphorbia sect. **Tenellae** 28, 43, **581–588**

Euphorbia sect. *Tetracanthae* 821
Euphorbia sect. **Tirucalli** 7, 9, 20, 23, 28, 32, 41, 43, **868–892**
Euphorbia sect. *Triacanthae* 641
Euphorbia sect. *Trichadenia* 60
Euphorbia sekukuniensis 45, 643, 700, 711, **801–805**, 863
Euphorbia ser. **Hystrix 92–111**, 95
Euphorbia ser. **Meleuphorbia** 16, 27, 32, 37, **111–198**, 894
Euphorbia ser. **Rhizanthium 198–212**
Euphorbia ser. **Treisia 213–234**
Euphorbia sessiliflora 889
Euphorbia silenifolia 9, **199–208**, 211, 212
Euphorbia siliciicola 515
Euphorbia similis 718, 722
Euphorbia spartaria 8, 9, 42, 46, 306, 484, 498, 507, 531, **533–538**
Euphorbia spicata 517, 546
Euphorbia spinea 19, 20, 483, **538–541**, 880
Euphorbia squarrosa 811, 812
Euphorbia stapelioides 8, 10, 11, 43, 46, 482, 495, 510, **542–547**, 558, 815
Euphorbia steelpoortensis 641, 645, 695, **806–809**
Euphorbia stegmatica 455, 459
Euphorbia stellata 10, 29, 48, 65, 71, 139, 166, 345, 391, 642, 643, 711, 739, **810–819**, 826, 921
  subsp. ***micracantha*** 29, **816–818**
  subsp. ***stellata*** 29, 30, 31, **816–818**
Euphorbia stellispina 8, 16, 29, 32, 112, 146, 152, 155, 162–163, **183–188**
  var. *astrispina* 183
Euphorbia stenocaulis vii, 898, 945, 947
Euphorbia stenoclada 20
Euphorbia stolonifera 591, 594, 599, **601–606**
Euphorbia strangulata 754
Euphorbia striata 12, 37, 607, 617, 622, 627, **631–635**
  var. *brachyphylla* 629
  var. *cuspidata* 632
Euphorbia subfalcata 87
Euphorbia subg. **Athymalus** 7, 10–15, 12, 19, 20, 23, 26, 33, 35, 38, 40, 43, 44, **55–474**
Euphorbia subg. **Chamaesyce** 10–13, 19, 22, 23, 26, 28, 29, 32–34, 38, 41, **475–588**
Euphorbia subg. **Esula** 7, 10, 12, 13, 20, 23, 26, 28, 32, 33, 38, 39, 41, 54, 55, **589–637**, 636
Euphorbia subg. **Euphorbia** 10, 11, 13, 15, 20, 22, 29, 39, 40, 55, , **639–892**
Euphorbia subg. *Rhizanthium* 57
Euphorbia subg. *Trichadenia* 57, 60
Euphorbia submammillaris 130
Euphorbia subsalsa 641, 726, 768, **818–822**, 820
  subsp. *fluvialis* 765
  subsp. *otzenii* **819–822**
  var. *kaokoensis* 723
Euphorbia subsect. **Africanae 591–606**
Euphorbia subsect. **Dactylanthes 60–91**
Euphorbia subsect. **Florispinae** 20, **92–234**
Euphorbia subsect. **Hypericifoliae 477–482**
Euphorbia subsect. **Medusea** 16, **234–408**, 235
Euphorbia subsect. *Meleuphorbia* 111
Euphorbia subsect. **Platycephalae 409–413**
Euphorbia subsect. **Pseudeuphorbium** 16, **413–460**
Euphorbia subsect. *Pseudomedusea* 234
Euphorbia suffulta 23, 482, **547–549**
Euphorbia superans 349
Euphorbia suppressa 293, 295
Euphorbia susannae 34, 35, 111, **112**, 128, 166, **189–193**, 352
Euphorbia symmetrica 10

Euphorbia systeloides 461
Euphorbia taitensis 712
Euphorbia tenax 483, 497, 531, **549–554**
Euphorbia tenuispinosa 712
Euphorbia terracina 636, 637
Euphorbia tetragona 16, 21, 36, 643, 663, 709, 711, 712, 730, 804, **822–827**, 833, 836, 895, 928
Euphorbia tettensis 477, 481
Euphorbia tirucalli 6, 7, 41, 54, **889–892**, 893, 955
Euphorbia torrei 22, 955, **959–963**, 962
Euphorbia tortilis 54
Euphorbia tortirama 643, 670, 684, 696, **828–832**, 831, 832, 841, 896–897
Euphorbia tortistyla 791
Euphorbia transvaalensis 22, 570, **572–577**, 820
Euphorbia triangularis 21, 23, 24, 48, 54, 644, 681, 701, 709, 711, 730, 825, **832–837**, 895, 896, 898, 928
Euphorbia trichadenia 6, 61, **86–91**, 338, 459
  var. *gibbsiae* 56, 87
Euphorbia tridentata 70, 74, 75
Euphorbia truncata 277
Euphorbia tuberculata 9, 264, 267, 271
Euphorbia tuberculatoides 264
Euphorbia tuberosa 1, 12, 37, 43, 200–206, **208–212**, 438
Euphorbia tubiglans 34, 45, 111, 112, 166, 174, **193–198**
  var. *jansenvillensis* 193
Euphorbia tugelensis 214, 216–217
Euphorbia turbiniformis 10, 143–144
Euphorbia uhligiana 832
Euphorbia umfoloziensis 665, 668–670
Euphorbia uncinata 812
Euphorbia unicornis 16, 37, 898, 918, 937, 945, **954–957**
Euphorbia unispina 16, 21
Euphorbia vaalputsiana 500, 504
Euphorbia valida 135, 139
Euphorbia vallaris 692, 861
Euphorbia vandermerwei 27, 47, 643, 669, 670, 696, 697, 830, 831, **838–842**
Euphorbia venenata 650, 655, 754, 755
Euphorbia venteri 43, 641, 645, 676–677, 742, 744, 797, **842–847**
Euphorbia verdickii 471
Euphorbia verruculosa 18, 19, 483, **554–558**
Euphorbia versicolores 315, 318, 321
Euphorbia viminalis 893
Euphorbia viperina 357, 893
Euphorbia virosa 3, 4, 9, 21, 29, 32–34, 37, 48–52, 644, 651, 653–656, 724, 754, 757, 758, **847–854**, 880
  f. *caespitosa* 847
  f. *striata* 847
  subsp. *arenicola* 847, 850, 854
Euphorbia volkmanniae 650, 755
Euphorbia waterbergensis 27, 642, 645, 700, 715, **854–858**
Euphorbia whellanii 945
Euphorbia wildii 413
Euphorbia willowmorensis 236, 393, **404–408**
Euphorbia wilmaniae 83, 419, 422
Euphorbia woodii 321, 325
Euphorbia zambesiana 477, 481
Euphorbia zoutpansbergensis 28, 643, 692, 701, **858–863**

F
Faucaria 62
Fockea comaru 62
Fockea edulis 62
Fockea multiflora 933

G

Galarhoeus genistoides 613
Geranianceae 18
Gibbaeum 190
   G. album 153
   G. geminum 326
   G. heathii 153, 326
   G. nebrownii 169
   G. pachypodium 153
   G. petrense 153
   G. pubescens 168
Glottiphyllum 62, 65, 74, 200

H

Haworthia 74, 200, 391
Hereroa 71
Hyaenanche 6
Hydnora 49
Hydnora africana 50, 777, 884

K

Kalanchoe 43, 47, 715
Kalanchoe elizae 937, 940, 945
Kalanchoe humilis 945, 955
Kleinia 15
Kleinia longiflora 538, 598, 651, 676, 880, 893

L

Leipoldtia schultzei 226, 229
Lophophora williamsii 143
Lyciopsis 55, 57, 462

M

Medusea 55, 57, 234
Medusea globosa 64
Medusea hamata 432
Medusea major 264
Medusea patula 69
Medusea procumbens 389, 395
Medusea tessellata 264
Mesembryanthemum lignescens 848
Mestoklema 71
Monadenium 55, 639, 863
Monadenium guentheri 21
Monadenium pedunculatum 957
Monadenium spinescens 21, 962
Monadenium torrei 959
Monechma incanum 377
Monsonia camdeboensis 142
Montinia caryophyllacea 98
Mopane 410, 465, 472, 566, 571, 593, 647, 648, 724, 756, 757, 761, 762, 766, 795, 820, 844, 873, 933
Myrothamnus 724, 925

N

Naboom 719

P

Pachypodium 868
Pachypodium saundersiae 666
Passerina 160, 161

Pedaliaceae 43
Pedilanthus 6, 7, 54
Periplocoideae 47
Phyllanthus 6
Phytosciulus 49
Plectranthus 43, 47, 682, 866
Polymita albiflora 37
Portulacaceae 43
Portulacaria afra 132, 158
Portulacaria fruticulosa 848
Portulacaria namaquensis 651, 848
Pterodiscus 43
Pteronia 62, 176
Pteronia incana 136, 218, 222, 226
Pteronia paniculata 172

R

Raphionacme zeyheri 62
Rhigozum obovatum 45, 62, 142, 146, 153, 184, 194, 777
Rhigozum trichotomum 46
Rhinephyllum muirii 169
Rhombophyllum dolabriforme 62
Ruschia 71, 74
Ruschia pungens 176
Ruschia robusta 377
Ruschia spinosa 136

S

Sansevieria 679, 682, 866, 955
Sarcostemma 15, 679
Schlecteranthus albiflorus 37
Selaginella 47, 949, 955
Senecio radicans 71, 142, 176, 191
Sesamothamnus 43
Sesamothamnus benguellensis 849
Synadenium 6, 7, 35
Synadenium arborescens 868

T

Tapinanthus oleifolius 50
Tetracanthae 641, 712
Tetranynchus 49
Tirucalia vii, 55, 639, 868
Tirucalia aequoris 511
Tirucalia amarifontana 525
Tirucalia angrae 525
Tirucalia arceuthobioides 549
Tirucalia aspericaulis 517
Tirucalia berotica 592
Tirucalia brachiata 525
Tirucalia burmanni 484
Tirucalia caterviflora 525
Tirucalia chersina 525
Tirucalia cibdela 533
Tirucalia congestiflora 870
Tirucalia corymbosa 484
Tirucalia curocana 440
Tirucalia cymosa 550
Tirucalia damarana 874
Tirucalia dregeana 422
Tirucalia ephedroides 489
   var. *debilis* 490
   var. *imminuta* 495
Tirucalia gentilis 500

Tirucalia giessii 505
Tirucalia gregaria 879
Tirucalia gummifera 884
Tirucalia herrei 508
Tirucalia indecora 525
Tirucalia juttae 511
Tirucalia karroensis 484
Tirucalia lignosa 440
Tirucalia macella 484
Tirucalia mauritanica 595
Tirucalia mixta 550
Tirucalia muricata 517
Tirucalia negromontana 523
Tirucalia paxiana 596
Tirucalia perpera 525
Tirucalia rectirama 533
Tirucalia rhombifolia 525
Tirucalia rudolfii 525
Tirucalia spartaria 533
Tirucalia spicata 517
Tirucalia spinea 538
Tirucalia stapelioides 542
Tirucalia stolonifera 601
Tirucalia tenax 549
Tirucalia tirucalli 889
Tirucalia transvaalensis 572
Tirucalia verruculosa 554
Tithymalus aphyllus mauritaniae 601
Tithymalus apiculatus 614
Tithymalus attenuatus 204
Tithymalus bergii 204
Tithymalus brachypus 595
Tithymalus capensis 632
Tithymalus confertus 614
Tithymalus crispus 208

Tithymalus ellipticus 203
Tithymalus epicyparissias 624
Tithymalus genistoides 613
Tithymalus involucratus 624
Tithymalus longipetiolatus 204
Tithymalus mauritanicus 595
Tithymalus meyeri 618
Tithymalus multicaulis 628
Tithymalus ovatus 629
Tithymalus revolutus 613
Tithymalus silenifolius 203, 208
Tithymalus striatus 631
Tithymalus truncatus 618
Tithymalus tuberosus 208, 212
Tithymalus zeyheri vii, 56, 595
Treisia 55, 57, 213
Treisia clava 213, 221
Treisia erosa 130
Treisia hystrix 97
Treisia tuberculata 221, 264
Tribulocarpus 43
Trichodiadema 74, 622, 623
Tylecodon hallii 848
Tylecodon rubrovenosus 848

V
Viscum dielsianum 50
Viscum minimum 50, 51
Vitaceae 18

X
Xerophyta 47, 728, 898, 912, 920, 940, 945, 947, 952